SURFACTANT SCIENCE SERIES

ADSORPTION AND AGGREGATION OF SURFACTANTS IN SOLUTION

SURFACTANT SCIENCE SERIES

ADDITIONAL VOLUMES IN PREPARATION

ADSORPTION AND AGGREGATION OF SURFACTANTS IN SOLUTION

edited by

K. L. Mittal

Hopewell Junction, New York, U.S.A.

Dinesh O. Shah

University of Florida
Gainesville, Florida, U.S.A.

CRC Press
Taylor & Francis Group
Boca Raton London New York

CRC Press is an imprint of the
Taylor & Francis Group, an **informa** business

CRC Press
Taylor & Francis Group
6000 Broken Sound Parkway NW, Suite 300
Boca Raton, FL 33487-2742

First issued in paperback 2019

ISBN-13: 978-0-8247-0843-6 (hbk)
ISBN-13: 978-0-367-39567-4 (pbk)

Visit the Taylor & Francis Web site at
http://www.taylorandfrancis.com

and the CRC Press Web site at
http://www.crcpress.com

Preface

This volume embodies, in part, the proceedings of the 13th International Symposium on Surfactants in Solution (SIS) held in Gainesville, Florida, June 11–16, 2000. The theme of this particular SIS was "Surfactant Science and Technology for the New Millennium." The final technical program comprised 360 papers, including 96 poster presentations, which was a testimonial to the brisk research activity in the arena of surfactants in solution. In light of the legion of papers, to chronicle the total account of this event would have been impractical, so we decided to document only the plenary and invited presentations. The contributors were asked to cover their topics in a general manner; concomitantly, this book reflects many excellent reviews of a number of important ramifications of surfactants in solution.

Chapters 1–4 document the plenary lectures, including the written account of the special "Host Lecture" by one of us (DOS) and Prof. Brij Moudgil. Chapters 5–32 embody the text of 28 invited presentations covering many aspects of surfactants in solution. Among the topics covered are: surfactant-stabilized particles; solid particles at liquid interfaces; nanocapsules; aggregation behavior of surfactants; micellar catalysis; vesicles and liposomes; the clouding phenomenon; viscoelasticity of micellar solutions; phase behavior of microemulsions; reactions in microemulsions; viscosity index improvers; foams, foam films, and monolayers; principles of emulsion formulation engineering; nano-emulsions; liposome gene delivery; polymeric surfactants; and combinatorial surface chemistry.

As surfactants play an important role in many and diverse technologies, ranging from high-tech (microelectronics) to low-tech (washing clothes) applications, an understanding of their behavior in solution is of paramount importance. Also, as we learn more about surfactants and devise new surfactant formulations, novel and exciting applications will emerge.

The present compendium of excellent overviews and research papers provides a bounty of up-to-date information on the many and varied aspects of surfactants in solution. It also offers a commentary on current research activity regarding the behavior of surfactants in solution. We hope that anyone involved or interested—centrally or tangentially—in surfactants will find this book useful. Further, we trust that both veteran researchers and those embarking on their maiden voyage in the wonderful world of surfactants will find this treatise valuable.

To put together a symposium of this magnitude and quality requires dedication and unflinching help from a battalion of people, and now it is our pleasure and duty to acknowledge those who helped in many and varied manners in this endeavor. First and foremost, we express our heartfelt and most sincere thanks to Prof. Brij Moudgil, Director of the Engineering Research Center for Particle Science and Technology, University of Florida, for helping in more ways than one. He wore many different hats—as co-chairman, as troubleshooter, as local host—and he was always ready and willing to help with a smile. Next we are thankful to faculty members, postdoctoral associates, graduate students, and administrative staff of both the Center for Surface Science and Engineering and the Engineering Research Center for Particle Science and Technology, University of Florida.

We acknowledge the generous support of the following organizations: the Florida Institute of Phosphate Research, the National Science Foundation, and the University of Florida. Many individual industrial corporations helped us by providing generous financial support and we are grateful to them. We also thank the exhibitors of scientific instruments and books for their contribution and support.

We are grateful to the authors for their interest, enthusiasm, and contribution without which this book would not have seen the light of day. Last, we are appreciative of the efforts of the staff at Marcel Dekker, Inc. for giving this book a body form.

K. L. Mittal
Dinesh O. Shah

Contents

v

Contributors

Kazunari Akiyoshi Department of Synthetic Chemistry and Biological Chemistry, Kyoto University, Kyoto, Japan

Alexei A. Antipov Max-Planck-Institute of Colloids and Interfaces, Potsdam, Germany

Abraham Aserin Casali Institute of Applied Chemistry, The Hebrew University of Jerusalem, Jerusalem, Israel

Robert Aveyard Department of Chemistry, Hull University, Hull, United Kingdom

Núria Azemar Departament de Tecnologia de Tensioactius, Institut d'Investigacions Químiques i Ambientals de Barcelona, Barcelona, Spain

Soo Kyoung Bae Department of Chemical and Biomolecular Engineering, KAIST, Daejeon, Korea

Rahul Prabhakar Bagwe Department of Chemical Engineering, University of Florida, Gainesville, Florida, U.S.A.

Rajkumar Banerjee* Division of Lipid Science and Technology, Indian Institute of Chemical Technology, Hyderabad, India

---

**Current affiliation*: University of Pittsburgh, Pittsburgh, Pennsylvania, U.S.A.

F. Bautista Departamento de Ingeniería Química, Universidad de Guadalajara, Guadalajara, Mexico

Vance Bergeron Ecole Normale Superieure, Paris, France

Arun Kumar Chattopadhyay Corporate Research and Development, United States Bronze Powders Group of Companies, Haskell, New Jersey, U.S.A.

Arabinda Chaudhuri Division of Lipid Science and Technology, Indian Institute of Chemical Technology, Hyderabad, India

Sherril D. Christian[†] Institute for Applied Surfactant Research, University of Oklahoma, Norman, Oklahoma, U.S.A.

Per M. Claesson Department of Chemistry, Surface Chemistry, Royal Institute of Technology, and Institute for Surface Chemistry, Stockholm, Sweden

John H. Clint Department of Chemistry, Hull University, Hull, United Kingdom

L. Dähne Max-Planck-Institute of Colloids and Interfaces, Potsdam, Germany

Prasanta Kumar Das[*] Division of Lipid Science and Technology, Indian Institute of Chemical Technology, Hyderabad, India

F. Debuigne Laboratoire de Résonance Magnétique Nucléaire, Facultés Universitaires Notre-Dame de la Paix, Namur, Belgium

Imre Dékány Department of Colloid Chemistry, University of Szeged, Szeged, Hungary

Nikolai D. Denkov Laboratory of Chemical Physics and Engineering, Faculty of Chemistry, Sofia University, Sofia, Bulgaria

Surekha Devi Department of Chemistry, M. S. University of Baroda, Baroda, Gujarat, India

---

[†]Deceased.
[*]*Current affiliation*: Massachusetts Institute of Technology, Cambridge, Massachusetts, U.S.A.

Edwin Donath   Department of Biophysics, Institute of Medical Physics and Biophysics, University of Leipzig, Leipzig, Germany

William A. Ducker   Department of Chemistry, Virginia Tech, Blacksburg, Virginia, U.S.A.

Julian Eastoe   School of Chemistry, University of Bristol, Bristol, United Kingdom

Jordi Esquena   Departament de Tecnologia de Tensioactius, Institut d'Investigacions Químiques i Ambientals de Barcelona, Barcelona, Spain

Shmaryahu Ezrahi   Technology and Development Division, IDF, Tel Hashomer, Israel

Ana Maria Forgiarini   Departament de Tecnologia de Tensioactius, Institut d'Investigacions Químiques i Ambientals de Barcelona, Barcelona, Spain

Changyou Gao   Department of Polymer Science and Engineering, Zhejiang University, Hangzhou, People's Republic of China

Anilkumar G. Gaonkar   Research and Development, Kraft Foods, Inc., Glenview, Illinois, U.S.A.

María José García-Celma   Departament de Farmàcia, Facultad de Farmàcia, Universitat de Barcelona, Barcelona, Spain

Nissim Garti   Casali Institute of Applied Chemistry, The Hebrew University of Jerusalem, Jerusalem, Israel

Brian Grady   School of Chemical Engineering and Materials Science, University of Oklahoma, Norman, Oklahoma, U.S.A.

Michael Gradzielski   Physical Chemistry I, University of Bayreuth, Bayreuth, Germany

Asen D. Hadjiiski   Laboratory of Chemical Physics and Engineering, Faculty of Chemistry, Sofia University, Sofia, Bulgaria

Maria Häger   Institute for Surface Chemistry, Stockholm, Sweden

Norikatsu Hattori* Department of Applied Chemistry, Nagoya Institute of Technology, Nagoya, Japan

Richard K. Heenan ISIS Facility, Rutherford Appleton Laboratory, Chilton, United Kingdom

Heinz Hoffmann Physical Chemistry I, University of Bayreuth, Bayreuth, Germany

Krister Holmberg Department of Applied Surface Chemistry, Chalmers University of Technology, Göteborg, Sweden

Klaus Horbaschek Physical Chemistry I, University of Bayreuth, Bayreuth, Germany

Qun Huo Department of Polymers and Coatings, North Dakota State University, Fargo, North Dakota, U.S.A.

Ivan B. Ivanov Laboratory of Chemical Physics and Engineering, Faculty of Chemistry, Sofia University, Sofia, Bulgaria

Paqui Izquierdo Departament de Tecnologia de Tensioactius, Institut d'Investigacions Químiques i Ambientals de Barcelona, Barcelona, Spain

L. Jeunieau Laboratoire de Résonance Magnétique Nucléaire, Facultés Universitaires Notre-Dame de la Paix, Namur, Belgium

Eui-Chul Kang Department of Synthetic Chemistry and Biological Chemistry, Kyoto University, Kyoto, Japan

Dae-Wook Kim Department of Chemical Engineering and CUPS, Hanyang University, Seoul, Korea

Jin-Chul Kim Department of Chemical and Biomolecular Engineering, KAIST, Daejeon, Korea

Jong-Duk Kim Department of Chemical and Biomolecular Engineering, KAIST, Daejeon, Korea

Douglas T. Lai Development Center for Biotechnology, Taipei, Taiwan

Current affiliation: Chisso Corporation, Tokyo, Japan.

Dominique Langevin   Laboratoire de Physique des Solides, Université Paris Sud, Orsay, France

Roger M. Leblanc   Department of Chemistry, University of Miami, Coral Gables, Florida, U.S.A.

Eun-Ok Lee   Department of Chemical and Biomolecular Engineering, KAIST, Daejeon, Korea

Stefano Leporatti   Institute of Medical Physics and Biophysics, University of Leipzig, Leipzig, Germany

Ganzuo Li   Key Laboratory for Colloid and Interface Chemistry of State Education Ministry, Shandong University, Jinan, People's Republic of China

Heinz Lichtenfeld   Max-Planck-Institute of Colloids and Interfaces, Potsdam, Germany

J. Lyklema   Physical Chemistry and Colloid Science, Wageningen University, Wageningen, The Netherlands

Octavio Manero   Instituto de Investigaciones en Materiales, Universidad Nacional Autónoma de México, Mexico City, Mexico

C. Manohar   Department of Chemical Engineering, Indian Institute of Technology, Mumbai, India

Laura Márquez   Laboratory FIRP, School of Chemical Engineering, University of the Andes, Mérida, Venezuela

Isabel Mira*   Laboratory FIRP, School of Chemical Engineering, University of the Andes, Mérida, Venezuela

Helmuth Möhwald   Max-Planck-Institute of Colloids and Interfaces, Potsdam, Germany

Daniel Morales   Departament de Tecnologia de Tensioactius, Institut d'Investigacions Químiques i Ambientals de Barcelona, Barcelona, Spain

---

Current affiliation: Institute for Surface Chemistry, Stockholm, Sweden.

Brij M. Moudgil Engineering Research Center, University of Florida, Gainesville, Florida, U.S.A.

Sergio Moya Max-Planck-Institute of Colloids and Interfaces, Potsdam, Germany

J. B.Nagy Laboratoire de Résonance Magnétique Nucléaire, Facultés Universitaires Notre-Dame de la Paix, Namur, Belgium

Charmian J. O'Connor Department of Chemistry, The University of Auckland, Auckland, New Zealand

Seong-Geun Oh Department of Chemical Engineering and CUPS, Hanyang University, Seoul, Korea

Hirofumi Okabayashi Department of Applied Chemistry, Nagoya Institute of Technology, Nagoya, Japan

Szilvia Papp Department of Colloid Chemistry and Nanostructured Materials Research Group, Hungarian Academy of Sciences, University of Szeged, Szeged, Hungary

Rita Patakfalvi Department of Colloid Chemistry and Nanostructured Materials Research Group, Hungarian Academy of Sciences, University of Szeged, Szeged, Hungary

Alison Paul School of Chemistry, University of Bristol, Bristol, United Kingdom

Alejandro Peña* Laboratory FIRP, School of Chemical Engineering, University of the Andes, Mérida, Venezuela

Jeffrey Penfold ISIS Facility, Rutherford Appleton Laboratory, Chilton, United Kingdom

J. H. Pérez-López Departamento de Ingeniería Química, Universidad de Guadalajara, Guadalajara, Mexico

Naveen Kumar Pokhriyal Department of Chemistry, M. S. University of Baroda, Baroda, Gujarat, India

---

*Current affiliation: Rice University, Houston, Texas, U.S.A.

J. E. Puig Departamento de Ingeniería Química, Universidad de Guadalajara, Guadalajara, Mexico

Igor Radtchenko Max-Planck-Institute of Colloids and Interfaces, Potsdam, Germany

Nalam Madhusudhana Rao Centre for Cellular and Molecular Biology, Hyderabad, India

I. Ravet-Bodart Laboratoire de Résonance Magnétique Nucléaire, Facultés Universitaires Notre-Dame de la Paix, Namur, Belgium

Laurence S. Romsted Department of Chemistry and Chemical Biology, Rutgers, The State University of New Jersey, New Brunswick, New Jersey, U.S.A.

Emily Rumsey School of Chemical Sciences, University of East Anglia, Norwich, United Kingdom

Jean-Louis Salager Laboratory FIRP, School of Chemical Engineering, University of the Andes, Mérida, Venezuela

John F. Scamehorn Institute for Applied Surfactant Research, University of Oklahoma, Norman, Oklahoma, U.S.A.

Dinesh O. Shah Departments of Chemical Engineering and Anesthesiology, Center for Surface Science and Engineering, University of Florida, Gainesville, Florida, U.S.A.

Qiang Shen Key Laboratory for Colloid and Interface Chemistry of State Education Ministry, Shandong University, Jinan, People's Republic of China

Seung-Il Shin Department of Chemical Engineering and CUPS, Hanyang University, Seoul, Korea

Andrew R. Slagle Institute for Applied Surfactant Research, University of Oklahoma, Norman, Oklahoma, U.S.A.

Conxita Solans Departament de Tecnologia de Tensioactius, Institut d'Investigacions Químiques i Ambientals de Barcelona, Barcelona, Spain

J. F. A. Soltero Departamento de Ingeniería Química, Universidad de Guadalajara, Guadalajara, Mexico

Gollapudi Venkata Srilakshmi   Division of Lipid Science and Technology, Indian Institute of Chemical Technology, Hyderabad, India

David Steytler   School of Chemical Sciences, University of East Anglia, Norwich, United Kingdom

G. B. Sukhorukov   Max-Planck-Institute of Colloids and Interfaces, Potsdam, Germany

Cynthia Q. Sun*   Department of Chemistry, The University of Auckland, Auckland, New Zealand

Junzo Sunamoto   Advanced Research and Technology Center, Niihama National College of Technology, Ehime, Japan

Slavka S. Tcholakova   Laboratory of Chemical Physics and Engineering, Faculty of Chemistry, Sofia University, Sofia, Bulgaria

Edward E. Tucker   Institute for Applied Surfactant Research, University of Oklahoma, Norman, Oklahoma, U.S.A.

Eric Tyrode[†]   Laboratory FIRP, School of Chemical Engineering, University of the Andes, Mérida, Venezuela

Núria Usón   Departament de Tecnologia de Tensioactius, Institut d'Investigacions Químiques i Ambientals de Barcelona, Barcelona, Spain

A. Voight   Max-Planck-Institute of Colloids and Interfaces, Potsdam, Germany

Wolfgang von Rybinski   Corporate Research, Henkel KGaA, Düsseldorf, Germany

Matthias Wegener   Corporate Research, Henkel KGaA, Düsseldorf, Germany

Noelia B. Zambrano[‡]   Laboratory FIRP, School of Chemical Engineering, University of the Andes, Mérida, Venezuela

---

Current affiliation:
*Hort Research, Auckland, New Zealand.
†Royal Institute of Technology (KTH), Stockholm, Sweden.
‡M.W. Kellogg Ltd., Middlesex, United Kingdom.

Weican Zhang State Key Laboratory of Microbial Technology, Shandong University, Jinan, People's Republic of China

Li-Qiang Zheng Department of Chemistry, Shandong University, Jinan, People's Republic of China

Linda Zhu Institute for Applied Surfactant Research, University of Oklahoma, Norman, Oklahoma, U.S.A.

ADSORPTION AND AGGREGATION OF SURFACTANTS IN SOLUTION

1

Highlights of Research on Molecular Interactions at Interfaces from the University of Florida

DINESH O. SHAH and BRIJ M. MOUDGIL University of Florida, Gainesville, Florida, U.S.A.

ABSTRACT

An overview of research highlights of the past three decades from the University of Florida on molecular interactions at interfaces and in micelles is presented. This overview includes work on (1) the kinetic stability of micelles in relation to technological processes, (2) molecular packing in mixed monolayers and phase transition in monolayers, (3) microemulsions and their technological applications including enhanced oil recovery (EOR) processes and preparation of nanoparticles of advanced materials, (4) adsorption of polymers at solid–liquid interfaces and selective flocculation, and (5) the mechanical strength of surfactant films at the solid–liquid interface and its correlation with dispersion stability as well as the interfacial phenomena in chemical–mechanical polishing (CMP) of silicon wafers. Detailed results explaining the role of molecular interactions at interfaces and in micelles as well as pertinent references are given for each phenomenon discussed.

I. INTRODUCTION

It is a great pleasure and privilege for us to summarize the highlights of research on molecular interactions at interfaces from the University of Florida on this 13th International Symposium on Surfactants in Solution (SIS-2000). During the past 30 years at the University of Florida, we have had an ongoing research program on fundamental aspects as well as technological applications of interfacial processes. Specifically, this overview includes the kinetic stability of micelles in relation to technological processes, molecular packing in mixed monolayers and phase transition in monolayers, microemulsions and their technological applications including enhanced oil recovery (EOR) processes and preparation of nanoparticles of advanced materials, adsorption of polymers at solid–liquid interface and selective flocculation, the mechanical strength of surfactant films at the solid–liquid interface and dispersion stability, and the interfacial phenomena in chemical mechanical polishing (CMP) of silicon wafers.

II. MONOLAYERS

During the past quarter century, considerable studies have been carried out on the reactions in monomolecular films of surfactant, or monolayers. Figure 1 shows the surface pressure–area curves for dioleoyl, soybean, egg, and dipalmitoyl lecithins [1]. For these four lecithins, the fatty acid composition was determined by gas chromatography. The dioleoyl lecithin has both chains unsaturated, soybean lecithin has polyunsaturated fatty acid chains, egg lecithin has 50% saturated and 50% unsaturated chains, and dipalmitoyl lecithin has both chains fully saturated. It is evident that, at any fixed surface pressure, the area per molecule is in the following order:

Dioleoyl lecithin > soybean lecithin > egg lecithin > dipalmitoyl lecithin

It can be assumed that the area per molecule represents the area of a square at the interface. Thus, the square root of the area per molecule gives the length of one side of the square, which represents the intermolecular distance. Figure 2 schematically illustrates the area per molecule and intermolecular distance in these four lecithins. The corresponding intermolecular distances were calculated to be 9.5, 8.8, 7.1, and 6.5 Å, respectively, at a surface pressure of 20 mN/m [2]. Thus, one can conclude that a change in the saturation of the fatty acid chains produces subangstrom changes in the intermolecular distance in the monolayer.

In addition, it was desired to explore the effects that these small changes in intermolecular distance had on the enzymatic susceptibility of these lecithins to hydrolytic enzymes such as phospholipase A [3–5], a potent hydrolytic enzyme found in cobra venom. Thus, microgram quantities of enzyme were injected under this monolayer. By measuring the rate of change of surface potential, one can indirectly measure the rate of reaction in the monolayer. It is assumed that these quantities [i.e., change in surface potential $\Delta(\Delta V)$ and the extent of reaction] are proportional to each other. The kinetics of hydrolysis, as measured by a decrease in surface potential, were studied for each lecithin monolayer as a function of initial surface pressure and are shown in Fig. 3 [6]. It was found that initially the reaction rate

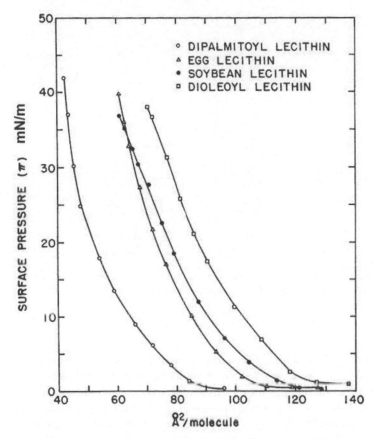

FIG. 1 Surface pressure–area curves of dipalmitoyl, egg, soybean, and dioleoyl lecithins.

increased as the surface pressure increased. Subsequently, as the surface pressure increased further, the reaction rate decreased until a critical surface pressure was reached at which no reaction occurred. The critical surface pressure required to block the hydrolysis of lecithin monolayer increased with the degree of unsaturation of fatty acid chains (Fig. 3). Thus, it appears that as the intermolecular distance increases because of the unsaturated fatty acid chains, a higher surface pressure is required to clock the penetration of the active site of the enzyme into the monolayer to cause hydrolysis. This also led to a suggestion that subangstrom changes in the intermolecular distance in the monolayer were significant for the enzymatic hydrolysis of the monolayers.

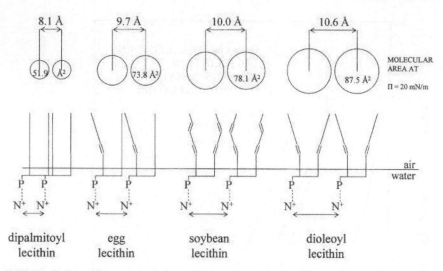

FIG. 2 Schematic representation of the area per molecule and intermolecular distance in dioleoyl, soybean, egg, and dipalmitoyl lecithin monolayers based on the data plotted in Fig. 1.

In addition to hydrolysis reactions, the enzymatic synthesis in monolayers was studied [7]. In this case, a steric acid monolayer was formed on an aqueous solution containing glycerol. After compression to a desired surface pressure, a small amount of enzyme lipase was injected under the monolayer. The lipase facilitated the linkage of glycerol with fatty acid and produced monoglycerides, diglycerides, and triglycerides in the monolayer [8–10], which could be detected by thin-layer chromatography (TLC) or high-performance liquid chromatography (HPLC).

Because the amount of product that can be synthesized using a monolayer is in microgram quantities, this method is not attractive for large-scale enzymatic synthesis. Therefore, studies of enzymatic reactions in monolayers were extended to studies of enzymatic reactions in a foam. A foam provides a large interfacial area, and by continuous aeration one can generate an even larger interfacial area. A soap bubble is stabilized by monolayers on both inside and outside surfaces of the bubble (Fig. 4). The glycerol and enzyme can be added into the aqueous phase before producing the foam. Thus, it was shown that almost 88% of free steric acid could be converted to di- and triglycerides in 2 h by reactions in foams (Fig. 5). For surface-active substrates (or reactants) and enzymes, reactions in foams offer a very interesting possibility to produce large-scale synthesis of biochemicals using a foam as a reactor.

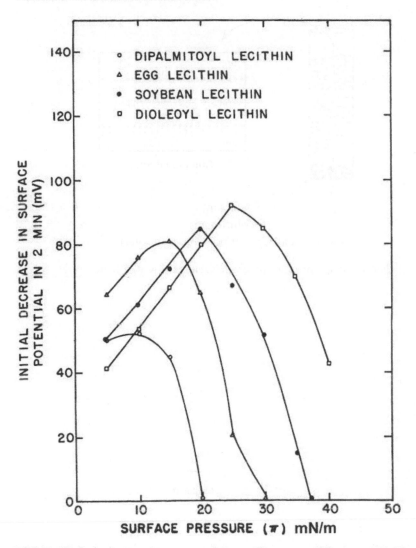

FIG. 3 Hydrolysis rate (as measured by surface potential) versus initial surface pressure of various lecithin monolayers.

Another interesting investigation at the Center for Surface Science and Engineering (CSSE) was focused on the possible existence of phase transitions in mixed monolayers of surfactants. Figure 6 shows the rate of evaporation from pure and mixed monolayers of cholesterol and arachidyl (C_{20}) alcohol, as well as their mixed monolayers [11]. It is evident that the pure

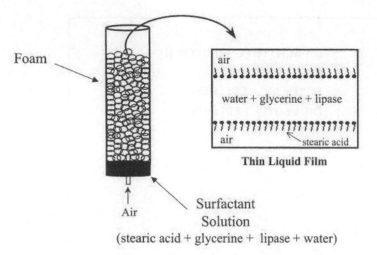

FIG. 4 Schematic diagram of a lipase-catalyzed reaction in a foam vessel.

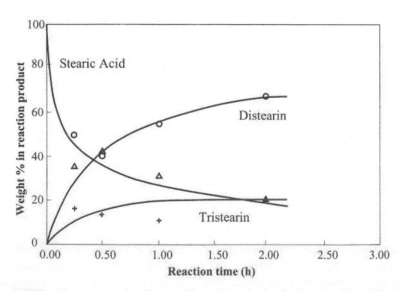

FIG. 5 Decrease in free fatty acids and synthesis of di- and triglycerides by lipase in foam.

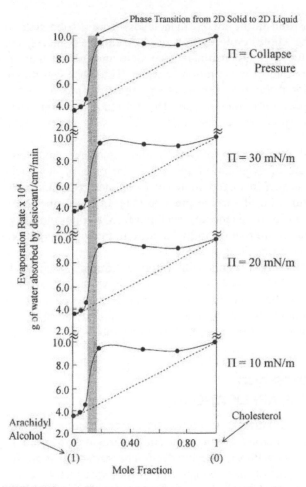

FIG. 6 Rate of evaporation from pure and mixed monolayers of cholesterol and arachidyl (C_{20}) alcohol.

C_{20} alcohol monolayer allows only one third of the water loss related to evaporation of the pure cholesterol monolayer. This is presumably due to the fact that the C_{20} alcohol forms monolayers that are in the two-dimensional *solid* state. In contrast, the cholesterol monolayers are in the two-dimensional *liquid* state. However, when cholesterol is incorporated into a C_{20} alcohol monolayer, the cholesterol mole fraction needs to be only about 20% to liquefy the solid monolayers of C_{20} alcohol. The abrupt increase in evaporation rate of water at 20–25 mol% cholesterol illustrates the two-dimensional phase transition in the mixed monolayers from a solid state to a liquid state. After the cholesterol fraction reaches about 25 mol%, the monolayer remains in the two-dimensional liquid state and, hence, there is no further change in the rate of evaporation of water. Thus, one can utilize the evaporation of water through a film as a very sensitive probe for observing the molecular packing in monolayers. The existence of a solid state or a liquid state for monolayers can be inferred from such experimental results.

It has been shown that mixed monolayers of oleic acid and cholesterol exhibit the minimum rate of evaporation at a 1:3 molar ratio of oleic acid to cholesterol. This is shown in mixed monolayers of oleic acid and cholesterol in Fig. 7 as a function of surface pressure [11]. In has further been shown that a 1:3 molar ratio in mixed fatty acid and fatty alcohol monolayers, one observes the maximum foam stability, minimum rate of evaporation, and maximum surface viscosity in these systems [12].

Monolayers are fascinating systems with extreme simplicity and well-defined parameters. During the past 35 years of research, we have found the studies on monolayers to be rewarding in understanding the phenomena occurring at the gas–liquid, liquid–liquid, and solid–liquid interfaces in relation to foams, emulsions, lubrication, and wetting processes.

III. MICELLE KINETICS AND TECHNOLOGICAL APPLICATIONS

It is well recognized that a surfactant solution has three components: surfactant monomers in the aqueous solution, micellar aggregates in solution, and monomers absorbed as a film at the interface. The surfactant is in dynamic equilibrium among all of these components. From various theoretical considerations as well as experimental results, it can be assumed that micelles are dynamic structures whose stability is in the range of milliseconds to seconds. Thus, in an aqueous surfactant solution, micelles break and reform at a fairly rapid rate [13–15]. Figure 8 shows the two characteristic relaxation times, τ_1 and τ_2, associated with micellar solutions. The shorter time, τ_1, generally of the order of microseconds, is related to the exchange of surfactant monomers between the bulk solution and the micelles, whereas the longer time, τ_2, generally of the order of milliseconds to seconds, is related to the formation or dissolution of a micelle after several molecular exchanges [13,14]. It has been proposed that the lifetime of a micelle can be approximated by $n\tau_2$, where n is the aggregation number of the micelle [15]. Thus, relaxation time τ_2 is proportional to the lifetime of the micelle. A large value of τ_2 represents high stability of the micellar structure.

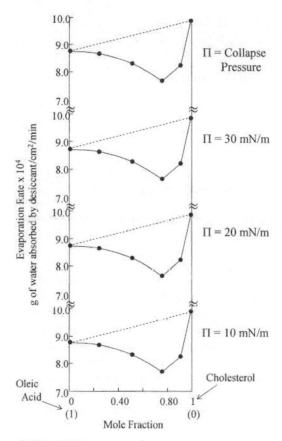

FIG. 7 Minimum rate of evaporation at a 1:3 molar ratio in mixed monolayers of oleic acid and cholesterol.

Figure 9 shows the relaxation time τ_2 of micelles of sodium dodecyl sulfate (SDS) as a function of SDS concentration [13,16,17]. It is evident that the maximum relaxation time of micelles is observed at an SDS concentration of 200 mM. This implies that SDS micelles are most stable at this concentration. For several years researchers at the CSSE have tried to correlate the measured τ_2 with various equilibrium properties such as surface tension, surface viscosity, and others, but no correlation could be found. However, a strong correlation of τ_2 with various dynamic processes such as foaming ability, wetting time of textiles, bubble volume, emulsion droplet size, and solubilization of benzene in micellar solution was found [18].

Fast relaxation time, microseconds

Slow relaxation time, milliseconds

FIG. 8   Two relaxation times of micelles, τ_1 and τ_2, and related molecular processes.

Figures 10 and 11 summarize the effects of SDS concentration on the phenomena mentioned as well as on other related phenomena. Figure 10 shows typical phenomena in liquid–gas systems, and Fig. 11 shows typical phenomena in liquid–liquid and solid–liquid systems. It is evident that each of these phenomena exhibits a maximum or minimum at 200 mM SDS, depending on the molecular process involved. Thus the "take-home message" emerging from our extensive studies of the past decades is that micellar stability can be the rate-controlling factor in the performance of various technological processes such as foaming, emulsification, wetting, bubbling, and solubilization [19].

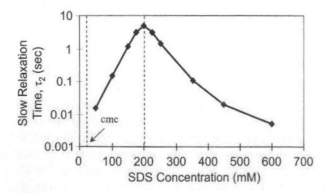

FIG. 9   Relaxation time, τ_2, of SDS micelles as a function of SDS concentration. Maximum τ_2 found at 200 mM (vertical line).

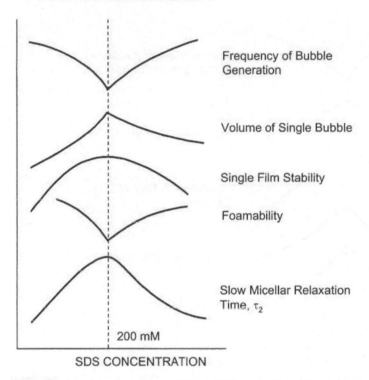

Frequency of Bubble Generation

Volume of Single Bubble

Single Film Stability

Foamability

Slow Micellar Relaxation Time, τ_2

200 mM

SDS CONCENTRATION

FIG. 10 Various liquid–gas system phenomena exhibiting minima or maxima at 200 mM SDS.

The currently accepted explanation for the effect of surfactant concentration on micellar stability was proposed by Aniansson and coworkers in the 1970s and expanded by Kahlweit and coworkers in the early 1980s [13–16]. Annianson's model [13–15] nicely predicts micelle kinetics at a low surfactant concentration based on stepwise association of surfactant monomers. Hence, the major parameters in this model are the critical micelle concentration (cmc) and the total concentration of the surfactant in solution.

At higher surfactant (and hence counterion) concentrations, experimental results begin to deviate from Aniansson's model. Kahlweit's fusion–fission model [16] takes into account the concentration and ionic strength of the counterions in these solutions and proposes that as the counterion concentration increases, the charge-induced repulsion between micelles and submicellar aggregates decreases, leading to coagulation of these submicellar aggregates.

In both of these models, the effects of intermicellar distance as well as the distance between submicellar aggregates have not been taken into ac-

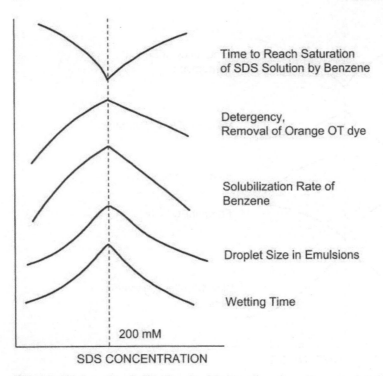

Time to Reach Saturation
of SDS Solution by Benzene

Detergency,
Removal of Orange OT dye

Solubilization Rate of
Benzene

Droplet Size in Emulsions

Wetting Time

200 mM

SDS CONCENTRATION

FIG. 11 Various liquid–liquid and solid–liquid system phenomena exhibiting minima or maxima at 200 mM SDS.

count. Researchers at the CSSE have been attempting to introduce the effect of intermicellar distance into micellar kinetic theory. As the SDS concentration increases, the number of micelles increases, and thus the intermicellar distance decreases. By knowing the aggregation number of the micelles, the number of micelles present in the solution can be calculated. The solution can then be divided into cubes such that each cube contains one micelle. From this, the distance between the centers of the individual cubes can be taken as the intermicellar distance.

Researchers at the CSSE determined that the concentration at which the SDS micelle was most stable (200 mM) coincided with an intermicellar distance of approximately one micellar diameter [19–23]. At this concentration, one would expect a tremendous coulombic repulsion between the micelles at such a short distance. A possible explanation for the observed stability is that there is a rapid uptake of sodium as a counterion on the micellar surface at this concentration, making the micelles more stable. Thus, the coulombic repulsion between micelles with the concomitant uptake of sodium ions allows the stabilization of micelles at this intermicellar distance and, hence, maximum τ_2 at 200 mM concentration.

By introducing this so-called intermicellar coulombic repulsion model (ICRM) into existing theoretical models, better agreement between theoretically calculated and experimental τ_2 values may be attained.

It should be mentioned that Per Ekwall proposed first, second, and third critical micelle concentrations for sodium octanoate solutions [22]. At the second cmc, he showed sudden uptake or binding of sodium ions to the micellar surface and he proposed that at the second cmc there was tight packing of surfactant molecules in the micelle. Thus, our 200 mM SDS concentration could be equivalent to the second cmc as proposed by Ekwall.

The phenomenon of a surfactant exhibiting a maximum τ_2 appears to be a general behavior, and perhaps other anionic or cationic surfactants may form tightly packed micelles at their own characteristic concentrations. As with the first cmc, this critical concentration may also depend upon physical and chemical conditions such as temperature, pressure, pH, salt concentration, and other parameters in addition to the molecular structure of the surfactant [24,25]. Work is currently in progress at the CSSE on identifying a similar critical concentration for nonionic surfactants as well as mixed surfactant systems.

In addition to this work, it has been shown that upon incorporation of a short-chain alcohol such as hexanol into the SDS micelles, the maximum τ_2 occurs at a lower SDS concentration [26–28]. Thus, it appears that in a mixed surfactant system, one can produce the most stable micelle at a lower surfactant concentration upon incorporation of an appropriate cosurfactant [29]. Also investigated was the effect of long-chain alcohols on the micellar stability. Results were similar to those for short-chain alcohols for all but dodecanol, which showed a significant increase in micellar stability over micelles containing only SDS because of the chain length compatibility effect [30].

Figure 12 shows the effect of coulombic attraction between oppositely charged polar groups as well as the chain length compatibility effect on the τ_2 or micellar stability of SDS plus alkyltrimethylammonium bromide (cationic surfactant) solutions. It shows that surface tension, surface viscosity, miscellar stability, foaming ability, and foam stability are all influenced by the coulombic interaction as well as the chain length compatibility effect [30,31]. It should be noted that the ratio (weight basis) of SDS to alkyltrimethylammonium bromide was 95:5 in this mixed surfactant system. However, even at this low concentration, the oppositely charged surfactant dramatically changed the molecular packing of the resulting micelles as well as the surfactant film absorbed at the interface.

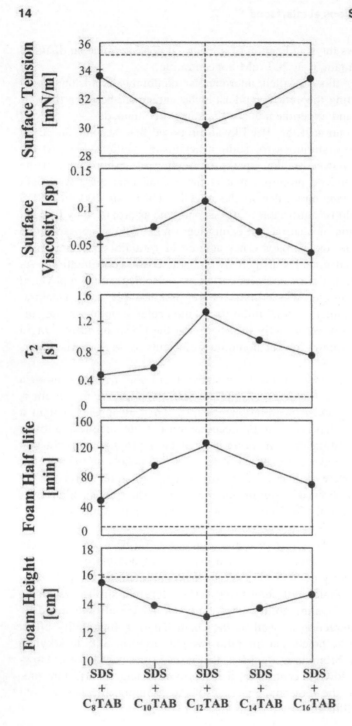

In view of the previous discussion, it is evident that micellar stability is of considerable importance to technological processes such as foaming, emulsification, wetting, solubilization, and detergency because a finely tuned detergent formulation can significantly improve the cleaning efficiency as well as reduce the washing time in a laundry machine, resulting in significant energy savings at a national and global level. Micellar stability is thus a critical issue in any application in which surfactants are present as micelles, and the subsequent monomer flux is utilized in the application.

IV. MICROEMULSIONS IN ENHANCED OIL RECOVERY AND SYNTHESIS OF NANOPARTICLES

As early as 1943, Professor J. H. Schulman published reports on transparent emulsions [32]. From various experimental observations and intuitive reasoning, he concluded that such transparent systems were microemulsions. Figure 13 illustrates the transparent nature of a microemulsion in comparison with a macroemulsion. He also proposed the concept of a transient negative interfacial tension to induce the spontaneous emulsification in such systems.

Considerable studies have been carried out on microemulsions during the past quarter-century, during which time it has been recognized that there are three types of microemulsions: lower-phase, middle-phase, and upper-phase microemulsions. The lower-phase microemulsion can remain in equilibrium with excess oil in the system, the upper-phase microemulsion can remain in equilibrium with excess water, and the middle-phase microemulsion can remain in equilibrium with both excess oil and water. As a result, the lower-phase microemulsion has been considered to be an oil-in-water microemulsion, the upper-phase microemulsion has been considered to be a water-in-oil microemulsion, whereas the middle-phase microemulsion has been the subject of much research and has been proposed to be composed of bicontinuous or phase-separated swollen micelles from the aqueous phase [33–44]. Figure 14 shows the lower-, middle-, and upper-phase microemulsions, as represented by the darker liquid in each tube [45–48].

The formation of lower-, middle-, and upper-phase microemulsions is related to the migration of surfactant from lower phase to middle phase to upper phase. Figure 15 illustrates that migration of the surfactant from the

---

FIG. 12 Effect of coulombic attraction between polar groups of surfactants and chain length compatibility on τ_2 or micellar stability and other interfacial properties of mixed surfactant solutions. The dashed line represents the specific property of 100 mM pure SDS solution.

FIG. 13 Illustration of the transparent nature of a microemulsion compared with a macroemulsion.

aqueous phase to the middle phase to the oil phase can be induced by changing a number of parameters, including adding salts (e.g., NaCl) to the system, decreasing the oil chain length, increasing the surfactant molecular weight, adding a cosurfactant, and decreasing the temperature [48,49]. However, it has been reported that in oil/water/nonionic surfactant systems, surfactant moves from lower phase to middle phase to upper phase as the temperature is increased [42,43], so each system must be carefully analyzed in order to determine the effects of certain parameters.

Microemulsions exhibit ultralow interfacial tension with excess oil or water phases. Therefore, the middle-phase microemulsion is of special importance to the process of oil displacement from petroleum reservoirs.

A.   Microemulsions in Enhanced Oil Recovery

Figure 16 shows a schematic view of a petroleum reservoir as well as the process of water or chemical flooding by an inverted five-spot pattern [33]. Several thousand feet below the ground, oil is found in tightly packed sand or sandstones in the presence of water as well as natural gas. During the primary and secondary recovery processes (water injection method), about 35% of the available oil is recovered. Hence, approximately 65% is left in the petroleum reservoir. This oil remains trapped because of the high interfacial tension (20–25 mN/m) between the crude oil and reservoir brine. It is known that if the interfacial tension between crude oil and brine can be reduced to around 10^{-3} mN/m, one can mobilize a substantial fraction of

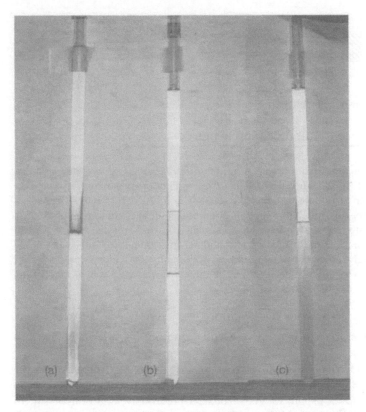

FIG. 14 Samples of (a) lower-, (b) middle-, and (c) upper-phase microemulsions in equilibrium with excess oil, excess water and oil, or excess water, respectively.

the residual oil in the porous media in which it is trapped. Once mobilized by an ultralow interfacial tension, the oil ganglia must coalesce to form a continuous oil bank. The coalescence of oil droplets has been shown to be enhanced by a very low interfacial viscosity in the system. The incorporation of these two critical factors into a suitable surfactant system for oil recovery was crucial in developing the surfactant–polymer flooding process for enhanced oil recovery from petroleum reservoirs. Conceptually, one injects a surfactant formulation in the porous media in the petroleum reservoir so that upon mixing with the reservoir brine and oil it produces the middle-phase microemulsion in situ. This middle-phase microemulsion, which is in equilibrium with excess oil and excess brine, propagates throughout the petroleum reservoir. The design of the process is such that the oil bank maintains

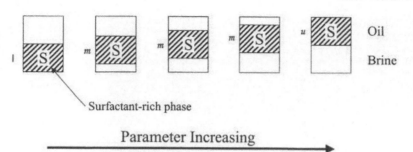

Surfactant-rich phase

Parameter Increasing

FIG. 15 The transition from lower- to middle- to upper-phase microemulsions can be brought about by the addition of salts or by varying other parameters. The transition from lower to middle to upper phase (I $\to$ m $\to$ u) occurs by (1) increasing salinity, (2) decreasing oil chain length, (3) increasing alcohol concentration (C_4, C_5, C_6), (4) decreasing temperature (for ionic surfactants), (5) increasing total surfactant concentration (for high-molecular-weight anionic surfactants), (6) increasing brine/oil ratio (for high-molecular-weight anionic surfactants), (7) increasing surfactant solution/oil ratio (for high-molecular-weight anionic surfactants), and (8) increasing molecular weight of surfactant.

ultralow interfacial tension with reservoir brine until it arrives at the production wells.

One parameter that has been discovered to be crucially important in the successful implementation of the surfactant–polymer flooding process is the salinity of the aqueous phase. As discussed previously, addition of salt to the microemulsion system induces the change from lower- to middle- to upper-phase microemulsion (Fig. 15) [33]. It was found that at a particular salt concentration, deemed the optimal salinity, a number of important parameters were optimized for the oil recovery process. The optimal salinity was found to occur when equal amounts of oil and brine were solubilized by the middle-phase microemulsion [50].

Figure 17 summarizes the various parameters that are important in the surfactant–polymer flooding process as a function of salt concentration [33,51–54]. It is evident that all of these parameters exhibit a maximum or a minimum at the optimal salinity. Thus, it appears that all of these processes are interrelated for the oil displacement in porous media by the surfactant–polymer flooding process. It also appears that the optimal salinity value is a crucial parameter for consideration of a system to be used in this process.

B. Formation of Nanoparticles Using Microemulsions

Another very interesting use of microemulsions that has been investigated in our laboratory over the past decade is in the production of nanoparticles.

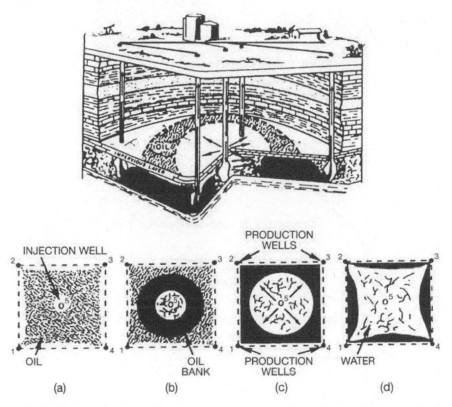

FIG. 16 Schematic view of a petroleum reservoir and the process of water or chemical flooding (five-spot pattern).

Figure 18 schematically illustrates the formation of nanoparticles using water-in-oil microemulsions. For this process, two identical water-in-oil microemulsions are produced, the only difference between the microemulsions being the nature of the aqueous phase, into which the two water-soluble reactants, A and B, are dissolved separately. Upon mixing the two nearly identical microemulsions, the water droplets collide and coalesce, allowing the mixing of the reactants to produce the precipitate AB. Ultimately, these droplets again disintegrate into two aqueous droplets, one containing the nanoparticle AB and the other containing just the aqueous phase [55–57]. Thus, a precipitation reaction can be carried out in the aqueous cores of water-in-oil microemulsions using the dispersed water droplets as nanoreactors. The size of the particles formed is physically limited by the reactant concentration as well as the size of the water droplets. In this way,

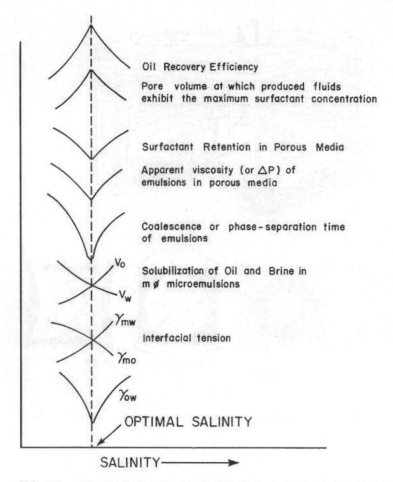

Oil Recovery Efficiency

Pore volume at which produced fluids exhibit the maximum surfactant concentration

Surfactant Retention in Porous Media

Apparent viscosity (or ΔP) of emulsions in porous media

Coalescence or phase-separation time of emulsions

V_o

Solubilization of Oil and Brine in m $\emptyset$ microemulsions

V_w

γ_{mw}

Interfacial tension

γ_{mo}

γ_{ow}

OPTIMAL SALINITY

SALINITY ⟶

FIG. 17 Various phenomena occurring at the optimal salinity in the surfactant–polymer flooding process for enhanced oil recovery.

monodisperse particles in the range 2–10 nm in diameter can be produced. The production of nanoparticles with homogeneity of particle size (i.e., small size range) has inherently been a problem with other conventional methods. This method of nanoparticle synthesis is an improvement over other methods for applications that require the production of monodisperse nanoparticles [58–60].

Superconducting nanoparticles have also been produced in our laboratory using the microemulsion method. Table 1 shows the composition of the two microemulsions used for synthesizing nanoparticles of YBCO (Yttrium Bar-

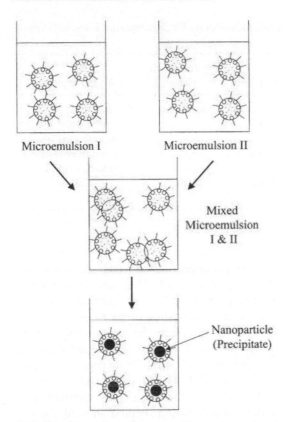

FIG. 18 Formation of nanoparticles using microemulsions (water-in-oil) as nano-reactors. The water droplets continually collide, coalesce, and break up upon mixing of two microemulsions containing reactants.

ium Copper Oxide) superconductor [61–63]. In this case, water-soluble salts of yttrium, barium, and copper were dissolved in the aqueous cores of one microemulsion and ammonium oxalate was dissolved in the aqueous cores of the other microemulsion. Upon mixing the two microemulsions, precursor nanoparticles of metal oxalates were formed. The nanoparticles were centrifuged and then washed with chloroform, methanol, or acetone to remove the surfactants and oil.

These nanopowders were then calcined at the appropriate temperature to convert the oxalate precursors into oxides of these materials. The oxides were then compressed into a pellet and sintered at 860°C for 24 h. The pellet was cooled, and the critical temperature of zero resistance was measured. It was found that this critical temperature did not show any change from crit-

TABLE 1 Composition of Two Microemulsions for Synthesizing Nanoparticles of YBCO Superconductors

	Surfactant phase	Hydrocarbon phase	Aqueous phase
Microemulsion I	CTAB + 1-butanol	n-Octane	(Y, Ba, Cu)nitrate solution, total metal concentration = 0.3 N
Microemulsion II	CTAB + 1-butanol	n-Octane	Ammonium oxalate solution, 0.45 N
Weight fraction (for both I and II)	29.25%	59.42%	11.33%

ical temperatures of superconductors produced by the traditional coprecipitation method. However, the fraction of the ideal Meissner shielding was strikingly different for the two samples prepared by different methods. It is the Meissner effect that is related to the levitational effect of the superconducting pellet on a magnetic field. Thus, it appears that the leakage of magnetic flux from the conventionally prepared sample was greater than that from the sample produced by microemulsion-derived nanoparticles. Figure 19 shows scanning electron microscope (SEM) images of sintered pellets produced by the two methods. It is evident that the pellets prepared from the nanoparticles produced by the microemulsion method showed 30–100 times larger grain size, less porosity, and higher density as compared with the samples prepared by conventional precipitation of aqueous solutions of these salts. A possible explanation for these effects is that nanoparticles, because of their extremely small size and large surface area, can disintegrate very quickly and allow diffusion of atoms to the site of the growing grains to support the growth process of the grains. Therefore, samples prepared from nanoparticles exhibit large grain size and low porosity [64]. Thus, it appears that these nanoparticles may be useful to produce high-density ceramics.

Since the pioneering work of the first 20 years on the formation of nanoparticles of heavy metals by the microemulsion method [65], we have added to the understanding of the mechanism and control of the reaction kinetics in microemulsions by controlling the interfacial rigidity of the microemulsion droplet. We have introduced the concept of a chain length compatibility effect observed for the reactions in microemulsions, in which the interfacial rigidity is maximized by a certain chain length combination of the surfactant, oil, and cosurfactant alcohol, causing a decreased reaction rate [66]. All of

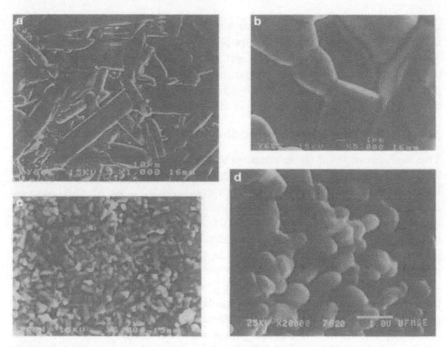

FIG. 19 SEM images of superconducting pellets prepared from nanopowders (a and b) using microemulsions and conventionally prepared powders (c and d) (by precipitation of aqueous solutions).

these contributions have led to a greater understanding of the method of nanoparticle production using microemulsion media.

V. CONTROL OF POLYMER ADSORPTION AT PARTICLE SURFACES

Polymers at particle surfaces play an important role in a range of technologies such as paints, polishing, filtration, separations, enhanced oil recovery, and lubrication. In order to optimize these technologies, it is important to understand and control the adsorption, conformation, and role of surface molecular architecture in selective polymer adsorption.

For a particular polymer functionality, the adsorption depends on the nature and energetics of the adsorption sites that are present on the surface. The adsorption of polymers via electrostatics, chemical bonding, and hydrophobic interaction is relatively well understood, and most of the unexpected adsorption behavior is attributed to hydrogen bonding, which is ubiquitous in nature. Research carried out at the University of Florida Engineering Research Center for Particle Science and Technology (UF-ERC) has focused on the role of surface chemistry and surface molecular architecture in hydrogen bonding of polymers to particle surfaces. In addition, practical applications of controlled polymer adsorption have been investigated for solid–solid separations via selective flocculation technology.

A. Control of Hydrogen Bonding

The surfaces of most oxides and minerals have two different kinds of acid sites, Brönsted and Lewis, on which hydrogen-bonding polymers can adsorb. Brönsted acids are defined as proton donors, such as the M—OH sites on oxide surfaces. The more electron withdrawing the underlying substrate, the greater is the Brönsted acidity. Lewis acid sites are defined as electron deficient or as having the ability to accept electrons. Examples of Lewis acid sites include $M^+(OH)_2$ groups on oxide surfaces.

For a hydrogen-bonding polymer such as poly(ethylene oxide) (PEO), whose ether oxygen linkage acts as a Lewis base, it was illustrated that not only did the *number* of hydrogen-bonding sites differ from one substrate to another but also the *energy* of the hydrogen-bonding sites varied [67]. Based on adsorption studies of PEO on silica and other oxides [67], it was determined that the amount of adsorbed polymer depended on the nature of the Brönsted acid (proton donor) sites on the particles. Hence, the more electron withdrawing the underlying substrate, the greater the Brönsted acidity, and thus the lower the point of zero charge (pzc) of the material. It is seen from Fig. 20 that SiO_2, MoO_3, and V_2O_5 strongly adsorb PEO whereas oxides with pzc greater than that of silica, such as TiO_2, Fe_2O_3, Al_2O_3, and MgO, did not exhibit significant adsorption of PEO. However, within a single system (silica) it has been found that PEO will adsorb onto sol-gel–derived silica but not onto glass or quartz at a pH of 9.5. This suggests that the strength of Brönsted acid sites (higher ability to donate protons), as determined by the surface molecular architecture, also influences the adsorption process. It was also shown that the nature and energetics of the surface sites could be modified by surface modification techniques such as calcination and rehydroxylation [68]. Upon calcination of a silica surface to 800°C, the number of isolated surface hydroxyl groups [determined from Fourier transform infrared (FTIR) spectroscopy] and three-membered silicate rings (determined from Raman spectroscopy) increased, resulting in higher surface acidity [determined from solid-state nuclear magnetic resonance (NMR) spectroscopy using triethyl phosphine oxide probe]. These changes led to higher adsorption of the PEO polymer (Fig. 21) [68].

Based on the polymer functionality, it may be possible to predict the surface molecular architecture or the surface sites that are required for the

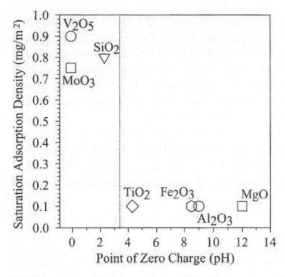

FIG. 20 Effect of surface Brönsted acidity on the adsorption of a hydrogen-bonding polymer, poly(ethylene oxide) (PEO). (From Ref. 31.)

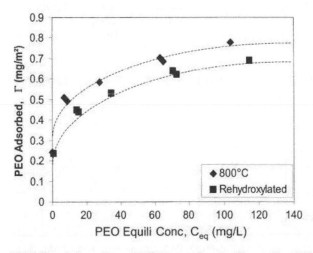

FIG. 21 Adsorption of PEO on sol-gel silica with different treatments (calcined 800°C, rehydroxylated). (From Ref. 32.)

polymer to adsorb. Studies of the adsorption of PEO, PAM (nonionic poly-
acrylamide), PAA [nonionic poly(acrylic acid)], PAH (polyallylamine hydrochloride), and PVA (poly (vinyl alcohol)) onto silica, alumina, and hematite were carried out, and a master table correlating functional group–
active surface site relationships was developed (Table 2) [69]. Strong
adsorption of PEO onto silica was observed, with none onto alumina and
hematite, in agreement with the earlier results [67].

No adsorption of PAM onto silica was observed; however, PAM was
found to adsorb onto hematite and alumina at pH 3.0. At pH 9.5, there was
no adsorption of PAM onto alumina. Given the lack of adsorption of PAM
onto silica, which has strong Brönsted acid sites, it was concluded that PAM

TABLE 2 Correlation of Polymer Functionality with the Surface Adsorption Sites

Polymer	Repeat unit	Functionality	Adsorbs onto	Adsorption sites
PEO	$\left[-CH_2-CH_2-O-\right]_n$	Ether	SiO_2	Brönsted
PVA	$\left[-CH_2-CH(OH)-\right]_n$	Hydroxyl	SiO_2	Brönsted
PAA	$\left[-CH_2-CH(C(=O)OH)-\right]_n$	Carboxylic acid	Fe_2O_3 Al_2O_3 TiO_2	Lewis
PAM	$\left[-CH_2-CH(C(=O)N(H)(H))-\right]_n$	Amide	Fe_2O_3 Al_2O_3 TiO_2	Lewis
PAH	$\left[-CH_2-CH(CH_2-NH_2)-\right]_n$	Amine	SiO_2	Brönsted

PEO, poly(ethylene oxide); PVA, poly(vinyl alcohol); PAA, poly(acrylic acid); PAM, poly-
(acrylamide) (nonionic); PAH, polyallylamine.
Source: Ref. 33.

adsorbed on Lewis acid sites. The lack of adsorption of PAM onto alumina at pH 9.5 was explained by "poisoning" of active Lewis sites due to preferential adsorption of hydroxide ions at pH values below the isoelectric point of alumina ($pH_{iep} = 8.8$).

Nonionic PAA was found to adsorb onto hematite and alumina but not onto silica at pH 3. Adsorption experiments were not conducted above pH 3 because PAA becomes ionized and the adsorption is dominated by electrostatic interactions.

Having developed the fundamental knowledge base on the role of surface molecular architecture in polymer adsorption and the surface site—polymer functional group correlation, the next logical step was to use these fundamentals in "real" particulate systems, which contain heterogeneous surfaces with impurities that in some cases may foil selective adsorption schemes.

B. Control of Polymer Adsorption for Selective Separation

Flocculation of fine particles using polymeric materials (flocculants) and separation of such aggregates from particles of the other component(s) in the dispersed phase is known as selective flocculation [67]. The competition between different surfaces for the flocculant must be controlled in order to achieve adsorption on the targeted component(s). The aggregates of the polymer-coated particles, or "flocs," thus formed are separated from the suspension by either sedimentation—elutriation or floc flotation.

The major barrier to further commercialization of the selective flocculation technology is the poor success in extension of single-component successes to mixed-component systems. The selectivity observed in single-component tests is often lost in mixed-component or natural systems. One of the significant reasons for this loss in selectivity is heteroflocculation, wherein a small amount of polymer adsorption on the inert material leads to coflocculation with the active material. One of the major advances at the University of Florida (UF) has been the development of the site-blocking agent (SBA) concept [70,71] to overcome heteroflocculation, thus achieving selectivity in particle separation. The SBA concept is illustrated schematically in Fig. 22. The concept involves blocking all the active sites for polymer adsorption on the inert material (component of the particle mixture not to be flocculated) by addition of the SBA. After the addition of the SBA, when the flocculant is added, it adsorbs only onto the active component (material intended to be flocculated), resulting in selectivity of separation. A lower molecular weight fraction of the same or a similar flocculant, which on its own is incapable of inducing flocculation in the active or the floc-

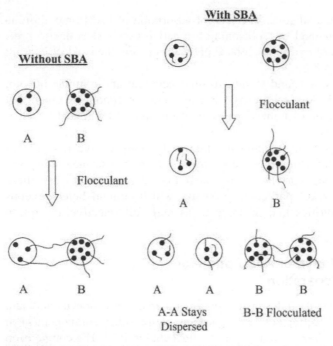

FIG. 22 Schematic illustrating the site-blocking agent (SBA) concept. (From Ref. 31.)

culating material, was successfully used as an SBA to minimize heteroflocculation. The concept has since been commercialized by Engelhard Corporation for removal of titania impurities from kaolin clays [72].

VI. SURFACTANT SELF-ASSEMBLY AT THE SOLID–LIQUID INTERFACE

Surfactants at the solid–liquid interface are used in various industrial processes ranging from ore flotation and paint technology to enhanced oil recovery [73]. Apart from the traditional uses of surfactants, surfactant structures are increasingly being investigated as organic templates to synthesize mesoscopic inorganic materials with controlled nanoscale porosity, which are expected to have applications in electronics, optics, magnetism, and catalysis [74]. Surfactant structures at the solid–liquid interface have also been utilized to stabilize particulate dispersions [75–77]. Recent work carried out at UR-ERC has shown that self-assembled surfactants can be utilized to prepare particulate dispersions under extreme conditions [78,79] in which traditionally used dispersing methods, such as electrostatics (surface charge), inorganic dispersants (sodium silicate, sodium hexametaphosphate), and polymers, may not result in a completely dispersed suspension.

Figure 23 depicts both the suspension turbidity—a measure of stability —and the surface forces present between the atomic force microscope (AFM) tip and silica substrate as a function of dodecyltrimethylammonium bromide ($C_{12}TAB$) concentration at pH 4 and 0.1 M NaCl. Under these conditions, the silica suspension without a dispersant is unstable. As the surfactant concentration is increased, the suspension remains unstable until a surfactant concentration of about 8 mM. Between surfactant concentrations of 8 and 10 mM, a sharp transition in the stability (unstable to stable) and forces (no repulsion to repulsion) is observed. A good correlation exists between the suspension stability and repulsive forces due to self-assembled surfactant aggregates. The repulsive force is an order of magnitude higher than electrostatic forces alone, indicating that the repulsion is steric in origin. It was proposed that the dominant repulsion mechanism was the steric repulsion due to the elastic deformation of the self-assembled aggregates when two surfaces approached each other.

The adsorption, zeta potential, and contact angle measurements on silica surfaces in 0.1 M NaCl at pH 4.0 as a function of solution $C_{12}TAB$ concen-

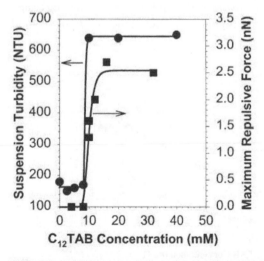

FIG. 23 Turbidity of silica particles after 60 min in a solution of 0.1 M NaCl at pH 4 as a function of $C_{12}TAB$ concentration, and the measured interaction forces between an AFM tip and silica substrate under identical solution conditions. (From Ref. 42.)

tration were measured and are presented in Fig. 24. Based on interface properties, the entire self-assembly process was divided into six stages, marked A–F in Fig. 24. At low concentrations (below 0.007 mM, region A in Fig. 24), individual surfactant adsorption takes place. The next structural transition in this system was the formation of hemimicelles, which is evidenced by a significant effect on both the zeta potential and hydrophobicity of the surface. At approximately 0.1 mM in region B, the sign of the zeta potential reverses but the contact angle continues to increase, indicating that the reversal in zeta potential is not due to formation of bilayers as suggested in the past [73,75–77]. This reversal in zeta potential while the hydrophobicity continues to increase was attributed either to hydrophobic association between the surfactant tails, resulting in formation of hemimicelles, or to some kind of specific adsorption.

In regions B and C, contact angle continues to increase, indicating increasing concentration of hemimicelles at the interface. Beyond a certain concentration (approximately 2.3 mM), in region D the hydrophobicity decreases, accompanied by a sharp increase in the zeta potential. This indicates the formation of structures with an increasing number of polar heads oriented toward the solution. The sharp increase in zeta potential and a corresponding decrease in contact angle were attributed to the transition of surfactant structure from hemimicelles to either bilayers, spherical aggregates (imaged at surfaces using AFM), or structures having semispheres on top of perfect monolayers (compact monolayer covering the entire surface), as suggested by Johnson and Nagarajan [80]. At higher surfactant concentrations beyond

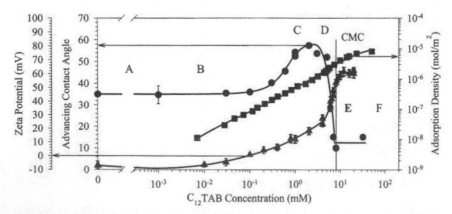

FIG. 24 Adsorption isotherm (squares), zeta potential (triangles), and contact angle (spheres) of silica surfaces in 0.1 M NaCl at pH 4.0 as a function of solution C_{12}TAB concentration. (From Refs. 42 and 43.)

the bulk cmc (regions E and F), based on AFM imaging, spherical aggregates or composite semispheres on top of perfect monolayers (it is not possible to distinguish between these structures on the basis of just AFM imaging) are known to exist [81]. Thus, based on the adsorption, zeta potential, and contact angle results, several plausible surfactant structures were proposed at the interface at different concentrations of the surfactant. To illustrate the exact structural transitions taking place at the interface, the FTIR–ATR (with polarized IR beam) technique, which can probe the adsorbed structures directly, was employed.

The FTIR–ATR technique relies on the fact that individual surfactant molecules, hemimicelles, monolayers, bilayers, and spherical or cylindrical aggregates at the interface will have different average orientations of the alkyl chains with respect to the surface normal. Different average orientations result in different absorptions of the plane-polarized IR beam and can thus be used to identify the surfactant structures at the interface [79].

Based on the FTIR–ATR study, the proposed surfactant structures in the different regions were verified. In region D, it was found that spherical aggregates were formed directly from hemimicelles, without the formation of bilayers. In fact, no evidence of bilayer formation was seen in this system even at very high surfactant concentrations.

Based on the contact angle FTIR, zeta potential, and adsorption results, the preceding structural transitions are summarized by the schematic shown in Fig. 25, which illustrates the structures present at the interface in the concentration regions A–F in Fig. 24.

A. Control of the Repulsion Barrier Using Cosurfactants

In bulk micellization processes, it has been proposed that oppositely charged surfactant incorporates itself into micelles and by reducing the repulsion between the ionic groups increases stability and lowers the bulk cmc [82,83]. A similar process can be expected to occur at the solid–liquid interface. As depicted in Fig. 26, very small additions of SDS were observed to have a dramatic effect on the formation of the surfactant surface structures. Figure 26 shows the correlation of suspension stability of silica particles with the maximum repulsive force measured against a silica plate in the presence of 3 mM C_{12}TAB and 0.1 M NaCl at pH 4 as a function of addition of SDS. At 5 μm SDS addition, no repulsive force is observed between the surfaces. However, at 10 μM SDS, strong repulsive force has developed and continues to increase with increase in SDS concentration. Correspondingly, over an identical range of SDS concentration, the initially unstable suspension becomes stabilized.

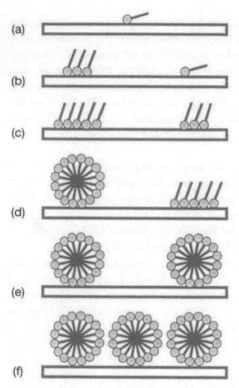

FIG. 25 Schematic representation of the proposed self-assembled surfactant films at concentrations corresponding to A–F in Fig. 24. (a) Individual surfactant adsorption, (b) low concentration of hemimicelles on the surface, (c) higher concentration of hemimicelles on the surface, (d) hemimicelles and spherical surfactant aggregates formed due to increased surfactant adsorption and transition of some hemimicelles to spherical aggregates, (e) randomly oriented spherical aggregates at onset of steric repulsive forces, and (f) surface fully covered with randomly oriented spherical aggregates. (From Ref. 43.)

The abnormally low concentration of SDS needed to form the surface surfactant structures may be explained by the preferential adsorption of SDS into self-assembled C_{12}TAB aggregates at the silica surface. Adsorption experiments, using total organic carbon analysis to determine the total amount of adsorbed surfactant and inductively coupled plasma spectroscopy to determine the SDS concentration through the sulfur emission line, have shown that nearly all the SDS added adsorbed at the solid–liquid interface. Hence, the molecular ratio at the interface was estimated to be on the order of 1:10 instead of 1:100 in bulk solution. This is particularly interesting because

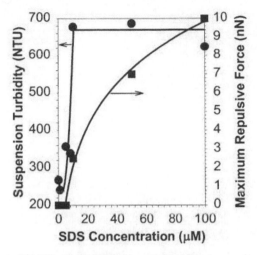

FIG. 26 Turbidity of a 0.02 vol% suspension of sol-gel–derived 250-nm silica particles after 60 min in a solution of 0.1 M NaCl at pH 4 with 3 mM C_{12}TAB as a function of SDS addition and the measured interaction forces between an AFM tip and silica substrate under identical solution conditions. (From Ref. 42.)

it was found that no measurable quantity of SDS adsorbed onto silica in the absence of C_{12}TAB. In addition, because little SDS is present in solution, the system is far below the bulk cmc and yet strong repulsive forces are once again observed.

The use of cosurfactants or other coadsorbing reagents is a critical factor in the utility of a surfactant dispersant in industrial processes. Not only can the concentrations for effective stabilization be reduced, but also many other options can become available to control the overall dispersion of single- and multicomponent suspensions. Availability of these engineered dispersant systems can enhance the processing of nanoparticulate suspensions for emerging specialized end uses, such as chemical–mechanical polishing of silicon wafers in microelectronics manufacturing.

VII. STABILIZATION OF CHEMICAL–MECHANICAL POLISHING SLURRIES UTILIZING SURFACE-ACTIVE AGENTS

Chemical–mechanical polishing (CMP) is a widely used technique in microelectronic device manufacturing to achieve multilevel metallization (Fig. 27). In the CMP process, the wafer surface (on which the microelectronic devices are built) is planarized by using a polymeric pad and a slurry com-

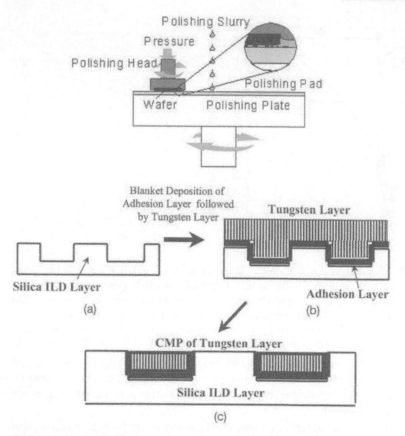

FIG. 27 (Top) Schematic representation of chemical–mechanical polishing (CMP) process. (From http://www.el.utwente.nl/tdm/mmd/projects/polish/index.html.) (Bottom) Review of tungsten CMP: (a) silica (interlayer dielectric) is etched, (b) tungsten is deposited onto silica ILD, and (c) CMP is applied to remove excessive tungsten layer and other levels are built on this level (multilevel metallization). (From Ref. 90.)

posed of submicrometer-size particles and chemical. The ultimate goal of CMP is to achieve an optimal material removal rate while creating an atomically smooth surface finish with a minimal number of defects. This can be accomplished by the combined effect of the chemical and mechanical components of the process. The mechanical action in CMP is mostly provided by the submicrometer-size abrasive particles contained in the slurries as they flow between the pad and the wafer surface under the applied pressure. The chemical effect, on the other hand, is provided by the addition of pH regulators, oxidizers, or stabilizers depending on the type of the CMP operation, which makes it easy for the reacted surface to be removed by abrasive particles.

As the rapid advances in the microelectronics industry demand a continuous decrease in the sizes of the microelectronic devices, removal of a very thin layer of material with atomically flat and clean surfaces has to be achieved during manufacturing [84]. These trends necessitate improved control of the CMP process by analyzing the slurry particle size distribution and stability effects on polishing. Past investigations suggest the use of monosized particles for the CMP slurries to achieve a planarized surface and to minimize the surface deformation [85]. However, in practical applications there may be oversize particles in the slurries in the form of hard-core larger particles (hard agglomerates) or agglomerates of the primary slurry particles because of slurry instability (soft agglomerates). Polishing tests conducted in the presence of hard agglomerates in the CMP slurries verified significant degradation in the polishing performance [86]. To remove the hard agglomerates, filtration of slurries is commonly practiced in industrial CMP operations. Nevertheless, even after filtering the slurries, the defect counts on the polished surfaces have been observed to be higher than desired [87]. This observation suggested the possibility of formation of soft agglomerates during the polishing operations. Indeed, it was reported that the commercial CMP slurries tended to coagulate and partially disperse during polishing [88]. Figure 28 shows AFM images of silica wafers polished with soft agglomerated baseline silica slurries of 0.2 μm monosize (at pH 10.5). It was observed that even the soft agglomerates resulted in significant surface deformations [89], indicating that the CMP slurries must remain stable to obtain optimal polishing performance.

A. Stabilization of Alumina Slurries Using Mixed Surfactant Systems for Tungsten Chemical–Mechanical Polishing

In CMP processes, polishing slurries have to be stabilized in extreme environments of pH, ionic strength, pressure, and temperature. In tungsten CMP, high concentrations of potassium ferricyanide are used to enhance surface oxidation of tungsten. These species reduce the screening length between the alumina particles of the CMP slurries to near zero, allowing for rapid coagulation of particles and destabilization of dispersions. It has been shown by Palla [90] that addition of a mixture of ionic (SDS) and nonionic (Tween 80) surfactants can stabilize alumina particles in the presence of high concentrations of charged species. Figure 29 illustrates schematically the mechanism of stabilization, which can be explained as enhanced ad-

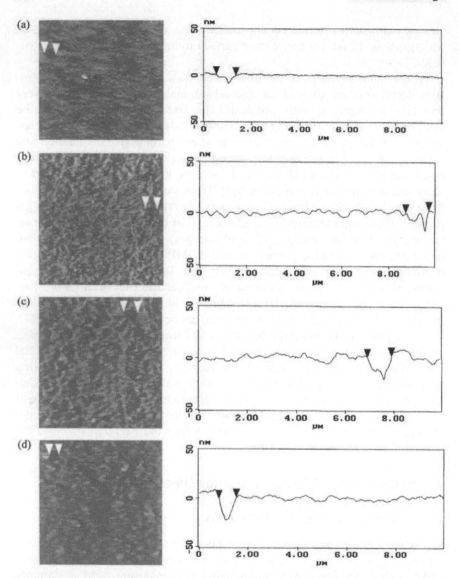

FIG. 28 Surface quality response of the silica wafers polished with (a) baseline 0.2-μm size, 12 wt% slurry; (b) dry aggregated slurry; (c) PEO flocculated slurry; (d) NaCl coagulated slurry. The inverted triangles in the AFM images (left) show the locations corresponding to the sample roughness plots (right). (From Ref. 89.)

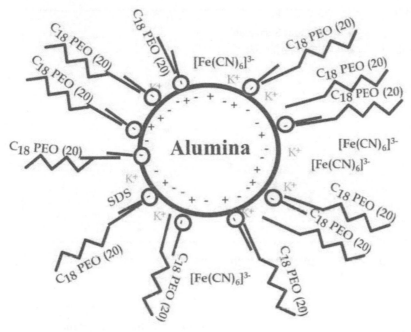

FIG. 29 Mechanisms of slurry stabilization with SDS (anionic) and Tween 80 (C_{18}PEO, nonionic) surfactants. (From Ref. 90.)

sorption of nonionic surfactant using the strongly adsorbing ionic surfactant as a binding agent. The stabilizing ability of the surfactant system was found to increase with increasing hydrophobicity of both the nonionic and ionic surfactants. The effect of surfactant concentration on stability is shown to have an optimal concentration range for a number of surfactants [90]. As shown in Fig. 30a and b, when the tungsten polishing was conducted using slurries stabilized by the described mixed surfactant system, 30% less material removal was obtained compared with the baseline slurry; however, much better surface quality was obtained [91].

B. Stabilization of Silica CMP Slurries Utilizing Self-Assembled Surfactant Aggregates: Role of Particle–Particle and Particle–Surface Interactions in CMP

As discussed in the previous section, self-assembled C_{12}TAB, a cationic surfactant, provided stability to silica suspensions at high ionic strengths by introducing a strong repulsive force barrier [78,79]. This novel concept has

Shah and Moudgil

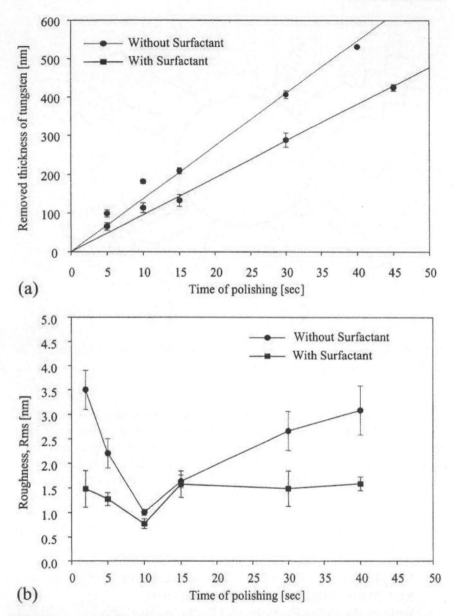

FIG. 30 (a) Material removal rate response of the tungsten CMP slurries stabilized with mixed surfactant systems (SDS and Tween 80). (b) Surface roughness response of the tungsten CMP slurries stabilized with mixed surfactant systems (SDS and Tween 80). (From Ref. 91.)

been applied to stabilize the coagulated silica CMP slurries in the presence of 0.6 M NaCl. The C_{12}TAB surfactant was used at 1, 8, and 32 mM concentrations according to previous findings reported by Adler et al. [78] and Singh et al. [79]. Figure 31 shows the mean particle size analyses of the baseline, 0.6 M NaCl, and 0.6 M NaCl + C_{12}TAB slurries [92]. Addition of 0.6 M NaCl destabilized the baseline CMP slurry by screening the charges around the silica particles at pH 10.5. Therefore, the mean size of the slurry increased to 4.3 μm from the baseline size of 0.2 μm. Addition of 1 mM C_{12}TAB further increased the mean particle size because the positively charged surfactant can screen more charges by adsorbing onto the silica particles. As described earlier, a jump was reported in the repulsive force barrier based on AFM force measurements at 8 mM C_{12}TAB, which was explained on the basis of the strength of the self-assembled surfactant aggregates as they formed between the AFM tip and the substrate [78,79]. As the repulsive force is increased, the slurry particles are expected to start stabilizing in the presence of 8 mM C_{12}TAB. In agreement with these findings, the mean size of the slurry with 8mM C_{12}TAB started to decrease, indicating that the stabilization had been initiated [92]. Finally, addition of 32 mM C_{12}TAB completely stabilized the 0.6 M NaCl–containing polishing slurry as enough repulsive force for particle–particle interaction was reached. Figure 32a summarizes the surface quality response in terms of

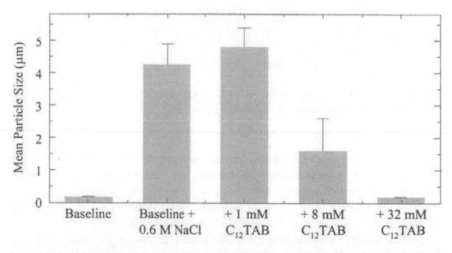

FIG. 31 Mean particle size analysis of the following slurries: baseline (Geltech 0.2 μm, 12 wt%, pH 10.5), baseline +0.6 M NaCl, and baseline +0.6 M NaCl + 1, 8, or 32 mM C_{12}TAB. The high-ionic-strength slurry is stable only at 32 mM C_{12}TAB addition. (From Ref. 92.)

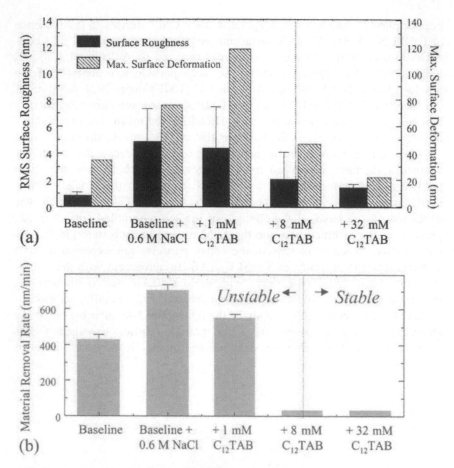

FIG. 32 (a) Surface quality response of $C_{12}TAB$ system. (b) Material removal rate response of $C_{12}TAB$ system. (From Ref. 92.)

surface roughness and maximum surface deformation (the maximum depth of the scratches or pits detected on the polished wafer surface) of the wafers polished with the preceding slurries. It is clear that the surface quality improves significantly for the wafers polished with the stable slurry (containing 32 and mM $C_{12}TAB$) as compared with the unstable slurries.

After stability was achieved for the high-ionic-strength CMP slurry by adding 32 mM $C_{12}TAB$, polishing experiments were conducted to measure the material removal rate response of the surfactant-containing slurries. Figure 32b illustrates the material removal rates obtained in the presence of $C_{12}TAB$ relative to baseline and 0.6 M NaCl–containing slurries. The material removal rate of the slurry containing only 0.6 M NaCl was higher than that with the baseline polishing. This is suggested to be due to the increased pad–particle–substrate interactions as a result of the screening of negative charges in the system by salt addition. This phenomenon was observed to enhance the frictional forces leading to more material removal [93]. As 1 mM C_{12}TAB was added to the slurry, the material removal rate response did not show a significant change. On the other hand, at 8 and 32 mM C_{12}TAB concentrations, the removal rate response decreased to 7 nm/min, compared with 430 nm/min with the baseline and 710 mn/min with the 0.6 M NaCl–containing slurries. Two reasons were suggested for the negligible material removal in the presence of 8 and 32 mM C_{12}TAB. First, it is known that the presence of surfactants results in lubrication between the abrasive and the surface to be polished and, therefore, decreases the frictional force [94]. Thus, the presence of C_{12}TAB in the polishing slurries at relatively high concentrations may result in negligible material removal by reducing the frictional forces. Indeed, it was shown that, in the presence of 32 mM C_{12}TAB at pH 4, the friction coefficient of silica–silica interaction was lower than the coefficient for bare silica surfaces at pH 4 [95]. The second alternative is that the high repulsive force barrier induced by the C_{12}TAB self-aggregated structures may prevent the particle–surface interaction and therefore result in a very low material removal rate. The concentration of C_{12}TAB at which the negligible material removal rate response was obtained coincided with the observation of the jump in maximum repulsive force as reported previously [78,79].

In order to distinguish the effects of lubrication and the repulsive force barrier on material removal response, it was planned to alter the magnitude of the repulsive force barrier. Accordingly, the force barriers of different chain lengths of the CTAB surfactant were measured by AFM above the cmc, where they formed the self-assembled aggregates. Figure 33a and b show the force–distance curves and the repulsive force barriers obtained for C_8TAB, C_{10}TAB, and C_{12}TAB surfactants at 140, 68, and 32 mM concentrations in the presence of 0.6 M NaCl at pH 9. It was observed that decreasing the chain length of the surfactant led to smaller repulsive forces between the abrasive particles and substrate. Figure 33b gives the magnitudes of the maximum repulsive force barriers per single 0.2-μm particle at the selected concentrations of C_8TAB, C_{10}TAB, and C_{12}TAB surfactants as 1.6, 2.5, and 4 nN, respectively. The force applied on a single 0.2-μm particle, on the other hand, was estimated to be close to 100 nN based on the assumption of hexagonal close packing at 100% surface coverage of particles at the contact points of the polishing pad [92]. Therefore, it is clear that the maximum repulsive force barriers introduced by the self-assembled surfactant aggregates are exceeded for all the chain lengths and the reduction

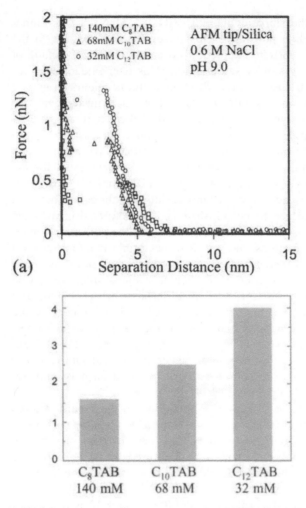

FIG. 33 (a) Force–distance curves for C_8TAB, C_{10}TAB, and C_{12}TAB surfactants at 140, 68, and 32 mM in the presence of 0.6 M NaCl at pH 9. (b) Maximum repulsive force response of C_8TAB, C_{10}TAB, and C_{12}TAB surfactants at 140, 68, and 32 mM in the presence of 0.6 M NaCl at pH 9 calculated for 0.2-μm particle. (From Ref. 92.)

in the material removal rates should be due to the lubrication effects introduced by the surfactant hydrocarbon chains.

The polishing tests were also conducted using the C_8TAB surfactant at 1, 35, and 140 mM exhibiting a relatively lower repulsive barrier. Figure 34a summarizes the surface quality response of the C_8TAB slurries. As the magnitude of the introduced repulsive barrier was very small, none of these slurries was stable. Therefore, the maximum surface deformation values were higher than desired. On the other hand, the surface roughness values

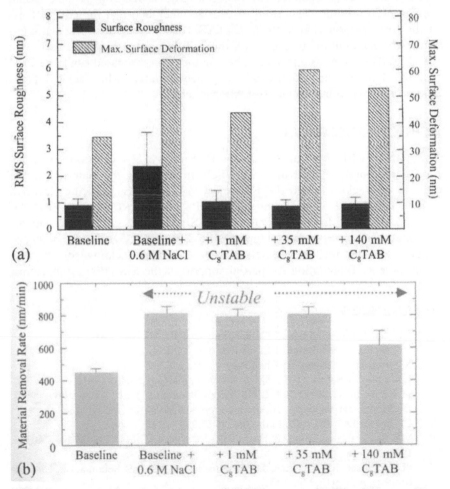

FIG. 34 (a) Surface quality response of C_8TAB system. (b) Material removal rate response of C_8TAB system. (From Ref. 92.)

were reasonable, which may be attributed to the decreased frictional forces due to the lubrication introduced by the surfactant. Most important, as can be seen in Fig. 34b, all the C_8TAB-containing slurries yielded high material removal rates. These results indicate that the surfactant chain length has a significant impact on not only the particle–particle interactions that affect the slurry stability but also the particle–substrate interactions, which alter the material removal mechanisms. To support this hypothesis further, polishing tests were conducted with 68 mM C_{10}TAB + 0.6 M NaCl slurries. The slurries were stable for this system, and the surface quality of the polished wafer was acceptable. The material removal rate, on the other hand, was only 50 nm/min. Therefore, it can be concluded that the compactness of the hydrocarbon chains of the C_{10}TAB system resulted in less particle–substrate interaction compared with the C_8TAB system but more than with the C_{12}TAB system, as expected. The phenomenon discussed can be utilized to control the surface quality of the polished wafers while achieving the desired material removal rate in CMP processes.

ACKNOWLEDGMENTS

The authors wish to thank Mr. James R. Kanicky, Ms. G. Basim, and Mr. Pankaj K. Singh for their invaluable help in preparing the manuscript. The authors also acknowledge the University of Florida Engineering Research Center for Particle Science and Technology (UR-ERC) (grant EEC 94-02989), UR-ERC Industrial Partners, CSSE affiliated companies, and the National Science Foundation (grant NSF-CPE 8005851) for the financial support provided for this research. Finally, the authors acknowledge the National Science Foundation for partial support of the SIS-2000 symposium.

REFERENCES

1. DO Shah, JH Schulman. J Colloid Interface Sci 25:107, 1967.
2. DO Shah, JH Schulman. J Lipid Res 6:311, 1965.
3. K Saito, DJ Hanahan. Biochemistry 1:521, 1962.
4. J Murata, M Satake, T Suzuki. J Biochem (Tokyo) 53:431, 1963.
5. M Kates. In: K. Bloch, ed. Lipid Metabolism. New York: Wiley, 1960, p 185.
6. JH Moore, JH Williams. Biochim Biophys Acta 84:41, 1964.
7. CP Singh, DO Shah. Colloids Surf A 77:219, 1993.
8. MB Stark, P Skagerlind, K Holmberg, J Carlfors. Colloid Polym Sci 268:384, 1990.
9. PDI Fletcher, RB Freeman, BH Robinson, GD Rees, R Schomacker. Biochim Biophys Acta 912:278, 1987.
10. SG Oh, CP Singh, DO Shah. Langmuir 8:2846, 1992.
11. YK Rao, DO Shah, J Colloid Interface Sci 137:25, 1990.

12. DO Shah. J Colloid Interface Sci 37:744, 1971.
13. EAG Aniansson, SN Wall. J Phys Chem 78:1024, 1974.
14. EAG Aniansson, SN Wall. J Phys Chem 79:857, 1975.
15. EAG Aniansson, SN Wall, M. Almgren, H Hoffmann, I Kielmann, W Ulbricht, R Zana, J Lang, C Tondre. J Phys Chem 80:905, 1976.
16. M Kahlweit. J Colloid Interface Sci 90:92, 1982.
17. T Inoue, Y Shibuya, R Shimozawa. J Colloid Interface Sci 65:370, 1978.
18. SG Oh, DO Shah. J Dispersion Sci Technol 15:297, 1994.
19. SG Oh, DO Shah. J Am Oil Chem Soc 70:673, 1993.
20. P Lianos, R Zana. J Colloid Interface Sci 84:100, 1981.
21. JB Hayter, J Penfold. J Chem Soc Faraday Trans I 77:1851, 1981
22. P Ekwall. In: JThG Overbeek, ed. Chemistry, Physics and Application of Surface Active Substances. New York: Gordon & Breach, 1967.
23. F Reiss-Husson, V Luzzati. J Phys Chem 68:3504, 1964.
24. V Luzzati. In: D. Chapman, ed. Biological Membranes. New York: Academic Press, 1968, p 71.
25. L Mandell, K Fontell, P Ekwall. In: Ordered Fluids and Liquid Crystals. Adv Chem Ser 63:89, 1967.
26. R Leung, DO Shah. J Colloid Interface Sci 113:484, 1986.
27. SY Shiao, A Patist, ML Free, V Chhabra, PDT Huibers, A Gregory, S Patel, DO Shah. Colloids Surf A 128:197, 1997.
28. T Inoue, Y Shibuya, R Shimozawa. J Colloid Interface Sci 65:370, 1978.
29. PDT Huibers, DO Shah. In: AK Chattopadhyay, KL Mittal, eds. Surfactants in Solution. New York: Marcel Dekker, 1996, p 105.
30. A Patist, V Chhabra, R Pagidipati, R Shah, DO Shah. Langmuir 13:432, 1997.
31. SY Shiao. PhD thesis, University of Florida, 1976.
32. TP Hoar, JH Schulman. Nature 152:102, 1943.
33. DO Shah. Proceedings of 1981 European Symposium on Enhanced Oil Recovery, Bournemouth, England, Elsevier Sequoia SA, Lausanne, Switzerland, 1981, pp 1–40.
34. LE Scriven. Nature 263:123, 1976.
35. LE Scriven. In: KL Mittal, ed. Micellization, Solubilization and Microemulsions. Vol 2. New York: Plenum Press, 1977, p 877.
36. ML Robbins. Paper 5839, presented at the SPE Improved Oil Recovery Symposium, Tulsa, OK, 1976.
37. RN Healy, RL Reed. Soc Pet Eng J 491, October 1974.
38. RN Healy, RL Reed. Soc Pet Eng J 147, June 1976.
39. CA Miller, R Hwan, WJ Benton, T Fort Jr. J Colloid Interface Sci 61:554, 1977.
40. C Ramachandran, S Vijayan, DO Shah. J Phys Chem 84:1561, 1980.
41. R Hwan, CA Miller, T Fort Jr. J Colloid Interface Sci 68:221, 1979.
42. K Shinoda. J Colloid Interface Sci 24:4, 1967.
43. K Shinoda, H Saito. J Colloid Interface Sci 26:70, 1968.
44. S Friberg, I Lapczynska, G Gillberg. J Colloid Interface Sci 56:19, 1976.
45. JC Noronha. PhD dissertation, University of Florida, 1980.
46. SI Chou. PhD Dissertation, University of Florida, 1980.

47. WC Hsieh. PhD dissertation, University of Florida, 1977.
48. KS Chan. PhD dissertation, University of Florida, 1978.
49. WH Wade, E Vasquez, JL Salager, M El-Emory, C Koukounis, RS Schechter. In: KL Mittal, ed. Solution Chemistry of Surfactants. Vol 2. New York: Plenum Press, 1979, p 801.
50. RL Reed, RN Healy. In: DO Shah, RS Schechter, eds. Improved Oil Recovery by Surfactant and Polymer Flooding. New York: Academic Press, 1977, p 383.
51. MY Chiang. PhD dissertation, University of Florida, 1978.
52. WC Hsieh, DO Shah. SPE 6594, International Symposium on Oilfield and Geothermal Chemistry, La Jolla, CA, 1977.
53. SJ Satter. SPE 6843, 52nd Annual Fall Conference and Exhibition of SPE-AIME, Denver, CO, 1977.
54. MC Puerto, WW Gale. SPE 5814, SPE Improved Oil Recovery Symposium, Tulsa, OK, 1976.
55. PDI Fletcher, AM Howe, BH Robinson. J Chem Soc Faraday Trans I 83:985, 1987.
56. HE Eicke, JCW Shepherd, A Steinemann. J Colloid Interface Sci 56:168, 1976.
57. C Minero, E Pramauro, E Pelizzetti. Colloids Surf 35:237, 1989.
58. CH Chew, LM Gan, DO Shah. J Dispersion Sci Technol 11:593, 1990.
59. MJ Hou, DO Shah. In: BM Moudgil, S Chander, eds. Interfacial Phenomena in Biotechnology and Materials Processing. Amsterdam: Elsevier, 1988, p 443.
60. M Dvolaitzky, R Ober, C Taupin, R Anthore, X Auvray, C Petipas, C Williams. J Dispersion Sci Technol 4:29, 1983.
61. P Ayyub, AN Maitra, DO Shah. Physica C 168:571, 1990.
62. P Kumar, V Pillai, SR Bates, DO Shah. Mater Lett 16:68, 1993.
63. V Pillai, P Kumar, MJ Hou, P Ayyub, DO Shah. Adv Colloid Iterface Sci 55: 241, 1995.
64. P Kumar, V Pillai, DO Shah. Appl Phys Lett 62:765, 1993.
65. M Boutonnet, J Kizling, P Stenius, G Maire. Colloids Surf 5:209, 1982.
66. V Pillai, DO Shah. In: V Pillai, DO Shah, eds. Dynamic Properties of Interfaces and Association Structures. Champaign, IL: AOCS Press, 1996, p 156.
67. S Mathur, PK Singh, BM Moudgil. Int J Miner Proc 58:201, 2000.
68. M Bjelopavlic, PK Singh, H El-Shall, BM Moudgil. J Colloid Interface Sci 226:159, 2000.
69. M Bjelopavlic, H El-Shall, BM Moudgil. In: V Hackley, P Somasundaran, J Lewis, eds. Polymers in Particulate Systems: Properties and Applications. New York: Marcel Dekker, 2001, p 105.
70. S Behl, BM Moudgil, TS Prakash. J Colloid Interface Sci 161:414, 1993.
71. S Behl, BM Moudgil, TS Prakash. J Colloid Interface Sci 161: 421, 1993.
72. S Behl, MJ Willis, RH Young. US patent 5,358,90, 1996.
73. MJ Rosen, Surfactants and Interfacial Phenomena. 2nd ed. New York: John Wiley & Sons, 1989.
74. N Kimizuka, T Kunitake. Adv Mater 8:89, 1996.
75. M Colic, DW Fuerstenau. Langmuir 13:6644, 1997.
76. AM Solomon, T Saeki, M Wan, PJ Scales, DV Boger, H Usui. Langmuir 15: 20, 1999.

77. LK Koopal, T Goloub, A de Keizer, MP Sidorova. Colloids Surf A 151:15, 1999.
78. JJ Adler, PK Singh, A Patist, YI Rabinovich, DO Shah, BM Moudgil. Langmuir 16:7255, 2000.
79. PK Singh, JJ Adler, YI Rabinovich, BM Moudgil. Langmuir 17:468, 2001.
80. RA Johnson, R Nagarajan. Colloids Surf A 167:31, 2000.
81. S Manne, HE Gaub. Science 270:1480, 1995.
82. JF Scamehorn. In: JF Scamehorn, ed. Phenomena in Mixed Surfactant Systems. Washington, DC: American Chemical Society, 1986, p 12.
83. A Patist, PDT Huibers, B Deneka, DO Shah. Langmuir 14:4471, 1998.
84. SP Murarka. Mater Res Soc Symp Proc 566:3, 2000.
85. LM Cook. J Noncryst Solids 120:152, 1990.
86. GB Basim, JJ Adler, U Mahajan, RK Singh, BM Moudgil. J Electrochem Soc 147:3523, 2000.
87. R. Ewasiuk, S Hong, V Desai. In: YA Arimoto, RL Opila, JR Simpson, KB Sundaram, I Ali, Y Homma, eds. Chemical Mechanical Polishing in IC Device Manufacturing III. PV 99-37. Pennington, NJ: Electrochemical Society, 1999, p 408.
88. Cabot Corporation Microelectronics Division. Material Safety Data Sheet for Semi-Sperse 12 and 25 Aqueous Dispersions, Aurora, IL, 2000
89. GB Basim, BM Moudgil. J Colloid Interface Sci, in press, (2002).
90. BJ Palla. PhD Dissertation, University of Florida, 2000.
91. M Bielman, U Mahajan, RK Singh, DO Shah, BJ Palla. Electrochem Solid State Lett 2:148, 1999.
92. GB Basim, I Vakarelski, PK Singh, BM Moudgil. Role of particle–particle and particle–substrate interactions in CMP. J Colloid Interface Sci, in press, (2002).
93. U Mahajan, M Bielman, RK Singh. Electrochem Solid State Lett 2:46, 1999.
94. J Klein, E Kumacheva, D Mahalu, D Perahia, LJ Fetters. Nature 370:634, 1994.
95. JJ Adler, BM Moudgil. J Colloid Interface Sci, in press, (2002).

2

Interaction Between Surfactant-Stabilized Particles: Dynamic Aspects

J. LYKLEMA   Wageningen University, Wageningen, The Netherlands

ABSTRACT

This chapter discusses the dynamics of particle interaction, that is, the rate dependence of the various subprocesses taking place when two particles meet. For the overall outcome, the extent to which subprocesses can relax during such an encounter is important. These subprocesses must be identified and their rates established relative to the rate of particle interaction. As an exercise, these ideas are elaborated for encounters between surfactant-covered particles. The dynamic differences between particle encounters in a sol and shelf stability in a sediment are briefly discussed. New insights into lateral transport by surface conduction are presented.

I. INTRODUCTION

Consider the interaction between a pair of surfactant-stabilized colloidal particles. When two such particles approach each other, a situation arises as sketched in Fig. 1. In the classical interpretation the Gibbs energy of interaction, $\Delta G(h)$, where h is the distance between the particle surfaces (or between the head groups of the surfactant), is obtained by estimating its contributions: $\Delta_{el}G$ for the double-layer part, $\Delta_{vdw}G$ for the van der Waals contribution, $\Delta_{entr}G$ accounting for changes in the entropy of the surfactant upon interaction, etc. Such interpretations are essentially *static*; at any moment the Gibbs energy is computed under the assumption that the structures of the surfactant layer and ionic double layers are at equilibrium. Typically, classical (statistical) thermodynamics is invoked for the analysis.

In *dynamic* approaches this assumption is relaxed. Now the rates at which the required structural changes take place are considered and compared with

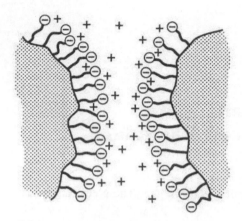

FIG. 1 Interaction between two surfactant-covered surfaces.

the rate of particle interaction. Quantitatively, this ratio is expressed through the *Deborah number De*, defined as

$$De = \frac{\tau(\text{process})}{\tau(\text{collision})} \tag{1}$$

where τ stands for the characteristic time of the phenomenon indicated. A variety of subprocesses can be recognized, depending on the complexity of the system. When for all these processes $De \ll 1$, the layers are continually at equilibrium during interaction and the static limit is attained.

In the following we shall discuss what might happen when this condition is not met.

II. DYNAMICS OF DLVO-TYPE INTERACTIONS

By way of introduction to the present issue, let us first consider the simple situation that the interaction is determined only by electrostatic repulsion and van der Waals attraction in the absence of adsorbed surfactants. This is commonly called the DLVO case.

The van der Waals part is of electromagnetic origin and is, therefore, virtually instantaneous: $De(\text{vdW}) \approx 0$. However, for the electrostatic part this is not so obvious. Verwey and Overbeek [1] discussed the difference between interaction at constant (surface) charge (density), σ^0, and at constant (surface) potential, ψ^0. Both cases lead to repulsion, although for different reasons. When σ^0 remains fixed upon reduction of h, the potential increases. The work to be done is of a *purely electrostatic* nature. However, when ψ^0 remains constant, the charge should diminish upon approach. Now work has to be done to desorb charge-determining ions from the surface or to compensate these charges by coadsorption of counterions. The work involved is now of a *purely chemical* nature. Figure 2 illustrates the difference in terms of potential-distance curves.

On closer reflection, it appears that the preceding distinction has a dynamic origin, the critical issue being whether or not σ^0 can be reduced during a particle collision. Interaction will take place at constant σ^0 when for the charge reduction process $De \gg 1$. So it depends on the system and on the rate of approach of the particles which case prevails. Polystyrene lattices, which carry covalently bound sulfate groups, and clay minerals, which exhibit a negative plate charge because of isomorphic substitution, are clearly constant-charge examples unless counterion adsorption can take place very rapidly. On the other hand, silver halides and oxides, for which σ^0 is determined by adsorption and desorption of charge-determining ions (Ag^+, I^- and H^+, OH^-, respectively) are constant-potential candidates provided De for the sorption processes is small enough. The distinction is not absolute: if we are capable of shooting AgI particles onto each other at such a high rate that the sorption processes cannot relax, we attain the constant-

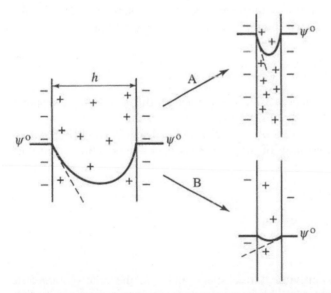

FIG. 2 Difference between interactions at constant charge (A) and at constant potential (B). Given is the potential as a function of distance. In case A, ψ^0 rises, whereas in case B, $\sigma^0 \to 0$. The dashed lines show $(d\psi/dx)_{x=0}$, which by virtue of Gauss' law is proportional to σ^0.

charge limit. Likewise, the charge inside clay particles can relax by solid-state diffusion over geological periods.

It is concluded that the distinction between the two cases is not absolute. Intermediate situations with partial relaxation may also occur; then the work to be done may be partly chemical, partly electrostatic. The analysis of these relaxations takes us to the domain of double-layer dynamics, and the prevailing issues are the identification of the relevant processes and the assessment of their rates.

III. RELAXATION OF SURFACTANT ADSORBATES

Let us next consider the more complicated case of an adsorbed ionic surfactant layer, as in Fig. 1. To keep things simple, we assume that the particle surface itself does not carry a charge. When two particles approach each other, for each adsorbate three relaxation mechanisms may be envisaged.

1. Desorption of surfactants
2. Relaxation of Stern ions
3. Relaxation of diffuse ions

Can we estimate the rates of these processes and compare them with the rate of interaction? For the interaction time $\tau(\text{int})$ we estimate

$$\tau(\text{int}) \approx \frac{1}{D\kappa^2} \approx 10^{-5} \text{ s} \tag{2}$$

for colloidal particles approaching each other by diffusion over a distance of order κ^{-1}. Here, D is the diffusion coefficient of the particles.

Let us first discuss relaxation mechanism 3 because relatively simple rules can be given. For relaxation of a space charge ρ by conduction, according to Maxwell

$$\rho(t) = \rho(t = 0)e^{-t/\tau(\text{diff})} \tag{3}$$

with the relaxation time obeying

$$\tau(\text{diff}) = \varepsilon_0 \varepsilon / K^{\text{L}} \tag{4}$$

where ε_0 is the permittivity of free space and ε is the relative dielectric constant of the solution and K^{L} its conductivity. For a derivation of Eq. (3) see Ref. 3. For typical electrolyte solutions (10^{-3}–10^{-1} M for monovalent salts) $\tau(\text{diff}) \approx O(10^{-9}$–$10^{-7}$ s), so that the diffuse double-layer part is always relaxed during interaction. Hence,

$$De(\text{diff}) = \frac{\tau(\text{diff})}{\tau(\text{int})} \ll 1 \tag{5}$$

where $De(\text{diff})$ is the Deborah number of the diffuse part of the double layer. There may be atypical cases, for instance, very narrow slits between two particles inside which the ions must be transported over long distances. Some situations have been considered by Lyklema et al. [4]. As our focus is on the surfactants, we shall not consider these here.

Consider now mechanism 1. Surfactants can be desorbed from the two surfaces if the particles attract each other so strongly that the binding Gibbs energy is overcompensated. Such situations can arise when the electrical repulsion is suppressed by electrolyte addition so that only the van der Waals attraction remains. Let us estimate the orders of magnitude for

$$\Delta G(\text{vdW}) = -\frac{A}{12\pi h^2} \tag{6}$$

where A is the Hamaker constant and h the distance between the surfaces. Of course, this equation can be refined in that the contributions of the surfactants are made explicit. This would lead to a more complicated equation, containing more than one h and more than one A, but for the moment we consider only orders of magnitude. Figure 3 holds for $A = 10\ kT$; h is in nm and it was assumed that the molecular cross section of the adsorbed

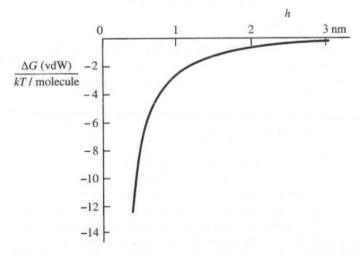

FIG. 3 Estimation of the order of magnitude of the van der Waals attraction between two surfactant-covered surfaces, expressed as Gibbs energies per molecule of surfactant.

surfactant was 2.5 molecules nm^{-2} to express the expelling Gibbs energy per molecule. This Gibbs energy must be compared with the molecular Gibbs energy of desorption, $-\Delta G(des)$, which for a surfactant with 10–15 CH_2 groups in the chain is about 12–18 kT, according to Traube's rule. From Fig. 3 we see that the driving force for desorption of single surfactants is marginal under the chosen conditions. The surfaces must come very close before substantial desorption sets in.

Even if the driving force is high enough, it is questionable whether the rate of desorption is fast enough to keep up with the rate of particle approach. For nonionic surfactants (where electrostatics does not play a role) desorption relaxation times from cellulose monolayers of the order of 10^{-4}–10^{-3} s have been reported [5,6]. In these experiments, desorption was initiated by diluting the solution. If we may use these data as representative for the particle interaction case, considering Eq. (2) we find

$$De(des) \sim 10\text{--}10^2 \qquad \text{(pair collision)} \tag{7}$$

meaning that desorption would not take place. The issue deserves further consideration. Questions to be addressed include the following: (1) Do the molecules desorb individually or collectively? (2) How do we account for the acceleration of particle approach because of particle attraction and deceleration because of hydrodynamic drag? (3) How do desorbed surfactant molecules leave the narrow gap between the approaching particles?

For the present purpose, we assume that no desorption will take place during particle encounter. However, under shelf stability conditions

$$De(des) \ll 1 \qquad \text{(shelf life)} \tag{8}$$

because then the particles will remain pressed against each other for a long time and, usually, under a pressure head. From practice it is known that shelf stability is quite different from the stability of dilute sols, and the dynamic differences are one of the reasons.

IV. RELAXATION OF STERN IONS

Having established that under many operational conditions $De(diff) \ll 1$ and $De(des) > 1$ or $\gg 1$, we now address mechanism 2, the relaxation of Stern ions. The issue of lateral mobility of ions that are closely associated with the head groups is related to electrokinetic problems and the dynamics of pair interaction in the absence of surfactants. As before, we must establish the driving forces and estimate the rates. These two processes are interdependent: the driving force depends on the extent of disequilibration of the double layer, but this extent depends on the rate of lateral transport and, hence, on the driving force.

Let us redefine the problem. We start from the premises that the diffuse part of the double layer is fully relaxed but that the surfactant adsorption is not relaxed at all. In other words, the surfactant charge σ^{surf} is fixed. We assume it to be negative, as in Fig. 1. Let us say that Na^+ ions are the counterions in the Stern layer, and let their surface charge density in this layer be σ^{Na^+}. Then the question is what happens to the sum

$$\sigma^i \equiv |\sigma^{surf} + \sigma^{Na^+}| \tag{9}$$

of the surfactant charge and the sodium ion charge if the particles approach each other. We have the following limiting conditions

$$De \equiv \frac{\tau(Na^+)}{\tau(int)} \gg 1 \qquad \sigma^i \text{ constant} \tag{10a}$$

$$De \equiv \frac{\tau(Na^+)}{\tau(int)} \ll 1 \qquad \sigma^i \to 0 \quad \text{for} \quad h \to 0 \tag{10b}$$

Here, $\tau(Na^+)$ is the residence time of sodium ions in the Stern layer and $\tau(int)$ is given by Eq. (2).

These are the classical conditions for interaction at constant charge and at constant potential, respectively, but now adapted for the present situation of surfactant-stabilized particles.

This problem has been solved for arbitrary De by Shulepov et al. [7] for a number of simplifying assumptions, including

1. Low potentials
2. Martynov double-layer model (it assumes that the Stern ions reside in a potential well)
3. For the mean particle flux

$$j_p = \langle v(h)c_p(h) \rangle \tag{11a}$$

the ensemble average is taken:

$$j_p = \langle v(h) \rangle \langle c_p(h) \rangle \tag{11b}$$

where $v(h)$ is the rate of approach and $c_p(h)$ the particle concentration. Both depend on h and, for that matter, on t

Figure 4 gives the variation of σ^i as a function of separation for various values of $De(\sigma^{Na^+})$. It illustrates the predicted limiting trends. For $De = \infty$, as in Eq. (10a), we have the constant-charge case, whereas for $De \to 0$ the charge decreases when the particles approach each other [eventually for $\kappa h \to 0$, $\sigma^i(h) \to 0$].

However, Fig. 4 also shows intermediate cases. We repeat that here charge should be interpreted as the sum of the charges attributable to the surfactant

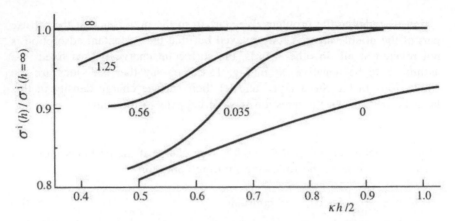

FIG. 4 Extent of variation of the sum of the surface charge and the Stern charge as a function of the dimensionless particle distance at various Deborah numbers for the Stern counterions.

ion and the counterions in the Stern layer. This sum can change with h only because the Stern charge varies; the charge of the surfactants remains constant because surfactants do not desorb.

However, as the theory leading to Fig. 4 is general, this figure would also apply when surfactant desorption was rate determining [after replacing σ^i by σ^{surf} and De by $De(\text{des})$].

This discussion indicates that the interaction dynamics can be phenomenologically elaborated with the residence times of Stern ions as the parameter. How can we measure these?

V. IONIC MOBILITIES IN STERN LAYERS

Over the past decade it has become possible to measure the lateral mobility of Stern ions and, in this way, establish their residence times. Basically, the procedure for obtaining these mobilities consists of measuring the surface conduction in that layer, $K^{\sigma i}$, and the Stern charge σ^i to obtain the required mobility of ion i in that layer, u_i^i, from

$$K^{\sigma i} = \sigma^i u_i^i \tag{12}$$

This equation holds for a Stern layer with only one ionic species, i, in it. Surface conduction in the Stern layer takes place parallel to that in the diffuse layer. Hence, the total surface conductivity is

$$K^\sigma = K^{\sigma d} + K^{\sigma i} \tag{13}$$

where $K^{\sigma d}$ and $K^{\sigma i}$ refer to the surface conductivity in the diffuse and the

Stern layer, respectively. Surface conductivities are measurable quantities. For instance, one can prepare a porous plug of the particles to be studied and measure its conductivity $K(\text{plug})$ as a function of the salt concentration. The $K(\text{plug})$ consists of two parts: a part determined by the interparticle fluid (which is proportional to c_{salt}) and a part contributed by surface conduction. The latter can be found from the intercept of $K(\text{plug})$ versus c_{salt} plots for $c_{\text{salt}} \to 0$. For $K^{\sigma d}$ a theory is available (by Bikerman); this quantity is determined by the electrokinetic potential ζ and c_{salt}. If ζ is available, say from additional electrophoresis measurements, $K^{\sigma d}$ can be computed and subtracted from K^{σ} to obtain $K^{\sigma i}$. One of the assumptions in this analysis is that the diffuse part of the double layer coincides with the part that is beyond the slip plane. The value of σ^i can be established as the difference between the surface charge and the diffuse charge, the latter being equated to the electrokinetic charge. For details of this and other procedures for establishing $K^{\sigma i}$ and hence u_i^i see the review in Ref. 8. That review also gives some feeling for the validity of the various assumptions that have to be made.

Lateral mobilities are now available for Stern layers on inorganic colloids and lattices. As far as the author is aware, such data have not yet been reported for surfactant-covered particles, although the required experiments are not difficult and might produce interesting information. The closest information we now have refers to phospholipid vesicles, where the (constant!) "surfactant" charge is caused by dissociation of the phosphate groups and where several counterions have been investigated [9]. In this study, electrophoresis experiments were combined with dielectric spectroscopy. For technical details we refer to the original paper, but it is very likely that the results are correct because they agree with independent data obtained by a quite different approach, namely from the difference between the isoelectric and isoconductic points [10]. Moreover, the trends are very similar to those on latices and inorganic surfaces, so the results will probably not be too different for surfactant adsorbates.

It is observed that

1. For monovalent counterions the mobility is, within experimental error, identical to that in the bulk.
2. The mobility ratio decreases drastically with the valence of the counterion.
3. Between counterions of the same valence there is no detectable ionic specificity.

These rules give rise to two considerations. First, if the dynamics of surfactant-stabilized particles are elaborated on the Gouy–Stern level, one can substitute bulk mobilities for those in the Stern layer, at least if the counterions are monovalent. For bivalent ones the average ratio is about

2/3, averaged over all systems studied so far. This conclusion, of course, also applies to such studies in the absence of surfactants.

The second consideration is of a more academic nature. Why is the mobility of, say, Na^+ ions in the Stern layer almost identical to that in the bulk whereas the viscosity in that layer is very high? Why is this ratio nonspecific? (On negatively charged polystyrene latices we found ratios of about unity even for protons, which migrate by a very different mechanism [11]). And why is there such a pronounced valence influence?

We are now beginning to get some grip on these matters by invoking help from molecular dynamics simulations for large numbers of molecules and ions [12]. Let us briefly summarize the main findings of these simulations.

1. It is generally known, and our simulations confirm, that liquids in contact with a hard surface exhibit density oscillations over the first few molecular layers, which can be described by the distribution function $\rho(z)$, where z is the distance from the surface. This function has a pronounced maximum not far from the sum of the radii of the surface atoms and those of the liquid molecules. After this maximum there is a minimum, then a more shallow second maximum, etc., the oscillations petering out after a few molecular cross sections.

2. The stagnant layer in electrokinetics, that is, the layer with an apparent infinitely high viscosity, is closely related to these density oscillations. Hence, the thickness of this layer is determined by the extension of these oscillations into the solution.

3. The electrokinetically bound ions also accumulate in this layer. At the same time, these ions are more or less identical to the Stern ions in double-layer electrostatics.

4. As we are talking about a general phenomenon, stagnant layers should be ubiquitous for all hard surfaces, be they charged or uncharged, hydrophilic or hydrophobic. They should also exist on the surfaces of surfactant-stabilized particles. However, they should not be observed for liquid–vapor interfaces because vapors are not hard. In fact, for such interfaces $\rho(z)$ follows a hyperbolic tangent profile rather than exhibiting oscillations.

5. When ions translate tangentially along hard surfaces, they reside for part of their time inside the stagnant layer and then can jump out, to move tangentially to the surface in the bulklike liquid, to return later to the stagnant layer, or to remain definitively in the diffuse part of the double layer after exchange against another ion. In other words, ionic conduction in the stagnant layer is short-circuited through the diffuse part.

6. The measured tangential mobility u_i^i in the stagnant layer [Eq. (12) and Table 1] is the average of the fractions of time spent in the stagnant

TABLE 1 Ratios $R \equiv u_i^s/u_i(\infty)$ Between Ionic Mobilities in the Stern Layer on Vesicles and the Corresponding Bulk Mobilities for Liposome Vesicles

| | Counterion | | | |
Reference	Na$^+$, Cs$^+$	Ca^{2+}	Cd^{2+}, Cu^{2+}	La^{3+}
Barchini et al. [9]	~1.0	~0.6	~0.6	
Verbich et al. [10]	~1.0	~0.8		~0.07

(bound) and diffuse (free) states. At present we are attempting to analyze these fractions by molecular dynamics simulations. It is at least in line with the observation (see points 1–3 in this section) that the ratio R is close to unity if the ions are readily exchangeable against the bulk. For monovalent counterions this is apparently the case. Such ions are readily short-circuited, or shunted, by the diffuse part. This observation is also in line with the absence of ion specificity. We do not yet have enough quantitative information for counterions of higher valence, but the fact that they are more strongly electrostatically bound to the surface may well help to explain their lower mobility ratio R.

7. An essential element in the analysis that we have learned to appreciate is the difference between self-diffusion of individual molecules and the collective diffusion in large collections of molecules. Figure 5 illustrates what we mean. The ion transport refers to the trajectories of individual species, whereas the displacement of the solvent reflects the viscosity, as obtained from the time correlation of the pressure tensor. See Ref.

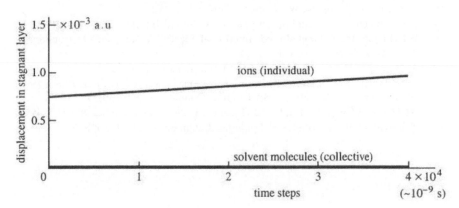

FIG. 5 Comparison of the lateral displacement of ions and solvent molecules in stagnant layers.

12 for details. Near the surfaces, this tensor is anisotropic; this surface viscosity is rather difficult to obtain, but because of the very large number of molecules that we could simulate, we were able to estimate the components of the tensor. The fluid transport simulated in this way typically represents the movement of large amounts of molecules, in which all molecular motions at any position are correlated.

In terms of phenomenological physical chemistry, the slopes of the two lines in Fig. 5 are measures of the ionic mobility in and the viscosity of the stagnant layer. The former is very much larger than the latter, meaning that ions do move in a hydrodynamically virtually stagnant layer.

VI. CONCLUSION

By analyzing the dynamics of surfactant-stabilized particles, a link was established with electrokinetics. In this way a cross-fertilization between two different domains of interfacial science was established. The potentials of this finding deserve further elaboration.

REFERENCES

1. EJW Verwey, JThG Overbeek. Theory of the Stability of Lyophobic Colloids. Amsterdam: Elsevier, 1948.
2. BV Derjaguin (=Deryagin). Acta Physicochim URSS 14:633, 1942.
3. For a derivation, see, e.g., J Lyklema. Fundamentals of Interface and Colloid Science. Vol I. 2nd printing. London: Academic Press, 1993, sec 6.6c.
4. J Lyklema, HP van Leeuwen, M Minor. Adv Colloid Interface Sci 83:33, 1999.
5. BV Zhmud, F Tiberg, J Kizling. Langmuir 16:2557, 2000.
6. LH Torn. PhD thesis, Wageningen University, 2000.
7. SYu Shulepov, SS Dukhin, J Lyklema. J Colloid Interface Sci 171:340, 1995.
8. J Lyklema. In: A Delgado, ed. Interfacial Electrokinetics and Electrophoresis. New York: Marcel Dekker, Chapter 3, p. 87, (2002).
9. R Barchini, HP van Leeuwen, J Lyklema. Langmuir 16:8238, 2000.
10. SV Verbich, SS Dukhin, H Matsumura. J Dispersion Sci Technol 20:83, 1999; H Matsumura, SV Verbich, SS Dukhin. Colloids Surf A 159:271, 1999.
11. M Minor, AJ van der Linde, J Lyklema. J Colloid Interface Sci 203:177, 1998.
12. S Rovillard, J de Coninck, J Lyklema. Langmuir 14:5659, 1998.

3

Solid Particles at Liquid Interfaces, Including Their Effects on Emulsion and Foam Stability

ROBERT AVEYARD and JOHN H. CLINT   Hull University, Hull, United Kingdom

ABSTRACT

Both hydrophilic and hydrophobic particles are "surface active" at air/water (a/w) and oil/water (o/w) interfaces. The free energy of attachment to an interface is simply related to the contact angle, θ, that the interface makes with the particle surface. Particles are most strongly anchored to a liquid surface when $\theta = 90°$.

Isolated hydrophobic particles resting at one surface of a thin aqueous film can cause the film to rupture (depending on θ) when the second surface engages the particle, causing the particle to bridge the film. For this reason, small hydrophobic particles can act as antifoams or foam breakers. Commercially, antifoams are often formulations in which hydrophobic particles are dispersed in a mineral or a silicone oil. Particles "adsorbed" in *close-packed* layers on droplet interfaces in an emulsion can stabilize the emulsion even in the absence of surfactant. Indeed, in some respects monolayers of particles behave similarly to adsorbed surfactants. Thus, hydrophobic particles [which are akin to surfactants with low hydrophile–lipophile balance (HLB)] can stabilize water-in-oil emulsions, whereas hydrophilic particles can be expected to stabilize oil-in-water emulsions (just as high-HLB surfactants do). In the first part of the chapter aspects of the current state of understanding of the effects that adsorbed particles have on foam and emulsion stability are reviewed.

The second part of the chapter describes some of our recent work on (1) possible effects of line tension on the wettability of particles at fluid interfaces and (2) lateral interactions between (charged) polystyrene particles in monolayers at a/w and o/w interfaces and the consequences for monolayer structure. It is shown how the contact angle θ influences the lateral interaction between particles in monolayers.

I. INTRODUCTION

When present in emulsions or foams, small solid particles can drastically influence the stability of the thin liquid films present and hence the stability of the whole system. Liquid droplets present in foams can have a similar effect [1–13]. For this to happen, the particles or droplets must be able to enter the surfaces of the films [14]. It is well known that once located within a liquid surface, a solid particle, for example, is usually strongly attached [15–17]. Thus, small particles are in some respects akin to surfactant and other surface-active molecules.

Studies of foam breaking by solid particles have usually focused on the behavior of individual particles at liquid interfaces. On the other hand, the ability of small particles to stabilize (Pickering) emulsions in the absence of a surfactant arises from processes involving particle assemblies (monolayers) at droplet surfaces [15]. Particle monolayers, like insoluble molecular monolayers [18], can be studied using a Langmuir trough (e.g., Refs. 19–24). "Adsorbed" particles exert a surface pressure, i.e., lower the interfacial tension, to an extent dependent on the surface concentration of particles. Monolayers can be compressed without particle loss to contiguous bulk phases because the energy of attachment to the surface is high.

The behavior of particle monolayers has not been widely studied, although significant work has been done, particularly over the last two decades [25–34]. We present here some of our own recent findings on the structure and interactions within particle monolayers and on the mode of collapse as the lateral pressure is increased [22–24]. The mode of collapse is of direct relevance to the measurement of particle wettability at interfaces, as will be seen. It is particle wettability that determines the ability or otherwise of particles to rupture thin liquid films and influence the stability of emulsions and foams.

In what follows we explain why particles are usually surface active at liquid interfaces. We also discuss mechanisms by which particles can rupture thin films in foams and emulsions. Because antifoam formulations are frequently dispersions of particle-containing oil droplets in an aqueous phase, we also allude to the way in which oil droplets can break foam films and to the synergy that exists between the oil and particles in this process. Finally, we illustrate some of our own work on the behavior of particle monolayers at (mainly) oil/water interfaces. Particularly important is the observation that very long range electrical repulsion between charge-bearing particles can occur through nonpolar oils at the oil/water interface.

II. ADSORPTION OF SOLID PARTICLES AT LIQUID INTERFACES

For simplicity, we consider spherical rather than irregularly shaped particles; general principles can be introduced in this way, and much of our own experimental work has involved the use of spherical polystyrene particles. Ideas introduced for solid particles apply to liquid droplets as well, but with the added complication that a droplet in bulk is deformed to a lens when it enters a liquid surface [15,35].

The wettability of a particle at a liquid interface, which is related to the energy of attachment to that interface, can be usefully expressed in terms of the contact angle θ that the liquid interface makes with the solid particle (Fig. 1). Here, we take θ to be the angle measured through the aqueous phase. Also, for simplicity we assume the liquid interface remains planar up to contact with the particle. Because we are concerned here with small particles (usually with diameters of a few μm), the latter assumption is not unreasonable. In Fig. 1 we show spherical particles immersed in an air/water or oil/water interface and similar systems where the solid surface (with the same surface characteristics as the particles) is planar and the liquid surface

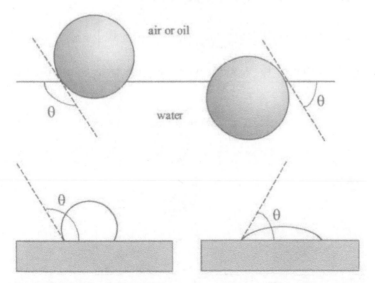

FIG. 1 Contact angles of water with solids in air or under oil. Top diagrams represent spherical particles in a plane liquid/fluid interface, and bottom diagrams show the equivalent angles for a drop of water resting on a plane solid surface. For angles less than 90° we refer to the solid as being hydrophilic in character and for angles greater than 90° we refer to the solid as being hydrophobic.

curved. The latter configuration is often the one used in the measurement of contact angles, which are then supposed to apply equally to the particle systems. Later, we discuss alternative methods for measurement of contact angles using particles directly.

In a general way we will refer to solids with contact angles of over 90° with the aqueous solutions as being *hydrophobic*; for angles < 90° we will call the solid *hydrophilic*. We note that solids having $\theta > 90°$ with water may well have $\theta < 90°$ with a surfactant solution. That is, a solid that is hydrophobic with respect to water may well be hydrophilic with respect to a surfactant solution.

The process of uptake of a particle, radius R, by a fluid/fluid (α/β) interface is illustrated in Fig. 2. Initially, the particle is completely immersed in the α-phase. The lower area A of the particle/α interface will, on uptake of the particle by the interface, be immersed in the β-phase (producing area A of solid/β interface) and a three-phase (α/β/solid) contact line is formed around the particle. Crucially, an area (designated S in Fig. 2) of plane α/β interface is lost. The various areas are each associated with an interfacial tension (γ) and the contact line with a line tension, τ, It is readily shown [16] that the free energy of attachment, $-G$, of the particle to, e.g., an oil/water (o/w) interface with tension γ_{ow}, is given by

$$G = -\pi R^2 \gamma_{ow} \left[(1 \pm \cos \theta)^2 - \frac{2\tau}{\gamma_{ow} R \sin \theta} (1 \pm \cos \theta) \right] \tag{1}$$

The free energy G is that of the system with the particle at equilibrium at

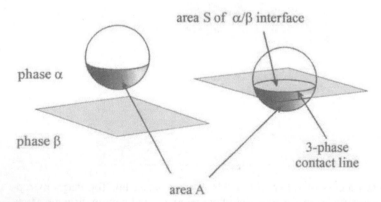

area S of α/β interface

phase α

phase β

3-phase
contact line

area A

FIG. 2 Uptake of a particle, originally in the α-phase, at the α/β interface. An area A of solid/α interface is lost, as is an area S of α/β interface. An area A of solid/β interface is gained and a three-phase contact line is formed.

the interface relative to the free energy of the system in which the particle is fully immersed in the more wetting bulk phase.

The dependence of G on the equilibrium contact angle is illustrated in Fig. 3; it is assumed here that the effects of line tension are absent (see later). Clearly, a particle of given radius is most strongly held to the interface when $\theta = 90°$. The dashed lines in the figure represent the free energy at the interface relative to that with the particle in the *less* wetting bulk phase. The changes in the free energy of a system as a particle is pushed vertically through the interface are shown in Fig. 4 [16,17]. The depth of immersion of a particle below the liquid interface is designated h so that the quantity $h/2R$ varies from zero (particle completely in the upper phase) to 1 (completely in the lower phase). The lower curve, for $\tau = 0$, exhibits a single minimum, corresponding to the equilibrium depth of immersion. This arises as a result of the way in which the areas A and S (Fig. 2) vary with the depth of immersion h.

In the cases where line tension is important, the curves of G versus $h/2R$ show two maxima and a single minimum. The maxima arise because of the way in which the contact line length varies with the extent of immersion. In the middle curve in Fig. 4, which is for an intermediate value of (positive) τ, the minimum lies below zero and there is an equilibrium configuration for the particle in the interface. For higher values of positive τ, however, the minimum can occur for $G > 0$. The minimum here corresponds to a

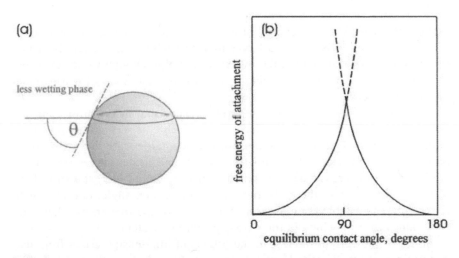

FIG. 3   (a) Spherical particle resting at a planar liquid/fluid interface. (b) Schematic diagram of the free energy of attachment $(-G)$ of the particle to the interface as a function of the equilibrium contact angle θ.

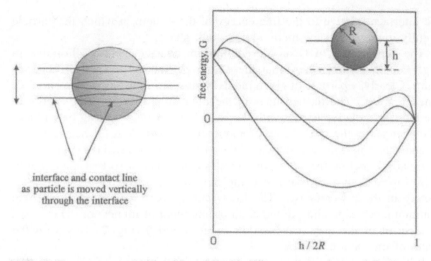

FIG. 4 Free energy as a function of depth of immersion of a particle through an interface. The areas of the particle/bulk phase contacts and area of liquid/fluid interface lost change with depth of immersion, as does the length of the three-phase contact line. The lowest curve on the graph corresponds to zero line tension. The two upper lines are for systems with positive line tensions (see text). The minima in the curves correspond to the equilibrium positions of the particles in the interface.

metastable state. The free energy of a system with the particle immersed in the more wetting bulk phase is lower than that where the particle is in the minimum energy state within the interface. For high enough values of τ, the minimum disappears and no stable or metastable state exists for a particle in the interface.

From what has been said, it can be appreciated that there are two special values of τ; there are a critical value, τ_c, for which G is zero and a maximum value, τ_m, above which no equilibrium or metastable state can exist for a particle in the interface. This situation is reflected in the curves of contact angle and free energy against assumed (positive) values of τ, shown in Fig. 5. For a given $\tau < \tau_m$ there are two values of contact angle, one of which represents an unstable configuration. Similarly, for the free energy there are two values, one of which is for a nonequilibrium system.

Some of the current interest in the effects of line tension arises from the uncertainty of the magnitude of line tension. First, we mention that, unlike the case for interfacial tensions, τ can be positive or negative. Positive values (considered earlier), which are normally observed for liquid/fluid/solid con-

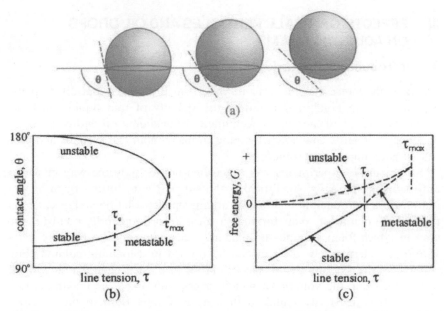

FIG. 5 Effects of line tension on contact angles and stability of particles at fluid/
fluid interfaces. (a) Effect of increasing positive line tension on a system with $\theta >$
90°; increase in τ causes θ to rise. (b) This effect is shown together with regions of
stability, metastability, and instability. (c) Corresponding effects of τ on the free
energy G (see text).

tact lines, correspond to a tendency of a contact line to shrink in size (just
as positive values of interfacial tension are associated with a tendency of a
surface to contract). Such line tensions tend to push the contact angle away
from 90°, forcing the particle toward the more wetting bulk phase. For a
given line tension, the effects are greater the smaller the particle size [see
Eq. (1)]. Negative line tensions, on the other hand, correspond to contact
lines that tend to expand; this forces the contact angle toward 90°. The
magnitudes of positive τ reported in the literature vary widely, from around
10^{-12} N (theoretical values) to the largest of experimental values in the range
10^{-6} to 10^{-5} N (see Ref. 17). These large values are expected to prevent
even large particles (say 1 mm in diameter) from entering a liquid interface
[17]. Whatever the subtleties of the sign, magnitude, and effects of line
tension may be, we are of the opinion that because particles with diameters
less than 10 μm can be readily incorporated into liquid surfaces, the high
values of line tension reported are probably in error.

III. EFFECTS OF SMALL PARTICLES AND OIL DROPS ON AQUEOUS FOAMS

A. Introductory Remarks

Much of the interest in the way in which small particles and oil droplets, separately or in combination, affect the stability of thin liquid films has arisen as a result of the use of formulations containing oil and particles as aqueous antifoams. Unwanted foaming in many industrial systems and processes is an important problem [1].

Some simple experiments can be carried out to study the way in which antifoams operate (Fig. 6). Probably the simplest is to form a foam by agitation (in a reproducible fashion) of an appropriate solution and observe (1) the initial volume of foam formed (a measure of *foamability*) and (2) the way in which foam volume subsequently falls with time (to measure *foam stability*). Alternatively, a foam can be formed in a foaming column (Fig. 6b). Clean gas, say nitrogen, is passed through a glass frit and foaming liquid to form a foam, the volume of which grows with time. The maximum rate of growth is obviously equal to the gas flow rate. Both of the methods

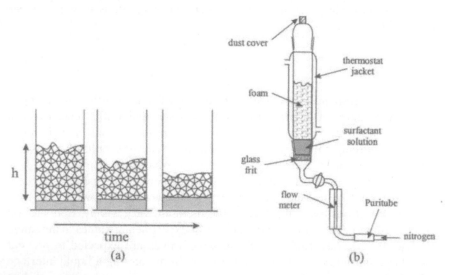

FIG. 6 (a) Foam volume (or height, h) as a function of time; initial volume is a measure of foamability and the change with time a measure of foam stability. (b) A foaming column for creation of foam and measurement of its volume (see text).

mentioned measure the total volume of gas trapped within a foam and give no direct information on the rupture of films within the bulk of the foam.

B. Some Experimental Results

The effectiveness of particles as aqueous foam breakers depends strongly on the wettability of the particles by the foaming liquid. In general, particles need to be hydrophobic (see earlier) in order to be effective. To illustrate this, we show in Fig. 7a the percent reduction of the initial foam volume of 0.2 mM aqueous cetyltrimethylammonium bromide (CTAB) solutions caused by the presence of spherical ballotini beads [7]. The beads were treated to varying extents with octadecyltrichlorosilane to yield a range of contact angles with the CTAB solution. It is clear that the higher the contact angle, the greater the initial foam reduction.

Contact angles can change as a result of variation of surfactant concentration as well as surface treatment of solid. The results shown in Fig. 7b relate to foam breaking of aqueous CTAB solutions by particles of ethylene-bis-stearamide (EBS), a waxy solid used in commercial antifoams [8]. In this case, changes in θ result entirely from changes in CTAB concentration in the foaming solutions. As in the case of surface treatment discussed before, the percent foam reduction increases with increase in contact angle (decrease in CTAB concentration). We note that the foamability of the CTAB solutions did not vary with surfactant concentration in the absence of the particles.

Commercial antifoams often rely on the synergistic effects of particles and water-insoluble oil droplets (e.g., hydrocarbon or silicone oil) in reducing foam volume. This is illustrated in Fig. 7b, where it is seen that EBS particles in combination with dodecane are much more effective in causing foam reduction than EBS particles alone.

The kind of results obtained using a foaming column, such as that illustrated in Fig. 6, can be seen in Fig. 8 [36]. Here, foam volume is plotted as a function of time for foams produced by passing nitrogen through 1 wt% aqueous solutions of the nonionic surfactant $C_{12}H_{25}(OCH_2CH_2)_5OH(C_{12}E_5)$. The surfactant solutions exhibit a cloud point (i.e., they phase separate on heating) at about 31°C. The phases formed are an aqueous phase rich in surfactant (the droplet phase) and a surfactant-lean aqueous phase. It is seen in Fig. 8 that foam volumes at a given time fall with increasing temperature. The gas flow rate used was the same for all temperatures. The inset shows the foam volume after 30 min as a function of temperature. Clearly, the volume falls rapidly up to about the cloud point (31°C) and then levels off. This may result from the antifoam effect of the phase-separated surfactant-rich droplets.

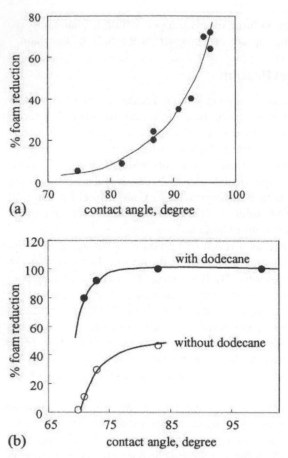

FIG. 7 (a) Foam reduction (see test) of 0.2 mM cetyltrimethylammonium bromide (CTAB) solutions caused by hydrophobized glass beads (average diameter 45 μm). The hydrophobicity of the beads and of glass plates (used for contact angle measurements) was adjusted by treatment with octadecyltrichlorosilane. (b) Foam reduction of aqueous CTAB solutions caused by ethylene-bis-stearamide (EBS) particles as a function of the contact angle of the surfactant solutions with flat EBS-coated plates in air. Open circles represent systems without dispersed dodecane drops and filled circles systems with dodecane. Here, changes in contact angle are brought about by changing the concentration of the surfactant solution.

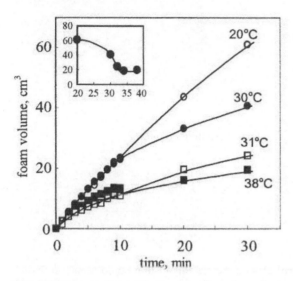

FIG. 8 Foam volume (in the foaming column) as a function of time for 1 wt% aqueous $C_{12}E_5$ over a range of temperature spanning the cloud point, which is around 31°C. The inset shows the foam volume formed after 30 min as a function of temperature. It is seen that the volume drops up to the cloud point and then becomes independent of temperature up to at least 38°C, the highest temperature studied.

C. Mechanisms of Foam Breaking

There is now a reasonable consensus concerning the modes of operation of hydrophobic particles, alone or in combination with oil droplets, in reducing foam volume [1–13]. With reference to Fig. 9a, consider a particle resting on one of the surfaces of a thin liquid foam film, exhibiting a contact angle > 90°. Films within foams drain and thin with time due to gravity and to suction of liquid into plateau borders (see later). Ultimately, the lower surface touches the surface of the particle (Fig. 9b) and the liquid rapidly "dewets" the solid surface. The resulting curvature of the film in the vicinity of the particles causes a local increase in pressure within the film. This excess of pressure (i.e., the Laplace pressure) forces the liquid away from the particle (as shown by the arrows in the curved liquid surfaces within the plateau borders).* When the two contact lines around the particle meet, the

---

*As a result of the action of surface tension, γ, there exists across a curved liquid surface a pressure difference, Δp, that depends on the curvature of the surface. The pressure is greater on the "inside" of the surface. If the principal radii of curvature of the surface are r_1 and r_2, then the Laplace equation can be written $\Delta p = \gamma[(1/r_1) + (1/r_2)]$.

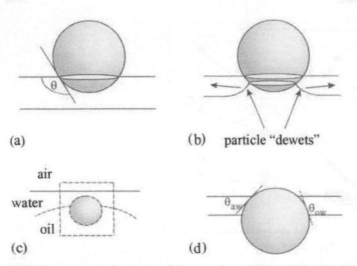

(a)   (b)   particle "dewets"

(c)   (d)

FIG. 9 Effect of spherical particles on the stability of thin liquid films. (a) Particle resting at equilibrium at the air/water surface. (b) As the film thins through drainage, the lower surface of the film comes in contact with the particle surface and the particle dewets. The lower contact line approaches the upper contact line, and the film ruptures when the contact lines meet. The arrows show the liquid flow generated by the Laplace pressure. (c) A particle at the surface of an oil droplet immersed in an aqueous film. (d) A magnified picture showing the two contact angles θ_{aw} and θ_{ow}; based on equilibrium behavior, if the sum of these angles exceeds 180°, the thin asymmetric water film will rupture, aiding the entry of the oil drop into the water surface (see text).

film ruptures. If the contact angle was < 90° the two contact lines would rest at equilibrium, one on each side of the equator, and rupture would not be expected to occur.

When particles are present together with oil, it is supposed that the foam-breaking entities are the oil droplets, whose entry into a film surface is facilitated by the particle. This is illustrated in Fig. 9c and d. The oil droplet, with a particle at its surface (Fig. 9c), approaches the film surface from within the film and the particle enters this surface. It can be appreciated that there are now two different contact angles to consider, that (θ_{aw}) of the air/water interface with the particle and θ_{ow} of the oil/water interface with the particle, as shown in the figure. If the sum ($\theta_{ow} + \theta_{aw}$) exceeds 180°, then the two 3-phase contact lines will tend to overlap so that the thin asymmetric oil/water/air film (Fig. 9d) becomes unstable. This leads to entry of the oil drop into the film surface, and subsequent rupture of the foam film is effected by the oil in ways to be discussed later. The facilitation of droplet entry by the presence of the particles leads to the observed synergy between the oil and the particles in foam breakdown. In practice, equilibrium contact angles are not the only important quantities; dynamic effects associated with rapid movement of contact lines are likely to be equally important so that sums of angles of less than 180° may well enhance drop entry [15]. It has also been shown that particles with suitable wettabilities can prevent drop entry into a liquid surface [15].

When an oil droplet within a liquid film arrives at one of the surfaces of the film, it may or may not enter the surface (see Fig. 10) [14,15,37]. It is possible that a metastable thin aqueous film is formed between the oil drop and air. Assuming that this film, if present, can be ruptured in some way, then in an equilibrated system it is thermodynamically feasible for the drop to enter the surface if the sum of the equilibrium oil/water and air/water interfacial tensions (γ_{ow} and γ_{aw} respectively) exceeds that of the oil/air interface (γ_{oa}). For this reason it is usual to define an equilibrium spreading coefficient, $S_{w,oa}$ for water spreading on the oil/air interface as

$$S_{w,oa} = \gamma_{oa} - (\gamma_{ow} + \gamma_{aw}) \tag{2}$$

It can be shown (e.g., see Refs. 14 and 15) that if $S_{w,oa} = 0$ water spreads on oil. Clearly, if this is the case an oil drop cannot enter the air/water interface. If the spreading coefficient is negative, however (positive values are not possible in equilibrated systems), the aqueous phase cannot spread on the oil and drop entry into the film surface is feasible.* An entry coefficient, $E_{o,aw}$, for the entry of an oil drop into a water surface can be defined as $-S_{w,oa}$. In this case drop entry in an equilibrated system is thermodynamically feasible if $E_{o,aw}$ is positive. Entry is not possible if the entry coefficient is zero, and negative values are not possible.

Assuming a droplet has entered a liquid surface (being transformed into a lens in the process—Fig. 10b), one or more of several processes may occur that can cause film rupture. The lens may remain in situ without any spreading. If so, then ultimately, as the foam film thins, the lens will form a bridge across the film (Fig. 10c) [1]. The bridge then stretches, as a result of uncompensated capillary pressures [10], and ultimately breaks (Fig. 10d–f).

Usually, the material from the lens (Fig. 11a) spreads along a foam lamella surface in one form or another. There may be macroscopic spreading of the whole lens (Fig. 11b) to give a thick (duplex) film. The spreading oil

---

*In a system at phase and adsorption equilibria, it is a thermodynamic result that (assuming, for example, γ_{oa} is the largest of the three tensions) $\gamma_{oa} \leq \gamma_{ow} + \gamma_{aw}$. This being the case, it can be appreciated from Eq. (2) that $S_{w,oa}$ must be zero or negative.

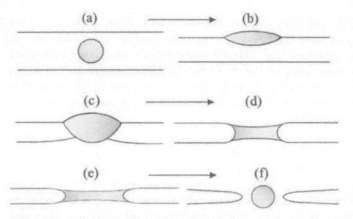

FIG. 10 Entry of an oil droplet (a) into a thin aqueous film to form a lens (b) and subsequent events (c) leading to bridge formation (d), bridge stretching (e), and ultimately film rupture (f).

drags along the underlying liquid (Marangoni effect), which causes local thinning of the film and possibly rupture. Surfactant monolayers (such as those stabilizing a foam film) can be regarded in the context as planar micelles; just as micelles in a bulk solution solubilize oil molecules, so often do planar surfactant monolayers (Fig. 11c) [38]. As a result, when a drop enters a lamella surface (Fig. 11a), tension gradients are produced. As the solubilization spreads out from the lens, the underlying liquid is dragged along, producing film thinning and maybe rupture, as with macroscopic

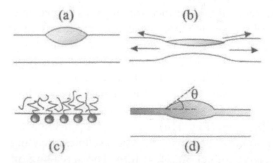

FIG. 11 (a) A drop (lens) of oil resting on an aqueous film surface. (b) Macroscopic spreading of an oil lens. (c) Two-dimensional solubilization of oil molecules in a surfactant monolayer. (d) A thin oil film in equilibrium with a macroscopic oil lens. Spreading of oil macroscopically or as a solubilized monolayer or a thin film can drag along the underlying aqueous phase, causing film thinning and rupture.

spreading. Finally, spreading can also occur as a thin multilayer film that is ultimately in equilibrium with a thick (duplex) film. When this situation exists, the tensions on the two sides of the oil film differ from the macroscopic values of oil/air and oil/water tensions as a result of interactions across the film. This gives rise to a finite contact angle between the thin and thick oil films, as illustrated in Fig. 11d. In any event, spreading of oil can again in principle lead to rupture. Summarizing then, spreading either macroscopically or as a monolayer or a thin multilayer can cause foam film rupture by Maragoni thinning of the films. For such spreading, however, obviously oil drops within foam films must be able to enter the film surfaces.

IV. PARTICLE MONOLAYERS AT LIQUID SURFACES

A. Emulsions Containing Particles

We have considered the behavior of single particles at liquid surfaces. Now we discuss assemblies of particles in the form of monolayers. It has long been known that emulsions can be stabilized by such monolayers in the absence of a surfactant; such emulsions are usually referred to as Pickering emulsions [15]. The behavior of Pickering emulsions in relation to particle wettability brings out remarkable parallels between wettability and the hydrophile–lipophile balance (HLB) of surfactants. The parallel is remarkable because, although surfactant molecules are amphiphilic, the particles to be considered here are not.*

In Fig. 12 we depict two approaching oil droplets whose surfaces contain close-packed monolayers of spherical particles. For an emulsion to be stable the monolayers must remain intact on droplet contact. For the two droplets to coalesce, either the particle monolayers must be compressed (if the particles remain in the liquid/liquid interface) or particles must be displaced from the interfaces. Monolayer compression of already close-packed particle layers is unlikely (see later). For colliding drops, particle displacement must presumably be into the droplets. If the particles are hydrophobic, i.e., θ in Fig. 12a is greater than 90°, displacement of particles into the oil droplets is relatively easy. If, on the other hand, $\theta < 90°$, as in Fig. 12b, the particles must surmount an energy barrier corresponding to the contact angle passing through 90° before they can enter the oil phase, which is the less wetting phase for the particles. This can be readily appreciated by inspection of Fig.

---

*Spherical particles in which one half of the surface is hydrophilic and the other half hydrophobic have been prepared, although their properties have not been widely studied. The particles are referred to as *Janus* beads, after the Roman god (of doorways, passages, and bridges!) who is represented by two heads facing in opposite directions.

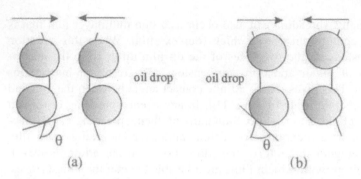

FIG. 12 Approach to two oil drops, coated with particle layers, in water. (a) The contact angle of the particle with the oil/water interface is greater than 90°. (b) The contact angle is less than 90°.

3. Thus, hydrophilic particles tend to stabilize oil-in-water (o/w) emulsions, whereas hydrophobic particles stabilize water-in-oil (w/o) emulsions. On this basis, hydrophilic particles are analogous to high-HLB surfactants and hydrophobic particles to low-HLB surfactants.

It is well known that the HLB of surfactants can be modulated by mixing high- and low-HLB surfactants, and emulsions can be inverted from o/w to w/o or vice versa in this way. It has been shown [39] that a similar inversion can be effected by mixing particles of differing wettabilities. We show results for stabilization and inversion of toluene + water emulsions (in the absence of a surfactant) in Fig. 13. The emulsions were stabilized by mixtures of hydrophobic and hydrophilic silica particles (particle diameters in the range 15 to 30 nm). The hydrophilic silica had 100% silanol groups on its surface and the hydrophobic sample half of this, the remainder of the groups having been reacted with dimethyldichlorosilane. Emulsion types obtained, and hence emulsion inversion, were determined by conductivity measurements. A low concentration of NaCl added to the aqueous phase rendered oil-in-water emulsions conducting, whereas water-in-oil emulsions had very low conductivities.

Particles that are small relative to emulsion drop size are also expected to have an effect on emulsion properties in systems stabilized by surfactants when present in only small amounts (say a sufficient number of particles to give 10% coverage of droplet surfaces). We reported elsewhere [40] on a preliminary study of the ways in which the stability to flocculation and coalescence of water-in-oil emulsions stabilized by the anionic surfactant Aerosol OT are modified by polystyrene latex particles. There is evidence that the particles bridge droplets to give weak flocs, which slow down droplet coalescence.

Clint and Taylor [42] devised a method that they applied to extremely small particles (overbased detergent particles used in engine oils), which had diameters of only a few nm. The principle is to form a monolayer of the (monodisperse) particles at the required interface (air/water in the systems studied by Clint and Taylor) on a Langmuir trough. The monolayers behave in a fashion analogous to insoluble molecular monolayers, and surface pressure (Π) versus surface area (A) plots can be generated (see later). If compressed sufficiently, the particle monolayer will collapse, the collapse point being evident from the shape of the Π–A curve. Clint and Taylor supposed that at and beyond collapse, particles are ejected from the surface. On this basis they equated the collapse pressure (Π_c) to the free energy of removing (into the more wetting phase) the particles present in a unit area of interface. They obtained a simple relationship between collapse pressure, contact angle, and the tension γ of the free interface (between particles):

$$\cos \theta = \pm\left(\sqrt{\frac{\Pi_c 2\sqrt{3}}{\pi \gamma}} - 1 \right) \tag{3}$$

Although this approach has been used with apparent success, it is not clear whether the assumption involving the mode of collapse is likely to be correct. For this and other reasons (mainly scientific curiosity), we have undertaken studies of the behavior of monolayers of spherical monodisperse particles at both the air/water and oil/water interfaces, and we will now describe some of this work.

C. Compression, Structure, and Collapse of Particle Monolayers

The particles studied [22,23] were monodisperse, surfactant-free spherical polystyrene latex particles with sulfate groups on the surface. When these groups are fully ionized in water, the particles have a surface charge density of around 8 μC cm^{-2}, equivalent to 1 sulfate group per 2 nm^2. Unless otherwise stated, the particle diameter was 2.6 μm.

The particles are described as hydrophobic by the makers (Interfacial Dynamics, Portland, OR), but in the present context they are better described as hydrophilic. We have measured the contact angles of 6-μm-diameter particles, directly by microscopy, at the air/water and oil (octane)/water interfaces, and some images are shown in Fig. 14. Although we cannot give the angles very precisely, they are clearly less than 90° (measured as before into the water phase), the angle for the air/water surface being about 30° and that for the octane/water interface around 75°. As will be seen later, this difference in contact angles and the concomitant large difference in particle

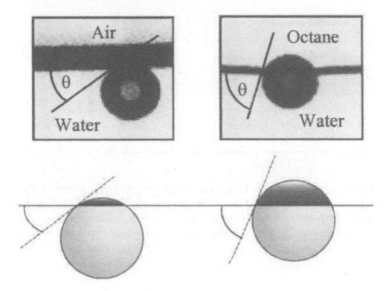

FIG. 14 Microscopically observed contact angles of polystyrene latex particles, diameter 6 μm, at (a) the air/water surface and (b) the octane/water interface. The lower diagrams give an impression of the great difference in the areas of the interfaces between particle and air and particle and octane.

surface areas exposed to the air phase and to the octane phase are crucial in determining the way in which particles interact laterally and form surface structures.

The Langmuir trough used in the study was a modified miniature trough obtained from Nima (Coventry, UK) and could be placed on a microscope stage; the salient features of the trough are illustrated in Fig. 15. Particle monolayers were observed with objectives designed for use with reflected light, and simultaneously the II–A curve was observed on the monitor of the computer controlling the operation of the Langmuir trough.

A typical II–A curve for a particle monolayer at the octane/water interface is shown in Fig. 16; we include in the figure images of the monolayers in different states of compression prior to collapse. We observe finite surface pressures out to high monolayer areas, indicative of strong repulsion between the particles. The possible origin of this repulsion is discussed in detail later. From the figure it is obvious that this repulsion leads to highly ordered structures even at large particle separations. As the monolayer is compressed, the high degree of structure is retained up until monolayer collapse, i.e., at the collapse pressure Π_c indicated in the figure. There is, however, some

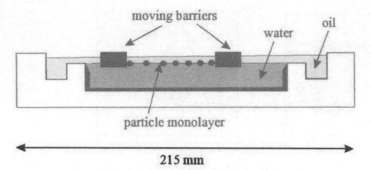

FIG. 15   Schematic of the miniature Langmuir trough used to study particle mono-layers at oil/water and air/water interfaces. The trough, machined from Teflon, has a steel lining with sharp edges to pin the oil/water or air/water interface. For study of oil/water systems the oil is added to the water surface and overflows into the channels around the trough. The barriers enclosing the monolayer are made of steel.

distortion of the hexagonal structure to a rhombohedral configuration for surface pressure in excess of about 25 mN m^{-1}.

Two possible modes of monolayer collapse at high surface pressure are illustrated in Fig. 17. One possibility, as assumed by Cling and Taylor [42], is that particles are ejected from the monolayer into the more wetting bulk phase (Fig. 17a). An alternative collapse process could involve folding the monolayer (Fig. 17b). The images shown in the figure clearly indicate that in the system studied, folding rather than particle ejection occurs, even well beyond collapse. Care must, therefore, be taken before using the collapse pressure to calculate contact angles according to Eq. (3).

D.   Repulsive Interactions Between Particles Within Monolayers; the Surface Equation of State

The polystyrene latex particles carry a charge in aqueous solution resulting from the surface sulfate groups, and as mentioned the particle dispersions are stable. Particle monolayers can be formed at air/solution surfaces as well as at oil/solution interfaces. There is obviously an electrical double layer formed around the parts of the particle surface in contact with the aqueous phase. Lateral repulsions in the monolayer due to double-layer interactions can be screened by addition of inert electrolyte to the aqueous phase. Images of relatively dilute monolayers on 10 mM aqueous NaCl are depicted in Fig. 18. Monolayers at the air/0.01 M NaCl solution surface are much less ordered than those at the octane/solution interface, although there is very little particle aggregation. Monolayers at the octane/solution interface retain

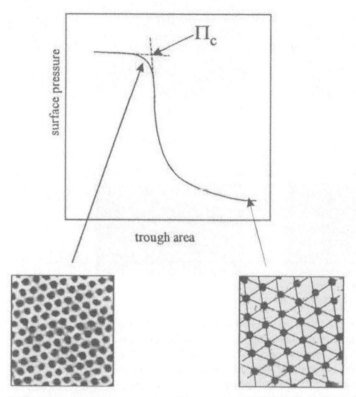

FIG. 16 Typical shape of a pressure–area isotherm for a particle monolayer at the oil/water interface. The images shown correspond to the regions of the isotherm as indicated. At high trough area the particles are well spaced and, due to strong lateral repulsions, form a highly ordered array; the lines have been drawn on the right-hand image simply to give an impression of the high order. Order is retained when the particles are compressed up to collapse, although it becomes distorted from hexagonal (see text).

very high order. When the NaCl concentration in the aqueous subphase is increased to 1 M, particles at the air/solution surface are completely aggregated, while those at the octane/solution interface retain a remarkable degree of order with little particle aggregation taking place.

These observations taken together can be interpreted as follows [22,23]. Aggregation caused by salt at the air/solution interface is presumably a result of screening the double-layer repulsion through the aqueous phase. Because order is retained in monolayers at the octane/solution interface up to very high electrolyte concentrations, it appears that the major component of the

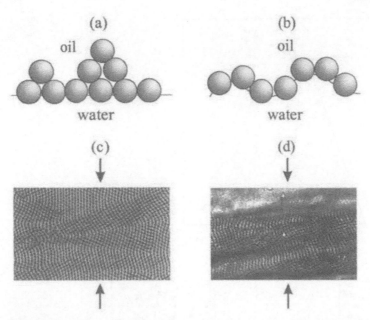

FIG. 17 Possible modes of monolayer collapse when the trough area is less than the close-packed area of the planar monolayer. (a) Particles are ejected into the more wetting phase; (b) the monolayer folds. (c and d) Images of monolayers of polystyrene latex particles compressed beyond collapse in the direction indicated by the arrows. (c) Monolayer just beyond collapse; (d) monolayer well beyond collapse. No ejection of particles from the monolayer is observed.

repulsion giving this order must be acting through the oil phase. Ordering occurs in monolayers with very large particle separations (Figs. 16 and 18), so the repulsion through the oil must be extremely long range.

From knowledge (or assumption) of interparticle interactions within a monolayer, it is possible to obtain a theoretical surface equation of state (i.e., a relationship between the surface pressure and the area available to particles in the monolayer) that can be compared with the experimental π–A curves. The state of charge of a polystyrene latex particle (with sulfate groups at its surface) is illustrated in Fig. 19. In the aqueous phase, an asymmetrical double layer is set up that leads to a net dipole normal to the interface. The magnitude of this dipole is expected to depend on the concentration of the inert electrolyte (e.g., NaCl) in the aqueous subphase, and the dipoles lead to lateral repulsion between particles. The sulfate groups also carry permanent dipoles, and this leads to a net dipole normal to the interface within the oil phase and hence to another contribution to lateral repulsion. It turns

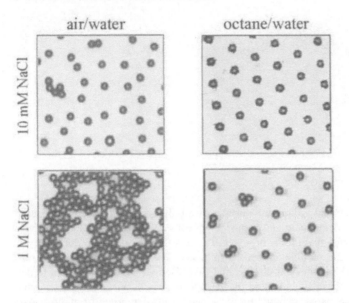

FIG. 18 Images of monolayers of polystyrene latex particles at air/water (left) and octane/water (right) interfaces. The top images are of systems in which the aqueous phase contains 10 mM NaCl; bottom images are for 1 M NaCl.

out, however, that repulsion from these dipoles is far too small to give the long-range order observed experimentally. We are led to suppose, therefore, that there is some small amount of free charge at the solid/octane interface, possibly arising from some retention of (hydration?) water and concomitant ionization of sulfate groups. Such charge could not be screened by a water-soluble electrolyte such as NaCl. It might be argued that such charge should also be present at the particle/air interface and hence lead to strong, long-range repulsion in monolayers at the air/solution interface. We recall, however, that the contact angle of particles at the air/water interface is 30° or less while at the oil/water interface it is around 70 to 80°; this leads to large differences in the solid/air and solid/oil interfacial areas (hence charge), as can be appreciated from Fig. 20. Further, the possibility exists that any water trapped at the air/solution interface could easily evaporate.

The charge on the particle/oil interface can be treated as an equivalent point charge, q, suitably placed (at vertical distance ζ from the interface). With reference to Fig. 20, the charge in the oil phase (relative permittivity ε_o of about 2) gives rise to an image charge of opposite sign and very similar magnitude a distance ζ into the aqueous phase, relative permittivity ε_w about

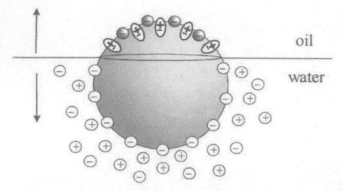

oil

water

FIG. 19 Charges on and around the surface of a polystyrene latex particle resting at the interface between octane and water or aqueous electrolyte. The latex surface contains sulfate groups that ionize in water. There is thus an electrical double layer around the part of the particle immersed in water. The sulfate groups at the oil/particle interface have permanent dipoles that give rise to a net dipole normal to the interface in the oil phase, as indicated by the upper arrow. Likewise, there is a net dipole in the aqueous phase, denoted by the lower arrow, that arises from the asymmetric double layer around the particle surface in contact with water. Because the particles are strongly repulsive at the oil/water interface but not at the air/water interface, we propose that there is a (small) negative charge arising from hydrated sulfate groups at the solid/oil interface.

80.* The lateral repulsion between charges q at separation D (see Fig. 20) arises from charge–dipole interactions involving the image charges and from coulombic interaction through the oil phase. It transpires that at small particle separations the two types of interactions are of comparable magnitude, whereas at larger separations the coulombic interaction through the oil is dominant. The equation of state for the particle monolayer at the oil/water interface can be expressed as

$$\Pi = \frac{q^2}{2\sqrt{3}\varepsilon_o R^3 x^{2/3}} \left[1 - \frac{1}{a} + \ln\left(\frac{1+a}{2}\right) \right] \tag{4}$$

where

$$a = \left(1 + \frac{16\zeta^2}{D^2} \right)^{1/2}$$

---

*The image charge in water, e_i, due to charge e_o in oil is given by $e_i = e_o((\varepsilon_o - \varepsilon_w)/(\varepsilon_o + \varepsilon_w))$.

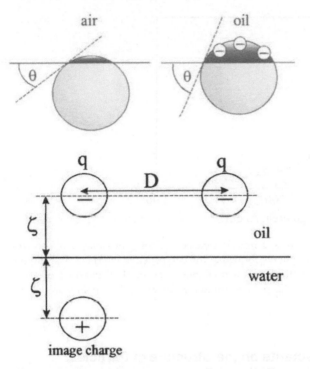

FIG. 20 Electrical interactions between particles at the oil/water interface. The upper diagram shows that the area of solid/upper phase contact depends strongly on the contact angle θ. The lower diagram shows the relevant charges and distances. The charge at the solid/oil interface can be taken as equivalent to a suitably placed point charge, q, a distance ζ from the oil/water interface. Adjacent charges are separated by distance D. A charge in the oil phase gives rise to an image charge (approximately $-q$ in the case of the octane/water interface) of opposite sign in the aqueous phase a distance ζ from the interface.

and $x = A/A_h$, A being the area occupied by the monolayer at surface pressure Π and A_h the area at hexagonal close packing of the monolayer. In order to effect a fit of theory to an experimental isotherm, it is necessary only to choose a value for the charge q. The fit shown in Fig. 21 has q equal to 1% of the total nominal charge on a fully ionized particle in water. The shape of the theoretical isotherm closely matches that of the experimental isotherm, and it can be appreciated that little charge is needed at the oil/particle interface to give the very long range repulsion observed.

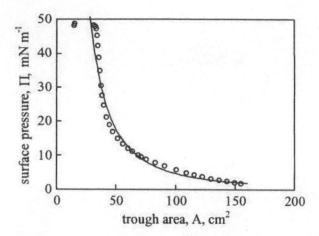

FIG. 21 Surface pressure versus trough area curves for a monolayer of 2.6-μm-diameter polystyrene latex spheres formed at the octane/water interface. The full line is from theory (see text) with an assumed charge at the solid/oil interface of 1% of the total nominal charge of the particle immersed in water. The points are experimental data.

E. Effects of Surfactants on the Structure of Particle Monolayers and on Collapse Pressures

In general, it appears that the surfactant present in the aqueous phase leads to a reduction in monolayer structure at the oil/aqueous solution interface [23]. At concentrations close to the critical micelle concentration (cmc) the particles are almost completely aggregated. We show images of such aggregated monolayers in Fig. 22. Because the particles carry a negative charge

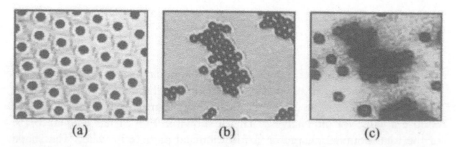

(a)   (b)   (c)

FIG. 22 Effects of surfactants, close to the cmc, on monolayer order for polystyrene spheres (diameter 2.6 μm) at the octane/aqueous solution interface. Monolayers (a) on water, (b) on 1 mM CTAB solution, and (c) on 10 mM aqueous SDS.

(ionized sulfate groups) it is tempting to suppose that the effect that CTAB has on the monolayer is a result of charge neutralization. That this is unlikely to be the case can be appreciated from the observation that the anionic surfactant sodium dodecyl sulfate (SDS) has the same effect as CTAB. At present, these observations remain unexplained.

The addition of a surfactant to systems with particle monolayers at oil/water interfaces allows us to determine the way in which the monolayer collapse pressure, Π_c, varies with the interfacial tension, γ_{ow}, of the (particle-free) oil/water interface [23]. This, in turn, can give us some feel for the physical origin of particle monolayer collapse. Quite remarkably, we find that there is a very close correspondence between Π_c and γ_{ow}, as seen in Fig. 23. The tensions of the oil/water interface have been varied by addition of a range of concentrations of four different surfactants (including anionic, cationic, and nonionic), namely CTAB, SDS, cetylpyridinium chloride (CPC), and the pure sugar surfactant decyl β-glucoside (DBG).

With reference to Fig. 24, the uncovered oil/water interface between particles in a monolayer tends to contract, the tendency being determined by the value of γ_{ow}. The repulsive particles, on the other hand, tend to cause the film-covered interface to expand, this tendency being governed by Π_c.

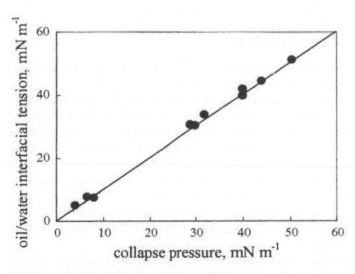

FIG. 23 Graph showing the near equivalence of the monolayer collapse pressure and the tension of the oil/water interface in the absence of particles. Tensions (and hence collapse pressures) have been adjusted by addition of different concentrations of a range of surfactants including SDS, CTAB, DBG, and CPC (see text for surfactant abbreviations).

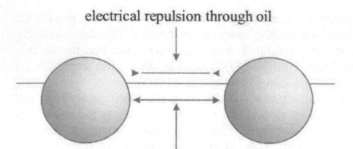

electrical repulsion through oil

contractile force due to interfacial tension

FIG. 24 Origin of equality of collapse pressure and interfacial tension. Particles repel through the oil phase, and the oil/water interface tends to contract. When the forces balance, the effective interfacial tension of the particle-covered interface is zero.

It appears that, from the equality of Π_c and γ_{ow}, monolayer collapse occurs when the effective tension of the particle-covered interface is equal to zero (for which $\Pi_c = \gamma_{ow}$).

V. CONCLUDING REMARKS

The behavior of particles at interfaces undoubtedly has important practical consequences, not the least in determining the stability of foams and emulsions containing particles. Whereas aqueous foam stability and breakdown in the presence of hydrophobic particles have been reasonably well studied in recent years, there are still a number of problems to be resolved, including how particle irregularities, particle size, and dynamic wetting effects can be adequately treated. When it comes to emulsion stability in the presence of particles, far less systematic experimental work has been done. There are intriguing similarities between particle wettability (in particle-stabilized emulsions) and surfactant HLB (in surfactant-stabilized emulsions). The study of such similarities, both experimentally and theoretically, will no doubt prove rewarding.

The effects of particles on foam stability are usually discussed in terms of individual particles because a single particle can rupture a thin liquid film. Emulsion stabilization by particles alone presumably involves close-packed monolayers around the emulsion drops. However, less than close-packed layers can, in principle, have important effects on emulsion stability in systems stabilized by surfactants, and this area warrants further study.

In addition to any practical applications that there might be (and the bureaucrats controlling science usually want to be told there are potential practical benefits in a research program), there are a number of areas regarding the behavior of particle monolayers that require fundamental study. For example, the forces giving rise to strong repulsion and high order within monolayers and the effects of surfactants on this warrant further investigation. Little is known about structures of monolayers containing mixtures of particle size, type, and charge. The behavior of monolayers of particles at liquid surfaces, where the particles cannot be observed by optical microscopy, also needs to be probed. The hope would be that in the not too distant future, the understanding of colloids in two dimensions will equal that of bulk colloidal systems.

REFERENCES

1. PR Garrett. In: PR Garrett, ed. Defoaming. Surfactant Science Series. Vol 45. New York: Marcel Dekker, 1993 p 1.
2. S Ross. J Phys Colloid Chem 54:429, 1950.
3. PR Garrett. J Colloid Interface Sci 76:587, 1980.
4. K Koczo, LA Lobo, DT Wasan. J Colloid Interface Sci 150:492, 1992.
5. PR Garrett. J Colloid Interface Sci 69:107, 1979.
6. A Dippenaar. Int J Miner Process 9:1, 1982.
7. R Aveyard, BP Binks, PDI Fletcher, CE Rutherford. J Dispersion Sci Technol 15:251, 1994.
8. R Aveyard, P Cooper, PDI Fletcher, CE Rutherford. Langmuir 9:604, 1993.
9. GC Frye, JC Berg. J Colloid Interface Sci 54:130, 1989.
10. ND Denkov, P Cooper, JY Martin. Langmuir 15:8514, 1999.
11. ND Denkov. Langmuir 15:8530, 1999.
12. ES Basheva, D Ganchev, N Denkov, K Kasuga, N Satoh, K Tsujii. Langmuir 16:1000, 2000.
13. ND Denkov, KG Marinova, C Christova, A Hadjiiski, P Cooper. Langmuir 16: 2515, 2000.
14. R Aveyard, BP Binks, PDI Fletcher, T-G Peck. J Chem Soc Faraday Trans 89: 4313, 1993.
15. R Aveyard, JH Clint. J Chem Soc Faraday Trans 91:268, 1995.
16. R Aveyard, JH Clint. J Chem Soc Faraday Trans 92:85, 1996.
17. R Aveyard, BD Beake, JH Clint. J Chem Soc Faraday Trans 92:4271, 1996.
18. GL Gaines Jr. Insoluble Monolayers at Liquid–Gas Interfaces. New York: Interscience, 1966.
19. H Schuller. Kolloid Z 216–217:389, 1967.
20. E Sheppard, N Tcheurekdjian. J Colloid Interface Sci 28:481, 1968.
21. A Doroszkowski, R Lambourne. J Polym Sci C 34:253, 1971.
22. R Aveyard, JH Clint, D Nees, VN Paunov. Langmuir 16:1969, 2000.
23. R Aveyard, JH Clint, D Nees, N Quirke. Langmuir 16:8820, 2000.

24. R Aveyard, JH Clint, D Nees. Colloid Polym Sci 278:155, 2000.
25. P Pieranski. Phys Rev Lett 45:569, 1980.
26. AJ Hurd. J Phys A 18:L1055, 1985.
27. DJ Robinson, JC Earnshaw. Phys Rev A 46:2045, 1992.
28. DJ Robinson, JC Earnshaw. Phys Rev A 46:2055, 1992.
29. DJ Robinson, JC Earnshaw. Phys Rev A 46:2065, 1992.
30. DF Williams, JC Berg. J Colloid Interface Sci 152:218, 1992.
31. DJ Robinson, JC Earnshaw. Langmuir 9:1436, 1993.
32. J Stankiewicz, MAC Vilchez, RH Alvarez. Phys Rev E 47:2663, 1993.
33. Z Horvalgyi, M Mate, M Zrinyi. Colloids Surf A 84:207, 1994.
34. D Goulding, J-P Hansen. Mol Phys 96:649, 1998.
35. R Aveyard, JH Clint, D Nees, VN Paunov. Colloids Surf A 146:95, 1999.
36. S Bird. Unpublished work, University of Hull.
37. V Bergeron, ME Fagan, CJ Radke. Langmuir 9:1704, 1993.
38. R Aveyard, P Cooper, PDI Fletcher. J Chem Soc Faraday Trans 86:3623, 1990.
39. BP Binks, S Lumsdon. Langmuir 16:3748, 2000.
40. R Aveyard, BP Binks, JH Clint, PDI Fletcher. In: JF Sadoc, N Rivier, eds. Foams and Emulsions. Dordrecht: Kluwer Academic Publishers, 1999, p 21.
41. Z Horvolgyi, S Nemeth, JH Fendler. Colloids Surf A 71:207, 1999.
42. JH Clint, SE Taylor. Colloids Surf A 65:61, 1992.

4

From Polymeric Films
to Nanocapsules

HELMUTH MÖHWALD, HEINZ LICHTENFELD, SERGIO MOYA, A. VOIGHT, and G. B. SUKHORUKOV Max-Planck-Institute of Colloids and Interfaces, Potsdam, Germany

STEFANO LEPORATTI University of Leipzig, Leipzig, Germany

L. DÄHNE, IGOR RADTCHENKO, and ALEXEI A. ANTIPOV Max-Planck-Institute of Colloids and Interfaces, Potsdam, Germany

CHANGYOU GAO Zhejiang University, Hangzhou, People's Republic of China

EDWIN DONATH University of Leipzig, Leipzig, Germany

ABSTRACT

A method to prepare well-defined responsive microcapsules and nanocapsules with engineered walls is reported. A decomposable colloidal template is coated by polyelectrolyte multilayers, and after core removal a hollow capsule, refillable with drugs, is obtained. The release of the capsule as well as mechanical properties can be tuned by adjusting pH and temperature.

I. INTRODUCTION

Much has been learned about the structure of organic films and interfaces, and techniques to prepare these films in a controlled way have been developed. These have been concerned mostly with planar systems, but there is no obvious reason why this knowledge cannot be transferred to curved interfaces, which would offer many advantages:

One would be able to obtain systems with a high specific surface area. Not only would this allow many applications requiring controlled surfaces (e.g., chromatography, enzyme technology, separation technology), but also other techniques to study interfaces would be employed. Our motivation resulted from the application of methods typically used for bulk samples: nuclear magnetic resonance (NMR), differential scanning calorimetry and flash spectroscopy.

Coating colloids in a defined way is also a prerequisite to understanding colloidal solutions because the interparticle interactions are determined by their interfaces.

Having succeeded in coating colloids, there is an obvious next step: to template colloids by dissolving the core and thus obtaining hollow capsules. This route is discussed in the following, including the interesting properties and possible applications of these capsules.

II. PREPARATION OF COATED COLLOIDS AND CAPSULES

A technique to prepare polymeric films with nanometer precision has been introduced by Decher [1] and is called layer-by-layer adsorption. With this technique a charged surface can be coated by dipping it into a solution of an oppositely charged polyelectrolyte. The latter is adsorbed, reversing the surface charge under suitable conditions, and thus a polyelectrolyte with the opposite charge can again be adsorbed. Repeating the process leads to polymeric films of low roughness (<1 nm), and the thickness can be controlled with nanometer accuracy by the number of dipping cycles and ionic conditions. Functional molecules can be integrated into the films oriented along the surface normal and controlled with nm precision. Depending essentially on electrostatic interactions, the technique is applicable to many multiply charged systems: synthetic or natural polymers, proteins, colloidal particles, and dyes. It also does not require planar and flat surfaces. Instead, the surface can be rough, porous, or strongly curved. Consequently, we have developed various protocols to coat colloidal particles by this technique [2] (Fig. 1).

These protocols consist of either incubating the colloidal particles in a solution containing more polyelectrolyte than needed for adsorption and then removing the excess polyelectrolyte by centrifugation, or filtering before adding a polyelectrolyte with the opposite charge, or adding just as much polyelectrolyte as needed for saturation coating. The conjecture that each adsorption step leads to reversal of the surface charge is verified by electrophoretic mobility measurements.

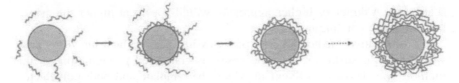

FIG. 1 Scheme of the layer-by-layer adsorption of polyelectrolytes on colloidal particles. The excess polyelectrolyte in the supernatant has to be removed prior to adsorption of the next polyion species.

The coating can be followed quantitatively by single-particle light scattering. Figure 2 shows that the distribution of the scattering intensity per particle shifts to higher values depending on the number of coated layers. This shift can be quantified by applying the Rayleigh–Debye–Gans or Mie theory, knowing the refractive index of the polymer [3]. One thus derives thickness increases of about 1 nm per coating step, which is close to the value determined by X-ray reflectivity for planar surfaces. Also, there is little broadening of the distribution with coating, proving that the deposit thickness is the same for different particles. This does not prove that the coating is uniform. However, if this were not the case one would observe particle aggregation because there would be coulombic attraction between oppositely charged areas on different colloids. This is not observed by light

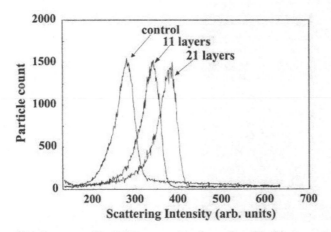

FIG. 2 Normalized light scattering intensity distributions (single-particle light scattering) of PAH/PSS-coated polystyrene sulfate latex particles (ϕ 640 nm). Particles with 11 and 21 layers are compared with uncoated (control) ones.

scattering. A dimer or higher aggregate would appear at higher intensities, in contrast to our findings.

An alternative method to coat particles in a less defined but faster way, which may be sufficient in some cases, is controlled precipitation. In this method one chooses a solvent in which the template and both polyelectrolytes are soluble. Then the quality of the solvent is decreased and the polyelectrolytes precipitate on the templates. Examples in which this idea has been verified are (1) a suitable water/acetone/brine mixture as a good solvent and reduction of solvent quality by evaporating acetone or by adding water and (2) precipitation of DNA by addition of methanol to a DNA solution in water. The confocal micrograph in Fig. 3 shows that this leads to a rather uniform coating.

Because the basic force for the layer buildup is the electrostatic attraction, the wall material can be of many different but multiply charged entities. We have demonstrated this versatility using synthetic and natural charged polymers, proteins, inorganic colloids, DNA, charged dyes, and trivalent metal ions [4]. Also, the colloidal core to be coated can be of many different types, and this advantage will be apparent in the following. We have successfully coated lattices, biological cells, inorganic particles, and precipitates of enzyme, DNA, or dyes, and even hydrophobic particles and oils would be coated after charging their surfaces by adding low- or high-molecular-weight amphiphiles [5].

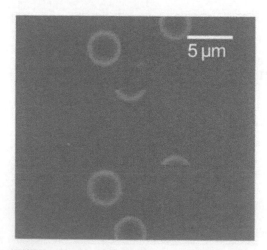

FIG. 3 Confocal fluorescence microscopy image of melamine formamide (MF) particles covered with DNA labeled with acridyl orange precipitated by dropping ethanol in aqueous solution.

These cores have become very important because they define the conditions under which they can be removed. This process then yields hollow capsules because generally the polyelectrolyte films are permeable to small molecules but impermeable to macromolecules with molecular weights of some thousands. In the case of weakly cross-linked melamine formaldehyde, the core could be removed by going to pH 1 [6], biological templates were removed with a deproteinizer solution, inorganic particles could be removed by milder pH changes, and organic cores were dissolved by adding a proper solvent [7].

All these different procedures lead to hollow capsules with the following main features:

Sizes and size distribution, depending on the template, between 70 nm and 10 μm
Wall thickness tunable in the nm range
Wall composition tunable along the surface normal
Variable inner and outer surfaces
Variable wall material

This unique variability should lead to wide tunability of properties, and examples of this are given in the next section.

III. PROPERTIES OF HOLLOW CAPSULES

A. Encapsulation and Release

Figure 4 shows confocal fluorescence micrographs of a hollow capsule after adding a fluorescently labeled polyelectrolyte to the outside aqueous medium [8]. The left image was taken at a pH close to that during preparation. The dark interior demonstrates that the wall is impermeable to the polyelectrolyte, and this holds over times of at least days. However, on reducing the pH to 6 (middle), the wall becomes permeable as evidenced by the fluorescence from the interior. On increasing the pH again to 8 and removing the outside polyelectrolyte, one observes that the polyelectrolyte is entrapped; i.e., the wall has become impermeable again. This intriguing finding also held for proteins and labeled dextrans with molecular weights from 70,000 to 2,000,000 and can be explained as follows [8].

The multilayer is composed of a strong polyelectrolyte, polystyrenesulfonic acid sodium salt (PSS), for which all dissociable groups are charged, and a weak polyelectrolyte, polyallylamine hydrochloride (PAH), for which, depending on the pH, only a portion of the side groups is charged. At the pH of preparation, the negatively and the positively charged groups compensate and the film interior is nearly neutral. This is the case for preparation

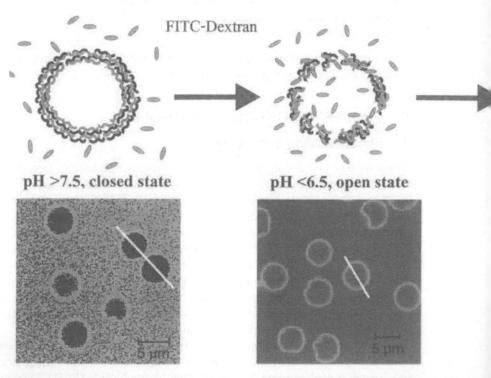

G. 4 Permeation and encapsulation of fluorescein isothiocyanate (FITC)–dextran (molecular ultilayer capsules. (Left) pH 10, (center) pH 3, (right) pH increased to 10 after the capsules ł 3. The bulk FITC–dextran was removed by washings at pH 10. (Top) scheme; (bottom) cc at a pH between 7 and 8. On reducing the pH, more amine groups of PAH become charged and the interior of the film is charged up. The internal coulombic repulsion will be partly compensated by the counterions from solution. This yields an internal osmotic pressure that results in a force destabilizing the film. This destabilization may result in disruption of the film or in pore formation, which is obviously the case here. The finding of reversible pore formation is accidental, and we would expect this reversibility not to be repeated many times. This is, however, not needed for many practical applications. The preceding arguments indicate that this release threshold can be varied through preparation parameters and the type of polyelectrolyte, and the ionic strength yields another control parameter. This has indeed been found. The pH- or ionic strength–controlled pore formation can be seen in the scanning force microscopy (SFM) images in Fig. 5. One realizes the presence of holes after treatment to obtain conditions in which macromolecules permeate. This indicates that these holes correspond to the pores postulated earlier, and they are not due to a drying artifact because the images were not taken under water. They might have been much smaller before drying, which probably increased their size. This experiment yields parameters for controlled encapsulation and release of macromolecules including proteins. For small molecules (molecular weight $<10^3$), however, the films prepared up to now have been largely permeable as measured by confocal microscopy with dye probes in experiments similar to those in Fig. 4.

Then one may encapsulate by controlled precipitation in an experiment as illustrated in Fig. 6 [9]. In this case the negatively charged dye car-

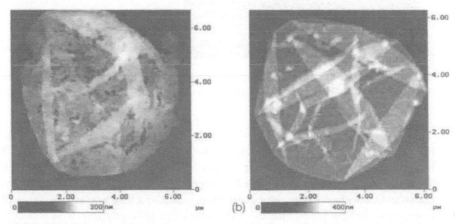

FIG. 5 SFM images of capsules prepared on MF particles and treated at pH 3.5 (a) and pH 12 (b) before drying.

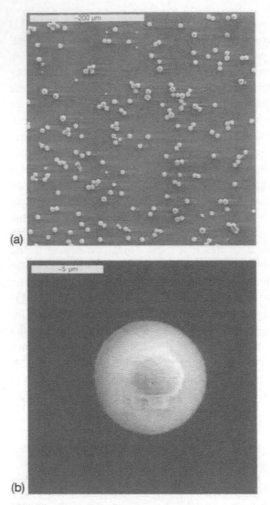

FIG. 6 Scanning electron micrographs of precipitates of 6-carboxyfluorescein in the presence of shells based on discocytes at different magnifications.

boxyfluorescein was used as a model drug. The capsule was prepared with the inner surface positively charged and the outer one negatively charged. The dye in solution is distributed nearly uniformly inside and outside the capsule but with a higher concentration near the inner surface because of coulombic forces. Lowering the pH causes dye precipitation, which occurs predominantly near the inner surface. Then the crystal growth proceeds toward the center and more dye is sucked into the inside until the capsule is filled with precipitate. The images in Fig. 6a and b show that crystallization occurs exclusively in the inside (for weak supersaturation and if the outside is positively charged). Even more complicated shapes such as the discoidal erythrocytes can be templated (Fig. 6b).

To study drug release, one may either reverse the conditions for inside precipitation or just use the precipitated particles and coat them in a defined way as described earlier. This has been done in the experiment leading to Fig. 7. The fluorescein dye does not fluoresce in a precipitated form but fluoresces as a monomer in solution. Hence, release from a particle into the solution outside the capsule causes a fluorescence increase, and the intensity is proportional to the amount of released material. The curves in Fig. 7 show qualitatively the expected delay of the release with increasing coating thickness. Although the different phases of the curves are not all well understood, one may derive quantitative conclusions from the parts one understands.

For intermediate times one expects a stationary (saturation) concentration of monomeric dye inside and a nearly zero concentration outside. Hence,

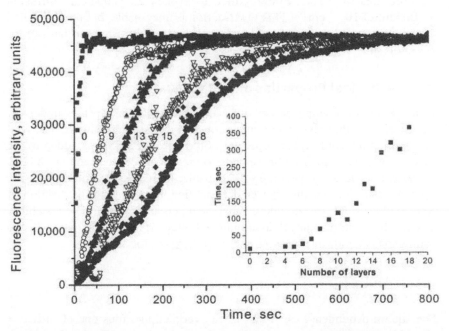

FIG. 7 Fluorescence increase with time obtained by dissolving fluorescein particles covered with shells of different thicknesses (9, 13, 15, and 18 layers) compared with naked (0) fluorescein particles. (Inset) Dependence of core decomposition time on the number of layers in the shells.

the concentration gradient as the driving force and thus the flux across the wall are time independent. This leads to a linear increase of monomeric dye concentration outside and explains the linear (middle) region of the fluorescence–time dependence. On increasing the wall thickness, the concentration gradient should decrease in an inversely proportional manner, and indeed the inset of Fig. 7 shows the corresponding dependence of the slopes of the linear region versus wall thickness. The permeability coefficient thus derived can be converted into a diffusion coefficient, and one obtains $D \sim 10^{-11}$ cm^2/s. This value is about four orders of magnitude larger than measured previously for similar dyes with planar polyelectrolyte films. This difference may be due to differences in preparation procedures, but it may also be due to the fact that the permeation process cannot be described as a diffusion process. Although we could describe the findings quantitatively with Fick's law, an alternative description may be to assume small pores through the whole wall but with a density or conductivity of pores inversely proportional to the wall thickness. To estimate numbers, if a dye diffusing through a pore had the same diffusion coefficient as in water ($\sim 10^{-5}$ cm^2/s) a fraction of the pore area of $\sim 10^{-6}$ would suffice to explain an "apparent" diffusion coefficient of 10^{-11} cm^2/s. This low fraction is imaginable, but more detailed experiments will be needed to elucidate the proper diffusion mechanism for low-molecular-weight compounds.

B. Mechanical Properties and Stability

Figure 8 shows a sequence of confocal micrographs of capsules under osmotic pressure that enable measurement of the elastic modulus μ. The wall was stained by a fluorescent dye, the polyelectrolyte PSS was added to the outside, and the sequence corresponds to increasing PSS concentration [10]. Because PSS does not penetrate the capsule, it creates an osmotic pressure from the outside, and this causes deformation. By counting the fraction of deformed capsules, one can define a critical pressure P_c where the instability sets in. In a simple model assuming a hollow sphere with radius R and homogeneous wall thickness δ, one would expect the relation

$$P_c = 4\mu \left(\frac{\delta}{R}\right)^2$$

The square dependence on R and δ was verified, and thus one obtains $\mu = 500$ MPa. This value is somewhat less than that for a glassy polymer such as poly(methyl methacrylate), and thus we may consider the wall like a glassy polymer. Although the preceding experiment is elegant and meaningful, one would not consider μ typical for shell materials. We mentioned

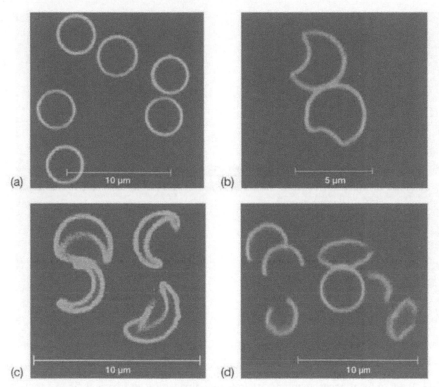

FIG. 8 Deformation of polyelectrolyte capsules consisting of 10 layers of PSS/PAH as a function of the PSS (molecular weight 70.000) bulk concentration. (a) Control, (b) 2.5 wt% PSS, (c) 5 wt% PSS, and (d) 10 wt% PSS. Capsule radius 2 μm.

earlier that a shell could be composed of many different materials including inorganic particles, and thus μ may vary considerably.

However, μ may vary even for the same material, as expected from the preceding pH variation but also as a function of temperature as revealed in Fig. 9 [11]. In this case, a capsule shrinks as the temperature is increased to 70°C. A detailed analysis of the height profiles from the AFM measurements reveals that this shrinkage is accompanied by a wall thickness increase.

Hence, the individual layers, initially rather stratified, would tend to coil for entropic reasons. A thickness increase at constant density would cause a surface area decrease and thus lateral shrinking. The remaining question now is, How is it possible, in view of the many electrostatic bonds in the film, to have a molecular reorientation? Indeed, if it is assumed that the reorien-

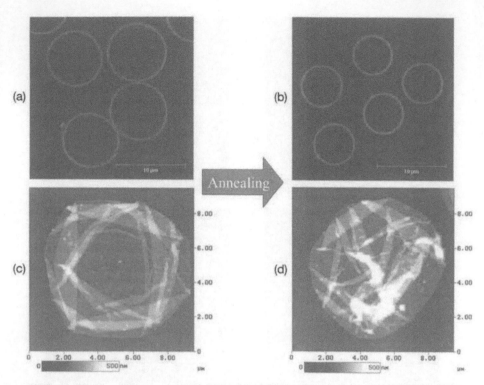

FIG. 9 Confocal laser scanning microscopy images (a and b) and SFM (contact mode) top view images (c and d) of 10-layer (PSS/PAH)₅ polyelectrolyte capsules prepared on melamine formaldehyde resin latex (ϕ 8.7 μm) solution, (a and c) before and (b and d) after annealing at 70°C for 2 h.

tation requires simultaneous breaking of many bonds (say 15) and each bond is screened partly by water (dielectric constant $\varepsilon \sim 40$), one would estimate with Eyring's theory reorientation times between hours and months and a strong temperature dependence [11]. On the other hand, this estimate shows that by varying molecular parameters such as charge–charge distance or water content in the film, one may obtain stability over years or instantaneous instability.

Another aspect of stability against precipitation, the so-called colloidal one, can be tuned trivially. The preparation process yields charge-stabilized colloids that are stable in solution for at least months. On the other hand, reducing the surface charge, e.g., by adsorbing oppositely charged polyelectrolytes without reversing the charge, eventually causes flocculation, as known in many applications of polyelectrolytes as flocculants.

ACKNOWLEDGMENT

We acknowledge the skillful technical assistance of H. Zastrow and I. Bartsch in characterizing the colloids.

REFERENCES

1. G Decher. Science 277:1232, 1997.
2. E Donath, GB Sukhorukov, F Caruso, S Davis, H Möhwald. Angew Chem Int Ed 37:2201, 1998.
3. H Lichtenfeld, L Knapschinsky, C Dürr, H Zastrow. Prog Colloid Polym Sci 104:148, 1997.
4. IL Radtchenko, GB Sukhorukov, S Leporatti, GB Khomutov, E Donath, H Möhwald. J Colloid Interface Sci 230:272, 2000.
5. S Moya, GB Sukhorukov, M Auch, E Donath, H Möhwald. J Colloid Interface Sci 216:297, 1999.
6. C Gao, S Moya, H Lichtenfeld, A Casoli, H Fiedler, E Donath, H Möhwald. Macromol Mater Eng 286:355, 2001.
7. O Tiourina, AA Antipov, GB Sukhorukov, N Larionova, Yu Lvov, H Möhwald. Macromol Biosci 1:209, 2001.
8. GB Sukhorukov, AA Antipov, A Voigt, E Donath, H Möhwald. Macromol Rapid Commun 22:44, 2001.
9. V Dudnik, GB Sukhorukov, IL Radtchenko, H Möhwald. Macromolecules 34: 2329, 2001.
10. C Gao, E Donath, S Moya, V Dudnik, H Möhwald. Eur Phys J E 5:21, 2001.
11. S Leporatti, C Gao, A Voigt, E Donath, H Möhwald. Eur Phys J E 5:13, 2001.

5

Investigation of Amphiphilic Systems by Subzero Temperature Differential Scanning Calorimetry

SHMARYAHU EZRAHI Technology and Development Division, IDF, Tel Hashomer, Israel

ABRAHAM ASERIN and NISSIM GARTI Casali Institute of Applied Chemistry, The Hebrew University of Jerusalem, Jerusalem, Israel

ABSTRACT

In this chapter some problems connected with the utilization of subzero temperature differential scanning calorimetry (SZT-DSC) are discussed. Among them are the determination of hydration numbers of surfactants and organic compounds, the determination of the hydration shell thickness, the effect of alcohol on the distribution of water between free and bound states in nonionic surfactant–based systems, and some considerations regarding the problem of phase separation of such systems in subzero temperatures. The significance of SZT-DSC for some novel applications is also discussed.

I. INTRODUCTION

If one considers the use of subzero temperature differential scanning calorimetry (SZT-DSC) as an analytical tool, a striking fact becomes evident: whereas this technique has been used extensively for the study of aqueous solutions of polymers, biopolymers, and other organic compounds such as phospholipids, only a little attention has been given to the investigation of amphiphilic systems. Even less attention has been paid to the study of ternary (water + surfactant + oil) and quaternary (water + surfactant + co-surfactant + oil) systems.

The work done in this area has been reviewed [1–3]. We have suggested two reasons for this infrequent use of SZT-DSC that might be stated tersely as follows [2]: the careful work needed to elucidate the nature of water–amphiphile(s) interactions and, much more important, the possibility of phase separation (and, sometimes, of phase transformation). We will provide greater detail on this second essential point in the following. What is important to emphasize for the present is that this putative phase separation might prima facie entail the invalidation of the extrapolation of low-temperature data to ambient temperature.

The aim of this chapter is to present the basic considerations and concepts germane to the use of SZT-DSC for the investigation of water behavior in surfactant-based systems and to relate them to some already established applications of microemulsions.

II. TYPES OF WATER IDENTIFIED BY SZT-DSC

We can, broadly speaking, discern two classes in relation to the behavior of water near surfaces: nonfreezable (i.e., most tightly bound or, alternatively, confined within very small enclosures) and freezable water. This second class reflects various degrees of water–amphiphile(s) interactions and, accordingly, subsumes several "types" (or "states") of water. We may use the following classification of Senatra et al. [4]:

1.   Free or bulk water—melts at 0°C
2.   Interphasal (or interfacial [1]) water—melts at about −10°C
3.   Bound water—melts at temperatures lower than −10°C

Succinct as these operational definitions may be, one cannot minimize their importance, for they correspond to thermal peaks observed when using the endothermic (i.e., controlled heating of previously frozen samples) mode of SZT-DSC [2] (see Fig. 1). The melting, temperature, −10°C, is just an arbitrary (and sometimes blurred) limit between various grades of water–amphiphile(s) dipole–dipole (or dipole–ion) interactions. It lacks a theoretical basis, but it may well represent intermediate bound water.

The bound water freezes at a lower temperature than the interphasal water, which is just another way of saying that the surfactant–water interaction is stronger for the bound water. However, the interphasal water–surfactant "hydrate" is, of course, more stable because if the positive entropic contribution (to the formation of ordered crystals in the freezing process) is overcompensated in the case of bound water at a lower temperature than in the case of interphasal water, then the bound water will obviously melt at this lower temperature. For simplicity, we assume here that no supercooling occurs. Clearly, the argument also remains essentially the same when supercooling does occur.

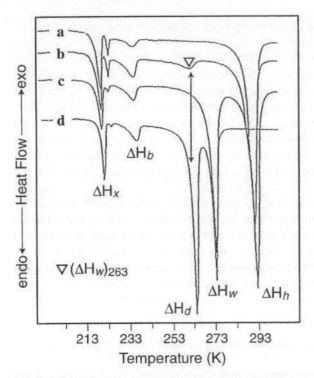

FIG. 1 Differential scanning calorimetric endotherms of the system K-oleate + hexanol 3:5 (wt/wt)–hexadecane (samples a, b, c) or dodecane (sample d)–water. In all samples, the surfactant/oil weight ratio was 0.68 and the water concentration C was expressed as the weight ratio of added water to total sample. $C_a = 0.071$, $C_b = 0.108$, $C_c = 0.290$, $C_d = 0.275$. $dT/dt = 4$ K/min. ΔH_x, ΔH_b, $(\Delta H_w)_{263}$, ΔH_d, ΔH_w, and ΔH_h are the enthalpies of fusion for hexanol, bound water, interphasal water, dodecane, free water, and hexadecane, respectively. Note that the dodecane peak at 263 K ($-10°C$) superimposes on the peak of interphasal water shown for sample b. (From Ref. 5.)

In contrast to the sharp peaks of Fig. 1, no clear distinction between free and bound water was observed in the water–lecithin [6] and water–dimyristoylphosphatidylcholine [7] systems. In these two cases the amount of water was adjusted by dehydration and the affinity between the water molecules and the binding sites gradually decreased toward that of free water [7].

We have discussed [2,8] a similar (but rather unusual in the case of microemulsions) behavior of some quaternary systems based on sugar esters, i.e., the superposition of the melting peaks of the free and interfacial water upon the addition of water.

It was shown [2] that the melting points of these types of water varied as a function of composition. For example, the system water–pentanol + dodecane 1:1 (by weight)–octaethylene glycol mono n-dodecylether [$C_{12}(EO)_8$] (hereafter designated as system A) was investigated by SZT-DSC along the water dilution line W5 for which the surfactant/alcohol/oil weight ratio was 2:1:1. An endothermic peak at about $-10°C$ was revealed and ascribed to interphasal water according to Senatra's definition [4]. Yet, in the related binary system water–$C_{12}(EO)_8$ interphasal water melts between -3 and $-4°C$ [9,10].

The somewhat vaguely defined nonfreezable water can be detected by SZT-DSC only in an indirect way: either by the absence of water-related SZT-DSC peaks or by a significant difference between the total water content of the system, W_T, and the amount of bound, W_B, and free, W_F, water— evaluated from their respective peaks [2]. In the first case, all water present should be regarded as nonfreezable. In the second case, the amount of non-freezable water, W_{NF}, can be evaluated from the equation

$$W_{NF} = W_T - (W_B + W_F) \tag{1}$$

The complications concerning the use of this equation were discussed in our previous review [2].

The water-related peaks can be identified by substituting D_2O for water with all other components and compositions being the same. The D_2O-relevant endothermic peaks typically shift toward higher temperatures [8,9,11– 20] (see Fig. 2).

As both dodecane and interphasal water melt at about $-10°C$ (with over-lapping of their fusion peaks), interphasal water peaks in dodecane-containing systems should be unambiguously identified. This can be done in two ways (besides using D_2O instead of water) [11]:

1. The contribution of interphasal water to the heat of fusion, $\Delta H_I(exp)$, measured at $-10°C$ is calculated [11] from the equation:

    $$W_I = \frac{\Delta H_I(exp) - \Delta H_D f_F}{\Delta H_I} \times 100 \tag{2}$$

    where W_I is the interphasal water concentration (in weight percent), ΔH_D is the heat of fusion of pure dodecane (191.6 J/g), f_D is the dodecane weight fraction, and ΔH_I is the heat of fusion of interphasal water (see Ref. 2 for more details).

2. If the fusion peak at $-10°C$ remains in the absence of dodecane or when dodecane is replaced by hexadecane [5,8,11,14,21,22], then we may safely attribute it to interphasal water (see Fig. 2).

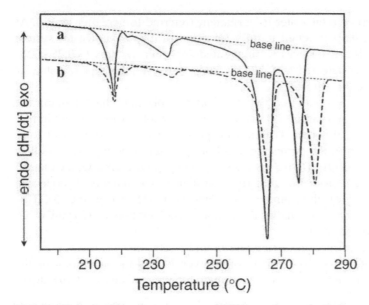

FIG. 2 Typical DSC thermograms of K-oleate–hexanol–dodecane–water micro-emulsion samples. Surfactant/oil = 0.2 g/mL; alcohol/oil = 0.4 mL/mL; water/(water + oil) = 0.222–0.4 g/g. Curve a, water-in-oil microemulsion sample; curve b, D_2O in oil microemulsion sample. Endothermic peaks due to the fusion of D_2O (277 K), "free" water (273 K), dodecane and "interphasal" water (263 K), "bound" water (233 K), and hexanol (220 K) were identified. (From Ref. 13.)

III. SATURATION OF THE SURFACTANT WITH WATER

Three types of water behavior in surfactant-based systems were observed.

1. No formation of free water [20,23–27]

      In some systems, inversion from water-in-oil to oil-in-water micro-emulsions is prevented by geometrical restrictions and the nature of the interfacial curvature. Thus, no core of free water is formed and when the surfactant becomes fully saturated, the addition of more water leads to phase separation.

2. Free water formation only after full hydration of the surfactant

      The distinction between the melting peaks of free and interphasal water enables the evaluation of their relative amounts in microemulsion samples. If the water content is gradually increased along known dilu-tion lines, it is usually observed that the concentration of the interphasal water increases until it reaches a constant value (relative to a fixed amount of the microemulsion sample) that corresponds to a specific molar ratio of total water to surfactant (referred to hereafter as W_o). At this point, the surfactant is fully (or almost fully, as shown for system A [9]) hydrated (as was demonstrated for other systems [8,9,28–30]). Free water is detected by SZT-DSC only above W_o; this type of behavior is manifested in Fig. 3.

It is also worth noting that the apparent plateau, characterizing the variation of interphasal water content with the total water content in Fig. 3, is absent (although it should be present if the surfactant was fully hydrated!) from the corresponding binary system (see Fig. 4) as well as from the $C_{12}(EO)_{23}$ (Brij 35)–water system [9]. Furthermore, the plateau is also absent from the quaternary system water–butanol + dodecane 1:1 (by weight)–the commercial sugar ester S1570 (see Fig. 5). This observation will be analyzed in more detail elsewhere (Ezrahi et al. paper in preparation).

3. Concurrent formation of free and bound water

In some cases, free water forms even before the saturation of the surfactant is achieved, as was shown for the gel phase of the dipal-

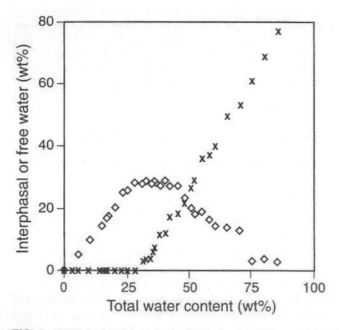

FIG. 3 Variation of free and interphasal water content with the total concentration of water for system A. Concentrations are calculated as a weight percentage relative to the weight of the microemulsion samples. (×) Free water; (◇) interphasal water. (From Refs. 11 and 20.)

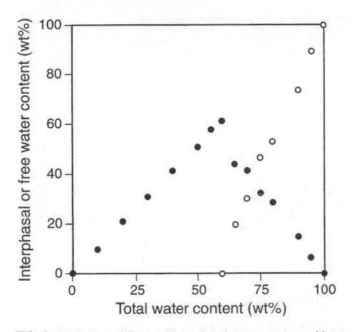

FIG. 4 Variation of free and interphasal water content with total water content for the system $C_{12}(EO)_8$–water [9]. (○) Free water; (●) interphasal water.

mitoylphosphatidylcholine (DPPC)–water system [31]. In the ternary water–sodium dodecyl sulfate-pentanol system [32] the interfacial (the authors prefer to call it "confined") water melts at $-10°C$. At low total water contents only this type of water is observed, whereas at higher water concentrations free water is also observed. We should realize that Fourier transform infrared (FTIR) [33], electron spin resonance (ESR) [34], time domain dielectric spectroscopy (TDDS) [35], and nuclear magnetic resonance (NMR) [36] measurements generally indicate that an equilibrium between free and bound water is established upon the addition of even a small amount of water (see Fig. 6).

On the other hand, we have already pointed out that using SZT-DSC, free water is detected in many cases only after the saturation of the surfactant. This apparent conflict is readily understood because water molecules that are not able to freeze (or melt) at about 0°C—and this is usually the case for water before the surfactant becomes fully (or almost fully) hydrated—are "considered" by SZT-DSC as bound [9,36]. In contrast, the spectroscopic distinction between these two types of water is based on the observation that the residence time of water mol-

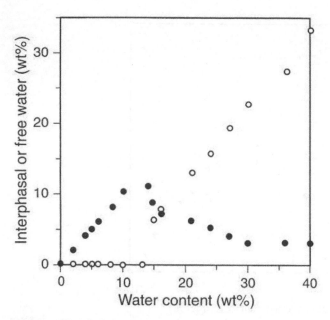

FIG. 5 Variation of free and interphasal water content with total water content for the system n-dodecane + 1-butanol + sucrose ester S1570 = 1:1:1.5 (by weight)/ water [8]. Concentrations are calculated as a weight percentage relative to the. weights of the microemulsion samples. (O) Bulk (free) water; (●) interphasal (bound) water.

ecules at a particular position in the bulk is shorter than the residence time of water molecules at the binding sites (in our case, most near the surfactant polar head groups) [2]. It should, however, be stressed that there is fast exchange between bound and free water molecules, and the motions of bound water, although reduced compared with the free water, are still fast [7,37].

IV. HYDRATION NUMBERS OF SURFACTANTS AND ORGANIC COMPOUNDS

Water interacts with specific binding sites on surfactants and organic compounds such as polymers, proteins, and lipids. The number of water molecules bound per substrate molecule is defined as the hydration number of the relevant compound. For ethoxylated surfactants (and polymers) it is convenient to define a quantity $N_{w/EO}$ as the number of bound water molecules per ethylene oxide (EO) group, usually at maximum hydration [2]. For ex-

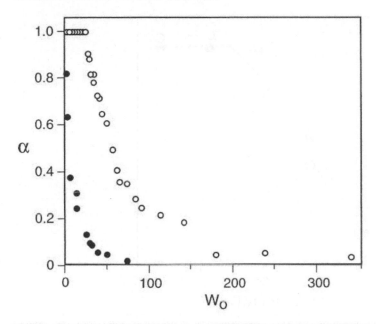

FIG. 6 Fraction of bound water (α) as a function of the water/surfactant molar ratio (W_o) for system A, along the water dilution line W5. (From Ref. 2.) (O) DSC data [9]; (●) NMR data [36]. Results based on chemical shift measurements were taken from Fig. 7 in Ref. 36 and reduced to scale. The decrease of α with the increase of W_o is slower when α is evaluated from T_1 relaxation time data (see Table 2 in Ref. 36).

ample, free water is first detected in system A at $N_{w/EO} \sim 3$ [corresponding to about 30 wt% of (total) water along the dilution line W5]. A similar value was obtained for the system water–dodecanol + polydimethylsiloxane 1:4 (by weight) [20,23], for binary systems based on such surfactants (see Fig. 7), and for aqueous solutions of polyoxyethylene (details are given in Ref. 2).

It should, however, be emphasized that, in general, the number of water molecules bound to each EO group (or to any other hydrophilic group for that matter) is a controversial subject because this number depends significantly on the method of measurement (as has been amply shown in Ref. 2) and, moreover, $N_{w/EO}$ may vary due to factors such as the composition of the system, the quantitative relation between its components, and the temperature [2].

Two instances (based on SZT-DSC results) will suffice to show our point: (1) $N_{w/EO}$ varies markedly at various surfactant-to-oil weight ratios in sys-

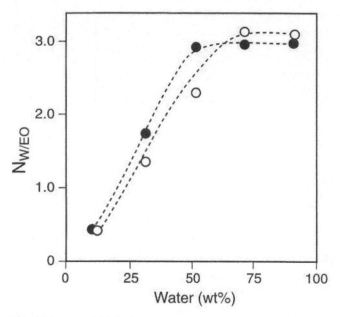

FIG. 7 Number of water molecules bound to each EO group, $N_{w/EO}$, as a function of solubilized water (in wt%) for two binary systems based on (●) Silwet L-7607 and (○) Silwet L-7600. (From Ref. 20.)

tems based on ethoxylated siloxanes (see Fig. 8); (2) for the binary system $C_{12}(EO)_8$–water we have evaluated [9,20] values of $N_{w/EO}$ up to ~5.7 (see Fig. 9) that are slightly higher than $N_{w/EO}$ = 3.7–4.3, determined by Andersson and Olofsson [10]. These values are in contrast to $N_{w/EO}$ ~ 3 obtained for the corresponding quaternary system A [11]. This difference may be interpreted as a result of the different microstructures assigned to these two related systems. The binary system consists of hexagonal liquid crystals [9,38]. A significant part of the interphasal water is trapped within the closed voids present between neighboring mesophase cylinders *without being bound to any specific site on the surfactant head group*. These trapped water molecules, together with those directly bound to EO groups, manifest themselves indistinguishably as one melting peak at about −3°C. In this binary system, $N_{w/EO}$ then includes not only the water that can be associated with each EO group of the surfactant but also the trapped water [9]. To explore this subject a bit more, we shall analyze the hydration behavior of other hexagonal liquid crystals. Thus, a value of $N_{w/EO}$ = 3.07 was determined [19,39] for the systems $C_{12}(EO)_{23}$ (Brij 35)–water and $C_{16}(EO)_{20}$ (Brij 58)–water. On the face of it, two arguments could now be suggested, on the

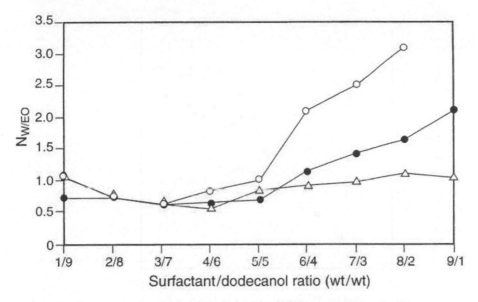

FIG. 8 Number of water molecules bound to each EO group, $N_{w/EO}$, in three surfactant systems at various surfactant-to-dodecanol ratios ($\triangle$, Silwet L-7605; $\bullet$, Silwet L-7600; $\circ$, Silwet L-7607) at the maximum solubilization point [20].

basis of this result, against our interpretation given before for the difference between the values of $N_{w/EO}$ in the $C_{12}(EO)_8$–water system and in the related system A:

1. We would have expected $N_{w/EO}$ for the systems based on Brij 35 and Brij 58 to be much higher as in the case of the $C_{12}(EO)_8$–water system.
2. The close agreement with $N_{w/EO} = 3$ obtained for the quaternary system A might prima facie lead us to conclude that the degree of hydration in surfactant-based systems depends on neither the length of the hydrophilic head group of the ethoxylated surfactant nor the presence of pentanol (or dodecane) [2,11].

The refutation of these arguments is based on the observation that water molecules could be trapped in coils formed by long head groups. This trapped water could be so strongly bound as to be considered as nonfreezable water [2,3,11] or, on the contrary, could just form a part of the kinetic unit in the water–surfactant system without a direct interaction [19]. Nonfreezable water has unequivocally been observed in the Brij 35–water [2,19] and Brij 58–water [19] systems. Thus, we have to add 1.4 molecules of nonfreezable water to the 3.07 molecules of interfacial water per EO group and then obtain about

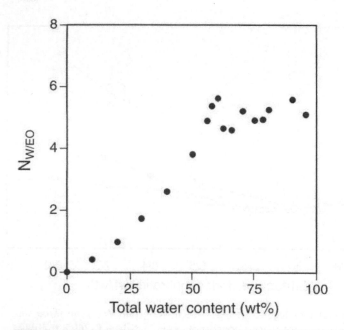

FIG. 9 Variation of $N_{w/EO}$ as a function of total water content for the $C_{12}(EO)_8$–water binary system [9].

4.5 bound water molecules per EO group [19] in good agreement with $N_{w/EO}$ = 5.05 evaluated for the system $C_{12}H_{25}(OCH_2CH_2)_{n=1-10}OSO_3Na$–water [40] and only a little lower than the value of $N_{w/EO}$ determined for the system $C_{12}(EO)_8$–water (see Fig. 9).

In contrast to the binary $C_{12}(EO)_8$–water system, the microstructure of the quaternary system A, at intermediate water contents along the dilution line W5, is considered to be local lamellar [38,41–43]. This structure would have remained just an erudite conjecture if it had not been sustained by the results of several experimental methods, including the use of small-angle X-ray scattering (SAXS) data and parameters derived therefrom [38,42], cryo-transmission electron microscopy micrographs (demonstrating the striated pattern of this structure for the first time) [42], NMR techniques [35,41] and electrical conductivity measurements [38]. This local lamellar structure is conceived as a type of ordered but highly obstructed bicontinuous micro-emulsion [9,38,43] (see Fig. 10). The water molecules that are not directly bound to the surfactant head groups can move rather freely (although they are more restricted than water molecules in ordinary bicontinuous micro-emulsions [38]) along the randomly oriented layered stacks featuring this structure. Using the SAXS data [9,42], we have shown [9] that for a mi-

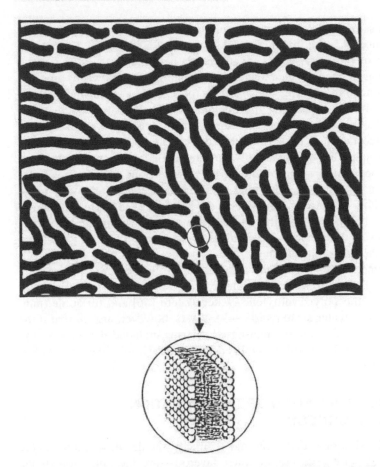

FIG. 10 Cartoon of the local lamellar structure. The black and white patches represent cross sections of the (surfactant + alcohol + oil) layers and the water layers, respectively. The relative thicknesses of the water- and non–water-filled regions were chosen approximately according to the values of d_o (the thickness of the layer of the nonwater components) and d (the measured periodicity) for the 55 wt% water sample on the water dilution line W5. The lamellae extend above and below the plane of the page. For simplicity, they are assumed to be perpendicular to the page, resulting in uniform cross section widths. (Inset) More detailed cartoon of the structure of the lamellar cross section as seen from above [38,40].

croemulsion sample of system A, containing 30 wt% of (total) water along the dilution line W5, the thickness of this water layer is about 1.5 nm. Now, the thickness of three consecutive water layers between surfaces (two water layers adjacent to each surface and an additional layer in between) amounts to only 0.75 nm [44], taking the diameter (or size) of a water molecule to be 0.25 nm [44,45]. Assuming the thickness of the hydration shell in our system to be about 0.5 nm (corresponding to two monolayers of interphasal water [11]; see later), it is obvious that there would still be enough space for "free" (i.e., having only water–water interactions) water molecules even for the microemulsion sample with 30 wt% water, and even more so for more water-rich samples as the total water layer becomes relatively much thicker, reaching 3.8 nm at 50 wt% of (total) water [11]. This motion is evidently less hampered than that of the water trapped within the hexagonal liquid crystals in the binary system. Thus, $N_{w/EO}$ is expected, and found, to be lower in the quaternary system.

Whereas the water–surfactant association implied by $N_{w/EO}$ is usually considered to be rather weak [11], the existence of definite stable hydrates was shown for the polyoxyethylene–water system [46] and for some (nonionic) surfactant–water systems [47–49]. It was, however, argued that if the phase diagrams of these surfactant-based systems included the putative hydrates, the phase rule would be violated [50]. Clearly, this issue merits further investigation.

V. THE DETERMINATION OF THE HYDRATION SHELL THICKNESS

Two interrelated topics that bear most directly on the description of the hydration shell—i.e., the bound water layer(s)—are the definition of the shell and its thickness. The problem of how the bound water can be sufficiently precisely defined is discussed elsewhere [11,37,51] and we shall not pursue it further here. It is clear, however, that the extent to which water is affected by a nearby surface is a function of the distance between them, namely the thickness of the hydration shell. Second-layer water (and, obviously, multilayer water) is much less perturbed than the water adjacent to the surface. We have used several methods to evaluate the thickness of the interphasal water layer in system A (as revealed by the low-temperature behavior of water) [2,11] and found it to be about 0.5 nm. Virtually the same value has been assessed for the thickness of the bound water layer on many organic and inorganic substrates [37,52–57]. As 0.5–0.6 nm is the thickness of two water molecules [45], we may envisage two monolayers of interphasal (or bound) water that are loosely associated with the substrate. We have shown that $N_{w/EO} = 3$ for system A at a total water content of 30 wt%.

Thus, three water molecules can, on average, bind per EO group with two of them hydrogen bonded directly to the EO oxygen atom and the third water molecule hydrogen bonded to both of these "first-layer" water molecules [11].

Of course, by defining characteristic properties of bound water other than its freezing, thicker layers of such water could be determined [2]. Thus, hydration forces may propagate through many layers of water and are detectable at 1–3 nm from the substrate [37]. Furthermore, the distance over which the cooperativity of hydrogen bonding extends is considered to be close to 10 nm (36 water molecules) [58]. This is also the estimated thickness of water layers that still maintain a distinct structure [59]. Somewhat shorter distances (3–5 nm) are suggested as the upper limit for distinguishing between free and bound water [1–3].

VI. THE ROLE OF ALCOHOL

The role of alcohol in nonionic surfactant–based systems has been a recurrent issue, especially in microemulsion research [60–65] (some works have specifically dealt with $C_{12}(EO)_8$, the surfactant used in system A [38,42,66–68]), but it has not been possible to date to clarify all the fine details concerning this topic (see Ref. 2 and references cited therein). The use of SZT-DSC could, in principle, shed some light on the interaction between alcohol and the other constituents in amphiphilic systems.

For system A we have shown [2,9,11] that although pentanol promotes water solubilization [38,65] and is present at the interface [38], its interaction with water or surfactant could not be detected in SZT-DSC measurements (see Table 2 in Ref. 2) [2,11]. This seemingly surprising observation has been interpreted [2,9] as the result of the bulkiness of the surfactant head group and the multiplicity of its binding sites (relative to the one hydroxyl group of the alcohol), endowing $C_{12}(EO)_8$ with a statistical advantage over pentanol.

Another contributing factor is the competition between water and alcohol for the binding sites of the surfactant, causing some pentanol to be rejected from the solvation shell upon the addition of water. This type of behavior has been exhibited for several systems [69,70]. The opposite effect, i.e., displacement of water upon the addition of alcohol, has also been evidenced. A spin probe technique has shown that $N_{w/EO}$ decreased from 3.3 to 0.8 upon the addition of pentanol or hexanol to an aqueous solution of Pluronic P-85 [poly(oxyethylene)$_{27}$/poly(oxypropylene)$_{39}$/poly(oxyethylene)$_{27}$] as hydration water was repelled away from the poly(oxyethylene)/poly(oxyethylene) border [71].

The addition of hexanol (or of anesthetics such as halothane) has assisted the release of unfreezable and bound water in partially hydrated phospholipid lamellar systems, as demonstrated by SZT-DSC. This dehydration effect was ascribed to the alcohol penetration into the interface [7].

The effect of alcohol on the distribution of (total) water between the free and bound states was exemplified [2,9] by the swelling of a fixed amount of a 1:1 (by weight) $C_{12}(EO)_8$–water mixture with increasing amounts of a 1:1 (wt/wt) solution of pentanol + dodecane. The relative amount of (total) water constantly diminished upon the addition of pentanol + dodecane, but when the alcohol concentration reached a threshold value, some of the (bound) water became free (Fig. 11). The same phenomenon was observed in the system phosphatidylcholine (25 wt%)–tricaprylin + alcohol (60 wt%; molar ratio of 1:5, respectively)–water (15 wt%). The alcohols used were ethanol, butanol, and hexanol, and the free water content increased in that order [72]. The same inverse dependence on the alcohol hydrophilicity was observed for sucrose ester–based microemulsions [29]. It seems that alcohol molecules adsorbed at the interface distort the three-dimensional network of

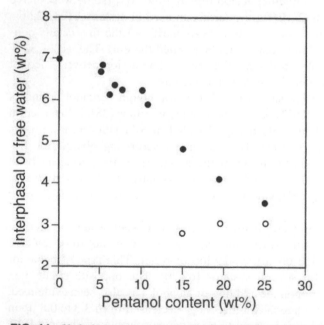

FIG. 11 Variation of interphasal and free water content as a function of pentanol content along a swelling line on which pentanol + dodecane 1:1 (w/w) solution is added in increasing amounts to a fixed amount of 1:1 (wt/wt) water–$C_{12}(EO)_8$ mixture [2]. (○) Free water; (●) interphasal water.

water about the surfactant head groups, thereby detaching the outer weakly associated water layers [2,9]. An additional, evidently related contributing factor is the thinning of the interfacial region by the alcohol, making it elastic and thus leading to undulations that disrupt the mesophase order [38,73,74].

For system A we may add another, plausibly related interpretation. We have shown [38,43] that a liquid crystal–to–isotropic transition is promoted by adding pentanol (and dodecane) to the binary $C_{12}(EO)_8$–water system. For a mixture of 50 wt% $C_{12}(EO)_8$ and 50 wt% water (designated as the 5PD swelling line in Ref. 38) it is possible to add up to 10 wt% of pentanol (+10 wt% of dodecane) before the hexagonal mesophase transforms to an isotropic nonviscous solution [38]. Yet, this is actually the threshold concentration of pentanol needed for the formation of free water in the same system (see Fig. 11)! Therefore, it is conceivable that the formation of free water is also connected with this structural transformation. Using SAXS measurements, we have shown that the thickness of the mantle within which the hydrated surfactant head groups are accommodated is slightly decreased whereas the radius of the nonpolar core and the available surface area per surfactant molecule at the interface increase (see Fig. 12) upon the addition of pentanol + dodecane. We have assumed that pentanol molecules penetrate into the interface and promote the liquid crystal–to–isotropic transition via a mild local disruption of aggregate packing [38,43]. At higher pentanol + dodecane concentrations, the structure of the isotropic solution was shown to be local lamellar [38,42,43]. On the basis of the linear dependence of $1/d^2$ (d is the periodicity) on the volume fraction ϕ_w of water, we have postulated a hypothetical local hexagonal microstructure as an intermediate between the hexagonal liquid crystals and the local lamellar solution [9]. Obviously, even in this stage the cylinders of the hexagonal mesophase must become widely separated, thereby freeing the trapped water (see earlier).

The shift of the melting peak of interphasal water toward more negative temperatures upon the addition of alcohol indicates that this compound also "strengthens" the water–surfactant interaction in the sense that the alcohol interferes with the water freezing at 0°C. The cryoscopic effect of the alcohol, which may also have a role in the freezing point depression, is expected to be quite small as the usual cosurfactants are only partially miscible with water. Thus, for the binary system $C_{12}(EO)_8$–water this endothermic peak is revealed at −3 to −4°C [2,9,10], which shifts to −10°C in the related quaternary system A [2,9,11]. Such a lowering of the freezing temperature cannot be accounted for by the cryosocopic effect of pentanol. Assuming that the solubility of pentanol in water is 2.19% [75], we have calculated the freezing point depression due to the cryoscopic effect to be only 0.2°C

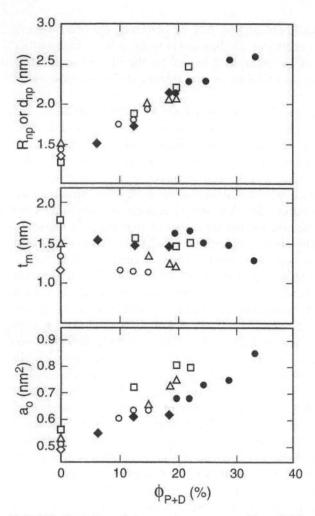

FIG. 12 Variation of structural parameters of model liquid crystal aggregates as a function of volume fraction of (pentanol + dodecane), ϕ_{P+D}, as derived from X-ray diffraction data. (Top) Size of the nonpolar compartment. R_{np} is the radius of the nonpolar core in the hexagonal mesophase H_I, and d_{np} is the thickness of the nonpolar layer in the lamellar mesophase L_α. (Middle) Thickness of the hydrophilic mantle, t_m. (Bottom) Available surface area per surfactant head group, a_o. Swelling lines: 4PD (□); 5PD (△); 6PD (○); 7PD (◆). These swelling lines correspond to the designation nPD, where $n \cdot (10\%)$ is the initial weight fraction of surfactant. Open symbols refer to H_I and filled symbols refer to the L_α phase. (From Ref. 38.)

for a microemulsion sample containing the following components (in wt%): water, 35; $C_{12}(EO)_8$, 35; pentanol, 15; dodecane, 15.

This depression of the melting point of the bound water is inversely proportional to the alcohol chain length. It was observed in sugar ester–based microemulsions [8], in system A [9], and in the system [9] water–butanol + dodecane 1:1 (by weight)–polyoxyethylene (10) oleyl alcohol $[C_{18:1}(EO)_{10}$; also known as Brij 97]. This system (hereafter designated as system B) was studied along the dilution line XB4, for which the surfactant/alcohol/oil weight ratio was 4:3:3. It was observed that the melting temperature of bound water in these systems leveled off at a certain alcohol chain length (see Fig. 13).

From this figure we may conclude [9] that:

1. The shorter the alcohol, the lower the bound water melting temperature.
2. Beyond a certain alcohol chain length, the melting temperature levels off.
3. The effect of alcohol is more pronounced in system A.

In Ref. 2 we pointed out the following three factors that contribute to the role played by Brij 97 in determining water binding using SZT-DSC:

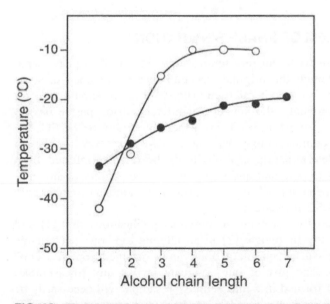

FIG. 13 Variation of the melting temperature of the bound water peak in systems A and B as a function of alcohol chain length. Total water in both systems is 40 wt%. (○) System A, along the dilution line W5. (●) System B, along the dilution line XB4. (From Ref. 9.)

1. Varying lengths of its head groups lead to trapping of water in the voids
    formed between them. Some of this water does not freeze even at the
    end of the cooling stage but rather when the system is heated back to
    ambient temperature, as manifested by the existence of exothermic
    peaks for system B [2,9].
2. The head group is long enough to form one coil within which water
    may be caged.
3. There is a relatively long and unsaturated hydrophobic chain.

The contribution of the alcohol in dictating the degree of hydration should
then be rather minor, as was indeed observed. Thus, even for hexanol, which
is scarcely soluble in water, the melting temperature of the bound water
peak is lower by 10 degrees than that of the interphasal water in system
A [9]!

In contrast, the structure of $C_{12}(EO)_8$ can hardly assist water binding and
thus the effect of alcohol is more prominent. Two mechanisms may be operative: the adsorption of the alcohol at the interface, which reaches the
plateau first at $n = 4$ (butanol; see Fig. 13), and the penetration of the alcohol
into the water layers enveloping the surfactant, which, obviously, depends
on the solubility of the alcohol in water [2,9].

VII. THE PROBLEM OF PHASE SEPARATION

Cooling a microemulsion to subzero temperatures may lead to phase separation [76–78]. Although the original structure may be retained in some
cases (especially for water-poor microemulsions) [76], we should take for
granted that, most probably, the concept of microemulsion per se has no
meaning for such low temperatures as are generally used in SZT-DSC. Because cooling a microemulsion may also cause phase transformations [2,78]
no microstructure determination can obviously be directly inferred from
SZT-DSC data. A more poignant question concerns the validity of any conclusion about the behavior of microemulsions at ambient temperature
reached using SZT-DSC measurements.

We have discussed this problem in previous publications [8,9,11] and
more systematically in our review [2] in an attempt to show that reliable
information on water–surfactant interactions can be obtained from SZT-DSC
experimental results. The gist of our argument is that any (measurable)
amount of free water formed in a microemulsion sample will necessarily be
detected as a melting peak at about 0°C in the endothermic mode of SZT-
DSC. *Thus, free water cannot be formed in the system and concurrently not
revealed by this peak!* The same conclusion should, of course, also be valid
if bound (or interphasal) water formed at ambient temperature was detached

from the surfactant during the cooling stage of the DSC freeze–thaw cycle and inevitably became free via phase separation. Our interpretation of SZT-DSC results, as with any other scientific suggestion should, in principle, be refutable. Therefore, in order to test our interpretation, we have to confront the question of whether SZT-DSC might ever be able to detect free water formed due to phase separation. The answer is simply yes. The existence of a *single* water melting peak at about 0°C *without any indication of the presence of bound water* [or, indirectly, using Eq. (1), of the presence of nonfreezable water, which is invisible to DSC; see earlier] could be conclusive proof that free water is formed from the bound water through phase separation, thereby rebutting our argument. However, such a single water melting peak at about 0°C is never observed. On the contrary, as shown before, when the water content of a microemulsion system is gradually increased, it is frequently observed that the surfactant interacts with the added water [as evidenced by the bound water melting peak(s)] and only after it becomes fully (or almost fully) hydrated does free water begin to appear. The claim that at low temperature a complete separation into ice and surfactant crystals takes place in water–surfactant binary systems [50,79–82] is, therefore, easily refuted. An additional argument, suggested by Nibu and coworkers [47–49], is that this phase separation implies a highly repulsive water–surfactant interaction in the solid state. Yet, many physical properties of ethoxylated nonionic surfactant–water binary systems depend on the strong (attractive) interaction between the polyoxyethylene chain and water molecules in liquid and liquid crystalline phases of these systems. It is then reasonable that such a strong interaction would also operate in a solid phase.

In conclusion, free water is formed not by detaching outer water layers through phase separation but rather by adding more and more water until its association with the surfactant is barely perceivable. Phase separation should then be regarded as a segregation of some (or all) of the free (pure) components (water, oil, or alcohol) of the microemulsion. Thus, the distribution of free and bound water in a microemulsion system (as well as in the related water–surfactant binary system) does not depend on the degree of phase separation [2,8,9,11].

The water behavior during phase transformation is a problem more difficult to tackle. Whereas the a posteriori absence of a free water melting peak before the hydration of surfactant has occurred may be regarded as a decisive answer in the case of phase separation, it might not suffice against the following argument: Although it is not reasonable that water, which was presumably free in the microemulsion sample at ambient temperature (before the saturation of the surfactant) would interact at subzero temperature with the hydrophilic head groups, such an interaction might occur *during the phase transformation*. [The possibility that (free) water might separate from the hydrated surfactant is, of course, still negated by the absence of a melting peak at 0°C.] This argument was refuted by comparing SZT-DSC results with the data obtained at ambient temperature by other experimental techniques (such as electrical conductivity [9,38,41], NMR spectroscopy [36] and viscosity [41]). For example, there is a sharp increase in the electrical conductivity of system A microemulsion samples beginning at about 30 wt% water, which is just the total water concentration at which free water is first detected by SZT-DSC. The same correspondence was observed for system B [9] and for sugar ester–based microemulsions [8].

We may also add that the calculated surface areas of surfactant molecules in liquid crystalline phases formed by the cooling of microemulsion samples fit well into the pattern set by the samples shown by SAXS measurements to be liquid crystals at room temperature [38]. In Ref. 2 we tried to show that the subzero temperature behavior of water was dictated a priori by defining the bound water in terms of its interference with the melting of frozen water at 0°C. This may lead to another problem. Suppose, for example, that the amount of "bound" water in system A (along the dilution line W5) was determined *at ambient temperature* by some experimental technique to be, say, 40 wt% for a fully hydrated surfactant, instead of 30 wt% as evaluated from DSC data obtained *at subzero temperature*. (As we have already seen, the opposite case is more prevalent; the bound water defined according to SZT-DSC is considered free by spectroscopic methods.) This excess of 10 wt% water would be patently revealed as free when using SZT-DSC because it does not perturb the melting of frozen water at 0°C. In light of the alleged evidence from other experimental techniques, one might doubt whether this 10 wt% water is "really" free. Yet, the freedom of this water in system A beyond the first two monolayers nearest to the surfactant molecules (the 0.5-nm thickness of the interphasal water layer; see earlier) is also manifested by electrical conductivity measurements performed at room temperature (see earlier). Obviously, SZT-DSC data for microemulsion systems should be individually analyzed according to these principles in order to interpret them. In Ref. 2 we discussed several secondary arguments in favor of the reliability of SZT-DSC results that might be applicable for certain systems.

VIII. RELEVANCE OF THIS WORK

Microemulsions are utilized in a wide variety of applications [83]. To provide even a short description of all these applications is not within the scope of this chapter. Instead, we concentrate on some applications that are related to the distinction between bound and free water. Obviously, this distinction is one of the most significant tasks of SZT-DSC measurements. We begin with system A, which serves as a model system for the development of fire-resistant hydraulic fluids based on microemulsions [38,84]. The loss of water by evaporation in open hydraulic systems is detrimental to the fire resistance capability of the hydraulic fluids. As the bound water should evaporate less than free water [85], it is important to be able to assess their relative concentrations [11]. An opposite example in which the formation of bound water would be disfavored is the dewatering of sludge [86]. Naturally, the energy cost for the removal of bound water from a given amount of sludge is higher than that needed for the removal of free water from the same amount of sludge. However, one of the most important applications of microemulsions is the utilization of their cores, more frequently water in oil, for conducting physical and chemical processes.

Several important advantages have been derived from the knowledge of the state of water obtained from these microemulsion systems: (1) better control of crystallization phenomena dictated by interfaces and water [87–90], (2) better control of organic reactions requiring free water such as hydrolysis, and (3) triggered enzymatic processes in which the enzymes are soluble in the water phase and can be activated upon turning the bound water into free water.

One can think about crystallization of amino acids, peptides, and other water-soluble active compounds in water-in-oil microemulsions. The water reservoir can be composed of bound water and, under given conditions, an excess of water can be added upon "demand." Thus, we may facilitate the mobility of the active matter within the core and control the size and shape of the crystallized matter. This is, therefore, a good method for controlling the formation of nanosized particles [91].

Similarly, it is possible to control polymorphism of drugs and other related pharmaceuticals or nutraceuticals by manipulating the amount of free and bound water in the microemulsions.

Organic reactions, mainly those that require free water, can also be controlled by the ratio of free and bound water in the microreactor of the reaction (the core of the microemulsion). Maillard reactions between amino acids (cysteine) and sugars (furfural) can be manipulated by the core water to improve the selectivity of products and the rates of reactions [92].

Finally, any enzymatic process using lipases or phospholipases can be triggered at will by changing the internal balance between the core water and the free water. On the shelf, the product will contain enzymes that will be immobilized and temporarily deactivated by the bound water, and when the microemulsion is further diluted, during use, the enzymes will become free and active and the reaction will be triggered.

We can conclude by saying that the control of free and bound water in the core of water-in-oil microemulsions provides us with new and important tools to reconsider old processes that were carried out in the bulk as well as to regenerate new reactions, new physical processes with improved selectivity, improved kinetic control, and controlled formation of nanosized particles.

IX. CONCLUDING REMARKS

This chapter has covered topics that bear most directly on the use of SZT-DSC in the identification of various water states supposed to be present in surfactant-based systems. The problem of phase transition (and phase transformation) has been discussed in more detail. If we have lingered in this discussion, it is because in solving this problem lies the justification of using SZT-DSC for the investigation of water behavior in surfactant-based systems. Finally, some novel applications, relating to the distribution of free and bound water were outlined.

ACKNOWLEDGMENTS

We thank Dr. I. Tiunova for her careful experimental work pertinent to many microemulsion systems described in this chapter and Dr. G. Berkovic for stimulating discussions and critical reading of the manuscript.

REFERENCES

1. PC Schulz. J Thermal Anal 51:135, 1998.
2. S Ezrahi, A Aserin, M Fanun, N Garti. In: N Garti, ed. Thermal Behavior of Dispersed Systems. Surfactant Science Series. Vol 93. New York: Marcel Dekker, 2001, pp 59–120.
3. PC Schulz, JFA Soltero, JE Puig. In: N Garti, ed. Thermal Behavior of Dispersed Systems. Surfactant Science Series. Vol 93. New York: Marcel Dekker, 2001, pp 121–181.
4. D Senatra, G Gabrielli, G Caminati, Z Zhou. IEEE Trans Electr Insul 23:579, 1988.
5. D Senatra, G. Gabrielli, G Caminati, GGT Guarini. In: KL Mittal, ed. Surfactants in Solution. Vol 10. New York: Plenum Press, 1989, pp 147–158.
6. VA Parsegian, N Fuller, RP Rand. Proc Natl Acad Sci USA 76:2750, 1979.
7. I Ueda, HS Tseng, Y Kaminoh, S-M Ma, H Kamaya, SH Lin. Mol Pharmacol 29:582, 1986.
8. N Garti, A Aserin, I Tiunova, M Fanun. Colloids Surf A 170:1, 2000.
9. S Ezrahi. Microemulsion systems as a basis for the fire resistant hydraulic fluids. PhD thesis, Hebrew University of Jerusalem, Jerusalem, 1997.
10. BA Andersson, G. Olofsson. Colloid Polym Sci 265:318, 1987.
11. N Garti, A Aserin, S Ezrahi, I Tiunova, G Berkovic. J Colloid Interface Sci 178:60, 1996.

12. D Senatra, L Lendinara, MG Giri. Can J Phys 68:1041, 1990.
13. D Senatra, GGT Guarini, G Gabrielli. In: V Degiorgio, M Corti, eds. Physics of Amphiphiles: Micelles, Vesicles and Microemulsions. Amsterdam: North-Holland, 1985, pp 802–829.
14. D Senatra, Z Zhou, L Pieraccini. Prog Colloid Polym Sci 73:66, 1987.
15. D Senatra, GGT Guarini, G Gabrielli, M Zoppi. J Phys (Paris) 45:1159, 1984.
16. D Senatra, G Gabrielli, GGT Guarini, M Zoppi. In: DO Shad, ed. Macro- and Microemulsions: Theory and Applications. ACS Symp Ser Vol 272. Washington, DC: American Chemical Society, 1985, pp 133–148.
17. D Senatra, R Pratesi, L Pieraccini. J Therm Anal 51:79, 1998.
18. FD Blum, WG Miller. J Phys Chem 86:1729, 1982.
19. PC Schulz, JE Puig. Colloids Surf A 71:83, 1993.
20. N Garti, A Aserin, I Tiunova, S Ezrahi. J Therm Anal 51:63, 1998.
21. D Senatra, G Gabrielli, GGT Guarini. Europhys Lett 2:455, 1986.
22. D Senatra, G Gabrielli, GGT Guarini. In: S Martellucci, AN Chester, eds. Progress in Microemulsions. New York: Plenum Press, 1989, pp 207–215.
23. N Garti, A Aserin, E Wachtel, O Gans, I Shaul. J Colloid Interface Sci 233: 286, 2001.
24. O Gans. Water solubilization within hydrophobic polymeric self assemblies. MSc thesis, Hebrew University of Jerusalem, Jerusalem, 1999.
25. DC Steytler, DL Sargeant, BH Robinson, J Eastoe, RK Heenan. Langmuir 10: 2213, 1994.
26. JC Ravey, M Bouzier, C Picot. J Colloid Interface Sci 97:9, 1984.
27. JC Ravey, M Bouzier. J Colloid Interface Sci 116:30, 1987.
28. C Boned, J Peyrelasse, M Moha-Ouchane. J Phys Chem 90:634, 1986.
29. N Garti, V Clement, M Fanun, ME Leser. J Agric Food Chem 48:3945, 2000.
30. T Silberstein. Microemulsion systems as a basis for phosphatidylcholine hydrolysis. PhD thesis, Hebrew University of Jerusalem, Jerusalem, 1999.
31. M Kodama, H Aoki. In: N Garti, ed. Thermal Behavior of Dispersed Systems. Surfactant Science Series. Vol 93. New York: Marcel Dekker, 2001, pp 247–293.
32. A Chahti, M Boukalouch, JP Dumas. J Dispersion Sci Technol 21:537, 2000.
33. G Giammona, F Goffredi, V Turco Liveri, G Vassalo. J Colloid Interface Sci 154:411, 1992.
34. H Caldararu, A Caragheorgheopol, M Vasilescu, I Dragutan, H Lemmetyinen. J Phys Chem 98:5320, 1994.
35. I Nir. Investigation of microemulsions by the TDS method. PhD thesis, Hebrew University of Jerusalem, Jerusalem, 1997.
36. D Waysbort, S Ezrahi, A Aserin, R Givati, N Garti. J Colloid Interface Sci 188:282, 1997.
37. JA Rupley, G Careri. Adv Protein Chem 41:37, 1991.
38. S Ezrahi, E Wachtel, A Aserin, N Garti. J Colloid Interface Sci 19:277, 1997.
39. JE Puig, JFA Soltero, EI Franses, LA Torres, PC Schulz. In: AK Chattopadhyay, KL Mittal, eds. Surfactants in Solution. Surfactant Science Series. Vol 64. New York: Marcel Dekker, 1996, pp 147–167.
40. E Tokiwa, K Ohki. J Phys Chem 71:1343, 1967.

41. S Ezrahi, A Aserin, N Garti. J Colloid Interface Sci 202:222, 1998.
42. O Regev, S Ezrahi, A Aserin, N Garti, E Wachtel, EW Kaler, A Khan, Y Talmon. Langmuir 12:668, 1996.
43. S Ezrahi, A Aserin, N Garti. In: P Kumar, KL Mittal, eds. Handbook of Microemulsion Science and Technology. New York: Marcel Dekker, 1999, pp 185–246.
44. J Israelachvili, J Mara. Methods Enzymol 127:353–360, 1986.
45. J Israelachvili, H Wennerström. Nature 379:219, 1996.
46. NB Graham. In: JM Harris, ed. Poly(ethylene glycol), Chemistry, Biotechnological and Biomedical Applications. New York: Plenum Press, 1992, pp 263–281.
47. Y Nibu, T Sulmori, T Inoue. J Colloid Interface Sci 191:256, 1997.
48. Y Nibu, T Inoue. J Colloid Interface Sci 205:231, 1998.
49. Y Nibu, T Inoue. J Colloid Interface Sci 205:305, 1998.
50. GG Chernik. Curr Opin Colloid Interface Sci 4:381, 2000.
51. ID Kuntz Jr, W Kauzmann. Adv Protein Chem 28:239, 1974.
52. M Kotlarchyk, JS Huang, S-H Chen. J Phys Chem 89:4382, 1985.
53. P Mazur. Ann NY Acad Sci 125:658, 1965.
54. A Banin, DM Anderson. Nature 255:261, 1975.
55. S Yariv. In: ME Schrader, GI Loeb, eds. Modern Approaches to Wettability—Theory and Applications. New York: Plenum Press, 1992, p 279.
56. EW Hansen, M Stocker, R Schmidt. J Phys Chem 100:2195, 1996.
57. R Schmidt, EW Hansen, M Stocker, D Akporiaye, OH Ellestad. J Am Chem Soc 117:4049, 1995.
58. PM Wiggins. Curr Top Electrochem 3:129, 1994.
59. BV Derjaguin, NV Churaev. Langmuir 3:607, 1987.
60. M Kahlweit, R Strey, G Busse. J Phys Chem 95:5344, 1991.
61. R Strey, M Jonströmer. J Phys Chem 96:4537, 1992.
62. M Jonströmer, R Strey. J Phys Chem 96:5993, 1992.
63. MHGM Penders, R Strey. J Phys Chem 99:6091, 1995.
64. MHGM Penders, R Strey. J Phys Chem 99:10313, 1995.
65. N Garti, A Aserin, S Ezrahi, E Wachtel. J Colloid Interface Sci 169:428, 1995.
66. H Kunieda, A Nakano, M Akimaru. J Colloid Interface Sci 170:78, 1995.
67. H Kunieda, A Nakano, M Angeles Pes. Langmuir 11:3302, 1995.
68. S Yamaguchi. Colloid Polym Sci 274:1152, 1996.
69. F Bastogne, BJ Nagy, C David. Colloids Surf A 148:245, 1999.
70. WO Parker Jr, C Genova, G Carignano. Colloids Surf A 72:275, 1993.
71. A Caragheorgheopol, H Caldararu, I Dragutan, H Joela, W Brown. Langmuir 13:6912, 1997.
72. Y Ikeda, M Suzuki, H Iwata. In: SP Rowland, ed. Water in Polymers. Washington, DC: American Chemical Society, 1980, pp 287–305.
73. A Martino, EW Kaler. Colloids Surf A 99:91, 1995.
74. SA Safran. Statistical Thermodynamics of Surfaces, Interfaces and Membranes. Reading, MA: Addison-Wesley, 1994.
75. PM Ginning, R Baum. J Am Chem Soc 59:1111, 1937.
76. PL Luisi, LJ Magid. CRC Crit Rev Biochem 20:409, 1986.

77. M Zulauf, H-F Eicke. J Phys Chem 83:480, 1979.
78. P-O Quist, B Halle, J Chem Soc Faraday Trans I 84:1033, 1988.
79. JS Clunie, JM Corkill, JF Goodman, PC Symons, JR Tate. Trans Faraday Soc 63:2839, 1967.
80. DJ Mitchell, GJT Tiddy, L Waring, T Bostock, MP McDonald. J Chem Soc Faraday Trans I 79:975, 1983.
81. K Kratzat, C Schmidt, H Finkelmann. J Colloid Interface Sci 163:542, 1994.
82. K Kratzat, H Finkelmann. J Colloid Interface Sci 181:542, 1996.
83. P Kumar, KL Mittal, eds. Handbook of Microemulsion Science and Technology. New York: Marcel Dekker, 1999.
84. N Garti, R Feldenkreiz, A Aserin, S Ezrahi, D Shapira. Lubr Eng 49:404, 1993.
85. K Pathananthan, GP Johari, J Chem Soc Faraday Trans 90:1143, 1994.
86. DJ Lee, SF Lee. J Chem Tech Biotechnol 62:359, 1995.
87. L Tunik, H Furedi-Milhofer, N Garti. Langmuir 14:3351, 1998.
88. H Furedi-Milhofer, N Garti, A Kamishny. J Cryst Growth 199:1365, 1999.
89. J Yano, H Furedi-Milhofer, E Wachtel, N Garti. Langmuir 16:9996, 2000.
90. J Yano, H Furedi-Milhofer, E Wachtel, N Garti. Langmuir 16:10005, 2000.
91. Y Berkovich, N Garti. Colloids Surf A 128:91, 1997.
92. S Vauthey, CH Milo, PH Frossard, N Garti, ME Leser, HJ Watzke. J Agric Food Chem 48:4808, 2000.

6

Aggregation Behavior of Dimeric and Gemini Surfactants in Solution: Raman, Selective Decoupling ^{13}C NMR, and SANS Studies

HIROFUMI OKABAYASHI and NORIKATSU HATTORI* Nagoya
Institute of Technology, Nagoya, Japan

CHARMIAN J. O'CONNOR The University of Auckland,
Auckland, New Zealand

ABSTRACT

The microstructures of the micelles formed by dimeric and gemini surfactants with a $(CH_2)_s$ spacer (s = 2, 3, 4, 6, 8, 10, and 12) or an aromatic spacer (o-, m-, or p-phenylenedimethylene) have been investigated. Raman and selective decoupling ^{13}C nuclear magnetic resonance (NMR) spectra have provided evidence for a conformational change within these surfactants. The aggregational behavior obtained from the small-angle neutron scattering (SANS) data is discussed with respect to the relationship between the structure of the micelle and its conformation.

I. INTRODUCTION

The packing property of self-assembling systems seems to depend on many factors. For a quantitative discussion of the relationship between the geometrical packing parameters and morphological characteristics, one must investigate the conformational change of surfactant molecules induced by

---

*Current affiliation: Chisso Corporation, Tokyo, Japan.

self-assembling because the packing parameters are affected by the conformational change of the surfactant molecules.

Bis(quaternary ammonium bromide) surfactants are composed of two n-alkyldimethylammonium bromide moieties whose polar head groups are chemically connected in pairs by an alkanediyl chain (referred to as a spacer). This dimerization brings about variations in physicochemical properties compared with those of the original surfactant molecules. In particular, it is well known that the length of the spacer markedly affects the morphology of an aggregate.

In this chapter, we discuss the relationship between the conformation of the dimeric and gemini surfactants and their micellar behavior, using Raman, selective decoupling ^{13}C nuclear magnetic resonance (NMR), and small-angle neutron scattering (SANS) data.

II. EXPERIMENTAL

A. Materials

Dimeric surfactants, bis(quaternary ammonium bromides) (DS, **I**), methonium bromides (**II**), and (phenylenedimethylene) bis(n-octylammonium) dibromides (geminis, **III**), were synthesized [1,2] and were identified by NMR and elemental analyses.

$$Br^-(CH_3)_2(C_mH_{2m+1})N^+\!-\!(CH_2)_s\!-\!N^+(C_mH_{2m+1})(CH_3)_2\,Br^- \tag{I}$$

($s = 2, 3, 4, 6, 8, 10,$ and 12; $m = 4, 10$) (abbreviations: sm2–4, sm2–10, sm3–4, sm3–10, sm4–10, sm6–10, sm8–10, and sm12–10).

$$Br^-(CH_3)_3N^+\!-\!(CH_2)_s\!-\!N^+\!-\!(CH_3)_3Br^- \tag{II}$$

($s = 4, 6, 8,$ and 10) [abbreviations: TMB ($s = 4$), HMB ($s = 6$), OMB ($s = 8$), and DMB ($s = 10$)].

$$Br^-(CH_3)_2(C_mH_{2m+1})N^+\!-\!CH_2\!-\!\!\bigcirc\!\!-CH_2\!-\!N^+(C_mH_{2m+1})(CH_3)_2Br^- \tag{III}$$

(o-, m-, and p-phenylenedimethylene, $m = 8$) (abbreviations: o-, oxy8; m-, mxy8; and p-, pxy8). The numbering of the three spacer portions of the geminis is as follows:

oxy8   mxy8   pxy8

B. Methods

Raman spectra were obtained on a Nicolet 950 Fourier transform Raman spectrometer (4000–100 cm^{-1}) using the neodymium:yttrium-aluminum-garnet (Nd:YAG) laser (CVI) excitation wavelength of 1064 nm. SANS measurements were made using the medium-angle neutron-scattering instrument (WINK, incident wavelength $\lambda = 0.1–1.6$ nm) installed at the pulsed neutron source KENS (Tsukuba, Japan). The intensity of the scattered neutrons was recorded on a position-sensitive two-dimensional detector. The ^{13}C NMR spectra were recorded on a Varian Unity-400 *plus* spectrometer operating at 100.58 MHz at 30°C, using an acquisition time of 2.560 s under a deuterium internal lock and 128,000 points in the time domain (sweep width 25,000 Hz). The decoupling power was optimized in order to decouple only for aromatic protons.

III. RESULTS AND DISCUSSION

A. Raman Spectra and Conformations of DS

1. DS Characteristic Bands

We have determined the molecular conformations of simple sm2–4 and sm3–4 by single-crystal X-ray diffraction analysis [3]. The results showed that the two *n*-butyl chains for sm2–4 were in the *trans* configuration with respect to the extended N—(CH$_2$)$_2$—N skeleton, and those for sm3–4 were in the *cis* configuration. In the Raman spectra of sm2–4 and sm3–4, we found Raman bands characteristic of the *trans* and *cis* forms in the skeletal deformation region [4], thus directly reflecting the conformation (Table 1).

The marked difference in the spectral features between the two simple DSs comes from the differences in their conformations. We note that the Raman band at 421–426 cm^{-1} of sm3–6 to sm3–12 disappears for sm2–6 to sm2–12 (Table 1) [4]. Therefore, the 421–426 cm^{-1} band may be regarded as a band characteristic of the *cis* form. Furthermore, the Raman bands at 505–509 and 542–544 cm^{-1} of sm2–6 to sm2–12 closely correspond to the bands at 509 and 546 cm^{-1}, respectively, for *trans* sm2–4, and those at 515–519 and 552–563 cm^{-1} of sm3–6 to sm3–12 also correspond to the bands at 516 and 553 cm^{-1}, respectively, of *cis* sm3–4, indicating that these bands may be used to distinguish between the *trans* and *cis* forms [4].

Because we found for the sm6–8 to sm6–12 series that the Raman bands at 423–425, 520, and 564–567 cm^{-1} were characteristic of the *cis* form [4], we concluded that the DS series with longer spacers also took up a *cis* form in the solid state. In the aqueous samples of sm6–8 to sm6–12, the 423–

TABLE 1 Raman Band Frequencies of *Trans* sm4–4, *Cis* sm3–4, and of Their Related DSs in the Skeletal Deformation Region

DS				Band frequencies (cm^{-1})				
Trans								
sm2–4		546		509	474		382	366
sm2–6		542		508	465		382	372
sm2–10		544		505	457		391	372
sm2–12		544		509	461		—	370
Cis								
sm3–4	553		536	519	463	426		376
sm3–6	552		530	515	453	421		376
sm3–10	562		530	515	457	422		374
sm3–12	563		534	515	459	422		377

424 cm^{-1} band of the solid samples disappeared, indicating that the *cis* form had broken down in aqueous solution. However, because the weak bands at 513–515 cm^{-1} were observed for the aqueous samples, the *trans* form could also coexist.

The Raman spectra of sm6–8 to sm6–12 in the gel and coagel states were very similar to those in aqueous solution [4], again suggesting that the *cis* form had almost completely disappeared in these states.

2. Accordion Bands as an Indicator of Hydrocarbon Ordering

We next examined the accordion modes of the *n*-alkyl-N segments and of the spacer portions for the solid and aqueous samples [4,5] so that we might discuss quantitatively the ordering of the hydrocarbon moieties in the aggregates.

The Raman spectra of *n*-hexyl, *n*-octyl, *n*-decyl, and *n*-dodecyl trimethyl-ammonium bromides in the 150–600 cm^{-1} region are shown in Fig. 1. Strong Raman bands are observed for all of these in the 160–300 cm^{-1} region, and these shift markedly to a lower frequency with an increase in chain length. We may assign the bands observed at 251, 208, 179, and 164 cm^{-1} to the accordion modes of the *n*-hexyl, *n*-octyl, *n*-decyl, and *n*-dodecyl chains, respectively. Figure 2 shows the linear relationship between the accordion band frequency and the reciprocal number of carbon atoms (1/C_N) for the four *n*-alkyltrimethylammonium bromides. We may assume that the accordion band frequencies of the *n*-octyl, *n*-decyl, and *n*-dodecyl chains,

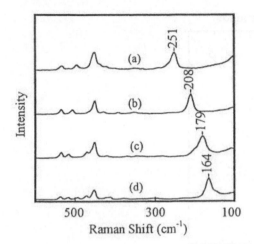

FIG. 1 Raman spectra for solid n-alkyltrimethylammonium bromides [(a) n-hexyl, (b) n-octyl, (c) n-decyl, and (d) n-dodecyl] in the skeletal stretching region.

for sm6–8 to sm6–12, are very close to those of the n-alkyltrimethylammonium bromides because the Raman bands for sm6–8, sm6–10, and sm6–12 were observed at 205, 178, and 160 cm^{-1}, respectively. For the Raman spectra of the n-paraffins (C$_1$–C$_{16}$), Mizushima and Shimanouchi [6] found that only one Raman band existed for each solid paraffin in the 150–300 cm^{-1} region and that its frequency was inversely proportional to the length of the chain (the number of carbon atoms). This result was accounted for by approximating the extended carbon chain as a continuous rod and assigning the Raman band to its longitudinal vibration, the frequency of which is given by $\nu = (1/2l)\sqrt{E/\rho}$, where E is Young's modulus, ρ the density, and l the length of the rod.

The Raman spectra of the spacer-related compounds TMB, HMB, OMB, and DMB have also been examined. The crystal structures of hexa- and decamethonium bromides [7] were elucidated by a single-crystal X-ray diffraction analysis, which showed that the molecular chain, which included the two nitrogen atoms and one CH$_3$ carbon atom of the terminal N(CH$_2$)$_3$ groups, was in a fully extended form. Thus, the methonium bromide chains should provide evidence for the accordion vibrational bands.

Because the plots of the band frequencies of the methonium bromides against the reciprocal number of carbon atoms are linear, these bands for TMB, HMB, OMB, and DMB are assigned to the accordion modes of the extended molecular chains [4,5].

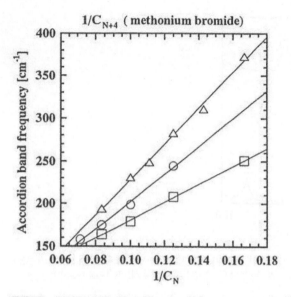

FIG. 2 Plot of the accordion band frequency against the reciprocal number of carbon ($1/C_N$) atoms* for ($\triangle$) n-paraffins (from n-hexane to n-dodecane), ($\square$) n-alkyltrimethylammonium bromide (n-hexyl, 251 cm^{-1}; n-octyl, 208 cm^{-1}; n-decyl, 179 cm^{-1}; and n-dodecyl, 164 cm^{-1}), and ($\circ$) methonium bromide (TMB, 244 cm^{-1}; HMB, 198 cm^{-1}; OMB, 175 cm^{-1}; and DMB, 156 cm^{-1}). [*Carbon number represents the total number (4) of two methyl carbon atoms and two nitrogen atoms for methonium bromide.]

The bands at 239 and 193 cm^{-1}, observed for the aqueous TMB and HMB solutions, respectively, were assigned to accordion bands arising from the all-*trans* form that existed in aqueous solution (spectra not shown). Thus, we may assume that the extended forms of TMB and HMB are preferentially stabilized in aqueous solution. However, for OMB and DMB, the extended forms must be stabilized in concentrated solutions, a decision based on the evidence from the Raman data in the skeletal stretch region [4,5] and the accordion mode data of simple soap molecules [8] and n-paraffins [6]. For the aqueous, gel, and coagel samples of sm6–8, weak and broad bands at 195–198 cm^{-1} were observed (spectra not shown), and these were assigned to the accordion mode of the hexamethylenediamine chain [4], which was superimposed on that of the n-octyl chain. For the coagel sm6–10, the band at 205 cm^{-1} was assigned to the accordion mode of the spacer [4], showing that the extended form of the shorter spacer chain was stabilized in all of the aqueous, gel, and coagel states.

B. Selective Decoupling ^{13}C NMR Spectra and Conformational Changes of Gemini Surfactants

Selective decoupling ^{13}C NMR spectra of oxy8, mxy8, and pxy8 have been investigated [9] and the results are summarized as follows.

1. pxy8

For the aqueous pxy8 sample, a conformational change should not be reflected in the selective decoupling ^{13}C NMR spectra because the *ipso*-carbon (^{a}C) is not influenced by the conformation of the *n*-octyl chain and, moreover, the four ^{b}C carbons are equivalent. Indeed, it was found that the resonance spectral features of the ^{a}C and ^{b}C carbons were characteristic of a triplet signal, and this feature did not change above and below the critical micelle concentration (cmc).

2. mxy8

We have found that for the signals of the ^{a}C, ^{c}C, and ^{d}C nuclei there exists a marked difference in the spectral features between monomeric and micellar solutions (Fig. 3). For the micellar solutions of mxy8, as listed in Table 2, the splitting pattern of the ^{a}C resonance characteristic of quintet lines (gg'gg') indicates that type I is preferentially stabilized upon micellization. Conversely, below the cmc, a spectral pattern of the ^{a}C resonance is found in which the quartet lines (a part of tggg') are superimposed upon the quintet lines (gg'gg'), implying that type I and type II coexist below the cmc.

For the ^{c}C resonance, the pattern characteristic of the gg' form becomes predominant upon micellization, although the two patterns characteristic of the different species of two gg' forms are superimposed upon each other. This result confirms the absence of types I' and II' and is consistent with the pattern obtained from a ^{a}C carbon.

From consideration of the ^{a}C and ^{d}C spectral intensities, we may assume that the high-field ^{d}C peak comes from a type I and the low-field peak comes from a type II. The difference may be caused by the environmental difference of the ^{d}C carbon in the solution, in addition to the difference in the charge density of the ^{d}C carbon, due to the conformational change about the two ^{c}CH$_2$—^{b}C single bonds.

3. oxy8

The dominant feature of the ^{a}C resonance was a quintet, characteristic of the gg' form, caused by spin–spin coupling through both two and three bonds between the ^{a}C nuclei and *o*-phenylenedimethylene CH$_2$ protons. There was no marked difference in the *J* values between the monomeric and micellar solutions. Consequently, we concluded that micellization of oxy8 did not result in a conformational change about the ^{s}CH$_2$—^{a}C single bonds,

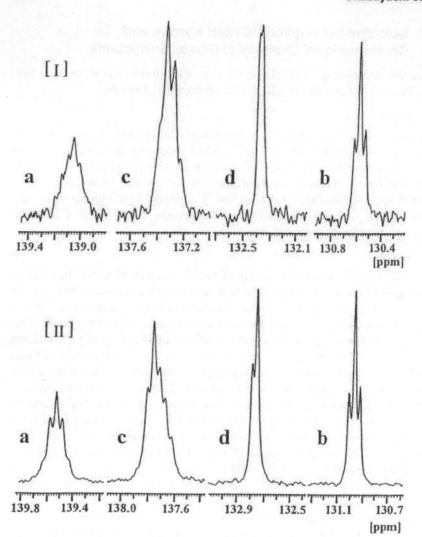

FIG. 3 Selective decoupling ^{13}C NMR spectra of aromatic ^{13}C carbons [^{a}C (a), ^{b}C (b), ^{c}C (c), and ^{d}C (d)] for mxy8 in monomeric (1.0 wt%, [**I**]) and micellar (3.0 wt%, [**II**]) solutions.

TABLE 2 Possible Skeletal Patterns of an mxy8 Molecule

Criterion: ^bC—^aC		Criterion: ^bC—^cC	
Type I	gg'gg'	Type I'	tgtg tg'tg
Type II	tggg'	Type II'	tggg'
Type III	tgtg tg'tg	Type III'	gg'gg'

showing that the protons of $^sCH_2$ did not take up the *trans* position with respect to the ^aC—^bC bond and that the two *n*-octyl-N segments were *trans* with respect to the ^aC—^aC bond.

The spectral feature of the ^bC resonance was characteristic of a triplet signal. However, micellization changed this feature to one in which the two triplet signals arising from two different conformers were superimposed. This observation indicates that these conformers are magnetically nonequivalent because of the conformational change about the two $^sCH_2$—^aC single bonds. Moreover, the exchange between these conformers occurs slowly compared with the NMR time scale.

Furthermore, it has been found for oxy8 and mxy8 that the stacking pattern of the aromatic rings varies upon micellization and that this variation is strongly reflected in the ^1H NMR spectral features of the aromatic protons (Fig. 4) [9].

For the oxy8 molecules, the micelles may be loosely packed owing to the presence of *gauche* forms in the skeleton of the rigid spacer because of preferential stabilization of a specific conformation upon micelle formation.

For the conformation of oxy8, which is stabilized upon micellization, the N—^sC—^aC—^aC—^sC—N skeleton, including *o*-phenylenedimethylene, is planar. Moreover, the conformation about the rigid ^aC—^sC single bond is *gauche* and the two *n*-octyl-N portions are in a *trans* configuration with respect to the ^aC—^aC double bond. Thus, when two *n*-octyl-N segments take up the extended form, the two octyl chains become nonparallel to each other and the packing feature of the *n*-octyl chains is lost.

Conversely, for the type I form of mxy8 stabilized upon micellization, the N—^sC—^bC—^aC—^bC—^sC—N skeleton takes up an extended form, resulting in a densely packed state of *n*-octyl chains within the micelles. This dense packing results in high-order (crystalline) self-assembly. The molecular shape of pxy8 probably favors the formation of a densely packed aggregated state, resulting directly in crystal formation from the premicelles.

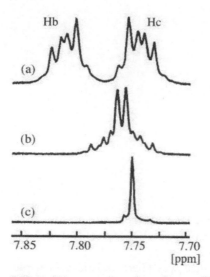

FIG. 4 Concentration dependence of ^{1}H NMR spectra in the aromatic protons (Hb and Hc) of oxy8 [(a) 4.0, (b) 3.0, (c) 1.0 wt%] in D_2O.

C. SANS Spectra and Micellar Structure of DS and Geminis

The dependence of the neutron-scattering intensity $d\Sigma/d\Omega$ on the magnitude of the scattering vector (Q) can be expressed as a function of both the micellar particle structure factor $P(Q)$ and the interparticle structure factor $S'(Q)$, that is, $d\Sigma/d\Omega = I_0P(Q)S'(Q)$, where I_0 is the extrapolated zero-angle scattering intensity. The single micellar particle structure factor $P(Q)$ was calculated by applying the micellar model. The factor $S'(Q)$, provided by Hayter and Penfold [10], is a function of the diameter, σ, the charge and number density of the micelles, and the dielectric constant of the solvent.

For a model of the DS micelle, the shape of a DS micelle with $m = 10$ was thought to be prolate, with a hydrophobic core with a major axis, a, and minor axis, b. The b value was set equal to the extended length of the n-decyl chain. The Stern layer, of thickness t, consists of the dimethylammonium head groups, associated with counterions and water; spacer methylene groups; and the hydrated portion of the n-decyl chain. Calculations were made for both prolate and oblate models. We found that the prolate model consistently provided a better fit to the observed SANS data than did the oblate model [11,12].

The SANS spectra of representative DS ($m = 10$, $s = 12$) micellar solutions are shown in Fig. 5 [11,12]. The observed SANS data were analyzed

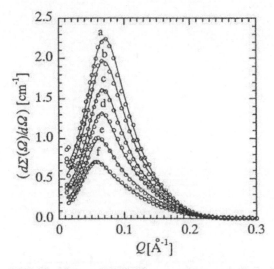

FIG. 5 Observed SANS spectra (open circles) of the DS ($m = 10$, $s = 12$)–D_2O system (23°C) [(a) 4.0, (b) 3.5, (c) 3.0, (d) 2.5, (e) 2.0, and (f) 1.5 wt%] and calculated intensity profiles (solid lines). The average percentage deviation per datum point was within ±3.8% for all spectra.

by using the aggregation number, n, the degree of micellar ionization, α, and the number of hydrated methylene group, n_{wet}, as the fitting parameters and by assuming monodispersity. The closeness of the fit between the observed and the calculated data is excellent. The extracted parameters for the representative DS ($m = 10$, $s = 12$) micelles are listed in Table 3.

A ladder model of micellar growth [13] could be applied to the micelles formed by DS. The slope in the linear plots of n versus $(X - X_{cmc})^{1/2}$ decreased with an increase in spacer length (Fig. 6), indicating that the extent of micellar growth decreased as the spacer lengthened. We should note that there is a marked difference in the slopes for the two DSs with $s = 2$ and 3 and that marked growth in the micelle occurred for the former DS. However, for the DS with $s = 6$, the slope of the linear plot approached zero, showing that the extent of micellar growth was either very small or nonexistent. The slopes for the DSs with $s = 8$ and 10 are the smallest in the series. Therefore, we assumed that for the DSs with $s = 8$ and 10 the extent of micellar growth was very small and that minimal micellar growth occurred. We should note that the slope for the DS with $s = 12$ again becomes larger, indicating the onset of micellar growth at this spacer length.

The magnitude of $(\Delta - n_0\delta)$, obtained from the slope, represents the difference in free energy between n_0 surfactants in the two end caps of a

BLE 3 Calculated Parameters for the DS ($m = 10$, $n_s = 12$)–D_2O System at 23°C[a]

%	n	α	n_{wet}	a (Å)	b (Å)	t (Å)	$(a + t)/(b + t)$
	21.1	0.44	1.6	13.6	13.6	7.5	1.00
	21.6	0.36	2.2	12.9	14.5	8.2	1.08
	22.9	0.31	2.2	12.9	15.5	8.3	1.12
	24.1	0.27	2.5	12.5	16.6	8.6	1.19
	25.6	0.22	2.2	12.9	17.2	8.2	1.20
	26.4	0.20	2.3	12.7	18.0	8.4	1.25

average aggregation number of the micelle; α, degree of ionization of the micelle; n_{wet}, number of hy
er; a, major axis of the prolate micelle given by $a = (4\pi m V_{tail})/(3b^2)$; b, minor axis of the prolate micelle
iickness of the Stern layer given by $t = 2.95 + 1.27(2 + n_{wet})$; N_s, number of water molecules associa
ye-Hückel screening length; σ, macroion diameter.

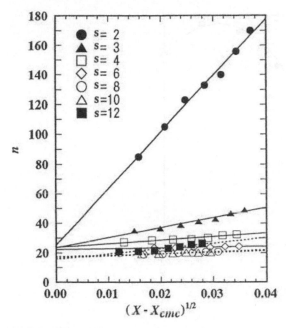

FIG. 6 Plot of the aggregation number, n, as a function of $(X - X_{cmc})^{1/2}$ (X, molar fraction) for the DS series ($m = 10$, $s = 2$, 3, 4, 6, 8, 10, and 12).

spherocylindrical micelle and the same number of surfactants in the cylindrical portion [13]. The value of $-(\Delta - n_0\delta)$ provides a measure of the rate of micellar growth. The calculated energy differences for the DS series are plotted against the spacer carbon number (Fig. 7). As the spacer carbon number increases, the parameter $-(\Delta - n_0\delta)$ increases exponentially with s in the range $s = 2-8$ and then decreases until $s = 12$. The variation in this energy difference indicates that an increase in spacer carbon number when $s \leq 10$ prevents micellar growth but that when $s = 12$ the micelle starts to grow again.

The SANS spectra of geminis in D_2O have also been analyzed [14]. For the oxy8 micelles, calculations have been made for both prolate and oblate models by assuming monodispersity. In the curves of $I(Q)$ against Q observed at concentrations less than 3.0 wt%, the peaks arising from the intermicellar interactions disappeared. For oxy8 sample solutions below the second cmc, therefore, the SANS spectra were analyzed by assuming $S'(Q) = 1$. For those above the second cmc, the so-called rescaled mean spherical approximation (RMSA) procedure for calculation of $S'(Q)$ was used [10],

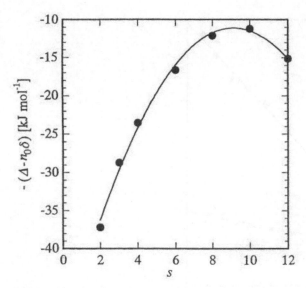

FIG. 7 Spacer chain length dependence of $-(\Delta - n_0\delta)$ for the DS series ($m = 10$, $s = 2, 3, 4, 6, 8, 10,$ and 12).

and these $S'(Q)$ calculations used the more accurately prescribed diameter σ [15]. The closeness of the fit between the observed and the calculated data for the prolate model was excellent.

Figure 8 shows the average aggregation number, n, plotted against $(X - X_{cmc})^{1/2}$ for the oxy8 micelles. The plot provides two straight lines in the concentration range measured, and there is a marked difference in the slopes of the two straight lines below and above the second cmc, showing that marked growth in the oxy8 micelles occurs above the second cmc. In the concentration region between the first and second cmc values, the aggregation number (n) obtained for the oxy8 micelles is 3–7, indicating that the micellar size in this region is relatively small and that formation of premicelles rather than micelles may occur at the first cmc.

The SANS data of oxy8 micellar solutions may suggest a "gemini → submicelle → assembly" mechanism for the growth of oxy8 micelles. We have analyzed the SANS spectra of myx8 and pxy8. As a consequence, it may be assumed that the mxy8 and pxy8 molecules form only premicelles in water. In fact, the aggregation numbers (n) calculated for both the mxy8 and pxy8 solutions were about 2.

Such a difference in the aggregation behavior between the gemini oxy8 and the geminis mxy8 and pxy8 may be related to their molecular conformations, which are stabilized in the micellar state.

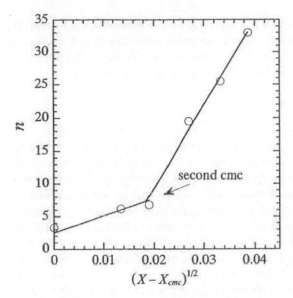

FIG. 8 Plot of aggregation number, n, as a function of $(X - X_{cmc})^{1/2}$ (X, molar fraction) for the oxy8 micelles calculated assuming monodispersity.

IV. CONCLUSION

For the DS molecules in the solid state, the molecular skeleton takes up a *cis* form, and their *n*-octyl, *n*-decyl, and *n*-dodecyl chains in the solid state as well as the hexamethylene spacer are in a fully extended state. In the aqueous and gel phases, the skeletal *trans* form, in addition to this skeletal *cis* form, may coexist.

The accordion modes in the skeletal deformation region and the skeletal stretch modes in the $1000–1100$ cm^{-1} region suggest that the all-*trans* form, for the *n*-alkyl chains and spacer segment with carbon number below 12, become abundant in the aqueous and gel phases.

For the DS samples, with $s = 2–12$, we may assume from the Raman data that a preferential stabilization of the fully extended form occurs upon micellization for $s \leq 10$ and that the populations of the conformers containing a *gauche* form become predominant with an increase in spacer carbon number to $s > 10$. This tendency probably causes the appearance of a maximum in the plots of the energy difference $[-(\Delta - n_0\delta)]$ versus s, as obtained from the SANS analyses.

The selective decoupling ^{13}C NMR results for the geminis, oxy8 and mxy8, showed that the specific conformations about the CH$_2$—aromatic carbon single bonds for mxy8 were preferentially stabilized upon micellization; for oxy8, only the conformation in which the aromatic ring was sandwiched between two octyl chains was stabilized in both the monomeric and micellar states.

Such a difference in the stabilized molecular conformation between the geminis oxy8 and mxy8 must affect their aggregational behavior.

V. FUTURE RESEARCH

The conformational change of surfactant molecules, which is induced by formation of micelles, brings about variation in the packing parameters, affecting micellar size and shape. In order to confirm the relationship between the molecular conformation and the morphology of a self-assembly system, cryo-transmission electron microscopy (TEM) work must be carried out in addition to studying the conformation of the surfactant molecules. The addition of a polymer to surfactant solutions probably induces variation in the micellar shape. In this case, the interaction between the polymer and surfactant molecules could possibly result in a conformational change of the surfactant molecules. The application of conformational change within surfactants will become more and more significant, particularly with respect to interpreting the morphology of self-assembly systems.

REFERENCES

1. F Devinsky, I Lacko, T Iman. J Colloid Interface Sci 143:336–342, 1991.
2. R Zana, M Benrraou, R Rueff. Langmuir 7:1072–1075, 1991.
3. N Hattori, H Masuda, H Okabayashi, CJ O'Connor. J Mol Struct 471:13–18, 1998.
4. N Hattori, M Hara, H Okabayashi, CJ O'Connor. Colloid Polym Sci 277:306–317, 1999.
5. N Hattori, H Yoshida, H Okabayashi, CJ O'Connor, R Zana. Vibr Spectrosc 18:83–90, 1998.
6. S Mizushima, T Shimanouchi. J Am Chem Soc 71:1320–1324, 1949.
7. K Lonsdale, HJ Milledge, LM Pant. Acta Crystallogr 19:827–840, 1965.
8. H Okabayashi, M Okuyama, T Kitagawa. Bull Chem Soc Jpn 48:2264–2269, 1975.
9. N Hattori, A Yoshino, H Okabayashi, CJ O'Connor. J Phys Chem 102:8965–8973, 1998.
10. JB Hayter, J Penfold. J Chem Soc Faraday Trans 177:1851–1863, 1981.
11. H Hirata, N Hattori, M Ishida, H Okabayashi, M Furusaka, R Zana. J Phys Chem 99:17778–17784, 1995.
12. N Hattori, H Hirata, H Okabayashi, M Furusaka, CJ O'Connor, R Zana. Colloid Polym Sci 277:95–100, 1999.
13. PJ Missel, NA Mazer, GB Benedek, CY Young, MC Carey. J Phys Chem 84:1044–1057, 1980.
14. N Hattori, H Hirata, H Okabayashi, CJ O'Connor. Colloid Polym Sci 277:361–371, 1999.
15. M Kotlarchyk, SH Chen. J Chem Phys 79:2461–2469, 1983.

7

Snared by Trapping: Chemical Explorations of Interfacial Compositions of Cationic Micelles

LAURENCE S. ROMSTED Rutgers, The State University of New Jersey, New Brunswick, New Jersey, U.S.A.

ABSTRACT

The logic of the chemical trapping method is described and used to estimate water and halide ion concentrations at the surfaces of cetyltrialkylammonium halide micelles, CTRAX (R = Me and *n*-Pr, X = Cl, Br), as a function of surfactant head group size, counterion type, and counterion concentration. The results show that interfacial counterion concentrations never reach a plateau but increase continuously with added counterion with a concomitant decrease in interfacial water. At the sphere-to-rod transition, interfacial counterion concentrations show a sharp increase and interfacial water shows a concomitant decrease. The transition depends on counterion type and is consistent with partial dehydration of the halide counterions and tight ion pair formation between the surfactant's head groups and counterions. The chemical trapping method provides new information on a variety of micellar properties including hydration numbers of nonionic micelles, distributions of additives such as alcohols and urea, and determining polypeptide topology at interfaces. These applications are described briefly.

I. INTRODUCTION

Cationic micelles are aggregates of cationic surfactant monomers (Fig. 1). Surfactants are schizophrenic molecules typically composed of linear hydrocarbon chains affixed to water-soluble head groups. In addition to cationic surfactants (with anionic counterions), micelles may be composed of surfactants with anionic (with cationic counterions), zwitterionic, or nonionic head groups. Micelles, like microemulsions, are association colloids and

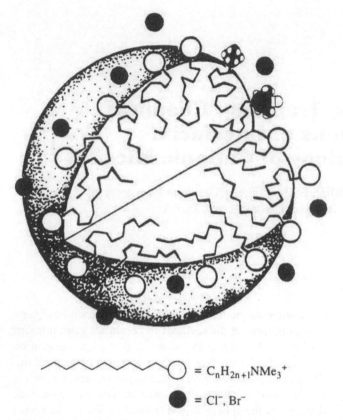

$$\text{\large\textcircle} \ = C_nH_{2n+1}NMe_3^+$$

$$\text{\Large\textbullet} \ = Cl^-, Br^-$$

FIG. 1 "Pacman" image of a cationic micelle in aqueous solution composed of alkyltrimethylammonium ions and halide counterions. The exposed sections illustrate the orientations of hydrocarbon tails in the micellar core and head groups and counterions in the interfacial region. The surrounding water is not shown. The image also illustrates the orientation of the arenediazonium ion probe with its tail buried in the core and its cationic head group oriented in the interfacial region. Space-filling representations show the relative sizes of the benzenediazonium ion and the α-methylenetrimethylammonium ion head group of the surfactant.

their solutions may also contain salts, polar organic molecules such as medium-chain-length alcohols, and oils such as benzene, octane, or chloroform. The aggregate structures formed in association colloids, e.g., spheroidal, rod, disk, cubic, or lamellar, depend on both the structures and concentrations of the components [1].

A delicate balance of forces controls the stability, size, and shape of cationic micelles and association colloids in general, and their properties are characterized by mass action or pseudophase models [2–4]. Aggregate formation is driven by the hydrophobic effect. Phase separation is prevented by the surfactant head groups and counterions retaining their contact with the surrounding aqueous phase. At equilibrium, dilute micellar solutions contain spheriodal aggregates, with aggregation numbers on the order of 10^2, that occupy only a small fraction of the total solution volume, typically $\leq 1\%$. Each micelle has a liquidlike hydrocarbon core filled with surfactant tails surrounded by an interfacial region containing the surfactant head groups and a large fraction of the surfactant counterions, about 50–90% (Fig. 1). The remaining counterions are "free" in the surrounding bulk water.

The stabilities, sizes, and shapes of the micelles depend on a balance of forces that is difficult to characterize, especially in structurally complex multicomponent micelles and microemulsions. Even in well-studied solutions of cationic surfactants and salts containing a common counterion, the factors contributing to this balance are still incompletely understood. The hydrophobic effect correlates well with the free energy of transfer of the surfactant tail from water to the micellar core and the number of methylenes in the chain. However, the interactions in the interfacial region surrounding the core must be more than coulombic in nature, i.e., the surfactant head groups and counterions interact only as point charges through an aqueous medium, because many micellar properties such as the critical micelle concentration (cmc) [5], micelle size [6], the sphere-to-rod transition [7], and specific counterion binding [8] depend on head group structure, counterion type, and surfactant and counterion concentrations. Adding more components further complicates the interpretation because of additional specific interactions. Solving this problem has proved difficult because quantitative treatments of specific interactions between surfactant head groups and their counterions (and other additives) in the interfacial regions of association colloids or with model compounds in bulk aqueous solutions are unavailable. Interfacial hydration is difficult to measure over wide ranges of solution composition because only a small fraction of the total water in solution is within the interfacial region.

This chapter describes the results of chemical trapping experiments in cetyltrialkylammonium halide (CTRAX) micelles. The primary focus is on the basic logic of the chemical trapping method and on its application to the experimental results in these systems. The method has broad application because it works with a wide variety of weakly basic nucleophiles over wide ranges of surfactant, counterion, and additive concentrations below pH 7 [9]. The goal of the work described here was to obtain new information on the relationships between surfactant head group structure, counterion type, and their concentrations on interfacial composition and thereby the balance of forces controlling aggregate structure. Particular attention was given to compositional changes at the sphere-to-rod transitions for CTMABr (R = Me, X = Br) and CTMACl (R = Me, X = Cl), which occur at about 0.1 M Br⁻ and 1.0 M Cl⁻, respectively. The atypical abbreviations for these two surfactants are necessary to differentiate them from those with larger head groups, here CTPABr (R = n-Pr). The results with CTEABr (R = Et) and CTBABr (R = n-Bu) are essentially the same as those with CTPABr and will be considered briefly in Section IV. A complete presentation of the results and the assumptions of the method is given in Ref. 10. The background information is in selected lead references. The current and potential applications of the method and their relevance to commercial and biological systems are discussed in Section VI.

II. LOGIC OF THE CHEMICAL TRAPPING METHOD

The chemical trapping method is grounded in the pseudophase ion exchange (PIE) model of association colloid effects on the rates of thermal reactions [11–13]. In this model, all the aggregates in homogeneous, thermodynamically stable, optically transparent micellar and other association colloid solutions are treated as separate phases, ergo pseudophases. Because the rates of transfer of organic reactants, surfactants, and other components between the micellar and aqueous pseudophases are generally orders of magnitude faster than the rate of a chemical reaction occurring in solution, the distributions of components and reactants are at equilibrium throughout the time course of the reaction. Thus, reactions occurring in each pseudophase are dependent not on the exit and entrance rates of the components but only on the concentrations of reactants in each pseudophase and on the rate constant for reaction within that pseudophase, and the observed rate of reaction is the sum of the rates in the micellar and aqueous pseudophases.

The distributions of organic reactants are generally described by a binding constant. Describing hydrophilic counterion distributions is more problematic. The fraction of the surfactant's counterions bound to the micelle, β, is generally assumed to be constant and independent of the concentrations of both the surfactant and added counterion. Values of β for many different counterions are generally in the range ~0.5–0.9 (50–90% bound), and the value obtained depends somewhat on the method used to measure it [14]. The rate constant for reaction in the aqueous pseudophase can be measured in the absence of micelles, but the rate constant for reaction in micelles must be obtained by fitting the change in observed rate constant as a function of the surfactant and salt concentrations. The PIE model has proved robust. It provides reasonable fits for a wide variety of chemical reactions up to about 0.1–0.2 M added surfactant and salt. At higher counterion concentrations or in the presence of high concentrations of hydrophilic counterions such as

OH$^-$ in cationic micelles, the model breaks down [12] and the results reported here are consistent with that breakdown (see later).

Scheme 1 and Figs. 1 and 2 illustrate the essential elements of the chemical trapping method and the assumptions required to interpret product yield results. The chemical trapping reagent is prepared as the 4-alkyl-2,6-dimethylarenediazonium tetrafluoroborate, z-ArN$_2$BF$_4$, and it is the z-ArN$_2^+$ ion that functions as the chemical probe. Water soluble 1-ArN$_2^+$ ($z = 1$, alkyl = CH$_3$) is used for studies in aqueous solution in the absence of micelles (Fig. 2). Water-insoluble 16-ArN$_2^+$ ($z = 16$, alkyl = C$_{16}$H$_{33}$) is used in micelles because its hydrocarbon chain binds the probe to the micellar core and its cationic head group orients the reactive diazonio group in the interfacial region of the micelles (Figs. 1 and 2). We use a preassociation model, Scheme 1, a modified form of the heterolytic dediazoniation mechanism proposed by

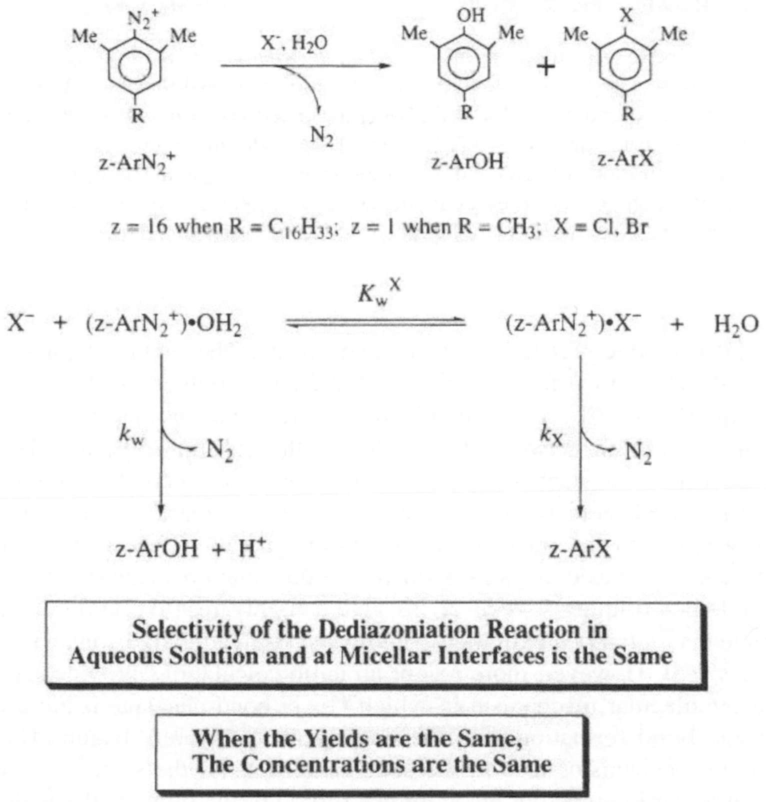

$z = 16$ when R = C$_{16}$H$_{33}$; $z = 1$ when R = CH$_3$; X = Cl, Br

$$X^- + (z\text{-ArN}_2^+)\bullet OH_2 \underset{}{\overset{K_w^X}{\rightleftharpoons}} (z\text{-ArN}_2^+)\bullet X^- + H_2O$$

$k_w \searrow N_2$ $\qquad\qquad\qquad$ $k_X \searrow N_2$

z-ArOH + H$^+$ $\qquad\qquad\qquad$ z-ArX

Selectivity of the Dediazoniation Reaction in Aqueous Solution and at Micellar Interfaces is the Same

When the Yields are the Same, The Concentrations are the Same

SCHEME 1 Basic assumptions of the chemical trapping method.

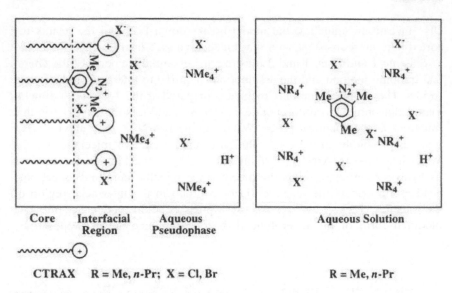

FIG. 2 The two boxes illustrate the mantra *when the yields are the same, the concentrations are the same* for a particular set of solution compositions. The box on the left contains the components for dediazoniation reactions in micelles, showing the location of the surfactant, $16\text{-}ArN_2^+$, X^-, and TMA^+. Product yields are proportional to the concentrations of H_2O and X^- in the interfacial region. The box on the right contains the components present in the aqueous reference solution. Product yields are proportional to the measured concentrations of water and TMAX.

Zollinger [15], to interpret the product yield results. The yields of phenol product, $z\text{-}ArOH$, from reaction with water and halo product, $z\text{-}ArX$, from reaction with Br^- or Cl^- are assumed to be determined primarily by the position of equilibrium between the ion–molecule and ion–ion pairs. The dediazoniation rate constant for each pair is probably of secondary importance for these dediazoniations because dediazoniation reactions are notoriously insensitive to the polarity of the reaction medium (see later). For several decades the basic consensus on the dediazoniation mechanism has been the rate-determining loss of N_2 to give a highly reactive aryl cation intermediate that is trapped extremely rapidly and competitively by available nucleophiles [15]. However, more recent ab initio calculations provide support for a bimolecular mechanism in which C—N bond cleavage is almost complete and bond formation with the nucleophile has barely begun [16]. Because both mechanisms lead to the same definition for the selectivity of the reaction toward competing nucleophiles [Eq. (1)], the bimolecular pathway for dediazoniation is not included in Scheme 1.

The crucial assumption in the chemical trapping method is that the selectivity of the dediazoniation reaction toward a pair of nucleophiles is the same for 16-ArN_2^+ in micelles (or other association colloids) as it is for 1-ArN_2^+ in a reference bulk aqueous solution of the same composition (Scheme 1 and Fig. 2). This assumption is expressed as

$$S_w^X = \frac{[H_2O]\{\%(1 - ArX)\}}{[TRAX]\{\%(1 - ArOH)\}} = \frac{H_2O_m\{\%(16 - ArX)\}}{X_m\{\%(16 - ArOH)\}} = \frac{K_w^X k_w}{k_X} \qquad (1)$$

where S_w^X is the selectivity of the trapping reaction toward a halide ion versus a water molecule; the second and third equalities are the definition of the selectivity in aqueous solution and in the interfacial region of micelles; % (percent) indicates measured yield of a reaction product that has been normalized to 100% of total halo and phenolic products; square brackets, [ ], indicate concentration in moles per liter of solution volume; [TRAX] is the molarity of tetraalkylammonium halide salt; and the subscript "m" indicates concentration in moles per liter of interfacial volume. The last equality describes the definition of the selectivity based on the mechanism in Scheme 1 [17]. As shown in the following, S_w^X for 1-ArN_2^+ in aqueous solutions of TRAX decreases with [TRAX] by about a factor of 2 up to about 3.5 M TRAX. To correct for the variation in S_w^X, the following corollary, stated succinctly in Scheme 1, is added to the selectivity assumption: when the product yield from reaction of 16-ArN_2^+ in CTRAX micelles is the same as the product yield from reaction of 1-ArN_2^+ in aqueous TRAX, i.e., %(16-ArX) = %(1-ArX) and %(16-ArOH) = %(1-ArOH), the concentrations of X and H_2O sensed by each probe are the same, i.e., $X_m = [X_w]$ and $H_2O_m = [H_2O_w]$ because S_w^X at the measured yields at those concentrations is the same. Because aqueous concentrations in the reference solutions are in molarity, interfacial concentrations are also in molarity, specifically in moles per liter of interfacial volume.

The relationships between the interfacial and aqueous concentrations of X and H_2O are given by

$$X_m = \frac{[X_m]}{([CTRAX] - cmc)V_m} = \frac{\beta}{V_m}; \quad H_2O_m = \frac{[H_2O_m]}{([CTRAX] - cmc)V_m} \qquad (2)$$

where V_m is the molar volume of reaction of 16-ArN_2^+, i.e., the volume of the interfacial region sampled by the probe as it reacts, which is assumed to be the same as the molar volume of the cationic head groups of CTRAX in the interfacial regions of their micelles, and ([CTRAX] − cmc) is the concentration of micelles in solution. The second equality on the right-hand side of the equation for X_m shows that X_m must be constant (and independent of surfactant and counterion concentrations) if β and V_m are constant. These two terms are generally assumed to be constant in applications of the PIE model, but the results here show that at high [TRAX] and [CTRAX] this assumption is not correct.

III. CHEMICAL TRAPPING IN AQUEOUS SOLUTIONS OF CTRAX MICELLES WITH ADDED TRAX: RESULTS

Figures 3–8 show chemical trapping results in TRAX solutions (Fig. 3), CTRABr solutions (R = Me, n-Pr) (Figs. 4, 6, and 7), and CTMACl solutions (Figs. 5 and 8). Experimental details and the complete results are given in Ref. 10. Figure 3 shows results for the reaction of 1-ArN$_2^+$ in aqueous TRAX solutions up to 3.5 M TRAX where R = Me and n-Pr and X = Br and Cl. The top two graphs in Fig. 3 show normalized product yields determined by HPLC for reaction with X$^-$, %(1-ArX), and H$_2$O, %(1-ArOH). Note that yields from reaction with Br$^-$ are greater than with Cl$^-$ and that there is an inverse change in %(1-ArOH) yields; i.e., when %(1-ArX) and X$_m$ increase, %(1-ArOH) and H$_2$O$_m$ must decrease because X and H$_2$O are the only two nucleophiles present. The third graph from the top shows the molarity of water for each salt in the solution, which was determined by weight. The bottom graph shows the resulting selectivities for each salt. The selectivities decrease by about a factor of 2 with added TRAX, the curves for S_w^{Br} in TPABr and TMABr solutions are almost the same, but the values of S_w^{Cl} are significantly lower, probably because 1-ArN$_2^+$ does not ion pair as strongly with Cl$^-$ as with Br$^-$. The ratio of the selectivities for Cl$^-$ and Br$^-$ is constant over the entire [TRAX] range and $S_{Cl}^{Br} = 1.9$. The %(1-ArX) yields and the S_w^X values are used to calculate X$_m$ and H$_2$O$_m$ in CTRAX micelles. The fact that S_{Cl}^{Br} is constant indicates that the variations in S_w^X are caused by different effects of the ionic strength on the stabilities of ion–molecule and ion–ion pairs in equilibrium with each other (Scheme 1) [9,18].

Figure 4 shows the interfacial concentrations of Br$_m$ and H$_2$O$_m$ in CTPABr and CTMABr micelles at several concentrations of added TMABr. Figure 5 shows parallel results in CTMACl solutions with added TMACl. Interfacial concentrations of Br$_m$ and Cl$_m$ were obtained by applying the mantra in Scheme 1: *when the yields are the same, the concentrations are the same.* For example, the %(1-ArCl) yield data in Fig. 3 were fitted with an exponential equation and used as a standard curve to calculate Cl$_m$ from %(16-ArCl) yields, by assuming that when %(16-ArCl) = %(1-ArCl), Cl$_m$ = [Cl$_w$]. The values of H$_2$O$_m$ were calculated from S_w^{Cl}, %(1-ArCl), %(1-ArOH), and [H$_2$O] M at each [TMACl] by using Eq. (1). Several patterns are apparent in these results. First, for the CTRABr surfactants, there is a sharp initial rise in Br$_m$ at low [CTRABr] when [TRABr] = 0 M (less visible for Cl$_m$ at

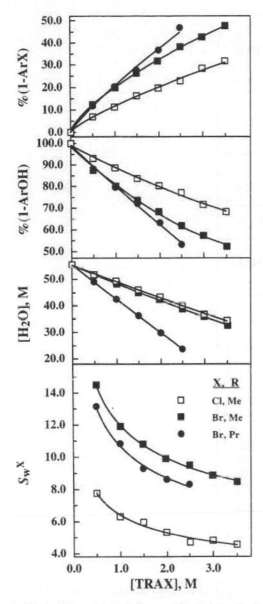

FIG. 3 Normalized dediazoniation product yields, water molarities, and halide ion versus water selectives S_w^X of Br^- and Cl^- in aqueous TRAX for R = Me and n-Pr at 40°C. The TMACl, TMABr, and TPABr solutions contained 0.01 M HCl and 0.01 M and 0.001 M HBr, respectively. Lines are fitted with an exponential equation. The lines in the %(1-ArX) graph serve as standard curves for estimating Br_m from %(16-ArBr) in CTPABr and CTMABr micelles and Cl_m in CTMACl micelles.

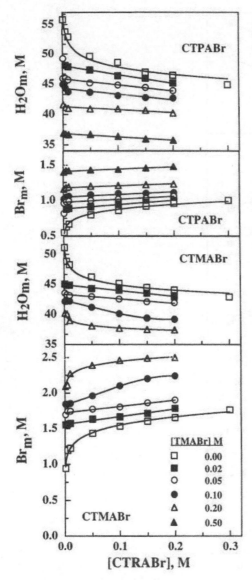

FIG. 4 Effect of increasing [CTRABr] and increasing [TMABr] on Br_m and H_2O_m for R = Me and n-Pr. Lines are drawn to aid the eye.

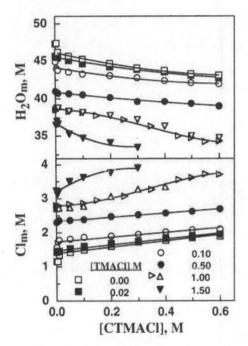

FIG. 5 Effect of increasing [CTMACl] and increasing [TMACl] on Cl_m and H_2O_m. Lines are drawn to aid the eye.

0 M TMACl in Fig. 5) that is absent in the presence of added salt. This rise is attributed to completion of binding of $16\text{-}ArN_2^+$ to the micelles with increasing surfactant concentration. This effect disappears in the presence of added salt because the added salt reduces the cmc and also enhances the binding of $16\text{-}ArN_2^+$ (a cosurfactant) to the micelles [19]. The first three points at [TMAX] = 0 M are not included in the following interpretations. Second, X_m increases gradually, essentially linearly with added CTRAX at each [TMAX], generating a series of parallel lines, except for the highest two [TMAX] in CTMABr and CTMACl solutions. Third, at the two highest [TMABr] and [TMACl], the Br_m and Cl_m molarities increase more rapidly and nonlinearly. For CTMABr, these changes occur at about 0.1 M TMABr, and for CTMACl, they occur at about 1.00 M TMACl. These solutions also become progressively more viscous (to the eye) with increasing TMAX. Fourth, the patterns for H_2O_m are the inverse of those for X_m. Interfacial water molarities range between about 30 and 55 M.

Figures 7 and 8 show plots of X_m and H_2O_m against $[X_w]$, the counterion concentration in the aqueous pseudophase surrounding the micelles, for

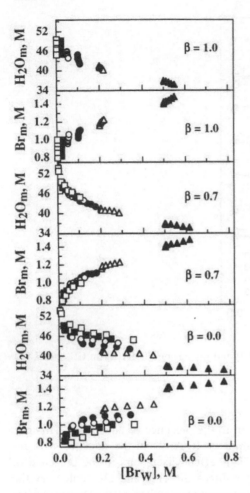

FIG. 6 Effect of increasing the degree of counterion binding, β, in CTPABr/TMABr solutions on plots of Br_m (data and symbols are in Fig. 4) versus $[Br_W]$. $[Br_W]$ was obtained by using Eq. (3) (see text).

each CTRAX/TRAX data set in Figs. 4 and 5. The X_m values fall on smooth curves except that the curves for CTMABr and CTMACl show discontinuities at about 0.1 M $[Br_w]$ and 1.0 M $[Cl_w]$, respectively. $[X_w]$ is given by

$$[X_w] = \frac{(1 - \beta)\{[CTRAX] - cmc\} + cmc + [HX] + [TMAX]}{1 - V\{[CTRAX] - cmc\}} \tag{3}$$

where the term $(1 - \beta)\{[CTRAX] - cmc\}$ is the concentration of counter-

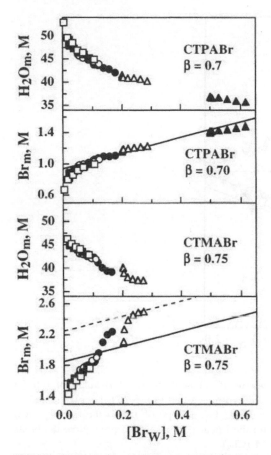

FIG. 7 Plots of Br_m and H_2O_m versus $[Br_w]$ at optimal β values shown in Fig. 6 for CTMABr and CTPABr with added TMABr. Data and symbols are in Fig. 3. $[Br_w]$ was obtained by using Eq. (3) (see text). Straight lines have a slope of 1, and each intercept was selected to give optimal contact with the linear portions of the curves. Note the break from the smooth curve for CTMABr above 0.1 M $[Br_w]$.

ions contributed to the aqueous pseudophase by the ionization of the micelles and [HX] is the concentration of the acid in solution, 0.01 M HBr for bromide surfactants and 0.1 M for CTMACl. The cmc was set equal to zero because the lowest surfactant concentration used was 0.01 M, at least 10 times greater than the cmc values of the surfactants, and the added TMAX lowers their cmc values even further. The V is the molar volume of the anhydrous surfactant and was included to correct for the excluded volume of the micelles. The greatest correction was 24% for CTMACl, and it was

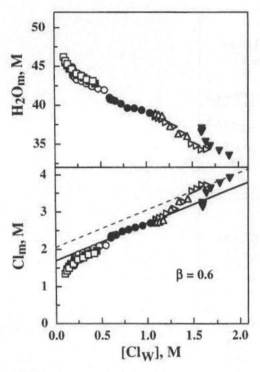

FIG. 8 Plots of Cl_m and H_2O_m versus $[Cl_w]$ at the optimal β value of 0.6 for CTMACl/TMACl solutions. Data and symbols are in Fig. 5. $[Cl_w]$ was obtained by using Eq. (3) (see text). The straight line has a slope of 1, and the intercept was selected to give optimal contact with the linear portion of the curve. Note the break from the smooth curve above 1.2 M $[Cl_w]$.

lower for the others. This correction has a negligible effect on the shapes of the curves in Figs. 7 and 8. For each surfactant β was treated as a disposable parameter and its optimal value was estimated from X_m versus $[X_w]$ plots with different β values. Figure 6 illustrates the effect of changing the value of β on Br_m and H_2O_m in CTPABr solutions at β = 1.0, 0.7, and 0.0. Note how dispersed the data points are in the plots at β = 1.0 and 0.0. The dispersion is minimal when β = 0.2, 0.3, and 0.4, and the value of 0.3 $\pm$ 0.1 was selected as the best. Similar plots were made for the other surfactants, and their optimal β values are listed in Figs. 7 and 8. The β values obtained in this way are in good agreement with those obtained by chemical trapping and with those published in the literature [10].

Figures 7 and 8 show plots of Br_m and H_2O_m versus $[Br_w]$ for CTPABr and CTMABr and Cl_m and H_2O_m against $[Cl_w]$ for CTMACl, respectively, that include all the data at each [TMABr] and [TMACl] in Figs. 5 and 6. Several patterns appear in these figures. Changes occurring in X_m with $[X_w]$ have concomitant inverse changes in H_2O_m with $[X_w]$, and the discussion at this point is focussed primarily on X_m. The Br_m and Cl_m increase relatively rapidly up to about 0.1 M $[Br_w]$ to 0.2 M $[Cl_w]$, respectively. Above these concentrations the patterns become different. CTPABr shows a linear increase in Br_m with a slope of 1. Its solutions remain fluid at all CTPABr and TMABr concentrations. The solid (and dashed) lines in both figures are drawn with a slope of 1 and the intercepts were selected to maximize contact with the data points. CTEABr and CTBABr exhibit the same pattern as CTPABr (data not shown). The Cl_m increases linearly up to about 1.0 M $[Cl_w]$. At higher $[Cl_w]$, CTMABr and CTMACl, respectively, show marked increases in Br_m and Cl_m (with concomitant decreases in H_2O_m) that appear to approach a new (dashed) line asymptotically with a slope of 1. As noted earlier, the solution viscosity increases in this region.

IV. DISCUSSION

The results in Figs. 4 and 5 show clearly that the values of X_m (and also H_2O_m) are not independent of surfactant and salt concentrations as required by Eq. (2) if both β and V_m are constant, as generally assumed in applications of the PIE model to micellar catalyzed reactions [20–22]. The increase in X_m with added salt is consistent with several sets of kinetic results in micelles in which the observed rate constants increase steadily with added reactive counterions and never reach a plateau. The linear regions of the results in Figs. 7 and 8 are qualitatively consistent with Eq. (4):

$$X_m = \frac{[X_m]}{([CTRAX] - cmc)V_m} + [X_w] = \frac{\beta}{V_m} + [X_w] \tag{4}$$

assuming V_m may also vary and that X_m increases incrementally with added $[X_w]$ and the coefficient of $[X_w]$ is 1. Note that Eq. (4) was obtained by adding the $[X_w]$ term to Eq. (2). Assuming β is constant, the initial rapid rise in X_m at low $[X_w]$ is caused by a decrease in V_m to a constant value; i.e., added salt shrinks the volume of the interfacial region without changing the ratio of interfacial counterions to head groups, β. The driving force for shrinking the interfacial volume is unknown, and there is no theory for predicting, even qualitatively, how the β/V_m ratio changes as a function of surfactant and salt concentrations. Above about 0.1 to 0.2 M $[X_w]$, the $[X_w]$ term becomes dominant. The incremental increase in X_m with added $[X_w]$ suggests that counterions from added TMAX are entering the interfacial region. For β to remain constant, both TMA^+ and X^- ions must enter the interfacial region together as a neutral pair. The ions displace water, but the net charge of the micellar surface remains constant.

Table 1 illustrates the changes in interfacial concentration of X_m and H_2O_m and their molar ratio, H_2O_m/X_m, for all four CTRABr surfactants and CTMACl at selected salt and surfactant concentrations. Several patterns are evident. The values of Br_m decrease with increasing size of the CTRABr head group at all [TMAX]; e.g., Br_m drops by about a factor of 2 at 0.02 M TMABr on going from trimethyl to tri-n-butyl head groups. At 0.02 M TMAX, i.e., in the absence of significant amounts of added TMAX, H_2O_m is essentially constant at 45.5 M $\pm$ 0.84 M $\pm$ 1.8% for all the surfactants. Indeed, the increase in Br_m approximately compensates for the decrease in

TABLE 1 Interfacial Compositions and Molar Volumes of CTRAX Micelles

[TMAX] (M)	[CTRAX] (M)	[X_W] (M)	X_m (M)	H_2O_m (M)	H_2O_m/X_m
	CTBABr				
0.02	0.01	0.024	0.71	44.9	63.3
0.1	0.01	0.105	0.91	41.2	45.3
0.2	0.2	0.289	1.13	36.5	32.3
0.5	0.2	0.622	1.59	27.9	17.6
	CTPABR				
0.02	0.01	0.024	0.89	47.9	53.8
0.1	0.01	0.104	1.06	43.7	41.2
0.2	0.2	0.287	1.23	40.2	32.7
0.5	0.2	0.616	1.48	35.8	24.2
	CTEABr				
0.02	0.01	0.024	1.32	44.7	33.9
0.1	0.01	0.104	1.57	42.4	27.0
0.20	0.2	0.284	1.78	39.1	22.0
0.5	0.2	0.611	2.05	35.3	17.2
	CTMABr				
0.02	0.01	0.024	1.57	44.9	28.6
0.1	0.01	0.104	1.85	42.2	22.8
0.2	0.2	0.271	2.50	37.4	15.0
0.5	—	—	—	—	—
	CTMACl				
0.02	0.01	0.124	1.47	45.5	30.9
0.1	0.01	0.205	1.78	43.8	24.6
0.5	0.2	0.736	2.48	40.2	16.2
1.0	0.2	1.26	3.07	37.3	12.1
1.5	0.2	1.80	3.77	34.1	9.05

head group size (volume) through the tri-n-butylammonium to trimethylammonium head group series. Note also that the values of X_m and H_2O_m for CTMABr and CTMACl are quite close at 0.02 and 0.1 M TMAX. The H_2O_m/X_m ratio correlates strongly with head group size. The ratio decreases rapidly with increasing [TMAX], and at low [TMAX] it is quite similar for CTMABr and CTMACl. The decrease in interfacial water concentration and the shrinkage of the interfacial volume permit tighter packing in the interfacial region with increasing concentration of added surfactant and salt, which is consistent with the known gradual increase in aggregation numbers of CTMABr and CTMACl [23]. Visual inspection shows that only CTMABr and CTMACl micelles undergo the sphere-to-rod transition when the H_2O_m/X_m ratios become small, i.e., above 0.1 M [Br_W] and 1.0 M [Cl_W], respectively. Together, these results show that micelle growth and the sphere-to-rod transition are governed by a balance of the hydrophobic effect against the free energy of head group and counterion hydration and not just coulombic interactions.

V. ORIGINS OF MICELLAR GROWTH AND THE SPHERE-TO-ROD TRANSITION

The equilibria in Scheme 2 offer an alternative basis for understanding the relationship between surfactant and salt concentrations, the shapes of the profiles in Figs. 7 and 8, and micelle structure. Important interactions and structural changes are attributed to the free energies of formation of solvent-separated and tight ion pairs. At low surfactant and salt concentrations micelles are small and spheroidal, the H_2O_m/X_m ratios are relatively high, and the interfacial region probably contains both free water and water of hydration around the head groups and counterions. The extent of ion pair formation increases as H_2O_m/X_m decreases with increasing surfactant concentration and added salt. However, ion pair formation also depends on the strength of the interactions between head groups and counterions. Ion pair

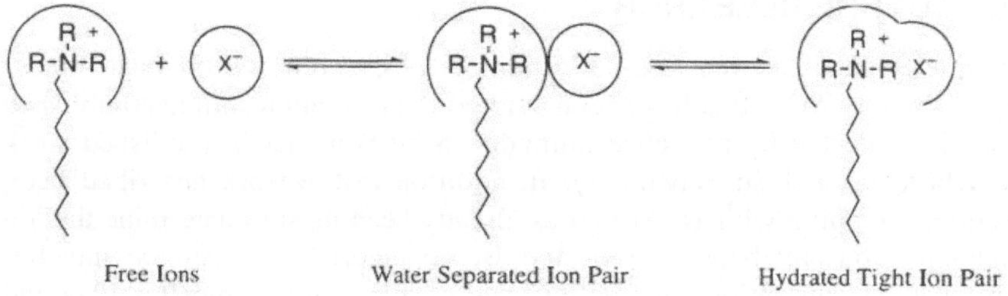

Free Ions   Water Separated Ion Pair   Hydrated Tight Ion Pair

SCHEME 2   Head group and counterion hydration.

formation should be greater for surfactants with small head groups because the counter anion interacts more strongly with the positive charge on smaller alkylammonium head groups than with the charges on the bulkier, more hydrophobic alkyl ammonium head groups. Ion pairs should also form more easily with counter anions that are less strongly hydrated.

The results in Figs. 7 and 8 show that as [TMAX] and/or [CTRAX] increases, water is removed from the interfacial region, the interfacial volume shrinks, and X_m increases relatively rapidly below about 0.2 M [X_W]. Note that the changes in X_m correlate with [X_W] and that added TMAX makes a larger contribution than added CTRAX because each equivalent of CTRAX adds only $(1 - \beta)$[CTRAX] counterions to the solution [Eq. (3)]. At higher [TMAX], significant quantities of TMA^+-X^- pairs enter the interfacial region and displace interfacial water. The decrease in H_2O_m also enhances the formation of water-separated ion pairs between surfactant head groups and counterions. Water around the three alkyl groups of the surfactant head groups is more easily removed than water of hydration around the halide ions, which have higher free energies of hydration [24]. The amount of water hydrating the water-separated ion pair is not known. The sphere-to-rod transition occurs when sufficient water has been removed from the interfacial region to permit the formation of hydrated tight ion pairs by partial dehydration of the halide ions. The sphere-to-rod transition of CTMABr and CTMACl micelles depends on counterion type, and the ease of partial dehydration of the counterion should correlate with its free energy of hydration, which is greater for Cl^- than Br^- [24]. Therefore, higher concentrations of TMACl/CTMACl than TMABr/CTMABr are required for partial dehydration of Cl^- than Br^-. Above the sphere-to-rod transition, increasing dehydration continues with increasing [X_W], the rods grow longer, and the solution viscosity increases. Had more TMABr been added to solutions of CTEABr and CTPABr, their micelles might also have undergone the sphere-to-rod transition. Alternatively, they might phase separate as does CTABABr [25] in 0.5 M TMABr after standing for several days [10].

VI. FUTURE RESEARCH

Scheme 3 shows the reactions of a number of different weakly basic nucleophiles with z-ArN_2^+ that have been studied in association colloids (published work is indicated by reference numbers in brackets and unpublished work by a bold asterisk in Scheme 3). In addition to the work described here, chemical trapping with 16-ArN_2^+ has already been used to determine the ion exchange constant between Cl^- and Br^- counterions in cationic micelles [26], Cl^- concentrations at the surfaces of zwitterionic micelles [27] and phospholipid micelles and vesicles [28], hydration numbers and terminal OH

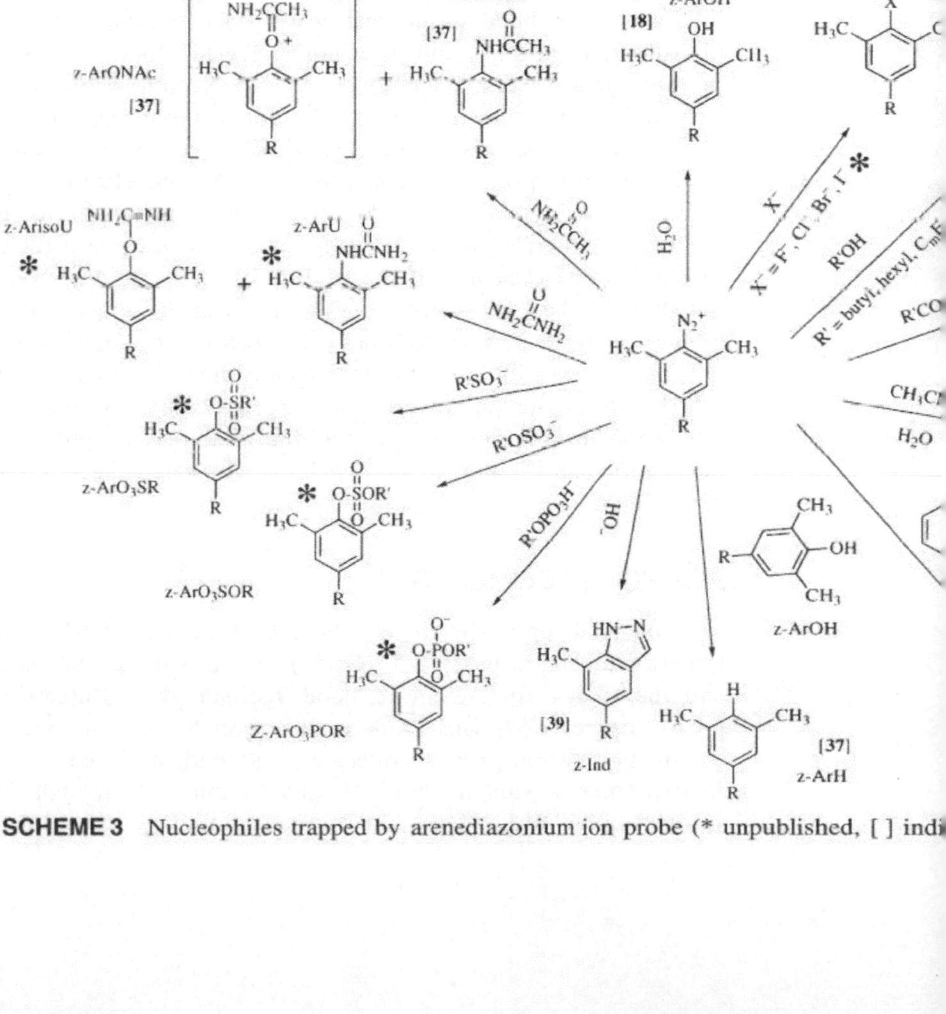

SCHEME 3 Nucleophiles trapped by arenediazonium ion probe (* unpublished, [ ] indi

group distributions in nonionic micelles [29], and alcohol distributions in oil-in-water and water-in-oil droplets and bicontinuous microemulsions [30]. The chemical trapping method has also been used to estimate the thickness of the interfacial layers of a reverse microemulsion [31], the counterion concentrations in the condensed volume of an anionic polyelectrolyte [32], Br^- concentrations in the interfacial regions of gemini micelles [33], coion (Cl^- and Br^-) concentrations in anionic micelles [34], and the degree of ionization, α, of cationic micelles [35]. The method has also been used in studies of the mechanism of reaction of $16\text{-}ArN_2^+$ with SDS head groups in SDS micelles [16]. A full review of this work has appeared as a book chapter [9]. Some unpublished work includes estimates of urea, aromatic counterion, and benzene concentrations at micellar surfaces.

Several important new directions are envisioned: monitoring the change in counterion concentration and interfacial water molarity of catanionic aggregates as they pass through the spontaneous micelle-vesicle transition [36]; trapping of phosphate groups of phospholipids, which should give information on the head group and water concentrations at the surfaces of phospholipid vesicles; and determining the topologies of polypeptides and proteins at aggregate interfaces. The $1\text{-}ArN_2^+$ ion is trapped by both the carbonyl oxygen and nitrogen of the amide group, and the mechanism of these reactions has been worked out [37]. The reaction with the carbonyl oxygen leads to products formed from C—N bond cleavage, i.e., the equivalent of cleaving the peptide bond at room temperature around pH 7. The reaction of $16\text{-}ArN_2^+$ with a polypeptide at an aggregate surface should lead to tagging and cleavage of the peptide bonds in the interfacial region but not those in the hydrocarbon region or out in the aqueous phase. Analysis of the fragments formed by liquid chromatography–mass spectrometry and conceptual reassembly of the polypeptide should provide a map of the peptide bonds of the polypeptide that are accessible to $16\text{-}ArN_2^+$ within the interfacial region. The results should provide new insight into polypeptide orientation and topology, e.g., α-helix or β-pleated sheet, at membrane-mimetic interfaces.

ACKNOWLEDGMENTS

The financial support from the Chemical Dynamics (CHE-952606) and International Programs (INT-97-22458) of the National Science Foundation and the Center for Advanced Food Technology at Rutgers University is greatly appreciated. This work would never have been completed without the many contributions of colleagues and students listed in the references. In particular, I want to thank Hernan Chaimovich, Iolanda M. Cuccovia, Jason Keiper, Valdir Soldi, Jihu Yao, and, as always, C. A. Bunton.

REFERENCES AND NOTES

1. B Jonsson, B Lindman, K Holmberg, B Kronberg. Surfactants and Polymers in Aqueous Solution. Chichester, UK: Wiley, 1998.
2. C Tanford. The Hydrophobic Effect: Formation of Micelles and Biological Membranes. 2nd ed. New York: Wiley, 1980.
3. K Holmberg, S-G Oh, J Kizling. Prog Colloid Polym Sci 100:281–285, 1996.
4. K Holmberg. Adv Colloid Interface Sci 51:137–174, 1994.
5. MJ Rosen. Surfactants and Interfacial Phenomena. New York: Wiley, 1978.
6. S Berr, RRM Jones, JSJ Johnson. J Phys Chem 96:5611–5614, 1992.
7. G Porte, J Appell. In: KL Mittal, B Lindman, eds. Surfactants in Solution. Vol 2. New York: Plenum, 1984, pp 805–823.
8. JD Morgan, DH Napper, GG Warr. J Phys Chem 99:9485–9465, 1995.
9. LS Romsted. In: J Texter, ed. Reactions and Synthesis in Surfactant Systems. New York: Marcel Dekker, 2001, pp 265–294.
10. V Soldi, J Keiper, LS Romsted, IM Cuccovia, H Chaimovich. Langmuir 16: 59–71, 2000.
11. CA Bunton, J Yao, LS Romsted. Curr Opin Colloid Interface Sci 2:622–628, 1997.
12. CA Bunton, LS Romsted. In: P Kumar, KL Mittal, eds. Handbook of Microemulsion Science and Technology. New York: Marcel Dekker, 1999, pp 457–482.
13. CA Bunton, F Nome, FH Quina, LS Romsted. Acc Chem Res 24:357–364, 1991.
14. Note that when two different counterions are present, one reactive and one nonreactive, their distribution is described by an empirical ion exchange constant and β is generally assumed to be the same for both ions.
15. H Zollinger. Diazo Chemistry I. Aromatic and Heteroaromatic Compounds. Weinheim: VCH Publishers, 1994.
16. IM Cuccovia, MA da Silva, HMC Ferraz, JR Pliego, JM Riveros, H Chaimovich. J Chem Soc Perkin Trans 2:1896–1907, 2000.
17. If the rate constants for the competing reactions are equal, Scheme 1, then the preassociation exchange constant equals the selectivity [10,18]. In cationic micelles with up to about 3 M added counterion, the average rate constant is about 25% less than the rate constant in water and both rate constants vary by about 10%. Consequently, S_w^X depends primarily on K_w^X. However, in sodium dodecyl sulfate micelles, the variation in the observed rate constant decreases by about a factor of 2 compared with water and S_w^X (where X is the dodecyl sulfate head group) may depend on K_w^X, k_w, and k_X [16].
18. A Chaudhuri, JA Loughlin, LS Romsted, J Yao. J Am Chem Soc 115:8351–8361, 1993.
19. L Miola, RB Abakerli, MF Ginani, PB Filho, VG Toscano, FH Quina. J Phys Chem 87:4417–4425, 1983.
20. LCM Ferreira, C Zucco, D Zanette, F Nome. J Phys Chem 96:9058–9061, 1992.
21. CA Bunton, JR Moffatt. Langmuir 8:2130–2134, 1992.

22. Z-M He, JA Loughlin, LS Romsted. Bol Soc Chil Quim 35:43–53, 1990.
23. R Ranganathan, LT Okano, C Yihwa, FH Quina. J Colloid Interface Sci 214: 238–242, 1999.
24. Y Marcus. Ion Solvation. Chichester, UK: Wiley, 1985.
25. SA Buckingham, CJ Garvey, GG Warr. J Phys Chem 97:10236–10244, 1993.
26. JA Loughlin, LS Romsted. Colloids Surf 48;123–137, 1990.
27. IM Cuccovia, LS Romsted, H Chaimovich. J Colloid Interface Sci 220:96–102, 1999.
28. MK Jain, J Rogers, B-Z Yu, J Yao, LS Romsted, OG Berg. Biochemistry 36: 14512–14530, 1997.
29. LS Romsted, J Yao. Langmuir 15:326–336, 1999.
30. J Yao, LS Romsted. J Am Chem Soc 116:11779–11786, 1994.
31. PK Das, A Chaudhuri, S Saha, A Samata. Langmuir 15:4765–4772, 1999.
32. BA McKernan, LS Romsted. In: Electrified Polymer/Solution Interfaces. ACS Symposium Series. Washington, DC: American Chemical Society, in press.
33. FM Menger, JS Keiper. Angew Chem Int Ed Engl 39:1906–1920, 2000.
34. IM Cuccovia, A Agostihno-Neto, CMA Wendel, H Chaimovich, LS Romsted. Langmuir 13:5302–5305, 1997.
35. IM Cuccovia, IN da Silva. H Chaimovich, LS Romsted. Langmuir 13:647–652, 1997.
36. O Soderman, KL Herrington, EW Kaler, DD Miller. Langmuir 13:5531–5538, 1997.
37. LS Romsted, J Zhang, L Zhuang. J Am Chem Soc 120:10046–10034, 1998.
38. A Chaudhuri, LS Romsted, J Yao. J Am Chem Soc 115:8362–8367, 1993.
39. R Banerjee, PK Das, A Chaudhuri. Biochim Biophys Acta 1373:299–308, 1998.

8

Effect of Surfactants on Pregastric Enzyme–Catalyzed Hydrolysis of Triacylglycerols and Esters

CHARMIAN J. O'CONNOR and CYNTHIA Q. SUN*
The University of Auckland, Auckland, New Zealand

DOUGLAS T. LAI Development Center for Biotechnology,
Taipei, Taiwan

ABSTRACT

A brief overview is provided of the function of lipases and their structural features and of their mechanism of action and uses. Specifically, details of these properties are given for pregastric lipases. Data are then presented for the reactivity of kid pregastric lipase against 4-nitrophenyl butyrate in the presence of the surfactants Triton X-100, Tween 20, sodium taurocholate, cetyltrimethylammonium bromide, sodium dodecyl sulfate, Aerosol-OT, and lecithin, all rates of hydrolysis being measured at pH 6.5, 25°C, and for the kid pregastric lipase–catalyzed hydrolysis of tributyrin in the presence of lecithin, sodium taurocholate, casein, and Tween 20 at pH 6.5, 35°C. A combination kinetic model has been used to explain the dual characteristics of the kinetic behavior of the lipase. The added surfactant serves both as a solubilizer, increasing the concentration of the soluble form of the substrate, and as an emulsifier, which creates mixed lipid–micelles in solution. The enzyme behaves as both a lipase and an esterase, and the catalyzed hydrolysis of both the free substrate and the aggregates takes place simultaneously, with the rate of reaction being largely dependent on the concentration of added surfactant.

---

*Current affiliation: Hort Research, Auckland, New Zealand.

I. INTRODUCTION

A. Definition and Function of Lipases

Lipases are a ubiquitous class of enzymes found in many organisms, including plants, seeds, fungi, microbes, and all animal life forms. These diverse origins are reflected in the great variety of lipases, in both their properties and specificities. Lipases are carboxylic ester hydrolases and are classified as triacylglycerol acylhydrolases (EC 3.1.1.3). They catalyze the hydrolysis of triacylglycerols to yield, usually sequentially, diacylglycerols, monoacylglycerols, and glycerol with free fatty acids being released at each step in the hydrolysis pathway [Eq. (1)].

$$
\begin{array}{c}
CH_2OCOR_1 \\
| \\
CHOCOR_2 \\
| \\
CH_2OCOR_3
\end{array}
\quad + 3\,H_2O
\xrightarrow{\text{Hydrolysis}}
\quad
\begin{array}{c}
CH_2OH \\
| \\
HO\!-\!C\!-\!H \\
| \\
CH_2OH
\end{array}
\quad + \quad
\begin{array}{c}
R_1COOH \\
R_2COOH \\
R_3COOH
\end{array}
\tag{1}
$$

triacylglycerol   **glycerol**   **fatty acids**

Lipases can catalyze other reactions, including acyl exchange and acyl transfer, and may catalyze the hydrolysis of a range of different carboxylate esters. Therefore, the distinction between lipases and other members of the carboxylic ester hydrolase family (EC 3.1.1) is not always clear. In kinetic terms, it was shown that an esterase enzyme (e.g., horse liver esterase) was active only against monomeric substrates, while the activity of a lipase (e.g., porcine pancreatic lipase) was significantly enhanced when the substrate was increased above its solubility limit [1], although this generalization did not always apply when tested across a range of lipases and esterases [2]. Instead, a more classical definition was suggested, whereby esterases are enzymes that act on water-soluble neutral esters and lipases act on water-insoluble neutral esters [2]. Lipases are, therefore, a special group of esterases that are active predominantly at a lipid–water interface.

1. Stereospecific Numbering of the Lipase Substrate

The glycerol molecule is prochiral and the two primary hydroxyl groups are enantiotropic. The modification of one of these groups (e.g., by esterification) leads to a chiral molecule. To describe the stereochemistry of glycerol (propan-1,2,3-triol) derivatives, the carbon atoms are stereospecifically numbered (sn). When the secondary hydroxyl is shown on the left of the central carbon in a Fisher projection, the carbon atoms are numbered 1, 2, 3 from top to bottom. The glycerol becomes sn-glycerol (i.e., stereospecifically numbered glycerol) as opposed to rac-glycerol. Using this nomenclature, the carbon numbering and absolute configuration of the glycerol spine are conserved regardless of substitution [Eq. (2)].

$$
\begin{array}{ccc}
\text{CH}_2\text{OH} & & \text{CH}_2\text{OCOR} \quad sn\text{-}1 \\
| & \xrightarrow{\text{esterification}} & | \\
\text{HO}-\text{C}-\text{H} & & \text{HO}-\text{C}-\text{H} \quad sn\text{-}2 \\
| & & | \\
\text{CH}_2\text{OH} & & \text{CH}_2\text{OH} \quad sn\text{-}3
\end{array} \tag{2}
$$

prochiral molecule   two enantiomers

B. Structural Features of Lipases

Lipases belong to the superfamily of proteins with a common structural framework of an α/β-hydrolase fold. This fold is made up mostly of a parallel β-pleated sheet, flanked on both sides by α-helices. The nucleophilic active-site serine residue is found at the C-terminal edge of the sheet, in a tight bend, between a strand and a helix. This region can be identified by the consensus sequence, Gly-X$_1$-Ser-X$_2$-Gly (where X$_1$ is Tyr or His and X$_2$ is any amino acid), which is unique to lipases and esterases. Most carboxylate ester hydrolases of known structure contain a Ser-His-Asp/Glu catalytic triad in the active site. In a lipase, this triad is analogous to that of serine proteases in terms of functional groups but is structurally distinct. Geometrically, the serine in a lipase is positioned on the opposite side of the catalytic pocket and, therefore, the tetrahedral center generated by attack of the serine hydroxyl, **1**, is a mirror image of the tetrahedral intermediate generated by a serine protease, **2**.

The investigation of the unique structural features of lipases has improved our understanding of the mechanism of action of this class of enzymes. In the early 1990s, the first lipase three-dimensional structures were elucidated by X-ray crystallography of the lipases: *Mucor miehei*, *Geotrichum candidum*, and the human pancreatic lipase. Studies revealed a characteristic surface loop (i.e., a lid domain) that covered the active site of the enzyme, rendering it inaccessible to solvent. This finding suggests that a conformational change is required to allow substrate entry. Cocrystallization of inhibitors (n-hexylphosphonate ethyl ester [3] and diethyl p-nitrophenyl phosphate [4]) with the M. miehei lipase revealed a structure that was distinctly different from the native form. In the region of a short α-helical segment or surface loop, significant opening of the lid was evident upon activation of the enzyme. Consequently, the hydrophilic side of the lid becomes partially buried and a large surface of hydrophobic residues surrounding the active site is exposed, strengthening the interactions with the lipid substrate. This unusual conformational adjustment of a lipase, required for a substrate to bind, was implicated in the phenomenon of interfacial activation and provided a possible criterion for defining lipases. The open and closed forms of a lipase could provide the structural basis for interfacial activation, which is characterized by the increase in its lipolytic activity at an insoluble surface.

1. Mechanism of Action

Because the lipase active site is similar to that of serine proteases, the hydrolytic mechanism may also be analogous [5]. It is suggested that the catalytic reaction begins with the formation of a noncovalent Michaelis complex between the enzyme and the substrate, which then reacts with the nucleophilic oxygen of the serine to form a covalent tetrahedral transition state. An acyl enzyme intermediate is then formed by cleavage of the substrate ester bond and dissociation of a protonated diacylglyceride. An activated water molecule then attacks the serine ester and forms a second tetrahedral transition state. Collapse of this transition state results in fatty acid release and the regeneration of the enzyme.

C. Uses of Lipases

Lipases are an important class of industrial enzymes and are used as chiral catalysts in a variety of reaction types. In organic synthesis, lipases are one of the most widely utilized and versatile biocatalysts. Lipases maintain activity in a nonaqueous environment, a property resulting in several advantages over other classes of enzymes. The water-insoluble enzymes and products can be readily recovered and the enzymes recycled. The absence of water can decrease the likelihood of side reactions and decrease product or substrate inhibition [6].

Applications of lipases include ester hydrolysis, regioselective acylation or deacylation, interesterification, and resolution of racemic mixtures. Such reactions can be employed in conjunction with a variety of substrate types, ranging from glycerol derivatives to organometallics [6]. Consequently, lipases can be utilized in a range of industries, including the manufacture of pharmaceuticals, perfumes, agrochemicals, polymer chemistry, peptide synthesis, biosurfactant production, and oleochemistry [6].

D. Pregastric Lipases

Pregastric lipases have a fundamental role in the fat metabolism of mammals. In conjunction with pancreatic and gastric enzymes, they are an essential part of the digestive system. Pregastric lipases are, however, distinct in their nature and source. They are secreted from the oral glandular structures (lingual and pharyngeal tissues) [7]. Pregastric lipases in ruminants are important because, in contrast to pancreatic lipases, they are able to penetrate milk fat globules to catalyze the hydrolysis of triacylglycerols.

1. Substrate Selectivity of Pregastric Lipases

Pregastric lipases of ruminants are found to exhibit a general preference for catalyzing the hydrolysis of short-chain fatty acids from substrates ranging from lipids in milk to naphthyl esters and synthetic triacylglycerols. A study of calf pregastric lipases [8] found that they preferred substrates with short- and medium-chain fatty acid (C_4–C_{12}) esters, and exhibited very low activity against triacylglycerols composed of acyl groups of chain length greater than 16 carbons. D'Souza and Oriel [9] undertook a similar study with purified lamb pregastric lipase. Maximal activity was found against the triacylglycerol with a four-carbon fatty acid, and they were unable to detect hydrolytic activity against lipids with fatty acid chains longer than eight carbons.

Kim Ha and Lindsay [10] reported that all the ruminant pregastric lipases had a selectivity for short-chain and branched fatty acids. It was also suggested that these pregastric lipases exhibited selectivity for certain glycerides, and this was apparent from the differences in their ability to release major n-chain fatty acids from various milk fats. De Caro et al. [11] found that purified lamb pregastric lipase released short-chain fatty acids more readily from tributyrin (C_4) than from trioctanoin (C_8).

A study by Villeneuve et al. [12] utilizing chiral and racemic triacylglycerols observed that the lipases of all three ruminants (kid, lamb, and calf) were typoselective for hydrolysis of lipids with short-chain fatty acids. An investigation of the enzyme-catalyzed hydrolysis of tributyrin, triolein, and 4-nitrophenyl acetate (4-NPA) by O'Connor et al. [13] found that both calf and lamb lingual lipases were more active against tributyrin than against the acyl ester 4-NPA (against the longer chain lipid, triolein, the activity was also reduced relative to tributyrin). The preference of ruminant pregastric lipases to catalyze the hydrolysis of short-chain fatty acids is clearly shown by the amount of these fatty acids released during the initial hydrolysis of anhydrous milk fat [14].

2.   Stereoselectivity of Pregastric Lipases

Investigations of the ruminant pregastric lipases have revealed a preference for the sn-1 and sn-3 positions of a triacylglycerol. However, interpretation of these results is made more difficult because a higher distribution of the shorter chain fatty acids is located at the primary acyl positions, which are more accessible to lipolysis. It was, therefore, necessary to clarify whether the stereoselectivity was due to the enzyme or to the fatty acid distribution in milk fat.

Racemic triacylglycerol (C16:O—C4:O—C16:0) was used to determine the selectivity of calf pregastric enzyme for esters of primary alcohols (sn-1 and sn-3), independent of its selectivity for short-chain fatty acids [15]. It was found that the undigested monoacylglycerol recovered from the hydrolysis reaction contained the four-carbon fatty acid. Kim Ha and Lindsay [10] measured the concentration of the volatile free fatty acids released from ruminants' milk fats and found that the kid, calf, and lamb pregastric lipases all preferentially catalyzed the hydrolysis of the sn-1 and sn-3 positions of the glycerides.

Villeneuve et al. [12] employed an optically active triglyceride and its racemic equivalent in combination with different fatty acid chain lengths to test for the dual selectivity of ruminant pregastric lipases. Not only did all the lipases exhibit typoselectivity for short-chain fatty acids, but they were reported to be sn-3 stereoselective. The kid lipase was the most typoselective and the calf the least. The kid lipase also had slightly stronger stereoselectivity with 65% of free fatty acid being released from the sn-3 position, compared with 60% for the lamb lipase and 57% for the calf lipase. The extent of the respective specificity was found to correlate with the composition of and short-chain fatty acid distribution in the mother's milk. Bovine (cow's) milk has the highest percentage of short-chain fatty acids (C_4–C_8), the majority of which are on the sn-3 position, although the calf lipase exhibited the least selectivity. In contrast, caprine (goat's) milk has the lowest proportion of short-chain triacylglycerides with only a small distribution located on the sn-3 position. Relative to the enzymes from calf and lamb, that from the goat had the highest typo- and stereoselectivity.

O'Connor et al. [16] utilized triacylglycerols of varying carboxylic acid chain length and found that lamb pregastric lipase released only one fatty acid per molecule of triacylglycerol, a result suggestive of selectivity. Lamb pregastric lipase stereoselectivity was confirmed by nuclear magnetic resonance (NMR) studies [17]. Products of hydrolysis, **3**, were identified by carbon-13 NMR, enabling the positional preferences to be distinguished from the products of acyl migration **4**.

These results led to the proposal that tributyrin is initially hydrolyzed at the *sn*-3 position. Nonenzymatic, intramolecular acyl migration results in a mixture of 1,2- and 1,3-diacylglycerols, which then undergo further hydrolysis [17].

3. Uses of Pregastric Lipases

In the dairy industry, lipases are used in the hydrolysis of milk fat. Applications include flavor enhancement of cheeses, acceleration of cheese ripening, manufacture of cheeselike products, and lipolysis of butterfat and cream. Sources of lipases for cheese enhancement are the pancreatic glands or pregastric tissues of lamb, calf, or kid. Each pregastric lipase leads to its own characteristic flavor pattern, and these enzymes are essential in the production of quality cheeses such as Romano and provolone [15]. Pregastric lipases have also been used for the treatment of calf diarrhea or scours [15] and have potential for the treatment of malabsorption syndrome in children.

E. Lipase Inhibitors

Inhibitors act either reversibly or irreversibly. Reversible inhibitors include various modes of action: competitive, noncompetitive, and uncompetitive [18]. A competitive inhibitor competes with the substrate for the active site of the enzyme and forms a nonproductive enzyme–inhibitor complex. A noncompetitive inhibitor binds to a site distinct from that which binds the substrate and, by changing the surface properties or conformation, inactivates the enzyme [19]. Alternatively, an uncompetitive inhibitor binds at a site distinct from the substrate but binds exclusively to an enzyme–substrate complex rather than to a free enzyme. In comparison, irreversible inhibitors are more specific, in that they must combine with or destroy a functional group on the enzyme that is essential for its activity.

Because of the similarity of the catalytic triad of serine proteases and lipases, the same protease inhibitors have been utilized in studies of lipases. Transition state analogues such as butylboronic acid and phenylboronic acid can be used [19]. However, the inhibition is short-lived and, therefore, a study of the transition states is difficult. Irreversible lipase inhibitors may be targeted to different residues of the catalytic pocket and may contain a range of different functional groups.

X-ray crystallography and kinetic studies, employing lipase inhibitors, have provided valuable information regarding the molecular basis of lipase activity. The structural characterization of a lipase–inhibitor complex and its comparison with the native enzyme have confirmed the identity of the active site residues and the organization of the catalytic pocket and have provided an understanding of the mechanism. This knowledge is fundamental in the design of modified enzymes by protein engineering. Mutation of selected amino acids permits the creation of a wide range of lipases with different selectivities and improved performance (e.g., stability toward temperature, pH ranges, etc.) [6].

II. MATERIALS AND METHODS

A. Chemicals

The purification of kid pregastric lipase was described earlier [20]. The surfactants cetyltrimethylammonium bromide (CTAB), sodium dodecyl sulfate (SDS), bis(2-ethylhexyl) sodium sulfosuccinate (Aerosol-OT) (AOT), soybean lecithin, Tween 20, sodium taurocholate (NaTC), and Triton X-100 (TX-100) were purchased from Sigma. Stock solutions of surfactants were prepared in 50 mM Bis-Tris buffer, pH 6.5, 25°C. The substrates 4-nitrophenyl butyrate (4-NPB) and tributyrin were also purchased from Sigma.

B. Pregastric Lipase–Catalyzed Hydrolysis of 4-NPB

Aliquots of 4-NPB (30 μL of different concentrations dissolved in acetonitrile) were added to 2.97 mL of surfactant solution, mixed well by sonication, and at zero time 30 μL of enzyme solution was added. Initial rates of hydrolysis were determined spectrophotometrically for 3 min at 25°C.

C. Pregastric Lipase–Catalyzed Hydrolysis of Tributyrin

Various amounts of tributyrin were added to 40 mL of surfactant solution, and the mixture was sonicated until it became monodisperse. The pH of the emulsion thus formed was adjusted and then allowed to equilibrate to 35°C, and at zero time the enzyme solution (variable volume) was added. The release of butyric acid was monitored titrimetrically for 6 min.

III. RESULTS AND DISCUSSION

A. Micellar Catalysis

In a system containing surfactant, S, substrate, X, and enzyme, E, the amount of substrate available for an enzyme-catalyzed reaction will be reduced if the substrate was hydrophobic and sequestered within the aggregates of surfactant molecules. This reduction will also depend on the nature of the surfactant. Moreover, the ability of the enzyme to perform at its best may be reduced if it is adsorbed onto the micellar surface. These effects will become more pronounced if the concentration of the surfactant is greater than the critical micelle concentration (cmc).

At concentrations of the surfactant below the cmc, the micellar concentration will be equal to zero and the surfactant will remain in solution as the free monomer, which is capable of forming a substrate–surfactant complex. If the substrate itself is rather hydrophobic, it will form substrate aggregates, X_n, in an aqueous solution.

If the added surfactant does not inhibit the enzyme and the enzyme does not catalyze the hydrolysis of the substrate aggregates, then reaction Scheme 1 will apply.

The total rate of hydrolysis will consist of the sum of the rates of hydrolysis of the substrate monomers, v_x, and of the substrate–surfactant complex, v_{xs}. As the concentration of substrate, [X], increases, v_{xs} increases and v_x decreases. The net effect on the rate of hydrolysis is an increase in the effective substrate concentration as [X] increases up to the cmc, followed by a decrease in the effective concentration as [X] increases further beyond the cmc. Consequently, the catalyzed rate should decrease in a micellar system with a high surfactant-to-substrate ratio.

B. Pregastric Lipase–Catalyzed Hydrolysis of 4-NPB in the Presence of Surfactants

We have previously shown that lamb pregastric enzymes are active only against monomeric 4-nitrophenyldecanoate and that the rate of catalyzed hydrolysis remained constant when the substrate concentration was increased beyond its cmc [21]. Because pregastric lipases have a preference for catalyzing the hydrolysis of a monomeric substrate rather than its aggregated form, the monomeric pathway, v_x, should be favored if the concentration of the monomer is increased by the addition of surfactant to the system.

1. Nonionic Surfactants

In order to restrict the interaction of the surfactant only to the substrate (or to the enzyme itself), we chose 1 mM 4-NPB, i.e., at a concentration below

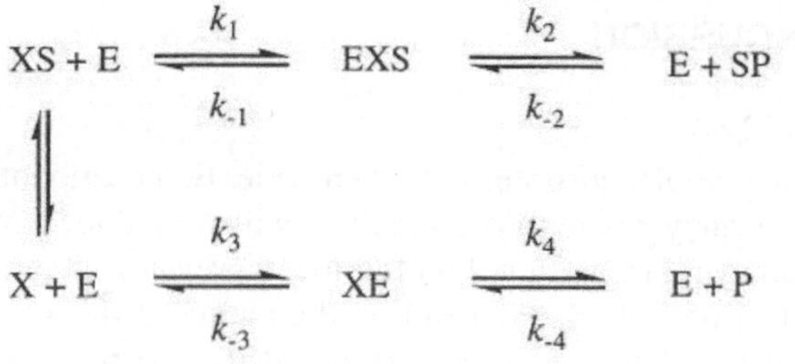

$$XS + E \underset{k_{-1}}{\overset{k_1}{\rightleftharpoons}} EXS \underset{k_{-2}}{\overset{k_2}{\rightleftharpoons}} E + SP$$

$$X + E \underset{k_{-3}}{\overset{k_3}{\rightleftharpoons}} XE \underset{k_{-4}}{\overset{k_4}{\rightleftharpoons}} E + P$$

SCHEME 1 Pathways for enzyme-catalyzed hydrolysis of substrate or substrate–surfactant complex: E, enzyme; X, substrate; S, surfactant; P, product.

its cmc, as the first substrate for our studies. Figure 1 shows the effect of the two nonionic surfactants, Triton X-100 and Tween 20, on activity, i.e., the rate of hydrolysis by the kid pregastric lipase.

Both surfactants first enhanced the activity, up to 100% at 0.7 mM Triton X-100 and 35% at 0.049 mM Tween 20, both of these concentrations being close to the respective cmc values. Beyond these concentrations, the activation effect decreased, with more than 80% of the maximal activity being lost in the presence of 4 mM Triton X-100 and even 10% inhibition being seen in the presence of 5 mM Tween 20. This loss in activity is probably due to the encapsulation of the substrate within the micelles of the added surfactant, which ultimately becomes unavailable to the enzyme for a catalyzed reaction.

2. Anionic and Cationic Surfactants

A similar effect was seen when NaTC and CTAB were used as surfactants (Fig. 2). The activity was enhanced by 55% in 5.6 mM NaTC and by 80% in 0.03 mM CTAB. Although only 10% of the maximal activity was lost in 25 mM NaTC, the activity quickly decreased as the concentration of CTAB increased until over 90% of the activity was lost at 0.5 mM CTAB.

Other ionic surfactants, particularly SDS and AOT, show evidence of much stronger binding to the lipase. At a concentration less than 25% of their cmc values, there was a very large inhibitory effect of these surfactants on the activity of kid pregastric lipase (Fig. 3). Complete inactivation was observed at 1.0 mM SDS and 0.7 mM AOT, values that are approximately 12% and 28% of their cmc values.

The final surfactant tested when using a fully solubilized form of the substrate was the phospholipid, lecithin. No inhibition was seen; rather, the catalysis initially increased threefold compared with the system without surfactant and then remained almost unchanged (Fig. 3).

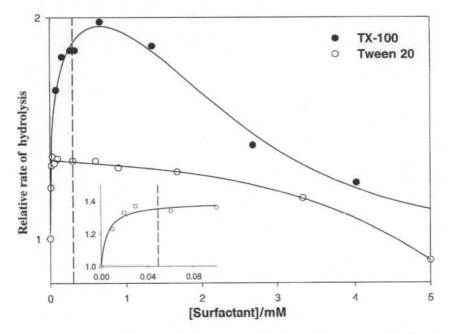

FIG. 1 Relative rates of hydrolysis of 4-NPB catalyzed by kid pregastric lipase at pH 6.5, 25°C in the presence of TX-100 and Tween 20. The cmc value for each surfactant is shown as a dashed line. Inset: Data for very small concentrations of Tween 20. (Adapted from Ref. 22.)

In all of these cases, it was not possible to observe an interfacial effect because the reaction medium was monophasic. The activation effects seen may be caused by a combination of activation of the enzyme itself by surfactant binding and interaction between the surfactant and the substrate to form a substrate complex that is more favored by the enzyme. The inhibition generally seen at high concentrations of surfactants suggests that the surfactants are unlikely to stimulate the enzyme itself to adopt a more highly active state.

C. Pregastric Lipase–Catalyzed Hydrolysis of Tributyrin in the Presence of Surfactants

In the absence of a surfactant, neither kid nor goat pregastric lipase catalyzed the hydrolysis of tributyrin. At concentrations of surfactants below their cmc values, the rate of hydrolysis increased with increasing surfactant concentration, but the data were scattered and nonreproducible in replicate experiments. Therefore, the remaining studies were carried out at concentrations

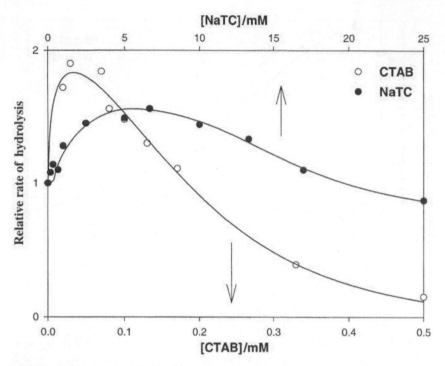

FIG. 2 Relative rates of hydrolysis of 4-NPB catalyzed by kid pregastric lipase at pH 6.5, 25°C in the presence of NaTC and CTAB. (Adapted from Ref. 22.)

of the surfactants lecithin, NaTC, casein, and Tween 20 above their cmc values. The results are given in Fig. 4, which shows values of apparent maximum velocity, V_{max}, that remain almost invariant with increasing concentration of NaTC, casein, and Tween 20 but increase slightly for increasing concentrations of lecithin.

Let us now assume that the pregastric enzyme has characteristics of both an esterase and a lipase. When it acts as an esterase, the catalyzed hydrolysis of tributyrin will depend largely on the concentration of the tributyrin–surfactant complex, XS, which, in turn, is dependent upon [S] below the cmc but remains constant as the micelle starts to form. The actual effective concentration of the substrate will be equal to that of the tributyrin–surfactant complex, [XS], which will, in turn, be dependent on the type of surfactant used.

When [S] is high, the concentration of [XS] will be close to zero because the substrate will be partitioned between the free monomers and the micelles and, at a given substrate concentration, the concentration of the substrate

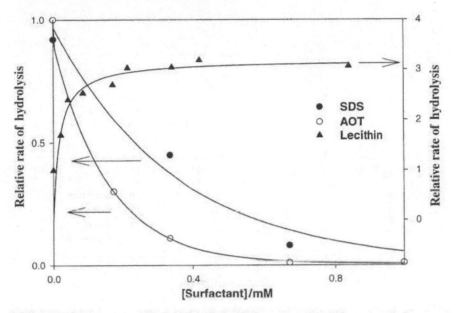

FIG. 3 Relative rates of hydrolysis of 4-NPB catalyzed by kid pregastric lipase at pH 6.5, 25°C in the presence of SDS, AOT, and lecithin. (Adapted from Ref. 22.)

actually solubilized will decrease as the surfactant concentration is increased. The enzyme will now behave more as a lipase, which has little or no activity against the soluble form of the substrate but whose activity is highly dependent upon surface area. Thus, the substrate concentration at the oil–water interface will dominate the rate of the catalyzed hydrolysis and will be influenced by the total surface area or micellar size. It is known, for example, that the average size of the casein–oil micelle decreases as the concentration increases and the micellar size remains constant at concentrations of casein greater than 1% (w/v) [23]. At low concentrations of surfactant, the reduction in the size of the micelle causes a significant increase in surface area and a corresponding increase in the rate of hydrolysis, as seen in Fig. 4.

If the enzyme reacts with the substrate only in the micellar phase, then the true value of V_{max} may be obtained from a double reciprocal plot of initial rate against surfactant concentration (Fig. 5). However, this relationship breaks down when the surfactant concentration approaches the cmc and the contribution from the esterase component becomes more dominant.

Figure 6 illustrates the interrelationship existing in the pregastric lipase–catalyzed hydrolysis of a lipid substrate. In the absence of an efficient catalyst, there is no activity. At low surfactant concentrations the rate of hy-

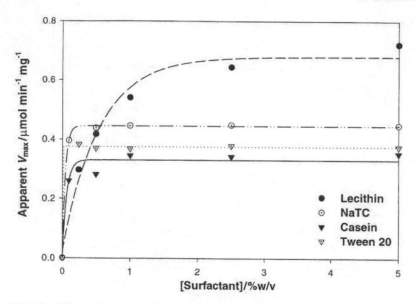

FIG. 4 Effect of surfactant concentration on the apparent V_{max} for kid pregastric lipase–catalyzed hydrolysis of tributyrin at pH 6.5, 35°C. (Adapted from Ref. 22.)

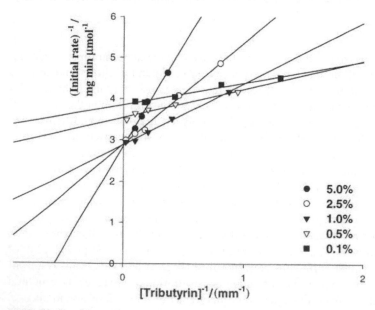

FIG. 5 Double reciprocal plots of initial rate of hydrolysis of tributyrin versus concentration of tributyrin in the presence of different concentrations of Tween 20. (Adapted from Ref. 22.)

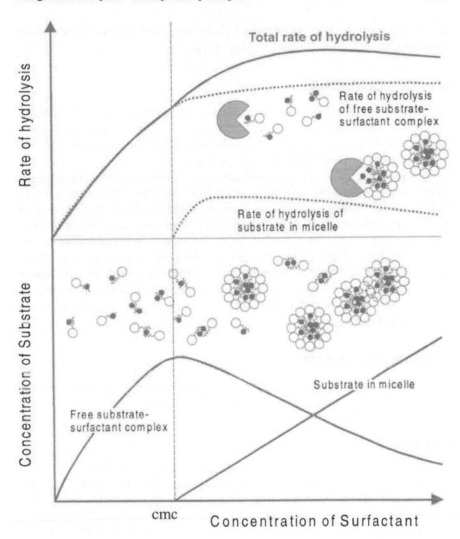

FIG. 6 Relationship between substrate distribution and substrate hydrolysis catalyzed by pregastric lipase and the concentration of added surfactant.

drolysis depends on the soluble form of the lipid–surfactant complex, XS, with little contribution from the aggregated substrate, X_n. As the concentration of S increases, more hydrolysis occurs on the surface of the micellized surfactant and the contribution from the homogeneous catalyzed pathway slowly disappears. The net effect of these two mechanisms bears a close similarity to a typical Michaelis–Menten plot.

IV. CONCLUSIONS

In conclusion, the behavior of the pregastric lipase reflects its dual functions, those of a lipase and an esterase, and which function is dominant is more likely to depend upon the substrate under attack and the environment of the reaction medium than upon the enzyme itself.

The presence of an efficient surfactant is essential for the pregastric lipase–catalyzed hydrolysis reaction of a lipid substrate. When the concentration of the substrate is greater than its solubility limit, the addition of surfactant to the solution, at a concentration either close to or above its cmc, will give rise to the formation of larger aggregates between substrate and surfactant. Under these conditions, the enzyme is behaving as a lipase. However, when the substrate is partially water soluble, the enzyme will behave as an esterase, and the purpose of the surfactant, even at low concentrations, is as a solubilizer that dissolves and increases the concentrations of the solubilized form.

ACKNOWLEDGMENT

We acknowledge financial assistance from the University of Auckland Research Committee and New Zealand Lottery Science.

REFERENCES

1. L Sarda, P Desneulle. Biochim Biophys Acta 30:513–516, 1958.
2. H Okuda. In: SA Kuby, ed. A Study of Enzymes. Vol 2. Boca Raton, FL: CRC Press, 1991, pp 580–593.
3. AM Brzozowski, U Derewenda, ZS Derewenda, GG Dodson, DM Lawson, JP Turkenburg, F Bjorkling, B Huge-Jensen, SA Patkar, L Thim. Nature 351:491–494, 1991.
4. U Derewenda, AM Brzozowski, DM Lawson, ZS Derewenda. Biochemistry 31:1532–1541, 1992.
5. FK Winkler, K Gubernator. In: SB Petersen, P Woolley, eds. Lipases: Their Structure, Biochemistry and Application. Vol 4. Cambridge: Cambridge University Press, 1994, pp 139–157.
6. G Gillis. J Am Chem Soc 65:846–852, 1988.
7. HA Ramsey, GH Wise, SB Tove. J Dairy Sci 39:1312–1321, 1936.
8. RK Hamilton, AM Raven. J Sci Food Agric 24:257–269, 1973.
9. TM D'Souza, P Oriel. Appl Biochem Biotechnol 36:183–198, 1992.
10. J Kim Ha, RC Lindsay. J Dairy Sci 76:677–690, 1993.

11. J De Caro, F Ferrato, R Verger, A De Caro. Biochim Biophys Acta 1252:321–329, 1995.
12. P Villeneuve, M Pina, J Graille. Chem Phys Lipids 83:161–168, 1996.
13. CJ O'Connor, RD Manuel, KW Turner. J Dairy Sci 76:3674–3682, 1993
14. CJ O'Connor, RH Barton, PAG Butler, AD MacKenzie, RD Manuel, DT Lai. Colloids Surf B 7:189–205, 1996.
15. JH Nelson, RG Jensen, RE Pitas. J Dairy Sci 60:327–362, 1977.
16. CJ O'Connor, AD MacKenzie, RH Barton, PAG Butler. J Mol Catal A 96:77–85, 1995.
17. RH Barton, CJ O'Connor. J Am Oil Chem Soc 75:967–976, 1998.
18. RA Copeland. Enzymes: A Practical Introduction to Structure, Mechanism and Data Analysis. New York: Wiley-VCH, 1996, pp 187–198.
19. S Patkar, F Bjorkling. In: SB Petersen, P Woolley, eds. Lipases: Their Structure, Biochemistry and Application. Vol 4. Cambridge: Cambridge University Press, 1994, pp 207–224.
20. DT Lai, RA Stanley, CJ O'Connor. J Am Oil Chem Soc 75:411–416, 1996.
21. CJ O'Connor, DT Lai, RH Barton. J Bioact Compat Polym 11:143–149, 1996.
22. DT Lai, CJ O'Connor. Langmuir 16:115–121, 2000.
23. Y Gargouri, A Bensalah, I Douchet, R Verger. Biochem Biophys Acta 1257:223–229, 1995.

9
Effect of Benzyl Alcohol on the Properties of CTAB/KBr Micellar Systems

GANZUO LI, WEICAN ZHANG, LI-QIANG ZHENG, and QIANG SHEN Shandong University, Jinan, People's Republic of China

ABSTRACT

The effect of benzyl alcohol on the microstructure of the micelle in the cetyltrimethylammonium bromide (CTAB)/KRr aqueous system and its rheological properties have been investigated by viscosity measurements, ^{1}H nuclear magnetic resonance (NMR), and laser light scattering (LLS) experiments. In the dilute surfactant solution, the addition of alcohol induces the viscosity of the 0.01 mol L^{-1} CTAB/KBr micellar system to go through a marked maximum. In 0.08 mol L^{-1} CTAB/0.8 mol L^{-1} KBr micellar solution, a small amount of alcohol can induce this system to have viscoelasticity. If more alcohol is added, the viscoelasticity of this micellar system decreases, In the dilute surfactant solution, the results of ^{1}H NMR and LLS experiments show that alcohol is located in the interfacial region of CTAB/KBr micelles and promotes these micelles to be larger at lower concentrations of alcohol. This process makes the viscosity of these systems rise. With continuously increasing alcohol content, it is solubilized in the palisades of the micelles and induces these rodlike micelles to be transformed into an oblate spheroid shape.

I. INTRODUCTION

There is now very strong evidence for a micellar phase in several aqueous cationic surfactant systems in the presence of added salt, such as cetyltrimethylammonium bromide or chloride (CTAB or CTAC) with added KBr or sodium salicylate (NaSal), cetylpyridinium bromide (CPyBr) + KBr, cetylpyridinium salicylate (CPySal) + NaSal, etc. The viscosity of these systems can be controlled between 0.001 and 100 Pa s at low values of surfactant concentration. In addition, such micellar systems are highly elastic. The last few years have seen growing interest in the structure and size of the micelles in salt solution and their rheological properties [1–14]. The numbers of papers and conferences dealing with this topic have considerably increased; however, there has been relatively little work concerning the effect of additives on the micellar properties of these systems [15]. In this chapter we will show the influences of benzyl alcohol on the shape and size of the micelles in the CTAB/KBr solutions and their rheological properties.

II. EXPERIMENTAL SECTION

All the reagents used in this study were analytical grade and water was doubly distilled.

A commercial laser light scattering spectrometer (ALV/SP-150, ALV Company, Germany) equipped with a solid-state laser (ADLAS DPY425 II, output power $\approx$ 400 mW at $\lambda = 532$ nm) as the light source and an ALV-5000 multi-τ digital correlator was used. The samples were filtered through a 0.5-μm Millipore filter into the cylindrical light scattering cell [16,17].

The specific refractive index increment (dn/dc) used in static light scattering was determined with a novel and precise differential refractometer [16,17].

The ^{1}H NMR spectra were recorded using a Varian Unity Inova-300 instrument. The ^{1}H chemical shifts are reported in δ units (ppm) relative to tetramethylsilane (TMS) at the external standard ($\delta = 0.00$).

The viscosity and viscoelasticity experiments were carried out on a HAAKE RS-75 rheometer with a cone and plate sensor system.

All measurements were conducted at a temperature of 30 ± 0.5°C.

III. RESULTS AND DISCUSSION

A. Effect of Benzyl Alcohol on the Viscosity of 0.01 mol L^{-1} CTAB Micellar Solution with Various Concentrations of KBr Salt

Figure 1 shows the influence of benzyl alcohol on the viscosity of 0.01 mol L^{-1} CTAB/KBr solution. Below 0.05 mol L^{-1} KBr salt, benzyl alcohol does not have a noticeable effect on the viscosity of this system, but above this salt content, the viscosity of this micellar system goes through a marked maximum upon addition of benzyl alcohol. The higher the KBr concentration, the higher the value of the maximum in viscosity and the lower the

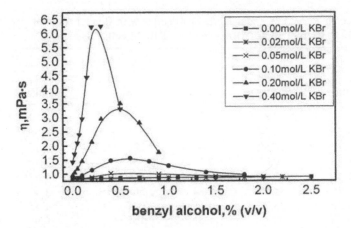

FIG. 1 Effect of benzyl alcohol concentration on the viscosity of the 0.01 mol L^{-1} CTAB/KBr micellar system.

alcohol concentration at which the viscosity of the system is highest. In Fig. 1, when the KBr concentration is 0.4 mol L^{-1}, the highest viscosity is obtained at 0.4% (v/v) benzyl alcohol.

B. Effect of Benzyl Alcohol Concentration on R_g, R_h, and M_w of Mixed Alcohol and CTAB Micelles in the Presence of KBr Salt

In the 0.01 mol L^{-1} CTAB/0.1 mol L^{-1} KBr micellar system, the micelles are rodlike [18]. The effect of benzyl alcohol concentration on the radius of gyration R_g, the hydrodynamic radius R_h, and the weight-averaged molar mass M_w of the micelles in this system is given in Table 1. In order to compare the LLS data with viscosity data, the latter are also listed in Table 1. It is seen that the R_g, R_h, and M_w of micelles go through marked maxima upon addition of benzyl alcohol in the CTAB/KBr system and the largest size of the micelles appears at the highest viscosity.

The value of R_g/R_h should be 0.775 for spheres and 2 for rigid rods according to theoretical prediction [2]. From Table 1, it is seen that before a benzyl alcohol concentration of 0.6% (i.e., the highest viscosity), the value of R_g/R_h increases up to 4 and M_w up to 4.52×10^6 with increasing benzyl alcohol content. If more benzyl alcohol is added, the ratio R_g/R_h decreases to 1.5 and M_w to 1.15×10^6, at which the benzyl alcohol content is 1.8%. These data suggest that the rodlike micelles grow gradually larger and longer with increasing amount of alcohol before the highest viscosity is attained. After that, the size of the micelles decreases and the long rodlike micelles

TABLE 1 Variance of R_g, R_h, M_w, and R_g/R_h of Micelles with Benzyl Alcohol Concentration in the 0.01 mol L^{-1}CTAB/0.1 mol L^{-1}KBr Micellar System

Benzyl alcohol, %(v/v)	R_g (nm)	R_h (nm)	R_g/R_h	M_w	η (mPa s)
0	—	7.8	—	—	0.82
0.1	49.76	19.34	2.57	7.31E + 5	0.96
0.2	65.72	24.59	2.67	1.22E + 6	1.15
0.4	82.28	30.07	2.73	1.55E + 6	1.48
0.6	134.4	33.43	4.02	4.52E + 6	1.57
0.8	57.76	31.87	1.81	1.56E + 6	1.45
1.2	45.22	28.77	1.57	1.03E + 6	1.15
1.8	30.88	20.15	1.53	6.15E + 5	0.96

change gradually to an oblate ellipsoid shape. Obviously, the change in the shape of the micelles causes the decrease in viscosity.

In order to elucidate the mechanism of the effect of alcohol on the size and shape of CTAB rodlike micelles in KBr solution, we measured the location of the benzyl alcohol in the CTAB/KBr micellar system using the NMR method.

C. Variation in the Solubilization Location of Benzyl Alcohol with Benzyl Alcohol Concentration in CTAB/KBr Micellar System

Generally, when an aromatic molecule is intercalated between the head groups or the long-chain methylenes of the surfactant in the micelles, the chemical shifts for some of the protons of the surfactant monomers are upfield because of the effect of the current of the aromatic ring [19]. We will utilize these changes to determine the location of benzyl alcohol in CTAB micelles and explain the mechanism of the influence of alcohol on the viscosity of the CTAB/KBr micellar system.

In the 0.01 mol L^{-1}CTAB/0.1 mol L^{-1} KBr system, the dependence of the chemical shifts and the shape of the signals of —CH$_3$, —(CH$_2$)$_{13}$—, and N—(CH$_3$)$_3$ groups of the CTAB molecule on benzyl alcohol concentration is shown in Fig. 2. In the ^{1}H NMR spectra, the resonance signals 1, 2, and 3 represent —CH$_3$, —(CH$_2$)$_{13}$—, and N—(CH$_3$)$_3$ groups, respectively, of the CTAB molecule. It can be seen from Fig. 2 that the peak due to the long-chain methylenes of the CTAB molecule shows marked line broadening when the benzyl alcohol concentration is more than 0.6%, at which the

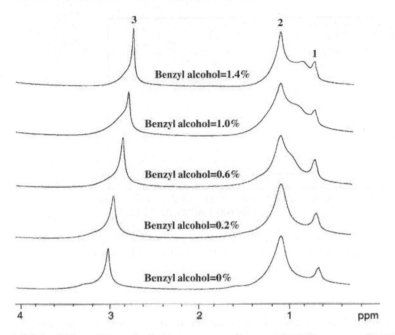

FIG. 2 Effect of benzyl alcohol concentration on ^{1}H NMR spectra of CTAB in 0.01 mol L^{-1}CTAB/0.1 mol L^{-1} KBr solution. 1, —CH$_3$; 2, —(CH$_2$)$_{13}$—; 3, N—(CH$_3$)$_3$.

viscosity of the CTAB micellar system is highest and the size of the micelles is largest.

The variations in the ^{1}H chemical shift ($\Delta\delta = \delta_0 - \delta$, where δ_0 is the ^{1}H chemical shift of CTAB molecules in 0.01 mol L^{-1}CTAB/0.1 mol L^{-1}KBr without benzyl alcohol and δ is the ^{1}H chemical shift of the same system with benzyl alcohol) directly indicate the effect of benzyl alcohol concentration on that of CTAB molecules. In the presence of 0.1 mol L^{-1} KBr salt, the variations in ^{1}H chemical shifts of the resolved signals of the CTAB molecule with benzyl alcohol concentration are shown in Fig. 3. A value $\Delta\delta > 0$ means an upfield shift of the ^{1}H chemical shift for the CTAB molecule.

The results in Fig. 3 show that the chemical shift of protons of N—(CH$_3$)$_3$ go upfield rapidly with the addition of benzyl alcohol, whereas the change in $\Delta\delta$ for —(CH$_2$)$_{13}$— and —CH$_3$ is very little.

The preceding results suggest that the aromatic ring of the benzyl alcohol molecules is located among N—(CH$_3$)$_3$ groups and its —OH groups are directed toward the bulk solution until a benzyl alcohol content 0.6%; below this concentration the electric field of the ring current of benzyl alcohol

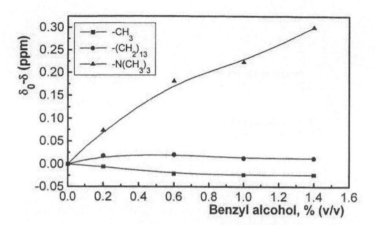

FIG. 3 Variations of chemical shifts of protons of CTAB in micelles with benzyl alcohol concentration in 0.01 mol L^{-1}CTAB/0.1 mol L^{-1} KBr solution.

induces upfield shifts in the ^{1}H chemical shift of N—(CH$_3$)$_3$ and does not have an apparent influence on the long-chain methylenes of CTAB molecules. If more benzyl alcohol is added, it is solubilized in the palissdes where the aromatic ring of the benzyl alcohol is located among long-chain methylenes and near the polar group of the CTAB molecule. Because of the effect of the ring current of benzyl alcohol, there is an upfield shift for the signals of protons of the long-chain methylenes near the aromatic ring and the ^{1}H NMR band of the long-chain methylenes starts to broaden and split. At this point, the rodlike micelles change to oblate in shape. In this case, the polar interfacial region can still solubilize alcohol molecules. The ^{1}H chemical shifts of N—(CH$_3$)$_3$— tend to be upfield continuously.

Israelachvili et al. [20] have considered the geometrical limitations that place restrictions on the allowed shape of a micelle. They gave a critical condition for the formation of rod and sphere micelles. The geometrical constant, f, is given as follows:

Rod micelles: $f = V/a_0 l_c = 1/3 - 1/2$ (1)

Sphere micelles: $f = V/a_0 l_c = 0 - 1/3$ (2)

where l_c is roughly equal to, but less than, the fully extended length of the hydrocarbon chain of the surfactant; a_0 denotes the optimal surface area per surfactant molecule, i.e., the area at which the free energy per surfactant molecule in a micelle is minimum; and V donates the hydrocarbon core volume per surfactant molecule in the micelle.

At the same time, these researchers believed that the oblate spheroid micelles formed by ionic and zwitterionic amphiphiles were unacceptable. As their peripheral regions have very high curvature while the central regions are too thick, the oblate spheroid micelle is energetically unfavorable due to electrostatic repulsion between the polar groups of surfactants. In the CTAB/KBr micellar system, the addition of neutral salt reduces the electrostatic interaction between the head groups in the micelle.

In the absence of salt, a medium-chain-length alcohol is solubilized in the palisades of micelles and brings about a decrease in the size of micelles [21]. In contrast, the benzyl alcohol is solubilized in the interfacial region of the CTAB rodlike micelles in KBr solution; the micelles become larger and longer as the structure is hardly disturbed by the alcohol. When the alcohol is in the palisades, a_0 will increase. In this process, V and l_c are almost constant. When $V/a_0 l_c \leq 1/2$, the rod micelles will undergo a transition to oblate spheroid. At this point, the size of the mixed alcohol and surfactant micelles will decrease. As a result of this transition, the viscosity of this system will decrease.

The results in Fig. 1 can be attributed to two reasons. First, the higher the KBr concentration, the smaller the amount of alcohol dissolved in the bulk solution. Second, the head groups of CTAB molecules in the aggregates approach each other and the total interfacial area of the micelles decreases with increasing KBr concentration so that the capacity for solubilizing the alcohol in the interfacial region decreases. Consequently, when the concentration of benzyl alcohol is low, it can be solubilized in the palisades of micelles and induce the change in the structure of the micelles.

D. Effect of Benzyl Alcohol on the Viscoelasticity of the 0.08 mol L^{-1} CTAB/0.8 mol L^{-1}KBr Micellar System

The preceding results show that a small amount of alcohol solubilized in the interfacial region of the aggregates induces rod micelles to be larger and longer. As a result, the viscosity of the system will increase. When the benzyl alcohol content becomes higher, it will be solubilized in the palisades of the micelles, the rod micelles gradually change into oblate spheroid ones, and the viscosity of the micellar system decreases. Generally, surfactants such as the classical soap do not have viscoelastic properties. In the present study, when the alcohol is solubilized in the interfacial region of CTAB micelles, the interfacial properties change. Thus, the viscoelastic properties of the CTAB/KBr system may also be modified. For this reason, we measured the effect of benzyl alcohol on the viscoelasticity of the 0.08 mol L^{-1}CTAB/0.8 mol L^{-1}KBr micellar system.

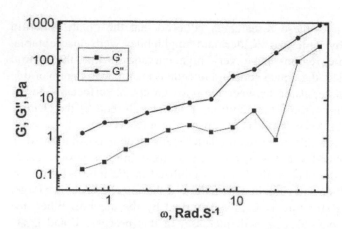

FIG. 4 Magnitudes of the storage modulus G' and the loss modulus G'' as a function of the angular frequency ω for the 0.08 mol L^{-1}CTAB/0.8 mol L^{-1} KBr micellar system.

Figures 4, 5, and 6 show the dependence of the storage modulus G' and loss modulus G'' on angular frequency (ω) in 0.08 mol L^{-1} CTAB/0.8 mol L^{-1} KBr, 0.08 mol L^{-1} CTAB/0.8 mol L^{-1} KBr/0.2% benzyl alcohol, and 0.08 mol L^{-1} CTAB/0.8 mol L^{-1} KBr/0.6% benzyl alcohol micellar systems, respectively.

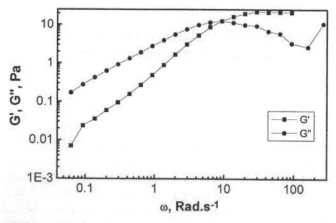

FIG. 5 Magnitudes of the storage modulus G' and the loss modulus G'' as a function of the angular frequency ω for the 0.08 mol L^{-1}CTAB/0.8 mol L^{-1} KBr/0.2% benzyl alcohol micellar system.

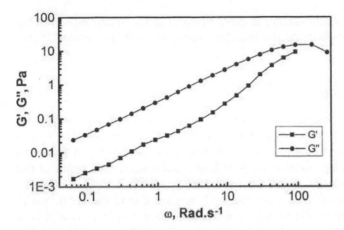

FIG. 6 Magnitudes of the storage modulus G' and the loss modulus G'' as a function of the angular frequency ω for the 0.08 mol L^{-1}CTAB/0.8 mol L^{-1} KBr/0.6% benzyl alcohol micellar system.

From Fig. 4, it can be seen that the rheological properties of 0.08 mol L^{-1}CTAB/0.8 mol L^{-1} KBr micellar system are very complicated.

For the 0.08 mol L^{-1}CTAB/0.8 mol L^{-1}KBr/0.2% benzyl alcohol micellar system, the rheological behavior can be described by a single relaxation time τ (11 s) and a single shear modulus G_0 (25 Pa) according to the Maxwell model at low frequency. Similar results have been obtained in 0.08 mol L^{-1}CTAB/0.8 mol L^{-1}KBr/0.1% benzyl alcohol and 0.08 mol L^{-1}CTAB/0.8 mol L^{-1}KBr/0.3% benzyl alcohol micellar systems. The G_0 of both systems is 25 Pa and τ values are 7 and 5 s, respectively. Obviously, the addition of a small amount of benzyl alcohol affects the relaxation process of these mixed micellar systems. The alcohol located in the interfacial region of the micelles enhances the attractive hydrophobic effects among the micellar surfaces and promotes the formation of a three-dimensional network of rodlike micelles similar to a polymer solution. At high frequency, the loss modulus passes through a minimum.

It is noted from Fig. 6 that if a large amount of benzyl alcohol is added to the CTAB/KBr micellar system, the viscoelastic properties disappear altogether. In this case, the network of rodlike micelles will be destroyed. Yiv et al. [22] showed that the micellar structure became very labile in the presence of a high alcohol concentration.

In conclusion, the alcohol plays an important role in the rheological properties of concentrated CTAB/KBr micellar systems. If it is added in small amounts, it induces this system to have viscoelasticity. When more alcohol is added, it destroys the rodlike micelles and breaks off the network structure.

IV. CONCLUSION

The results obtained in this study clearly show the complexity of the size and shape of CTAB micelles in KBr solution with the addition of benzyl alcohol. A small amount of alcohol solubilized in the interfacial region of the aggregates renders rodlike micelles larger and longer. As a result of this process, the viscosity of the dilute surfactant systems will rise and the viscoelasticity of concentrated solutions will increase due to the formation of a network structure of the micelles. When the alcohol content is higher, it will be solubilized in the palisades of the micelles and the rodlike micelles transform gradually into smaller oblate spheroid ones. Both the viscosity of dilute surfactant systems and the viscoelasticity of concentrated micellar systems will decrease.

ACKNOWLEDGMENT

The financial support of the National Natural Science Foundation of China (29973023) and Key Lab of Oil & Gas Reservoir Geology and Exploitation, Southwest Petroleum Institute is gratefully acknowledged.

REFERENCES

1. G Porte, J Appell, Y Poggi. J Phys Chem 84:3105–3110, 1980.
2. J Appell, G Porte. J Colloid Interface Sci 81:85–90, 1981.
3. J Appell, Y Poggi. J Colloid Interface Sci 87:492–499, 4981.
4. T Imae, R Kamiya, S Ikeda. J Colloid Interface Sci 108:215–225, 1985.
5. T Imae, S Ikeda. Colloid Polym Sci 265:1090–1098, 1987.
6. H Rehage, H Hoffmann. J Phys Chem 92:4712–4718, 1988.
7. ME Cates. J Phys Chem 94:371–375, 1990.
8. ME Cates, SJ Candau. J Phys Condens Matter 2:6869–6892, 1990.
9. PK Vinson, JR Bellare, HT Davis, WG Miller, LE Scriven. J Colloid Interface Sci 142:74–91, 1991.
10. F Kern, R Zana, SJ Candau. Langmuir 7:1344–1351, 1991.
11. F Kern, P Lemarechal, SJ Candau, ME Cates. Langmuir 8:437–440, 1992.
12. A Khatory, F Kern, F Lequeux, J Appell, G Porte, N Morie, A Ott, W Urbach. Langmuir 9:933–939, 1993.
13. A Khatory, F Lequexu, F Kern, SJ Candau. Langmuir 9:1456–1464, 1993.
14. JFA Soltero, JE Puig, O Manero, PC Schulz. Langmuir 11:3337–3346, 1995.
15. WJ Kim, SM Yang, M Kim. J Colloid Interface Sci 194:108–119, 1997.

16. M Li, Y Zhang, M Jiang, L Zhu, C Wu. Macromolecules 31:6841–6844, 1998.
17. M Li, M Jiang, L Zhu, C Wu. Macromolecules 30:2201–2203, 1997.
18. W Zhanag, GZ Li, JH Mu, Q Shen, LQ Zheng, HJ Liang, C Wu. Chin Sci Bull 45:1854–1857, 2000.
19. K Bijma, B Engberts. Langmuir 13:4843–4849, 1997.
20. JN Israelachvili, DJ Mitchell, BW Ninham. J Chem Soc Faraday Trans 2 72: 1525–1568, 1976.
21. R Zana, S Yiv, C Strazielle, P Lianos. J Colloid Interface Sci 80:208–223, 1981.
22. S Yiv, R Zana, W Ulbricht, H Hoffmann. J Colloid Interface Sci 80:224–236, 1981.

10

Vesicle Formation by Chemical Reactions: Spontaneous Vesicle Formation in Mixtures of Zwitterionic and Catanionic Surfactants

KLAUS HORBASCHEK, MICHAEL GRADZIELSKI, and HEINZ HOFFMANN   University of Bayreuth, Bayreuth, Germany

ABSTRACT

The influence of shear on the vesicle formation in surfactant systems was studied by preparing the liquid crystalline phases in the surfactant system tetradecyldimethylamine oxide (TDMAO)/sodium-3-hydroxy-2-naphthoate (SHNC)/formic acid (HCO_2H)/water with and without shear. The transition between the different liquid crystalline states in this system is controlled by the degree of protonation of the zwitterionic surfactant TDMAO and thus by the concentration of HCO_2H. Using the corresponding ester methyl formiate (MF) instead of HCO_2H, the transition between the different liquid crystalline phases can be studied without application of any shear. The phase transitions are induced by the change in the degree of charging during the hydrolysis of the ester that takes place on a time scale of several hours.

We started from a micellar solution of 100 mM TDMAO and 25 mM SHNC and added 100 mM MF. With time we observed the formation of a lamellar L_α phase, a vesicular L_v phase, and a vesicular precipitate. This precipitate dissolved to form a vesicular phase, an L_α phase, and finally again a micellar phase. The hydrolysis of the MF and the resulting phase transitions were observed by measurements of pH, conductivity, turbidity, and by means of freeze-fracture transmission electron microscopy. The microstructures of the phases obtained were compared with the microstructures of the corresponding phases when increasing amounts of HCO_2H were mixed with solutions of 100 mM TDMAO and 25 mM SHNC.

I. INTRODUCTION

The formation of vesicles has been observed in a great number of surfactant systems. They are formed in aqueous solutions of double-chained surfactants, in mixtures of various surfactants with cosurfactant, in mixtures of cationic surfactants with large hydrophobic counterions, and in mixtures of cationic and anionic surfactants [1–6]. It is generally assumed that the vesicle formation occurs spontaneously in these systems. However, there is evidence that the microstructures formed can be strongly affected by the preparation process. It has been found that the size distribution of the vesicles depends on the formation pathway as well as on preformed structures that are already present in the solution when the vesicles are built (matrix effect) [7–9]. In addition, vesicular solutions usually have to be mixed during the preparation process to homogenize the components. Therefore, they are exposed to shear during the preparation process. It is well known, however, that shear forces may have a great impact on the vesicle formation. To avoid the application of shear during the vesicle formation, chemical reactions, e.g., the hydrolysis of organic esters, can be used. With ester hydrolysis the composition of the surfactant aggregates changes in a way that enables the formation of vesicular phases. The hydrolysis of MF within phases containing TDMAO, for example, leads to protonation of the zwitterionic surfactant TDMAO, the degree of charging is altered, and transformations of one structural type to another can be triggered. If shear is avoided by doing so, it has been shown that the resulting microstructures in some systems are classical L_α phases of stacked bilayers instead of vesicular phases of multilamellar vesicles. Only if the L_α phases are exposed to shear are the stacked bilayers transformed to multilamellar vesicles [10–12].

Within the system TDMAO/SHNC/HCO_2H/water the phase behavior is determined by the degree of charging of the surfactant aggregates. This degree of charging can be varied by varying the concentration of added HCO_2H or in situ by hydrolysis of a corresponding amount of MF. Thus, the shear during the formation of the liquid crystalline phases can be avoided. The phase transformations have been studied by measuring the pH, conductivity, turbidity, and by means of freeze-fracture transmission electron microscopy (FF-TEM).

II. MATERIALS AND METHODS

TDMAO was a gift from Clariant AG Gendorf and was delivered as a 25% aqueous solution. It was purified by recrystallizing it twice from acetone and was characterized in terms of the melting point and critical micelle concentration (cmc). SHNC was produced by titration of alcoholic 3-hydroxy-2-naphthoate with alcoholic NaOH. 3-Hydroxy-2-naphthoate (>95%) was purchased from Fluka and recrystallized twice from acetone/water. Its purity was checked by gas chromatography. The MF and HCO_2H from Merck were of p.a. quality and used without further purification.

For freeze-fracturing, a freeze-fracture apparatus Bioetch 2005 of Ley-bold-Heraeus (Germany) was used. The replicas were examined with a CEM 902 electron microscope from Zeiss (Germany).

III. RESULTS AND DISCUSSION

The system TDMAO/SHNC/water is a mixture of zwitterionic and anionic surfactants. The addition of HCO_2H leads to partial protonation of TDMAO, and the zwitterionic surfactant TDMAO is transformed to a cationic surfac-tant $TDMAOH^+HCO_2^-$. The cationic surfactant interacts with the anionic surfactant SHNC and catanionic ion pairs $TDMAOH^+HNC^-$ are formed. At equal concentrations of anionic surfactant SHNC and cationic surfactant $TDMAOH^+HCO_2^-$, the system can be considered as a mixture of zwitterionic surfactant and anionic surfactant. As a consequence, the phase behavior of the system $TDMAO/SHNC/HCO_2H$ is similar to the phase behavior of catanionic surfactant systems.

Figure 1 shows a cut through the phase diagram at a concentration of 100 mM TDMAO and 25 mM SHNC. At equimolar concentrations of SHNC and HCO_2H, a voluminous precipitate is formed. Under this condition the bilayers are fairly uncharged and there is 25 mM excess salt in the solution. If the precipitate is charged either by an excess of HCO_2H or by an excess of the anionic surfactant SHNC, cationic-rich or anionic-rich vesicles are formed. If the charge density is increased further there is a transition from the vesicular L_v phase to a lamellar L_α phase and finally to a micellar L_1

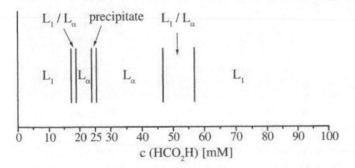

FIG. 1 A cut through the phase diagram of $TDMAO/SHNC/HCO_2H$/water at 25°C for c (TDMAO) = 100 mM and c (SHNC) = 25 mM.

phase. This phase sequence is exactly the same as that observed in many catanionic systems upon charging the equimolar mixtures with an excess of a cationic or anionic surfactant [13,14].

Starting from the micellar solution of 100 mM TDMAO and 25 mM SHNC, each of the preceding phases can be produced without applying any shear by hydrolysis of the appropriate amount of MF. If 100 mM MF is added, the ester hydrolysis results in time in solutions with the same composition as in the stationary system in Fig. 1. The additional MF and MeOH are at low concentrations and do not have any significant influence on the phase behavior of the solutions. The phase changes in a solution of 100 mM TDMAO/25 mM SHNC after the addition of 100 mM MF have been observed both with and without crossed polarizers (Fig. 2).

The starting solution is clear and isotropic (Fig. 2a). After 50 min the solution becomes slightly turbid, but there is no birefringence (Fig. 2b). The birefringence starts to develop about 10 min later (Fig. 2c). With time the solutions become more and more turbid (Fig. 2d), and after a reaction time of 4 h 30 min a strongly turbid whitish precipitate is formed (Fig. 2e). The precipitate dissolves 45 min later and before demixing into a rich-surfactant

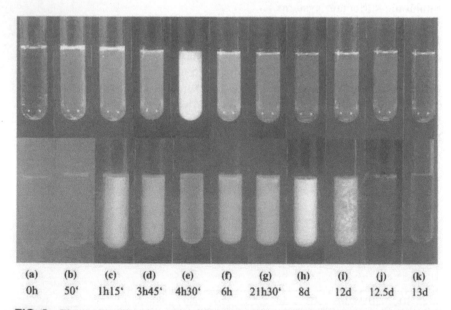

(a)	(b)	(c)	(d)	(e)	(f)	(g)	(h)	(i)	(j)	(k)
0h	50'	1h15'	3h45'	4h30'	6h	21h30'	8d	12d	12.5d	13d

FIG. 2 Phase transitions in 100 mM TDMAO/25 mM SHNC/100 mM methyl formiate (MF) during the hydrolysis of the methyl formiate. (Top row) without polarizers; (bottom row) with polarizers. The time is given after which the pictures were taken [days (d), hours (h), minutes (')].

and a low-surfactant solution. The solution stays birefringent and slightly turbid for several days. The solutions clear up more and more and the birefringence becomes brighter (Fig. 2f–h) and after some days shows a domainlike pattern (Fig. 2i). Finally, the solution again becomes isotropic but still shows little turbidity (Fig. 2j) that clears up with time (Fig. 2k).

As shown in Fig. 3, at the beginning of the experiment the pH of the solution is alkaline and the ester hydrolysis is fast. When the precipitate is dissolved ($t = 5$ h) the pH of the solution is slightly acidic and the ester hydrolysis is rather slow. Accordingly, the microstructure of the solution does not change very much for a long time (Fig. 2f–h). As the turbidity changes at the different phase transitions, the phase sequence during the ester hydrolysis can be followed very easily by observing the transmission of the sample with time (Fig. 4).

The two-phase area between the micellar and liquid crystalline phases is indicated by a sharp drop in the transmission after 50 min. However, even before this transition a slight turbidity develops within the solutions. Within the liquid crystalline phase the solutions become more and more turbid and the transmission drops to nearly zero when the precipitate zone is reached. When the precipitate dissolves, there is again a steep increase in transmission followed by a slight increase in transmission within the liquid crystalline phase.

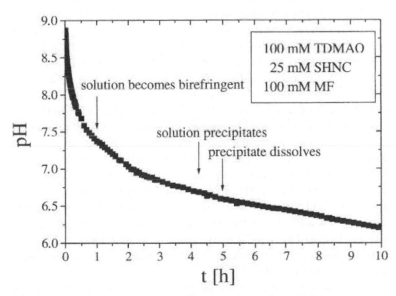

FIG. 3 The pH and phase transitions as a function of time, i.e., during the ester hydrolysis, in 100 mM TDMAO/25 mM SHNC/100 mM MF.

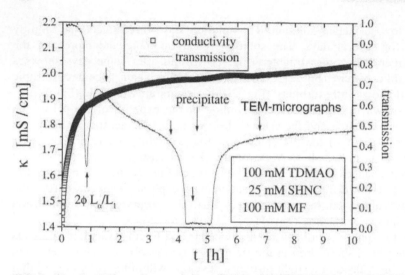

FIG. 4 Conductivity and transmission of 100 mM TDMAO/25 mM SHNC/100 mM MF as a function of time, i.e., during the hydrolysis reaction of MF (at 25°C).

The microstructures of the different liquid crystalline phases during the ester hydrolysis in 100 mM TDMAO/25 mM SHNC/100 mM MF were examined by FF-TEM. Small aliquots of the solution were prepared after 1 h 30 min, 3 h 45 min, 4 h 30 min, and 6 h 45 min (see arrows in Fig. 4). A micrograph of the birefringent, only slightly turbid solution after 1 h 30 min is shown in Fig. 5a. The structure of the solution is lamellar, but there is a strong tendency toward cross-fracturing. A large number of defects and perforations are visible. The interlamellar distance is about 65 nm. For comparison, a micrograph of a solution of 100 mM TDMAO/25 mM SHNC/20 mM HCO_2H is shown in Fig. 5b. The sample consists of strongly undulating bilayers and thus shows exactly the same structure as the phase produced by the chemical reaction. A micrograph of 100 mM TDMAO/25 mM SHNC/ 100 mM MF after 3 h 45 min is shown in Fig. 5c. The corresponding phase with 100 mM TDMAO/25 mM SHNC/22 mM HCO_2H is shown in Fig. 5d. Both phases are vesicular phases that consist of oligolamellar vesicles with a limited number of shells. The interlamellar distances vary strongly around a mean value of 65 nm. The vesicle membranes show undulations and some defects. In addition, lamellar fragments in a size range up to several hundreds of nanometers are present.

The microstructures of the precipitated phases are shown in Fig. 6a and b. Figure 6a shows the kinetic experiment after 4 h 30 min, and Fig. 6b shows the precipitate equimolar concentrations of SHNC and HCO_2H, i.e.,

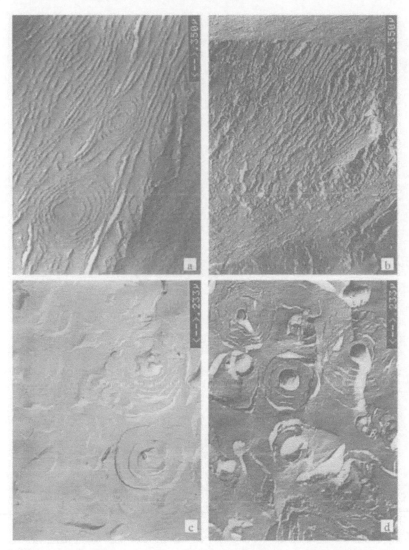

FIG. 5 FF-TEM micrographs of 100 mM TDMAO/25 mM SHNC/100 mM MF after $t = 1$ h 15 min (a) and after $t = 3$ h 45 min (c), of 100 mM TDMAO/25 mM SHNC/ 20 mM HCO_2H (b), and 100 mM TDMAO/25 mM SHNC/22 mM HCO_2H (d).

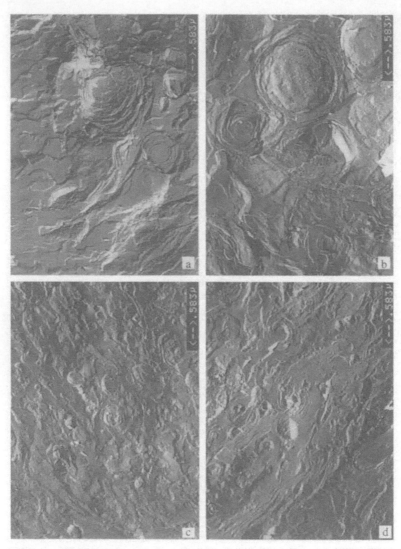

FIG. 6 FF-TEM micrographs of 100 mM TDMAO/25 mM SHNC/100 mM MF after $t = 4$ h 30 min (a) and $t = 6$ h 45 min (c), of 100 mM TDMAO/25 mM SHNC/ 25 mM HCO_2H (b), and of 100 mM TDMAO/25 mM SHNC/30 mM HCO_2H (d).

for 100 mM TDMAO/25 mM SHNC/25 mM HCO$_2$H. The precipitate is a liquid crystalline precipitate that consists of aggregated multilamellar vesicles. The size of the vesicles ranges up to several micrometers, but there also exist a number of smaller vesicles and lamellar fragments. The interlamellar distance in the vesicles varies strongly and ranges up to more than a hundred nanometers. There are some areas of the sample where there are no bilayers present, indicating that the sample was taken from a two-phase area. The micrographs in Fig. 6c and d show the liquid crystalline phases of 100 mM TDMAO/25 mM SHNC/100 mM MF at $t = 6$ h 45 min and of 100 mM TDMAO/25 mM SHNC/30 mM HCO$_2$H, respectively. The liquid crystalline phases in both samples are vesicular phases. The vesicles are oligolamellar and are embedded in a network of planar bilayers that coexist with the vesicles.

Although the liquid crystalline phases within 100 mM TDMAO/25 mM SHNC/100 mM MF have been produced by the hydrolysis of the ester and thus without applying any outer shear forces, they exhibit in each of the four cases examined exactly the same microstructural features as the corresponding samples that were prepared by direct mixing of the components. The transformations between the different microstructures are tuned only by the degree of charging of the surfactant aggregates at the given pH and thus by diffusive processes. In contrast to other systems examined so far, the vesicles within the solutions of TDMAO/SHNC/HCO$_2$H form spontaneously, and shear forces are not necessary for the vesicle formation.

IV. SUMMARY

The system TDMAO/SHNC/HCO$_2$H/water is closely related to catanionic systems. For equimolar mixtures of SHNC and HCO$_2$H, it can be regarded as mixture of zwitterionic surfactant TDMAO with cationic ion pairs TDMAOH$^+$HNC$^-$ and excess salt. At 100 mM TDMAO/25 mM SHNC/25 mM HCO$_2$H the precipitate formed is liquid crystalline and consists of aggregated multilamellar vesicles. If the precipitate is charged, a phase transition occurs: precipitate $\rightarrow$ L$_v$ $\rightarrow$ L$_\alpha$ $\rightarrow$ L$_1$/L$_\alpha$ $\rightarrow$ L$_1$.

The influence of shear forces on the vesicle formation in this system has been examined by preparing the liquid crystalline phases starting from the micellar phase of 100 mM TDMAO/25 mM SHNC by addition of MF and following the hydrolysis of the ester. In this way, the different liquid crystalline phases evolve with time without applying any outer shear. The phase sequence during the hydrolysis of 100 mM MF has been examined by measurements of pH and turbidity, and the microstructures of the phases have been determined by means of FF-TEM and compared with the microstructure in the system TDMAO/SHNC/HCO$_2$H. The phase sequence during the hydrolysis of MF is $L_1 \rightarrow L_\alpha \rightarrow L_v \rightarrow$ precipitate $\rightarrow L_v \rightarrow L_\alpha \rightarrow L_1$ and thus is exactly the same as in the samples produced by direct mixing of the components. Shear forces are not necessary for the formation of the vesicles; they form spontaneously depending on the degree of charging of the surfactant aggregates.

REFERENCES

1. H Kunieda, K Shinoda. J Phys Chem 82:1710, 1978.
2. BW Ninham, DF Evans, GJ Wei. J Phys Chem 87:5020, 1983.
3. H Hoffmann, C Thunig, P Schmiedel, U Munkert. Langmuir 10:3972, 1994.
4. EW Kaler, AK Murthy, B Rodriguez, JAN Zasadzinski. Science 245:1371, 1989.
5. BK Mishra, SD Samant, P Pradhan, SB Mishra, C Manohar. Langmuir 9:894, 1993.
6. M Gradzielski, M Müller, M Bergmeier, H Hoffmann, E Hoinkis. J Phys Chem B 103:1416, 1999.
7. J Oberdisse, C Couve, J Appell, JF Berret, C Ligoure, G Porte. Langmuir 12: 1212, 1996.
8. E Marques. Langmuir 16:4798, 2000.
9. S Lonchin, PL Luisi, P Walde, BH Robinson. J Phys Chem B 103:10910, 1999.
10. M Bergmeier, H Hoffmann, C Thunig. J Phys Chem B 101:5767, 1997.
11. H Hoffmann, M Bergmeier, M Gradzielski, C Thunig. Prog Colloid Polym Sci 109:13, 1998.
12. K Horbaschek, H Hoffmann, J Hao. J Phys Chem B 104:2781, 2000.
13. A Khan, E Marques. In: ID Robb, ed. Specialist Surfactants. London: Chapman & Hall, 1997, p 37.
14. EW Kaler, KL Herrington, AK Murthy, JAN Zasadzinski. J Phys Chem 96: 6698, 1992.

11

Mechanism of the Clouding Phenomenon in Surfactant Solutions

C. MANOHAR   Indian Institute of Technology, Mumbai, India

ABSTRACT

The possible mechanism for clouding in surfactant solutions of nonionic and mixed ionic–nonionic surfactants is reviewed. Semiquantitative arguments to predict the trends in changes of clouding temperatures in mixed systems are proposed. Results for Triton X-100 and additives, sodium dodecyl sulfate and salicylic acid, are presented. A methodology for systematic interpretation of light and small-angle neutron scattering (SANS) results in uncharged colloidal systems is proposed by taking advantage of the smallness of the ratio of the range of attractive interaction to the diameter. The use of these methods is illustrated by estimating the van der Waals depth of the inter-micellar potential in nonionic micellar systems. The experimental results seem to indicate that the well depth increases quadratically with temperature. The relation of this temperature dependence to hydration experiments is discussed. This procedure is generalized to include Coulomb interactions as a perturbation and is demonstrated by application to experiments on mixed nonionic and ionic surfactants. These results appear to show the phenomenon of charge condensation.

I. INTRODUCTION

The phenomenon of clouding of surfactant solutions is important in several industrial processes such as detergency and separation [1]. Understanding the mechanism underlying this phenomenon and developing methods for quantitative interpretation of experimental results have been challenging tasks. There have been attempts to develop systematic methods for analyzing this phenomenon [2,3]. The present chapter reviews these developments and suggests investigations to extract molecular level parameters characterizing the clouding phenomenon. The developments have become feasible for the following reasons.

1. There are semiquantitative arguments and experimental demonstration that this phenomenon is due to increased strength (U) of the short-range attraction between the micelles with temperature (T) [4].
2. The range (Δ) of the attractive interaction being smaller than the diameter (σ) can be used to advantage to formulate a rigorous and quantitative model [5] for the interpretation of small-angle neutron scattering (SANS) and light scattering (LS) results.

The next two sections discuss some of these developments, and the last section describes some experimental results.

II. SEMIQUANTITATIVE ARGUMENTS AND EXPERIMENTS

If one assumes that the interaction between micelles is responsible for clouding and phase separation, then the cloud point should be sensitive to the charge on micelles and should increase with surface charge. This has been confirmed by adding small amounts (subcritical micellar concentrations) of sodium dodecyl sulfate (SDS) to 1% Triton X-100 solution and observing that the cloud point increases [6]. The magnitude of the increase in the cloud point should be of the order of the Coulomb potential at the surface of micelle. This argument gives [4]

$$T_C = T_0 + \frac{Z^2 e^3}{(\varepsilon\sigma)} \frac{1}{(1 + 0.5\kappa\sigma)^2} \tag{1}$$

where T_C is the cloud point after the addition of ionic surfactant; T_0 is the cloud point of the pure nonionic surfactant; ε, κ, and σ are, respectively, the dielectric constant of water, Debye-Hückel screening parameter, and micellar diameter; and Z is the charge on the micelle. Two points are noteworthy. First is the *quadratic dependence* of the cloud point temperature on the charge Z and the effect of salt solution in reducing the cloud point (through κ). More important, the effect of salt *is independent of the valance of the salt and depends only on the ionic strength*. These features have been confirmed by experiments [7–9].

These arguments can be extended to systems in which the charge of the additive is pH dependent, and the cases of salicylic acid and aspirin have been investigated in detail [10,11]. These semiquantitative arguments give correct orders of magnitude of cloud point shifts and give confidence that the intermicellar interactions play a dominant role in the clouding phenomenon. These arguments need to be given a more concrete theoretical footing to interpret more rigorously the results of the SANS and LS experiments on nonionic surfactant solutions and their mixtures.

III. THEORETICAL MODELS

Theoretical models are discussed in Refs. 2 and 5. The picture that has to be modeled is that of the nonionic surfactant micelles interacting via an attractive (van der Waals) potential whose well depth increases with temperature. One of the simplest methods is to model this as a square well of depth U and width Δ along with steric repulsion at distances shorter than σ. In colloidal micellar systems σ is of the order of about 4–5 nm and Δ is the of the order of 0.1–0.4 nm. Therefore, $\Delta/\sigma < 1$ and this ratio becomes a convenient parameter to develop the statistical mechanics, especially for colloidal systems [5]. It has been shown that if one defines parameters

$$\tau = \left(\frac{\sigma + \Delta}{12\Delta}\right) \exp\left(\frac{-U}{kT}\right) \tag{2}$$

and

$$\eta = \frac{\phi(\sigma + \Delta)}{(\sigma - 2\Delta)} \tag{3}$$

where ϕ is the volume fraction of the micelles, then the system shows a gas–liquid transition in the phase space of (τ, η) with a critical point $\tau_c = 0.1213$. Below $\tau < \tau_c$ the system separates into two phases, one rich in micelles and the other lean in micelles. This describes the phenomenon of clouding in nonionic surfactant solutions.

This model can not only predict the phase separation but also give expressions for the structure factor obtainable from SANS and LS. The scattering intensity $I(Q)$ in both cases is given by

$$I(Q) = AP(Q)S(Q) \tag{4}$$

where Q is the scattering vector given by

$$Q = \frac{4\pi \sin(\theta/2)}{\lambda} \tag{5}$$

θ is the scattering angle, λ is the wavelength of the radiation (neutron or light) used, A is a constant dependent on the refractive index or scattering length density, and $P(Q)$ is the shape factor and for spherical particles of radius R is given by

$$P(Q) = \left[3 \, \frac{(\sin QR - QR \cos QR)}{(QR)^3} \right]^2 \tag{6}$$

The structure factor $S(Q)$ is a complicated expression, and for uncharged colloids such as nonionic micelles this expression is given in Ref. 5. In the present chapter, for convenience, the structure factor for uncharged micelles is designated by $S_0(Q)$. These expressions give quantitative agreement for the structure factors obtained in Monte Carlo computer simulations *without any adjustment of the parameters for both micellar and microemulsion systems*. This gives one confidence to try these models for nonionic micellar systems, and they have been applied to Triton X-100 and C_iE_j [2,12]. Apart from the good fits for the SANS and LS results, this analysis also gives the temperature dependence of U—the most important quantitative entity for the mechanism of clouding. From the analysis of the data available, it appears that U has a quadratic dependence on temperature T, namely

$$U = \alpha T + \beta T^2 \tag{7}$$

The phenomenon of clouding is related to hydration of the head groups, and measurements [13] show that the hydration numbers vary smoothly through the cloud point. For example, in the case of $C_{12}E_6$, the hydration number per EO group decreases smoothly from 4.2 to 2.8 as the temperature is raised from 20 to 60°C through the cloud point of 50°C. This is consistent with a smooth dependence of U on T. Quantitative relations between hydration and clouding do not exist at present.

Intuitively, one would expect that at the short distances where attractive forces become dominant, the micellar surface can no longer be regarded as smooth. Therefore, one would expect that clouding is related to the roughness of the surface and this would mean that the closest distance to which two micelles can approach is decided not by their diameters but by the highest bump created by the hydration of the surface. This would imply that the two micelles, in spite of a strong attraction, are never able to come close enough for binding to each other. When the temperature is raised, the hydration water molecules at the surface become more mobile and thus the surface roughness decreases, letting the two micelles approach closer and bind to each other. This drives the system to clouding and phase separation. This concept is illustrated in Fig. 1.

IV. EXTENSION OF THE THEORY TO CHARGED MICELLES

Scattering techniques have been used extensively for charged micellar systems and the results have been interpreted using the theories developed by

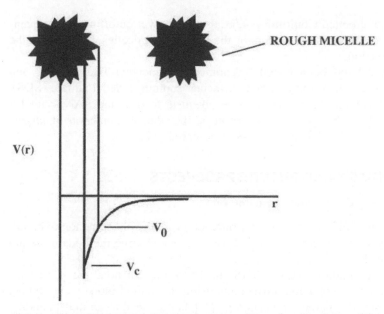

FIG. 1 Potential between two uncharged rough-surfaced micelles. Hydration of the surface contributes to roughness and this reduces the strength of adhesion from V_C to a lesser value V_0.

Hayter and Penfold [14,15]. This theory ignores the attractive interactions and uses only the Yukawa-type coulombic interactions, for which the Ornstein–Zernike equation can be solved analytically. Therefore, this model cannot explain the clouding phenomenon. Fortunately, most of the normal ionic micelles with strong coulombic interactions do not show the phenomenon of clouding. However, if the coulombic interactions are suppressed by addition of salt, the micelles show clouding and the Hayter–Penfold model becomes inapplicable [16]. In applications one considers the mixture of nonionic and ionic surfactants, and again for these low-surface-charge micelles there is a need to develop models that take into account both attractive and repulsive interactions.

One of these attempts which has shown some promise is the treatment of electrostatic interaction by random phase approximation [3] in which the expression for the structure factor becomes

$$S(Q) = \frac{S_0(Q)}{1 + [\rho v(Q)S_0(Q)/kT]} \tag{8}$$

where ρ is the number density of the colloids and $v(Q)$ is the Fourier transform of the screened Coulomb potential with a lower cutoff at σ. The quantity $S_0(Q)$ is the structure factor for the uncharged micelles discussed in the previous section.

This model has been tested for nonionic surfactants Triton X-100 and C_iE_j with addition of the charged surfactant sodium dodecyl sulfate (SDS), and the model has been found to be adequate to describe SANS and LS results [13,17]). In the case of Triton X-100, the phenomenon of charge condensation with addition of SDS is observed [17].

V. SUMMARY AND FUTURE PROSPECTS

The preceding developments enable one to

1. Check the role of intermicellar interactions in the phenomenon of clouding by simple experiments and use semiquantitative methods to obtain the trends.
2. Describe, quantitatively, SANS and LS results for nonionic surfactants with mixtures of ionic surfactants using theoretical models and extract the molecular parameters, such as U, that may lead to an understanding of the relation between hydration and clouding.

It remains to be seen whether the phenomenon of clouding in phase separation in microemulsions can also be described by these models. The role of surface roughness of the micelles in the phenomenon of clouding needs to be delineated by both experiments and theory.

REFERENCES

1. V Degiorgio. In: V Degiorgio, M Corti, eds. Physics of Amphiphiles: Micelles and Microemulsions. Amsterdam: North Holland, 1985.
2. SVG Menon, VK Kelkar, C Manohar. Phys Rev A 43:1130, 1992.
3. C Manohar, VK Kelkar. Langmuir 8:18, 1992.
4. C Manohar, VK Kelkar. J Colloid Interface Sci 137:604, 1991.
5. SVG Menon, KS Rao, C Manohar. J Chem Phys 95:9186, 1991.
6. BS Valaulikar, C Manohar. J Colloid Interface Sci 108:403, 1985.
7. L Marszall. Langmuir 4:90, 1988.
8. T Gu, S Qin, C Ma. J Colloid Interface Sci 127:586, 1989.
9. L Marszall. Colloids Surf 25:279, 1987.
10. C Manohar, VK Kelkar, BK Mishra, KS Rao, PS Goyal, BA Dasannacharya. Chem Phys Lett 171:451, 1990.
11. BS Valaulikar, BK Mishra, SS Bhagwat, C Manohar. J Colloid Interface Sci 144:304, 1991.
12. KS Rao, PS Goyal, BA Dasannacharya, VK Kelkar, C Manohar, SVG Menon. Pramana (J Phys) 37:311, 1991.

13. W Douglas, E Kaler. Langmuir 10:1080, 1994.
14. JB Hayter, J Penfold. Mol Phys 42:109, 1981.
15. J Hansen, JB Hayter. Mol Phys 46:651, 1982.
16. GG Warr, TN Zemb, M Drifford. J Phys Chem 94:3086, 1990.
17. VK Kelhar, BK Mishra, KS Rao, PS Goyal, C Manohar. Phys Rev A 44:8421, 1991.

12

Atomic Force Microscopy of Adsorbed Surfactant Micelles

WILLIAM A. DUCKER   Virginia Tech, Blacksburg, Virginia, U.S.A.

ABSTRACT

The many uses of surfactants have stimulated research on surfactant adsorption and on surfactant aggregation in bulk solution. For some time, evidence has suggested that surfactants also aggregate in the adsorbed state. In this chapter, we review the early evidence of surfactant aggregation at interfaces and then describe research in which an atomic force microscope (AFM) has produced real-space images of adsorbed micelles. The ability of the AFM to obtain these images has allowed researchers to examine the relationship between intermolecular forces and the shape of the adsorbed micelle. We describe how additional forces arising from the interface cause the structure of adsorbed micelles to differ from the structure of solution micelles. In particular, this chapter will focus on how surface charge and wettability affect micellar structure.

I. INTRODUCTION: EVIDENCE FOR THE EXISTENCE OF SURFACE MICELLES

A. Early Studies of Adsorbed Surfactant Aggregates

The concept of surface aggregation for surfactants was first introduced to explain abrupt changes in interfacial properties as a function of surfactant concentration. Gaudin and Fuerstenau [1] inferred the existence of surface aggregates from zeta potential measurements, which showed an increase in the gradient of surfactant adsorption at a particular concentration. This concentration was approximately a constant fraction of the critical micelle concentration (cmc), which suggested that the surface process was similar to bulk micellization [2]. Unlike simple monovalent ions, surfactant ions reversed the charge of solids even when they were not lattice ions. This adsorption against an electrostatic potential implied that the surfactants attracted each other. In early work, the attractive force was assumed to be van der Waals interactions, but later it was attributed to the hydrophobic effect. The surface density and wetting properties of the surfactant clusters were consistent with the surfactant adopting an orientation with the alkyl chains facing the solution. These small clusters were called "hemimicelles." Later, surface aggregates were proposed in a number of different systems and new names such as "admicelles" [3], surface micelles, and "solloids" [4] were proposed.

B. Adsorption Isotherms

Deviation from the Langmuir adsorption isotherm is often used to infer the aggregation of surfactants at interfaces. The Langmuir adsorption isotherm assumes that the adsorption energy is independent of surface coverage, and deviation is used to signal the presence of concentration-dependent intermolecular forces, e.g., electrostatic forces or the hydrophobic effect. The use of deviations from idealized isotherms to indicate aggregation is somewhat problematic because two interactions can produce opposing effects. For example, electrostatic interactions and hydrophobic interactions may combine to produce adsorption that resembles the Langmuir adsorption isotherm [5].

Somasundaran and Fuerstenau [6] measured the adsorption isotherm of sodium dodecyl sulfate (SDS) on alumina as well as the electrophoretic mobility of alumina. At low concentrations and low surface coverage (<1/50 monolayer), the surfactant adsorbed on the alumina by displacing counterions in a 1:1 ratio. At higher bulk concentrations and higher surface coverage (<1/10 monolayer), there was an increase in the rate at which the surface excess concentration increased with bulk concentration, even through adsorption was driven by a smaller positive potential. At even higher concentrations, adsorption continued against a negative potential. This adsorption was explained by invoking an attractive potential between adsorbed surfactants, much as for micellization in the bulk solution.

C. Models Describing Surfactant Adsorption

Zhu and Gu [7] used a simple mass action model to fit adsorption isotherms for nonionic surfactants. They noted that adsorption often increased with concentration in two steps, so they modeled adsorption with two binding constants. The first binding constant accounted for binding of the surfactant to the surface, and the second also accounted for attractive interactions between the adsorbed surfactant molecules. The latter are expected to be important at a high surface density of surfactant.

For charged surfactants, the surface electrical potential is an important component of the surfactant chemical potential in the adsorbed state. Thus, the calculation of surfactant adsorption density or aggregate shape must include a calculation of the surface potential. This is a difficult problem because the binding of the surfactant, its counterions, and other salt molecules all affect the surface charge, and interactions among these adsorbed species also affect the binding constants. Zhu and Gu used their two-state adsorption model to examine adsorption of cationic surfactants to silica [8]. This method is reasonably successful if the surface potential is either fitted to the data or measured experimentally.

Self-consistent field (SCF) theory provides a more powerful alternative for modeling the adsorption of surfactants. In SCF, surfactant molecules are regarded as small amphiphilic flexible-chain molecules, and each chain segment, solvent molecule, and ion fills a lattice site. The interactions between the segments are introduced as model parameters, so comparison between SCF and experiment can reveal which values of interaction parameters realistically reproduce adsorption behavior. Böhmer and Koopal [9] used SCF theory to examine the adsorption of nonionic surfactants onto both hydrophilic and hydrophobic surfaces. They used planar geometry and a mean-field approximation, so their work was not capable of reproducing surface aggregation, but they were able to make the following predictions about surface adsorption: (1) that regardless of the number of hydrophobic and hydrophilic groups on the surfactant, all surfactants will form a monolayer on a hydrophobic surface, and (2) that adsorption will exceed Langmuir adsorption for small hydrophilic head groups on both hydrophilic and hydrophobic surfaces. (The latter implies cooperative adsorption at the interface.) Tiberg et al. [10] used ellipsometry to measure accurately the adsorption isotherms and thicknesses of adsorbed layers of ethylene oxide surfactants on silica. They were able to confirm cooperative adsorption on hydrophilic silica but not on hydrophobic silica.

Goloub and Koopal [11,12] also used SCF calculations to examine the adsorption of cationic surfactants onto Aerosil silica. Experimentally, they found that, below a certain surfactant concentration, the addition of salt hindered adsorption, and above this concentration, the addition of salt aided adsorption. Using SCF calculations, they showed that the existence of this "common intersection point" was due to competition of salt ions for surface binding sites at lower surfactant concentrations and screening of electrostatic repulsions by salt at higher surfactant concentrations. Their calculations also suggested that the surfactant tails were attracted to Aerosil silica.

Johnson and Nagarajan [13,14] have also recently performed a theoretical treatment of the effect of surface charge and wettability on the shape of adsorbed surfactant aggregates.

D. Spectroscopic Investigations

The lateral organization of surfactants has also been probed by spectroscopic techniques. Levitz and coworkers [15–17] used fluorescence decay spectroscopy to show that nonionic surfactants aggregated on silica in much the same way as in bulk solution but at lower bulk concentrations. Somasundaran and Krishnakumar [18] used fluorescence quenching experiments to suggest that SDS formed aggregates of 50–350 molecules on alumina. Use of ^{2}H nuclear magnetic resonance (NMR) measurements for the same system showed that the surfactant did not form a continuous bilayer structure but instead formed either oblate, ribbon, or porous bilayer structures [19]. Although these measurements are useful in determining the surface structure, both NMR and fluorescence measurements are model dependent and suffer from the following experimental limitations: results from fluorescence measurements can be affected by the presence of the probe and quencher, and ^{2}H NMR spectroscopy depends on the availability of ^{2}H-surfactants and the presence of an invariant local quadrupole coupling.

E. Force Measurement and Contact Angles

The organization of surfactants in the direction normal to the solid–liquid interface has been examined using the surface force apparatus (SFA) (e.g., Refs. 20 and 21) and the atomic force microscope (AFM) [22]. Israelachvili and Pashley [23] used the SFA to measure the force between mica sheets in aqueous hexadecyltrimethylammonium bromide (C_{16}TABr) solutions [23]. At about 1/10 of the cmc the force was much less repulsive than the calculated double-layer plus van der Waals forces. Uncharged adsorbed layers of double-chain surfactants produced net attractive forces [24]. The authors explained these forces by invoking the adsorption of surfactant with an orientation that exposed a significant amount of the surfactant hydrocarbon tail to solution, thereby raising the energy of the film of water between the mica sheets. At this time the exact interpretation of these results remains controversial, but this model for adsorption is essentially a continuous layer of the hemimicelles originally proposed by Fuerstenau.

At higher concentrations of C_{16}TABr, the force was repulsive, with the appropriate decay length for charged and hydrophilic surfaces. The force was consistent with a degree of counterion binding for surface-adsorbed surfactant ($\sim$25%) similar to that previously observed in bulk solution [21]. This result was used to infer that the surfactant aggregated into a bilayer, consisting of an inner layer of surfactant with head groups facing the mica and an outer layer with head groups facing the solvent. The same evidence also supports the existence of spherical or cylindrical micelles. Contact angle [25] and flotation measurements have provided a similar picture of a hydrophobic surface (and thus hydrocarbon exposure) below about 1/10 cmc and a hydrophilic surface (head group exposure) at higher concentrations [26].

F. The Atomic Force Microscope

Although many techniques have been employed to determine both the concentration and the properties of adsorbed surfactants, most of these techniques do not directly probe that lateral organization of surfactants. In 1994, Manne et al. [27] showed that the AFM could be used to image directly the structure of surfactant aggregates at the solid–liquid interface above the surfactant cmc. This method is fast, does not require special sample preparation, is not restricted to particular surfactant molecules, and determination of the shape is not model dependent.

G. Neutron Reflection

Neutron reflectivity experiments provide the most direct method for probing the organization of surfactant micelles normal to the solid–solution interface [28]. The experimental data are used to select the most consistent model of the surface. Unfortunately, neutron reflectivity data often do not allow selection among proposed lateral arrangements of surfactants. Therefore, the AFM and neutron reflectivity are complementary; an AFM reveals the lateral structure and neutron reflectivity reveals the normal structure. The remainder of this chapter will focus on the use of the AFM.

II. ATOMIC FORCE MICROSCOPY

A. Contrast Generation in AFM Images

In an AFM, the surfactant aggregate is probed by measuring the force between a sharp tip and the micelle adsorbed at the interface between the solid and the solution. To obtain contrast there must be lateral variation in the force between the tip and the adsorbed micelle. This lateral variation in the force arises from both topography and chemical variations. The resolution is maximized in two ways: (1) by using a very sharp tip that receives a large contribution of total force from a small area of the solid and (2) by making measurements where the force has a very high gradient normal to the surface.

Sharp AFM tips typically have a radius of about 10 nm, which is similar to the smallest radius of curvature of adsorbed micelles. Therefore, the force between the tip and the micelle at any tip location includes large force contributions arising from a region of the micelle, not from a point on the micelle. Thus, AFM tips of today cannot be used to map accurately the shape of highly curved adsorbed micelles. The lateral extent of a small micelle can be estimated from an AFM image by examining the spacing between the centers of adjacent micelles. In other words, the AFM is good at examining the arrangement of micelles, which can then be used to infer the shape. For example, Fig. 1 shows an image of SDS adsorbed onto graphite. Clearly, there is one very long axis and one short axis of the micelle, so the structure is something like a cylinder or a hemicylinder. (Measurements of the thickness and density [29] of the adsorbed surfactant show that it is a hemistructure.) The long axis is easily resolved, and the short dimension parallel to the interface is inferred from the spacing between adjacent features. However, the determination of the exact cross section is usually beyond the resolution of the tip because the micelle curvature is too high compared with the radius of the tip. Although micelle shapes are often listed as spherical or cylindrical in the literature, the actual cross section could be

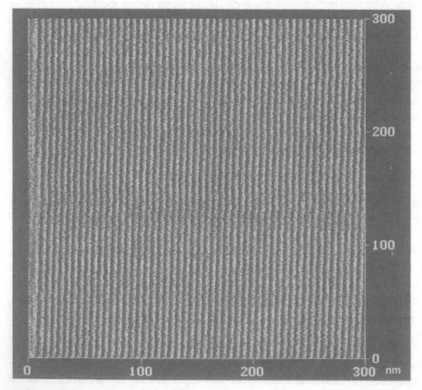

FIG. 1 AFM image of SDS adsorbed at the interface between graphite and a 2.8 mM SDS, 20 mM NaCl solution. (Reprinted with permission from Ref. 40. Copyright 1996 American Chemical Society.)

an ellipse or some other shape. The variation in height (or deflection) in the AFM signal while imaging an array of surface micelles is typically 0.1–0.5 nm, whereas the adsorbed micelles are typically one or two molecular lengths thick (~1.5 or ~3.0 nm). The difference in these two lengths arises because the relatively blunt AFM tip cannot fit in between adjacent aggregates.

AFM resolution also depends on the normal and lateral gradients in the force between the tip and the sample. Because the gradient of surface forces usually increases with decreasing surface separation, the highest gradients are usually obtained when the tip is close to the sample. Therefore, in order to understand the imaging of surfactants by AFM, it is necessary first to understand how the force between an AFM tip and a surfactant-coated substrate varies as a function of separation. Good reviews of intermolecular and surface forces are provided by Israelachvili [30] and by Evans and Wennerström [31].

The adsorption of charged surfactants often results in the generation of a charge on the sample and sometimes on the tip. These charges are compensated by counterions that inhabit the solution between the tip and the sample. Some of the counterions are closely associated with the surfactant head groups and some exist in a diffuse layer in solution. As the tip approaches the sample, work is required to confine these counterions to a smaller volume and, therefore, the tip experiences a repulsive force [32], which is known as the electrostatic double-layer force. The gradient of this force depends on the surface charge and on the concentration of electrolyte in solution. The gradient is larger for higher concentrations of electrolyte in solution. (For example, the double-layer force in a 1:1 electrolyte has a decay length of ~10 nm in a 1 mM solution and a decay length of ~1 nm in a 100 mM solution.) Therefore, the resolution of AFM imaging should be greater as an electrolyte is added to the solution. Of course, the addition of electrolyte may also affect the actual structure of the adsorbed micelles.

Adsorbed surfactants also generate surface forces due to the energy required to remove water from the surfactant head groups (hydration forces) and the energy required to confine the surfactant to a smaller volume on the surface (protrusion forces) [32]. These short-range forces (<1 nm) have high gradients that are ideal for AFM imaging. In contrast, the longer range double-layer force usually has a lower gradient, which decreases the resolution in the vertical direction and causes the interaction between the sides of the tip and the sample to make significant contributions to the total force. This effectively widens the tip. Therefore, it should be easier to obtain high-resolution images of net-uncharged surfactants or of charged surfactants in concentrated salt solutions.

The result of any measurement is influenced by the act of measurement, but in AFM imaging of adsorbed surfactants, the relationship is particularly simple. The adsorbed micelle owes its adsorption and cohesion to intermolecular forces, and the act of measurement introduces new forces. Sometimes the adsorption and structure are produced by a delicate balance of forces, and both can be altered by a small change in solution conditions. In order to determine the shape of the adsorbed micelle, the AFM tip must apply an additional force to the adsorbed micelle. This force may perturb the "native" structure, so it is important to weigh the desire to use a force that yields maximum contrast against the desire to measure the least perturbed structure. Along with the gradient, the magnitude of the force between the tip and the sample often increases with decreasing separation, so there is often a trade-off between obtaining high contrast and minimizing perturbation of the aggregate. This is particularly obvious for charged surfactants. For zwitterionic and nonionic surfactants, there is no net charge and therefore no long-range repulsive force, and so it is possible to image by sensing a very high force gradient when the force is zero [33].

The perturbations caused by the AFM tip can be used to the advantage of the experimentalist when desorption occurs in a layer-by-layer process. By imaging at a series of increasing forces, the adsorbate can be interrogated layer by layer, finally revealing the structure of the underlying substrate. This is particularly useful when trying to establish the relationship between the adsorbate and the adsorbent structures (see Section III.A). However, the adsorbate does not always adsorb in layers, and a major disadvantage when using the AFM is the inability to resolve adsorbate features below the surface of the adsorbed structure.

B. Stability of Adsorbed Surfactant Aggregates

To date, all images of surface micelles have been obtained above the cmc or just below the cmc (e.g., cmc/3 for SDS on graphite), where the density of adsorbed surfactant is high. There is no really convincing AFM evidence of aggregation at much lower surface concentrations. In particular, there is no AFM evidence for the hemimicelles that were originally postulated to form at 1/10 to 1/100 of the cmc. This lack of evidence does not necessarily mean that hemimicelles do not form. To form an image, the AFM must impart a force on the surfactant. This force must be less than the force required to disrupt the aggregate. Disruption can occur by fragmentation into monomers (either adsorbed or in solution), by desorption, or by lateral displacement. The forces probed are the cohesive energy of the aggregate, the adsorption energy, and the energy required to increase the area density of aggregates on the surface. It is notable that AFM images of surface micelles are always obtained when there is a dense layer of micelles covering the surface and a fairly monodisperse spacing between adsorbed micelles. Under these conditions, there must be large forces between the aggregates that hinder the lateral motion of an aggregate when it is subjected to stress from the tip. At low surface densities (where hemimicelles are postulated), the lateral interactions are usually smaller than at high densities so the aggregates may simply move sideways in response to stress from the tip. If the adsorbed aggregates are always moving away from the tip, they would be difficult to observe with an AFM.

Evidence for the lateral pressure among adsorbed micelles comes from images of mica in mixtures of strong acid and surfactant [34]. In the absence of surfactant, the acid etches the mica, but in the presence of surfactant and acid, the acid does not etch the mica, even when the tip is pushed hard against the mica. This suggests that the surfactant keeps a seal aound the tip and, therefore, that the surfactant is subject to a lateral pressure. The interactions among aggregates are considered further in Section III.E.

In addition to causing a redistribution of surfactant on the surface, the tip can cause desorption or adsorption. At equilibrium, this effect can be understood using a thermodynamic argument developed by Hall [35] and by Ash et al. [36]. When the force at a given separation increases as a function of surfactant chemical potential, the surfactant will desorb as the tip approaches. Conversely, when the force decreases at a given separation, the surfactant will adsorb. A recent paper describes changes in the adsorption from dodecyltrimethylammonium bromide ($C_{12}TAB$) solution that arise from the approach of a particle [37]. To date, there has been no systematic study of changes in micellar structure induced by an AFM tip, but Ducker and Wanless [34] did observe that micelles of hexadecyltrimethylammonium bromide ($C_{16}TABr$) transformed from cylindrical to approximately spherical geometry when the applied force was increased.

III. PARAMETERS AFFECTING THE SHAPE OF ADSORBED MICELLES

A. Solid Structure

All solid substrates studied to date concentrate the surfactant at the interface and confine the surfactant more or less to a single plane. Such concentration increases the importance of forces between the surfactant monomers (and micelles), and confinement to a plane increases the order of the micellar assembly. In addition, most early AFM work investigated the adsorption of surfactants onto *crystalline* substrates: mica and graphite. These substrates are ideal for AFM work because the perfect cleavage planes of these minerals allow the facile preparation of clean and reproducible surfaces that are smooth on an atomic scale. It is much easier to obtain images of small and delicate aggregates on substrates that are very smooth and robust. However, we see in the following section that the crystal structure of the substrate is a strong influence on the adsorbed-micelle structure.

1. Graphite

Manne and coworkers [27] investigated the adsorption of the cationic surfactant C_{16}TABr. Although C_{16}TABr formed approximately spherical micelles in solution, they found that C_{16}TABr formed hemicylindrical micelles at the interface between the basal plane of graphite and a solution above the cmc. The basal plane of graphite consists of a sheet of hexagonal rings of carbon atoms with a threefold axis of rotation. The axis of each adsorbed cylinder was aligned perpendicular to one of the graphite lattice axes. Hemicylindrical micelles also adsorbed onto graphite from C_{12}TABr, C_{14}TABr [38], C_{18}TABr [39], C_{20}TABr, SDS [40], (zwitterionic) dodecyldimethylammoniopropylsulfonate (DDAPS) [41], and several (nonionic) polyethylene oxide surfactant [41,42] solutions at concentrations at which most of these surfactants formed spherical or globular micelles.

Figure 1 shows SDS micelles adsorbed onto graphite. Clearly, graphite has a strong effect on the absorbed structure. Scanning tunneling microscope (STM) images of alkanes and other alkane derivatives on graphite show that alkyl chains adsorb in an all-*trans* configuration with their chain directors parallel and that two methylene units fit within one hexagonal ring of the graphite [43]. Presumably this close geometric fit allows the establishment of strong van der Waals forces between the surfactant and the graphite when the surfactant lies in this position. Thus, as orignally proposed by Manne, the orientation of the first layer of adsorbed surfactant is directed by the graphite to lie in an organized pattern with parallel chain directors. Additional adsorption is templated by the first layer: the surfactant completes a constant-curvature shape while minimizing exposure of the hydrocarbon to the aqueous solution. This results in a hemicylindrical micelle.

2. Mica

Mica is a sheet aluminosilicate. Substitution of aluminum for silicon in the cleavage plane leaves a net negative charge that is compensated by potassium [44]. In an aqueous solution, the electrostatic interaction between the mica lattice charge and the potassium is weaker than in air, and the potassium ions are distributed between the solution and the surface sites. These potassium ions can be displaced by other cations, for example, quaternary ammonium surfactants. Hence, the adsorption of surfactants is driven by an attractive electrostatic interaction.

Many experiments have shown that single-tail quaternary ammonium ions form cylinders and flat sheets at the interface between mica and aqueous solutions containing approximately spherical micelles [38,39,45]. Adsorption onto mica causes a reduction in the curvature of the micelle. It is well known that surfactant micelles in bulk solution adopt a lower curvature in concentrated salt solution than in surfactant-only solution. This effect occurs because the addition of counterions screens the repulsive interactions between the charged head groups. It has been hypothesized that the lower curvature for adsorbed micelles on mica occurs because the repulsions between the positive head groups are diminished when the head groups are in close proximity to the negative lattice charge [46]. In essence, the mica acts like a very concentrated set of counterions. Thus, the structure on the anionic surface may resemble the structure in a concentrated salt solution. These counterions are not only concentrated, they are also confined to a plane, which provides a template for planar adsorption.

3. Gold

Surfactants adsorbed onto gold surfaces form regular arrays that are similar to those observed on graphite [47]. Long cylinders or hemicylinders form in straight parallel lines on flat terraces. When these lines meet the steps in the crystal, the micelle prefers to deviate and remain on a single plane rather than step up or down to a parallel plane.

B. Wettability of the Solid Substrate

The adsorption of surfactant molecules from an aqueous solution to a solid is an exchange reaction that necessitates desorption of water. Therefore, the energy of adsorption of water to the solid is important in determining the adsorption of surfactant molecules [9,40]. If all other surface characteristics are equal, the energy of adsorption of a surfactant to a hydrophilic surface should be greater (less favorable) than the energy of adsorption to a hydrophobic surface, and therefore the surfactant should be more densely packed on the hydrophobic surface. Unfortunately, it can be experimentally difficult to change the wettability of the solid surface systematically without changing other properties, including the affinity of the surfactant for the solid.

The fraction of a surface that is covered by surface micelles depends on the shape of the micelle, so wettability should influence the shape of the adsorbed aggregate. When one considers a series of adsorbed aggregates in order of decreasing curvature (e.g., spherical micelle, cylindrical micelle, bilayer) the least curved aggregate (the flat layer) will generally cover the most solid surface [40]. Thus, adsorbed micelles should be perturbed from their equilibrium solution geometry to a less curved geometry at a hydrophobic surface.

Figure 2 shows the interfacial energies involved when micelles adsorb at interfaces. We will assume that the interfacial energy of the micelle (γ_{MW}) is approximately the same as for the adsorbed micelle and therefore that this makes no contribution to the energy. (This will not be true if the head group area changes; see later.) Compared with a micelle in solution, an adsorbed micelle has created a new interface between the micelle and the solid with interfacial energy γ_{SH}. The magnitude of γ_{SH} will depend on the orientation of the surfactant, but for now we will assume the surfactant tails are in contact with the solid. Adsorption reduces the area of the solid that is exposed to water, so the micelle is attracted to the surface with a wetting energy of $\gamma_{SW} - \gamma_{SH}$. This wetting energy will act to make the micelles cover the surface. This can be achieved in two ways: (1) by adsorbing more micelles or (2) by the spreading of each individual micelle. The latter is achieved by increasing the average packing parameter, P [48] and the aggregation number. The relationship between the packing parameter and spreading is shown in Fig. 3. The perturbation of the micellar shape from the free shape in solution is determined by the balance between the wetting energy and the energy required to perturb the shape of the micelle from its "free" curvature. This perturbation can be achieved by decreasing the average head group area or by decreasing the average alkyl chain extension.

The micelles can also cover a greater area of the surface by increasing the density of adsorbed micelles. In this case, the adsorption density is dictated by the balance between the wetting force and the repulsive micelle–micelle forces. This is considered further in Section III.E. The effects of wettability and charge on adsorbed surfactant shape are considered further in two papers by Johnson and Nagarajan [13,14].

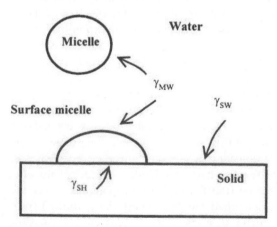

FIG. 2 Interfaces affected by micellar adsorption.

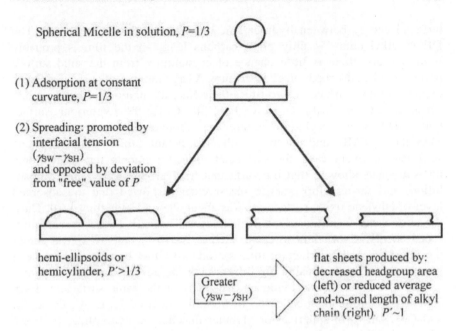

Spherical Micelle in solution, $P=1/3$

(1) Adsorption at constant curvature, $P=1/3$

(2) Spreading: promoted by interfacial tension $(\gamma_{SW}-\gamma_{SH})$ and opposed by deviation from "free" value of P

hemi-ellipsoids or hemicylinder, $P'>1/3$

Greater $(\gamma_{SW}-\gamma_{SH})$

flat sheets produced by: decreased headgroup area (left) or reduced average end-to-end length of alkyl chain (right). $P'\sim1$

FIG. 3 Effect of interfacial tension on the adsorbed micelle shape. The adsorption process has been artificially broken into two stages: (1) adsorption at constant aggregate curvature and (2) readjustment of the curvature in response to the microscopic solid–liquid interfacial tension. The P is the packing parameter in the free micelle; P' is the packing parameter in the adsorbed micelle. In general, the two packing parameters are not the same because different forces are present. The local forces can change both the effective head group area and the average length of the hydrocarbon chain.

The effect of solid wettability on the shape of adsorbed aggregates was first proposed to explain the formation of hemicylindrical SDS micelles on graphite [40]. The strong correlation between the crystallographic axis and the axis of the adsorbed micelle suggests that, for graphite, the main effect causing hemimicelle formation is actually the epitaxial adsorption of alkyl chains (see Section III.A). However, the effect of wettability has also been tested on a substrate that does not direct epitaxial adsorption.

We have studied the adsorption of many surfactants on amorphous silica surfaces that were made hydrophobic by chemisorption of diethyloctylchlorosilane (DEOS) [49]. DEOS is a silanating agent that produces an advancing water contact angle of 105° and a receding angle of 95°. The interfacial energy between the alkyl chain of surfactants and the silane chain is very small, so the spreading pressure is approximately the same as the interfacial energy between hydrocarbon and water ($\sim$50 mJ m^{-2}) [50]. The DEOS alkyl chain is only eight carbons long, so the film is probably liquidlike and there is little chance of crystallinity from the solid surface perturbing the adsorbed micelle structure. A large number of surfactants were exposed to this surface at two times their cmc: the nonionic ethylene oxide surfactants, $C_{10}E_5$, $C_{10}E_6$, $C_{12}E_5$, $C_{12}E_8$, $C_{14}E_6$, $C_{16}E_6$; the zwitterionic surfactants, DDAPS and decyl ammoniopropane sulfonate; the charged surfactants, SDS and C_{16}TABr; and the fluorocarbon surfactant, lithium perfluorooctane sulfonate. In every case, the surfactant formed a smooth flat sheet. Force measurements showed that the surfactant head groups were facing the solution, and earlier ellipsometric measurements showed that the adsorbed layer of ethylene oxide surfactants was about one molecule thick [10]. Thus, the surfactants form an oriented monolayer. The high DEOS–water interfacial energy is sufficient to cause each of the surfactants to spread across the interface. A flat surfactant film is produced either by reducing the head group area or by decreasing the thickness of the adsorbed surfactant film.

There is experimental evidence that many of the same surfactants form spherical or globular micelles on hydrophilic surfaces. $C_{10}E_5$, $C_{10}E_6$, $C_{12}E_8$, and $C_{16}E_6$ all form spherical or globular micelles on hydrophilic ($\theta \sim 0°$) silica [49]. The micelles have a very small contact area with the silica and are difficult to image. On hydrophilic silica there is a competition between water and the ethylene oxide head groups (rather than the tail) for the surface silanol groups. Water has a strong affinity for the silica silanol groups, and the micelle is unable to displace much water from the surface. C_{16}TABr also forms spherical micelles on hydrophilic silica up to about five times its cmc. For C_{61}TABr, the comparison is complicated by electrostatic interactions with the negatively charged silica. The zwitterionic surfactant forms spherical micelles on hydrophilic ($\theta \sim 0°$) silicon nitride [41]. In these cases, the solid surface has a strong affinity for the solvent water, the spreading energy is small, and so the adsorbed micelle has a shape similar to that observed in bulk solution.

Grant et al. [51] have extended the study of the effect of wettability on adsorbed surfactant shape by observing the shape of the nonionic surfactant $C_{12}E_8$ on a series of gold surfaces that were modified by adsorption of thiols. The most hydrophobic surface was produced by adsorbing alkane thiols, and increasingly hydrophilic surfaces were produced by adsorbing thiols from solutions containing an increasing ratio of ω-OH to ω-CH$_3$ groups. Consistent with earlier work, they found that the surfactant formed a flat monolayer on a pure ω-CH$_3$ surface and spherical micelles on a pure ω-OH surface. The micelles were only weakly bound to the pure ω-OH surface. An interesting new effect was observed for the surfactant adsorbed onto the surface that was 25% hydroxylated. The results are consistent with the adsorption of a bilayer. A bilayer structure has a similar effect to a monolayer in displacing the solvent water from the surface, but compared with a monolayer, the surfactant head group has displaced the alkyl chain at the thiol–surfactant interface. Presumably this is because the ethylene oxide head groups can hydrogen bond to the 25% of thiols that are —OH terminated.

C. Counterions

It is well known that the shape of micelles composed of charged molecules depends on the concentration and nature of the counterions present in solution. Counterions near the surfactant head groups reduce the electrostatic repulsion between the head groups, thereby reducing the effective head group area and increasing the mean curvature of the micelle [30]. Micelles adsorbed onto solid–solution interfaces should also be subject to the same influence but perhaps with diminished importance because of the control exerted by the substrate. For example, we have spent a great deal of effort examining the influence of various counterions on the adsorption of dodecyl sulfate onto graphite. The counterions affected the spacing between the adsorbed hemimicelles (Section III.E) but did not have a large effect on the aggregate shape. This is because the graphite strongly directs the formation of hemimicelles (Section III.A).

In order to observe the effect of counterions, it is necessary to use a solid substrate that has a weak effect on the surfactant. Silica is a good substrate because surfactants adsorbed to silica tend to retain a structure which is similar to that observed in bulk solution. In Section III.B we argued that this was because the silica was very hydrophilic, so water was difficult to displace from the interface. Patrick et al. [52], Velegol et al. [53], and Subramanian and Ducker [54] have investigated the effect of various counterions on the shape of $C_{16}TA^+$ micelles adsorbed onto silica. The addition of Br^- causes a transition to cylindrical (wormlike) micelles (Fig. 4), whereas Cl^- does not. The same phenomenon occurs in bulk solution and has been explained on the basis of the difference in polarizability of the ions [55]. An increased concentration of anions in solution leads to an increase in concentration of ions at the surface of the micelle and therefore to increased screening of the head group repulsions. The Br^- has a greater excess polarizability over the surrounding water than Cl^-, so it experiences greater attractive van der Waals forces from the micelle. Thus, Br^- is more concentrated at the micelle surface than Cl^- at the same bulk concentration.

We have investigated the effect of a variety of anions and found that the more polarizable ions are capable of effecting the sphere-to-cylinder transition on silica. (See Table 1.) In the table, we use the traditional physical

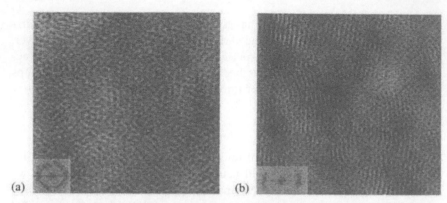

(a)   (b)

FIG. 4 AFM images at 300 nm of C_{16}TABr adsorbed at the interface between hydrophilic silica and (a) 10 mM C_{16}TABr solution and (b) 10 mM C_{16}TABr and 300 mM NaBr solution. The insets are Fourier transforms of the main images. (Reprinted with permission from Ref. 54. Copyright 2000 American Chemical Society.)

chemical descriptions of "soft" and "hard" to represent molecules that have high and low polarizability, respectively. The soft ions are able to effect the sphere-to-cylinder transition, but the hard ions are not. In particular, we have investigated the effect of substituting the more polarizable S for O in oxyacids. For example, CTA^+ forms spheres at the interface between silica and concentrated (0.4 M) CO_3^{2-} solutions but forms cylinders at the interface of

TABLE 1 Shape of CTA^+ Micelles Adsorbed onto the Interface Between Silica and Aqueous Salt Solution as a Function of Counterion Polarizability

Anion	Polarizability	Micellar shape
Cl^-	Hard	Oblate spheroid
Br^-	Borderline	Cylinder
Ac^-	Hard	Sphere
HSO_3^-	Borderline	Sphere
HS^-	Soft	Cylinder
SO_4^{2-}	Hard	Sphere
$S_2O_3^{2-}$	Soft	Cylinder or short rod
CO_3^{2-}	Hard	Sphere
CS_3^{2-}	Soft	Cylinder
SO_3^{2-}	Borderline	Sphere

Source: Adapted from Ref. 54.

even dilute (0.01 M) CS_3^{2-} solutions. The sulfite and hydrogen sulfite anions (SO_3^{2-} and HSO_3^-) are exceptions to this generalization. Although they are relatively polarizable, they do not cause the formation of cylinders. This may be because the most polarizable atom, S, has a partial positive charge. Apparently, to effect the sphere-to-cylinder transition, it is necessary to have a polarizable anionic section of the ion. One assumes that this group would also need to be accessible to the CTA^+ head group and not buried within the ion.

There has been some discussion of whether divalent ions should be better than monovalent ions at effecting the sphere-to-cylinder transition because of the increased electrostatic attraction [56]. Table 1 contains no evidence to support differences between monovalent and divalent ions. For example, the borderline soft HSO^{3-} ion forms spheres and continues to form spheres when the charge is increased by removal of a proton.

D. Surface Charge

The electrical interaction between charged surfactants and charged surfaces is clearly important in determining the arrangement of adsorbed molecules. Attractive interactions between the surfactant and oppositely charged groups at the interface lead to concentration of the surfactant at the interface and, therefore, enhance the opportunity for surfactant tail–tail interactions. The interactions with charged surface groups can act to diminish the repulsive interactions between charged surfactant head groups and, therefore, lead to a decrease in the curvature of the aggregate. This mechanism is the same as the mechanism that occurs on the addition of salt (Section III.C), except that the surface counterions are constrained to the surface, and sometimes the arrangement is dictated by the crystallography of the solid (e.g., for mica). The effect of surface charge has been investigated on silica, mica, and gold.

Manne et al. [38] examined the adsorption of $C_{14}TABr$ on silica as a function of pH. At twice the cmc, $C_{14}TABr$ forms approximately spherical micelles at pH 2.9, 6.3, and 9.9. In the absence of surfactant, silica is known to have a small surface charge near pH 2, and the overall surface charge becomes more negative as the pH increases [57]. Thus, the density of anionic surface sites on silica increases with pH. The presence of surfactant at the interface will probably increase the density of negative charges, but one would still expect to have more negative sites at high pH. One might have expected this increase in density of negative sites at high pH to lead to cylindrical micelles or even a flat bilayer, but this was not observed. We repeated their experiments for $C_{16}TABr$ at twice the cmc, extending the range to pH 11 (where there is probably already dissolution of the silica), and found that the aggregates were still spherical. Cylindrical $C_{16}TABr$ aggregates have been observed on silica at slightly higher concentrations [53,54], so there is no intrinsic reason why cylindrical aggregates cannot form. We believe that the surfactant micelles are not cylindrical at low Br^- concentration simply because the silica surface is so hydrophilic that water still occupies most of the surface in preference to the surfactant. Approximately spherical micelles displace fewer water molecules than cylindrical micelles when they adsorb.

We have investigated the effect of surface charge by measuring the arrangement of $C_{16}TABr$ molecules on mica. Mica is less hydrophilic than silica: it has a small ($\sim 5°$) water contact angle. The lattice charge on mica is fixed, but the net surface charge of mica can be regulated by the adsorption of cations, including the proton and cationic surfactants. All cations cause a reduction in surface charge, but some cations are more concentrated at the surface because of nonelectrostatic forces. For example, more polarizable ions have higher binding constants due to their lower affinity for water [58]. We have measured the structure of adsorbed CTA^+ micelles as a function of the concentration of H^+, Li^+, Na^+, K^+, and Cs^+ ions [34,59]. In the absence of salt, the CTABr at twice the cmc forms approximately spherical micelles in bulk solution and a flat (bilayer) sheet at the interface between the micellar solution and mica. The high density of mica lattice anions explains the difference between the bulk and surface structures. These ions screen the electrostatic repulsion between the charged cationic head groups, which allows a smaller head group area and a less curved aggregate.

When salt is added to solution, the flat, continuous structure breaks up into cylindrical or spherical micelles (see Fig. 5). Note that the increase in the curvature on addition of salt is the opposite of the effect that occurs in bulk solution, so clearly the change in curvature is controlled by the mica–solution interface. The rival cations compete with the quaternary ammonium head groups for the mica lattice charges. This reduces the density of anionic sites available to the surfactant. Fewer adsorbing surfactant head groups will associate directly with a lattice charge, releasing them from the plane of the mica surface and producing a larger repulsion between head groups. Both of these factors lead to more curved aggregates, i.e., spherical or cylindrical micelles. Salts differ in their ability to effect the planar-to-curved transition. The Cl^- salts are more effective at lowering the curvature than Br^- salts. This is simply a manifestation of the tighter binding of Br^- to the surfactant head groups, as discussed in the previous section. Furthermore, Cs^+ is more effective than Li^+ at converting the bilayer into spherical or cylindrical micelles. This is expected because Cs^+ is less strongly hydrated and so has a higher affinity for the mica in the absence of the surfactant. When surfactant displaces water at the surface, the Cs^+ should be even better than Li^+ at

FIG. 5 AFM images at 300 nm of $C_{16}TACl$ adsorbed at the interface between mica and a solution containing 2.7 mM CTACl and various concentrations of CsCl: (a) 0 mM CsCl; (b) 34 mM CsCl; (c) 100 mM CsCl. (Reprinted with permission from Ref. 59. Copyright 1998 American Chemical Society.)

replacing CTA^+ at the mica surface because the excess polarizability of Cs^+ should lead to a greater van der Waals interaction with the adsorbed micelles.

The effects of surface charge on adsorbed aggregate shape have also been examined on gold. Burgess et al. [60] performed an interesting study of the adsorption of SDS onto a gold (111) surface that was under potential control. When there was no applied potential, the SDS formed hemicylindrical surface micelles, as previously observed [47]. When a negative potential was applied, the surfactant desorbed. There is a potential at which the repulsive electrostatic force between the gold and the negatively charged surfactant is sufficient to overcome the attractive interaction between the surfactant alkyl chain and the gold atoms. When a large positive potential was applied, the hemimicelles transformed into a flat layer. The transformation occurred when the charge on the gold was equal to the total charge on the adsorbed surfactant. The flat structure is consistent with the formation of a surfactant bilayer. Thus, flat layers form on both mica and gold when the solid has a high surface-charge density.

E. Interactions Among Adsorbed Micelles

We investigated the effect of electrostatic interactions between hemicylindrical micelles adsorbed onto graphite. The electrostatic interaction was modified by varying the electrolyte concentration. Graphite strongly dictates that the overall geometry must remain hemicylindrical (Section III.A) even though the interaction between the head groups is affected by salt (Section III.C). An increase in salt concentration can reduce the repulsive force between the aggregates by reducing the Debye screening length. If counterions adsorb to the micelle, they also reduce the force by reducing the magnitude of the effective charge on the micelle.

We measured the separation between the long axes of the hemicylindrical micelles (the period) as a function of NaCl, $MnCl_2$, or $MgCl_2$ ions in a fixed surfactant concentration [40,61] (see Fig. 6). Note that the period is the sum of the micellar diameter and the intermicellar separation. The period increases approximately linearly with the Debye length for both NaCl and $MnCl_2$ solutions. The limiting period is slightly longer than 5 nm, which probably represents the diameter of the micelle, including water of hydration and counterions.

We can think of the equilibrium separation between the aggregates as being determined by the balance between a repulsive and an attractive force. The repulsive force is the electrostatic repulsion between micelles,and the attractive force is the wetting energy (described in Section III.B). The wetting energy decreases when the graphite is covered by micelles, so this promotes adsorption, and the electrostatic free energy increases with increased adsorption of the charge aggregates. The addition of an electrolyte to the solution reduces the electrostatic free energy, which allows more aggregates to adsorb. The average spacing is reduced by this additional adsorption. This picture is complicated somewhat by the known ability of salt solutions to alter the curvature of surfactant aggregates both in bulk solutions and at interfaces (Section III.C). In $MgCl_2$ solutions, the period does not continue to decrease with concentration but reaches a plateau at about 6 nm for a 2–4 mM solution. This suggests that the reduction in the separation between the adsorbed micelles is compensated by an increase in the diameter of the micelles.

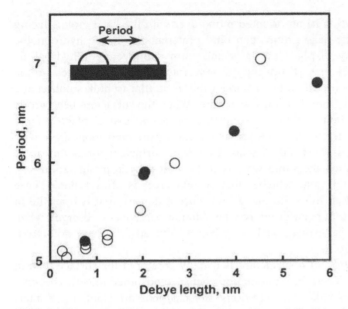

FIG. 6 Effect of solution Debye length on the period of hemicylindrical micelles adsorbed at the interface between SDS solution and graphite. (Closed circles) 2.8 mM SDS and NaCl; (open circles) 1 mM SDS and $MnCl_2$. (Adapted from Refs. 40 and 63.)

A similar argument can be made for the change in the period as a function of SDS concentration: the addition of sodium decreases the electrostatic repulsion and allows a smaller spacing between the adsorbed micelles. The addition of SDS also increases the chemical potential of SDS, allowing higher energy states to be reached on the surface.

IV. SUMMARY AND FUTURE DIRECTIONS

A variety of surface aggregates form at interfaces between micellar solutions and solid surfaces. The most common structures are spherical, globular, and cylindrical micelles; bilayers; and their corresponding hemistructures. The factors that affect micellar shape in solution are still important at interfaces, but additional effects arising from the solid often produce a perturbed structure at the interface. The final adsorbed structure depends on the competition for surface sites among the surfactant head group, the surfactant tail group, other solutes, and the solvent. When the solvent has a low affinity for the solid (hydrophobic surfaces), the surface micelles flatten out to reduce the fraction of the solid that is exposed to water. For very hydrophobic surfaces, this causes formation of an oriented monolayer with the head groups facing the solution. If the head groups can bind preferentially to the hydrophobic surface, they may displace the tails and form a bilayer. For hydrophilic surfaces, the water may still occupy a large fraction of surface sites, so the adsorbed micelles are in an environment that is similar to bulk solution and are less perturbed from their bulk structure. When the surfactant head group can form strong bonds with the solid, the surfactant can displace a large amount of water from the interface and form an oriented monolayer with head groups facing the solid. A second layer of surfactant forms on top of this surface in much the same way as it does on a hydrophobic solid. (At this point, it is not clear whether this second layer is always flat.) These bilayers occur when the solid has a high charge density that is opposite in sign to the surfactant head group charge. Similar behavior is observed when a covalent bond is formed, such as when a thiol adsorbs onto gold from ethanol solution [62].

The concentration of a surfactant is usually greater at the surface than in the bulk solution. Therefore, interactions between adsorbed micelles become important at lower bulk concentrations. For charged surfactants these interactions will be large and repulsive, but the addition of salt decreases the repulsion between the charged surface micelles. The final shape of the adsorbed aggregate depends on all the intermolecular interactions, but because the surfactant is usually more concentrated at the interface, the adsorbed structure is often more similar to structures formed in a bulk solution at a higher concentration that the solution from which the micelle actually adsorbs.

To date, almost all AFM work has focused on adsorption from micellar solution or from solutions just below the cmc. Below the cmc, the chemical potential of the surfactant is a much stronger function of solution concentration, so it should be possible to form a larger variety of surfactant structures. There is hope that new structures might be observed in this more dilute regime.

REFERENCES

1. AM Gaudin, DW Fuerstenau. Trans AIME 202:958–962, 1955.
2. DW Fuerstenau. J Phys Chem 60:981–985, 1956.
3. JH Harwell, JC Hoskins, RS Schechter, WH Wade. Langmuir 1:251–262, 1985.
4. JT Kunjappu, PJ Somasundaran. J Colloid Interface Sci 175:520–521, 1995.
5. M Bjelopavlic, J Ralston, GJ Reynolds. J Colloid Interface Sci 208:183–190, 1998.
6. P Somasundaran, J Fuerstenau. J Phys Chem 70:90–96, 1966.

7. BY Zhu, T Gu. Adv Colloid Interface Sci 37:1–32, 1991.
8. Y Gao, J Du, T Gu. J Chem Soc Faraday Trans 1 83:2671–2679, 1987.
9. MR Böhmer, LK Koopal. Langmuir 6:1478–1484, 1990.
10. F Tiberg, B Jönson, Z Tang, B Lindman. Langmuir 10:2294–2300, 1994.
11. TP Goloub, LK Koopal, BH Bijsterbosch. Langmuir 12:3188–3194, 1996.
12. TP Goloub, LK Koopal. Langmuir 13:673–681, 1997.
13. RA Johnson, R Nagarajan. Colloids Surf A 167:21–36, 2000.
14. RA Johnson, R Nagarajan. Colloids Surf A 167:37–46, 2000.
15. P Levitz, A El Miri, D Keravis, JH van Damme. J Colloid Interface Sci 99: 484–492, 1984.
16. P Levitz, JH van Damme. J Phys Chem 88:2228–2235, 1984.
17. P Levitz, JH van Damme. J Phys Chem 90:1302–1310, 1986.
18. P Somasundaran, S Krishnakumar. Colloids Surf A 93:79–95, 1994.
19. P-O Quist, E Söderlind. J Colloid Interface Sci 172:510–517, 1995.
20. P Kékicheff, HK Christenson, BW Ninham. Colloids Surf 40:31–41, 1989.
21. RM Pashley, PM McGuiggan, RG Horn, BW Ninham. J Colloid Interface Sci 126:569–578, 1988.
22. MW Rutland, TJ Senden. Langmuir 9:412–418, 1993.
23. J Israelachvili, RM Pashley. J Colloid Interface Sci 98:500–514, 1984.
24. RM Pashley, PM McGuiggan, BW Ninham, DF Evans. Science 229:1088–1089, 1985.
25. ZM Zorin, VP Romanov, NV Churaev. Colloid Polym Sci 257:968–972, 1979.
26. RW Smith, S Akhtar. In: MC Fuerstenau, ed. Flotation: AM Gaudin Memorial Volume. New York: AIME, 1976, pp 87–116.
27. S Manne, JP Cleveland, HE Gaub, GD Stucky, PK Hansma. Langmuir 10: 4409–4413, 1994.
28. G Fragneto, RK Thomas, AR Rennie, J Penfold. Langmuir 12:6036–6043, 1996.
29. FG Greenwood, GD Parfitt, NH Picton, DG Wharton. In: Adsorption from Aqueous Solution. Washington, DC: American Chemical Society, 1968.
30. J Israelachvili. Intermolecular and Surface Forces. 2nd ed. San Diego: Academic Press, 1991.
31. DF Evans, HK Wennerström. The Colloidal Domain. 2nd ed. New York: Wiley, 1999.
32. JN Israelachvili. Intermolecular and Surface Forces. 2nd ed. San Diego: Academic Press, 1991, p 304.
33. WA Ducker, DR Clarke. Colloids Surf A 93:275–292, 1994.
34. WA Ducker, EJ Wanless. Langmuir 15:160–168, 1999.
35. DG Hall. J Chem Soc Faraday Trans II 68:2169–2182, 1972.
36. SG Ash, DH Everett, C Radke. J Chem Soc Faraday Trans II 69:1256–1277, 1993.
37. V Subramanian, W Ducker. J Phys Chem B 105:1389–1402, 2001.
38. S Manne, HE Gaub. Science 270:1480–1482, 1995.
39. JF Liu, WA Ducker. J Phys Chem B 103:8558–8567, 1999.
40. EJ Wanless, WA Ducker. J Phys Chem 100:3207–3214, 1996.
41. LM Grant, WA Ducker. J Phys Chem B 101:5337–5345, 1997.

42. HN Patrick, GG Warr, S Manne, IA Aksay. Langmuir 13:4349–4356, 1997.
43. DM Cyr, B Venkataraman, GW Flynn. Chem Mater 8:1600–1615, 1996.
44. JGL Gaines. Nature 178:1304–1306, 1956.
45. HN Patrick, GG Warr, S Manne, IA Aksay. Langmuir 15:1685–1692, 1999.
46. WA Ducker, LM Grant. J Phys Chem 100:11507–11511, 1996.
47. M Jaschke, HJ Butt, HE Gaub, S Manne. Langmuir 13:1381–1384, 1997.
48. J Israelachvili, DJ Mitchell, BW Ninham. J Chem Soc Faraday Trans II 72: 1525–1568, 1976.
49. LM Grant, F Tiberg, WA Ducker. J Phys Chem B 102:4288–4294, 1998.
50. JN Israelachvili. Intermolecular and Surface Forces. San Diego: Academic Press, 1991, p 315.
51. LM Grant, T Ederth, F Tiberg. Langmuir 16:2285–2291, 2000.
52. HN Patrick, GG Warr, S Manne. Langmuir 15:1685–1692, 1999.
53. SB Velegol, BD Fleming, S Biggs, EJ Wanless, RD Tilton. Langmuir 16:2548–2556, 2000.
54. V Subramanian, WA Ducker. Langmuir 16:4447–4454, 2000.
55. BW Ninham, V Yaminsky. Langmuir 13:2097–2108, 1997.
56. HK Wennerström, A Khan, B Lindman. Adv Colloid Interface Sci 34:433–449, 1991.
57. WA Ducker, TJ Senden, RM Pashley. Langmuir 8:1831–1836, 1992.
58. RM Pashley. J Colloid Interface Sci 83:531–546, 1981.
59. RE Lamont, WA Ducker. J Am Chem Soc 120:7602–7607, 1998.
60. I Burgess, CA Jeffrey, X Cai, G Szymanski, Z Galus, J Lipkowski. Langmuir 15:2607–2616, 1999.
61. EJ Wanless, WA Ducker. Langmuir 13:1463–1474, 1997.
62. CD Bain, GM Whitesides. Science 240:62–63, 1988.
63. EJ Wanless, TW Davey, WA Ducker. Langmuir 13:4223–4228, 1997.

13

A Simple Model to Predict Nonlinear Viscoelasticity and Shear Banding Flow of Wormlike Micellar Solutions

**J. E. PUIG, F. BAUTISTA, J. H. PÉREZ-LÓPEZ,
and J. F. A. SOLTERO** Universidad de Guadalajara,
Guadalajara, Mexico

OCTAVIO MANERO Universidad Nacional Autónoma de
México, Mexico City, Mexico

ABSTRACT

A model consisting of the codeformational Maxwell constitutive equation coupled to a kinetic equation for breaking and re-formation of micelles is presented to reproduce most of the nonlinear viscoelastic properties of wormlike micelles. This simple model is also able to predict shear banding in steady shear and pipe flows as well as the long transients and oscillations that accompany this phenomenon. Even though the model requires six parameters, all of them can be evaluated from single and independent rheological experiments, and then they can be used to predict other flow situations. The predictions of our model are compared with experimental data for aqueous micellar solutions of cetyltrimethylammonium tosilate (CTAT).

I. INTRODUCTION

Wormlike micellar solutions exhibit a rich and complex rheological behavior because of their dynamical nature (continuous breaking and re-formation) and their ability to form entanglements similar to those of polymer solutions [1–8]. A crucial parameter in understanding the dynamics of wormlike micelles is the ratio of the breaking and the reptation times, ζ ($=\tau_{break}/\tau_{rep}$)

[9,10]. Depending on this ratio, the response at low frequencies can range, in the linear viscoelastic regime, from near-Maxwell behavior with a single relaxation time to a polymer-like behavior with a spectrum of relaxation times [2,4,9,10]. At high frequencies, because of the occurrence of the Rouse and other relaxation mechanisms, deviations from Maxwell behavior are observed even in the fast breaking regime [10].

The nonlinear rheological behavior of these systems is even more intriguing and interesting. One of the most outstanding features is the shear banding flow or "spurt effect," first detected in polymer solutions [11]. Here, under steady shear flow, a stress plateau is detected at a critical shear rate due to the development of nonhomogeneous flow where two bands supporting different shear rates coexist [12–14]. At and above this critical rate, stress oscillations and long transients are detected. Also, a metastable branch is followed under certain flow conditions [13,14]. At shear rates higher than a second critical value, the flow becomes homogeneous again and steady conditions are achieved quite rapidly, i.e., within a few relaxation times of the sample [15]. Shear banding flow has been visualized by nuclear magnetic resonance (NMR) velocimetry [16–19], by small angle neutron scattering [20,21] and by optical methods [22,23].

Several models have been forwarded to explain shear banding flow in wormlike micellar solutions [12,16,23]. The nonlinear extension of the reaction–reptation model predicts a stress saturation of $0.67G_0$ at a critical shear rate equal to $2.6/\tau_d$, where G_0 is the plateau elastic modulus and τ_d is the main relaxation time of the sample [12]. Other models predict a Newtonian behavior followed by shear thinning flow and then by an unstable flow region [16,23]. However, none of the models in the literature appears to be capable of predicting, in addition to the mentioned features, the long transients, the stress oscillations, the presence of a second metastable branch at high shear rates, which has been observed experimentally [18]; and a homogeneous flow region above the second critical shear rate.

Here we present a simple (although phenomenological) model that is able to predict all these features in both steady shear [15] and pipe flows [24]. Moreover, the model also predicts the nonlinear viscoelastic flow behavior of wormlike micelles [14].

II. THE MODEL

The model consists of the codeformational Maxwell equation coupled to a kinetic equation to account for the breaking and re-formation of micelles [15,24]. For simple shear flow, the model simplifies to the following system of ordinary differential equations:

$$N_1 = \frac{1}{G_0\varphi}\left[\frac{dN_1}{dt} - 2\sigma\dot\gamma\right] = 0 \tag{1}$$

$$\sigma + \frac{1}{G_0\varphi}\frac{d\sigma}{dt} = \frac{\dot\gamma}{\varphi} \tag{2}$$

$$\frac{d\varphi}{dt} = \frac{\varphi_0 - \varphi}{\gamma} + k_0(1 + \mu_1\dot\gamma)(\varphi_\infty - \varphi)\sigma\dot\gamma \tag{3}$$

In these equations, σ is the shear stress, $\dot\gamma$ is the shear rate, N_1 [$\equiv\sigma_{11} - \sigma_{22}$, with σ_{ii} the first ($i = 1$) and the second ($i = 2$) normal stresses] is the first normal stress difference, φ is the fluidity or inverse of the shear viscosity (η), φ_0 and φ_∞ are the fluidities at zero and infinite shear rates, respectively, G_0 is the shear plateau modulus, λ is a structure relaxation time, k_0 is a kinetic constant for structure breakdown, and μ_1 is the shear banding intensity parameter. In the derivation of these equations, the third normal stress, σ_{33}, was neglected and the subscripts of the shear stress and shear rate were dropped for simplicity.

Thus, for a steady shear flow, the time derivatives in Eqs. (1) to (3) are equal to zero. For this flow situation, one obtains

$$N_1 = \frac{2\sigma\dot\gamma}{G_0\varphi} \tag{4}$$

$$\varphi^2 - \varphi_0\varphi - k_0\lambda(\varphi_\infty - \varphi)(1 + \mu_1\dot\gamma)\dot\gamma^2 = 0 \tag{5}$$

For the fully developed isothermal flow of an incompressible fluid in a pipe of radius R and length L where the pressure drop along the length of the tube is ΔP, the shear stress distribution is given by $(\Delta P/L)(r/2)$; hence, the shear stress at the tube wall is given by this equation with the radial position $r = R$. Also, the apparent shear rate, $\dot\gamma_{app}$, is given by $4Q/(\pi R^3)$, where Q is the volumetric flow rate, which can be calculated from the fluidity data from

$$Q = \frac{\pi\Delta P}{L}\int_0^R \rho\left\{\int_0^\rho \xi\varphi(\xi)\,d\xi\right\}d\rho \tag{6}$$

The model needs six parameters (φ_0, φ_∞, G_0, λ, k_0, and μ_1) to predict the experimental data. However, as already described, each of these variables can be determined from independent rheological measurements, and then they can be used to predict and, in some cases, reproduce other flow situations [24]. The details of the measurements to obtain these parameters are given elsewhere [14,15].

III. RESULTS AND DISCUSSION

Figure 1 shows a schematic representation of the $\sigma-\dot\gamma$ relationship for fluids that exhibit shear banding under steady shear flow [25]. In this scheme, six flow regions are detected. At very low shear rates ($\dot\gamma \leq \dot\gamma_N$), the flow is Newtonian with $\eta = \eta_0$ (region I). For $\dot\gamma_N < \dot\gamma < \dot\gamma_{c1}$, shear thinning behavior is observed (region II). However, at the critical shear rate, $\dot\gamma_{c1}$, and up to $\dot\gamma_{c2}$, a stress plateau forms. Notice that the unstable region (region IV), ($d\sigma/d\dot\gamma < 0$) is bounded by two metastable branches (regions III and V). Once $\dot\gamma_{c2}$ is surpassed, the flow becomes homogeneous again (region VI) but with a smaller Newtonian viscosity (η_∞). Both the low- and the high-shear rate branches, by the way, have been observed experimentally [13,14,18].

The predicted plots of σ versus $\dot\gamma$ in a steady shear flow as a function of μ_1 or of the product of k_0 and λ are shown in Fig. 2. In both cases, these plots are remarkably similar to that shown in Fig. 1. In the first case (Fig. 2a), the shear banding region becomes wider and the stress plateau appears at lower values as μ_1 increases. Notice that as μ_1 goes to zero, the sigmoidal shape tends to vanish and when μ_1 is equal to zero, the sigmoid collapses into a monotonically increasing curve with an inflection point (labeled C in Fig. 2a); i.e., the stress plateau disappears when the shear banding intensity parameter becomes zero. On the other hand, under steady-state conditions

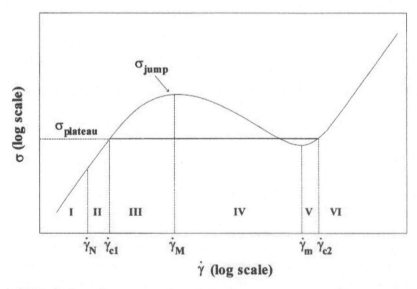

FIG. 1 Schematic representation of the shear stress σ versus shear rate $\dot\gamma$ relationship for materials that exhibit the spurt effect.

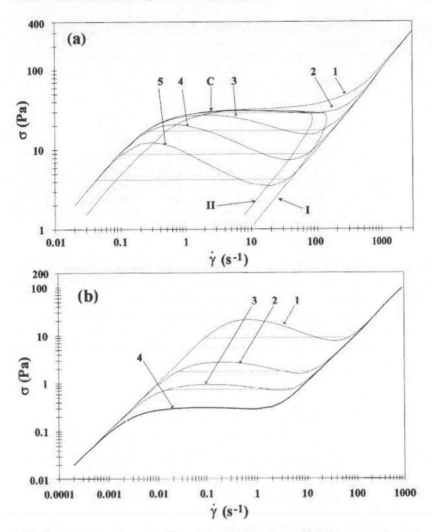

FIG. 2 Shear stress versus shear rate as a function of (a) the parameter μ_1 in s [0 (1); 0.01 (2); 0.1 (3); 1 (4); 10 (5)]; (b) the product $k_0\lambda$ in Pa^{-1} s [10^{-6} (1); 10^{-5} (2); 10^{-4} (3); 10^{-3} (4)].

[see Eq. (5)], the parameters k_0 and λ do not play separate roles. Here, as $k_0\lambda$ increases, $\dot{\gamma}_{c1}$, $\dot{\gamma}_{c2}$, and $\sigma_{Plateau}$ move to larger values. As shown in Table 1, the value of λ increases whereas that of k_0 diminishes with increasing CTAT concentration. However, the product of k_0 and λ diminished with increasing surfactant concentration and both $\sigma_{Plateau}$ and $\dot{\gamma}_{c1}$ shifted to larger

TABLE 1 Values of the Parameters of the Model for Various Concentrations of CTAT Micellar Solutions

C_{CTAT} (wt%)	G_0 (Pa)	φ_0 (Pa^{-1} s^{-1})	φ_∞ (Pa^{-1} s^{-1})	λ (s)	k_0 (Pa^{-1})	μ_1 (s)
1.5	2.5	0.152	80	0.058	0.015	0.0035
2	5.4	0.083	50	0.068	0.005	0.004
3	12	0.055	28	0.107	0.002	0.0045
5	41.5	0.0275	22	0.17	0.00021	0.005
10	176	0.0061	15	0.33	0.00003	0.007
15	380	0.005	12.6	0.38	0.00001	0.01
20	620	0.0042	12	0.42	0.0000042	0.017

values (see Fig. 2 of Ref. 14), in agreement with our predictions (Fig. 3). Unfortunately, $\dot{\gamma}_{c2}$ could not be determined in our measurements because the sample was expelled from the cone-and-plate geometry at high shear rates, so the predictions of the model could not be corroborated. Nevertheless, $\dot{\gamma}_{c2}$ and the development of a homogeneous flow for $\dot{\gamma} > \dot{\gamma}_{c2}$ have been determined in pipe flow experiments followed by NMR imaging velocimetry [18].

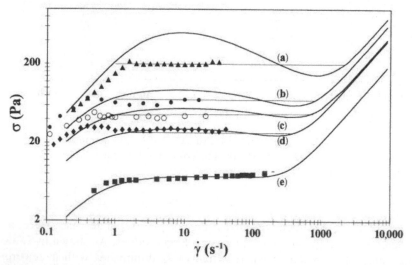

FIG. 3 Shear stress versus shear rate measured under steady-state conditions at 30°C for CTAT micellar solutions in wt% (■) 2; (◆) 3; (○) 5; (●) 10; (▲) 20.

Figure 3 shows the experimental steady shear measurements (symbols) and the predictions of the model (solid lines) for wormlike micellar solutions of various CTAT concentrations. The predictions were performed with the rheologically estimated parameters reported in Table 1. For low CTAT concentrations (line e), our model predicts a monotonically increasing relationship between σ and $\dot{\gamma}$ with a near-zero slope at intermediate shear rates that follows the experimental data closely. This is the result of the small value of μ_1 for this sample (Table 1). However, for higher CTAT concentrations (lines a to d), our model reveals a sigmoid whereas the experimental data show an increasing region of σ versus $\dot{\gamma}$ and then a stress plateau. However, the line that joins the critical shear rates $\dot{\gamma}_{c1}$ and $\dot{\gamma}_{c2}$ passes through the experimental data. The criterion employed to set the position of the dividing line is that the areas below and above this line and the sigmoid should be equal, which implies that the bands at $\dot{\gamma}_{c1}$ and $\dot{\gamma}_{c2}$ are supporting equal amounts of extended Gibbs free energy; otherwise, there would be a driving force to restore a homogeneous flow, contrary to experimental observations. Analysis of the shear banding process by irreversible thermodynamics indicates that this criterion is indeed valid [24,26]. Moreover, the criterion sets the value of the shear banding intensity parameter and establishes a *unique* relationship between σ_{Plateau} and μ_1.

Our model also predicts quite well the flow behavior of wormlike micelles in a pipe [24]. Here, because the shear stress is not constant across the tube, it is possible to have simultaneously both homogeneous and unstable flow regions. Figure 4 depicts normalized velocity profiles as a function of normalized radial position for different wall shear stresses (or pressure gradient). When $\sigma_w < \sigma_{\text{Plateau}}$, the parabolic profile of the Poiseuille flow of a Newtonian fluid (for $\dot{\gamma} \leq \dot{\gamma}_N$) or a shear-thinning fluid (for $\dot{\gamma}_N < \dot{\gamma} < \dot{\gamma}_{c1}$) is observed (Fig. 4). Also, steady conditions are achieved very rapidly (Fig. 5a, curve 1). However, once $\sigma_w \approx \sigma_{\text{Plateau}}$, a band supporting a higher shear rate appears near the tube wall with the development of flow instabilities and long transients (Fig. 5c); also, a nearly discontinuous shear stress profile close to the pipe wall is predicted (Fig. 4). The model, moreover, indicates a near-Newtonian profile at the center of the tube. On the other hand, when $\sigma_w > \sigma_{\text{Plateau}}$, the parabolic profile of a Newtonian fluid appears again but with a smaller shear viscosity, η_∞ (Fig. 4). Nevertheless, a small discontinuous region remains near the pipe center as a result of the shear stress dependence on radial position (inset of Fig. 4). Again, steady conditions are achieved very quickly (Fig. 5a, curve 2).

Two important predictions of our model are the long transients and the oscillations associated with shear banding flow. In steady shear flow, our model predicts overshoots and oscillations in steady shear flow when the shear rate is within the shear banding region (Fig. 5b and c). Moreover, it

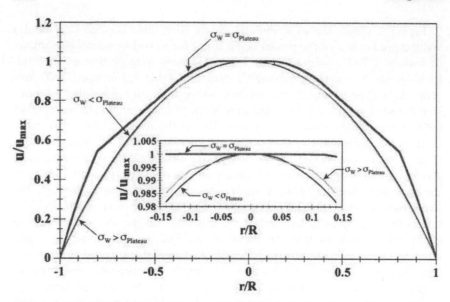

FIG. 4 Normalized velocity profiles as a function of normalized radial position for different wall shear stresses: $\sigma_W < \sigma_{Plateau}$; $\sigma_W = \sigma_{Plateau}$; $\sigma_W > \sigma_{Plateau}$. Inset: Enlargement of the profiles near the center of the tube.

predicts that for shear rates smaller than $\dot{\gamma}_{c1}$ or larger than $\dot{\gamma}_{c2}$, steady conditions are achieved very rapidly, i.e., a few relaxation times of the sample (Fig. 5a). However, as the shear rate approaches $\dot{\gamma}_{c1}$, our model indicates that the time required to achieve steady conditions increases rapidly and even very long times can be required for both steady shear and pipe flows (Fig. 6). These predictions are consistent with the pressure drop variations reported for the shear banding regime in a pipe flow of CTAT micellar solutions [27]. In this case, fluctuations in the pressure drop were noticed for over 2 h without any sign of ever achieving steady flow. Also, it was not possible to record meaningful data by NMR velocimetry imaging because of fluctuations in velocity profiles at shear rates near the critical value [18].

Besides describing the most important features of shear banding flow, our model can reproduce many of the details of the nonlinear rheological behavior of wormlike micelles such as inception of shear flow, instantaneous stress relaxation, and interrupted shear flow [14]. Figure 7 depicts the experimental data (symbols) and model predictions (solid lines) for stress relaxation after cessation of flow for different shear strain levels. In this case,

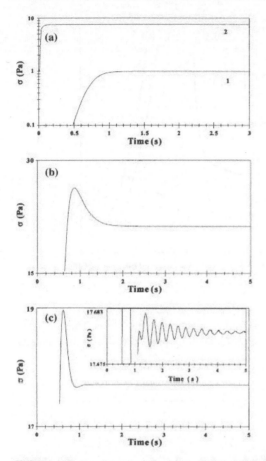

FIG. 5 Shear stress as a function of time after inception of shear flow: (a) curve 1 ($\dot{\gamma} < \dot{\gamma}_N$) and curve 2 ($\dot{\gamma} > \dot{\gamma}_{c2}$); (b) $\dot{\gamma}_N < \dot{\gamma} < \dot{\gamma}_{c1}$; (c) $\dot{\gamma}_{c1} < \dot{\gamma} < \dot{\gamma}_{c2}$. Inset in part (c) shows an enlargement of the scale.

when the shear strain is within the linear region, a single-exponential stress relaxation is observed, which is faithfully followed by our model. However, when the strain is within the nonlinear region, two relaxation times are dominant: one at very short times and another at long times. The latter one, by the way, is identical to the linear viscoelastic relaxation time (Fig. 7). At intermediate times, there is a transition region. Our model predicts quite well both the linear and the nonlinear regions as well as the transition regime up to moderate strain levels (10%). For higher values, our model overpredicts the experimental data. The Wagner approach, consisting of the generalized Maxwell constitutive equation with a damping function [$\exp(-k\gamma)$],

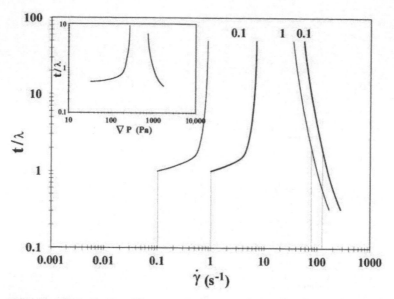

FIG. 6 Dimensionless time required to reach steady state (t/λ) as a function of shear rate for steady shear flow of wormlike micellar solutions. Inset: Dimensionless time as a function of the applied pressure gradient for pipe flow of wormlike micellar solutions.

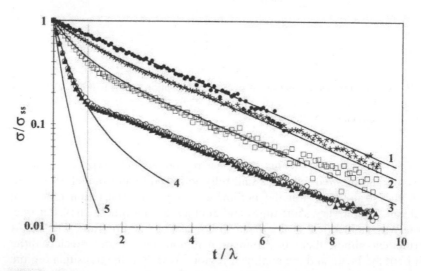

FIG. 7 Stress relaxation after cessation of flow as a function of shear rate at 30°C for a 10 wt% CTAT micellar solution: (●) 1 s^{-1}; (∗) 2 s^{-1}; (□) 3 s^{-1}; (○) 4 s^{-1}; (▲) 5 s^{-1}. The parameters used for the predictions of the model were taken from Table 1.

where k is the damping coefficient and γ is the shear strain [28], also fails to predict our data for strain levels higher than 10%.

IV. SUMMARY

A simple model with no adjustable parameters was used here to reproduce many of the features of the nonlinear viscoelastic behavior of wormlike micelles, including shear banding flow. It is noteworthy that shear banding flow is rarely observed in polymer solutions due to the typical broad molecular weight distribution observed in these systems, which masks the shear banding phenomenon. Wormlike micellar solutions also have a broad size distribution. However, the micelle size distribution is averaged by the chain breaking and recombination processes, so micellar solutions behave as highly monodisperse systems. Hence, shear banding flow is easily detected in wormlike micellar solutions, which makes them model systems for studying complex flow behavior.

ACKNOWLEDGMENT

This work was supported by the National Council of Science and Technology of México (CONACYT grant 3343-P-E9607).

REFERENCES

1. SJ Graveshold. J Colloid Interface Sci 57:575, 1976.
2. H Rehage, H Hoffmann. Rheol Acta 21:561, 1982; J Phys Chem 92:4712, 1988; Mol Phys 74:033, 1991.
3. T Shikata, Y Sakaiguchi, H Urakami, A Tamura, H Hirata. J Colloid Interface Sci 119:291, 1987.
4. T Shikata, H Hirata, T Kotaka. Langmuir 3:1081, 1987; Langmuir 4:354, 1988.
5. A Rauschter, H Rehage, H Hoffmann. Prog Colloid Polym Sci 84:99, 1991.
6. F Kern, R Zana, SJ Candau. Langmuir 7:1344, 1991.
7. JFA Soltero, JE Puig, O Manero, PC Schulz. Langmuir 11:3337, 1995.
8. JFA Soltero, JE Puig, O Manero. Langmuir 12:2654, 1996.
9. ME Cates. Europhys Lett 4:497; Macromolecules 20:2289, 1987; J Phys (France) 49:1593, 1988.
10. ME Cates, SJ Candau. J Phys Condens Matter 2:6869, 1990.
11. GV Vinogradov. Rheol Acta 12:357, 1973.
12. NA Spenley, ME Cates, TCB McLeish. Phys Rev Lett 71:939, 1993.
13. C Grand, J Arrault, ME Cates. J Phys II (France) 6:551, 1997.
14. JFA Soltero, F Bautista, JE Puig, O Manero. Langmuir 15:1804, 1999.
15. F Bautista, JFA Soltero, JH Pérez-López, JE Puig, O Manero. J Non-Newtonian Fluid Mech 94:57–66, 2000.

16. PT Callaghan, ME Cates, CF Rofe, JBFA Smelders. J Phys II (France) 6:375, 1996.
17. RW Mair, PT Callaghan. J Rheol 41:901, 1997.
18. MM Britton, RW Mair, RK Lambert, PT Callaghan. J Rheol 43:897, 1999.
19. LF Berret, D Roux, G Porte, P Lindner. Europhys Lett 25:521, 1994.
20. V Schmitt, F Lequeux, A Pousse, D Roux. Langmuir 10:955, 1994.
21. JP Decruppe, R Cressely, R Makhloufi, E Cappelaere. Colloid Polym Sci 273: 346, 1995.
22. R Makhloufi, JP Decruppe, A Ait-Ali, R Cressely. Europhys Lett 32:253, 1995.
23. NA Spenley, XF Yuan, ME Cates. J Phys II (France) 7:1071, 1996.
24. F Bautista, JFA Soltero, ER Macias, JE Puig, O Manero. J Non-Newtonian Fluid Mech, submitted.
25. TCB McLeish, RC Ball. J Polym Sci Polym Phys Ed 24:1735, 1986.
26. D Jau, J Casas-Vazquez, M Criado-Sancho. Thermodynamics of Fluid Under Flow. Berlin: Springer-Verlag, 2000.
27. S Hernández-Acosta, A González-Alvárez, O Manero, AF Méndez-Sánchez, J Pérez-González, L De Vargas. J Non-Newtonian Fluid Mech 85:229, 1999.
28. MH Wagner. Rheol Acta 15:136, 1976.

14

Preparation and Stabilization of Silver Colloids in Aqueous Surfactant Solutions

DAE-WOOK KIM, SEUNG-IL SHIN, and SEONG-GEUN OH   Hanyang University, Seoul, Korea

ABSTRACT

Silver colloids were prepared in the presence of various surfactants by the reduction of silver nitrate with hydrazine. Because of the positively charged hydrophobic nature of Ag nanoparticles, the Ag colloids prepared in aqueous surfactant solutions of sodium dodecyl sulfate (SDS) and Tween 20 showed good stability. But poor colloidal stability was observed in solutions of cetyltrimethylammonium bromide (CTAB) and NP-9. The stabilization of Ag colloids by surfactant molecules was explained on the basis of the electrostatic interaction between the Ag particles and surfactants and a stabilization model was proposed. The particle size distribution was investigated by ultraviolet (UV) absorption spectroscopy measurements. The UV absorption spectra showed different patterns depending on the nature of the stabilizers (i.e., surfactants). In the case of Tween 20 as a stabilizer, the smallest particles, about 11.6 nm in average diameter, were obtained. In the case of CTAB, pearl formation was observed because of the formation of relatively large particles about 300 nm in size.

I. INTRODUCTION

Colloidal dispersions of metals, semiconductors, and polymers have attracted considerable interest because of their photochemical [1], photocatalytic [2], and nonlinear optical properties [3]. Various methods for their preparation have been studied: controlled chemical reduction [4,5], photochemical or radiation-chemical reduction [6,7], photocatalytic reduction [8], etc.

It is well known that surfactants form several types of well-organized assemblies that provide specific size, geometrical control, and stabilization to particulate assemblies formed within the organized surfactant assemblies. The host surfactant assemblies that are available for the formation of nanoparticles are summarized in Table 1. The aqueous micellar solutions, reverse micelles, microemulsions, vesicles, monolayers, Langmuir–Blodgett films, and bilayer lipid membranes are typical surfactant assemblies that are often employed to prepare nanoparticles [9,10].

In this chapter, the preparation of Ag colloids in aqueous surfactant solutions and their stability were investigated. Depending on the nature of the materials prepared, two types of adsorption onto the surface of particles may be considered. If the surface of particles is hydrophobic, the hydrophobic part of the surfactant will adsorb onto the surface of particles and form a monomolecular film in an aqueous solution. If the surface of particles is hydrophilic, the hydrophilic part will adsorb onto the surface of particles and a bilayer surfactant film will form at the surface of particles in the aqueous solution because the hydrophobic part of the surfactant cannot be oriented toward the aqueous solution.

Furthermore, depending on the charge of the particle and surfactant head groups, two kinds of binding behaviors are shown. In the case of an ionic surfactant of opposite charge to the particle, the hydrophilic head groups of the surfactant bind to the particle surface, which then becomes hydrophobic [11]. Such surfactant-covered particles cannot be kept in an aqueous environment unless a double layer of surfactant molecules is formed, which is difficult to achieve with very small colloidal particles [12]. With an ionic surfactant of the same charge as the particle surface, binding of the surfactant head groups does not occur [13].

Our choice of silver metal was based on several properties of silver. First, silver is the cheapest noble metal. Second, it has a narrow intense plasmon absorption band in the visible region that is very susceptible to surface–interface effects [14].

The aim of this study was to prepare colloidal solutions of silver nanoparticles in surfactant aqeuous solutions and to investigate the effects of surfactant molecules on the particle size formed in surfactant solutions.

In this work, a positively charged Ag colloid was prepared in both the absence and presence of surfactants. The particle size distribution and UV absorption of Ag colloid were investigated. Furthermore, a model for stabilization of Ag colloid was proposed.

BLE 1 Properties of Organized Surfactant Assemblies

	Aqueous micelle	Reverse micelle	Microemulsion	Monolayers	Bilaye mem
hod of prepaation	Dissolving appropriate (above the cmc) amount of surfactant in water	Dissolving appropriate amount of surfactant in an apolar solvent and adding small amounts of water	Dissolving appropriate amount of surfactant and cosurfactant in water or oil	Spreading the surfactant (or a dilute solution of it in an organic solvent) on water surface	Painting surfac Teflo
ght average ıolecular /eight	2000–6000	2000–6000	10^4–10^7	Depends on area covered and density of coverage	Depends cover densi cover
lrodynamic iameter (nm)	4–10	4–10	5–500	Depends on area covered and density of coverage	Depends cover densi cover
ıe scale of ıonomer agregate formaon, breakown	10^{-4}–10^{-6} s	10^{-4}–10^{-6} s	10^{-4}–10^{-6} s	Monomer to subphase, minutes–hours	Monom teau, hours
›ility ıtion by /ater	Months Destroyed	Months Water pools, enlarges: water-in-oil microemulsion formed	Months Depends on the phase diagram	Days, weeks	Hours
nber of reacınts	Few	Few	Large	Large	Large
ıbilization ites	Distributed around and within the Stern layer, no deep penetration	Aqueous inner pool, inner surface, surfactant tail	Aqueous inner pool, inner surface, surfactant tail	Intercalation and surface	Either o sides bilaye the b

ce: Ref. 9.

II. METHODS

A. Materials

NP-9 [polyoxyethylene (9) nonyl phenol ether] was kindly provided by Ilchil Chemicals (Korea). Tween 20 [polyoxyethylene (20) sorbitan monolaurate, Aldrich], SDS (sodium dodecyl sulfate, Aldrich, 99.5%), CTAB (cetyltrimethylammonium bromide, Acros, 99%+ thin-layer chromatography grade) were used as received as surfactants. $AgNO_3$ (silver nitrate, Kojima Chemicals, Japan, 99.9%) as the starting material and $N_2H_4 \cdot H_2O$ (hydrazine monohydrate, Aldrich, 98%) as a reduction agent were used as received to prepare Ag nanoparticles. Water was double distilled using a Millipore system (Milli-Q).

B. Formation of Ag Colloids

Silver colloids were prepared by reduction of $AgNO_3$ solution (0.05 M) using hydrazine solution (0.1 M) in both the presence and absence of a surfactant. The Ag nanoparticles would be formed according to the following reaction:

$$N_2H_4 + 4Ag^+ + 4OH^- \rightarrow 4Ag^0 + 4H_2O + N_2$$

First, 20 g of the aqueous surfactant solution (0.01 M) was added in a glass vial. Then, 0.5 g of hydrazine solution was added and mixed thoroughly using a magnetic stirrer for 1 min. Finally, 0.5 g of $AgNO_3$ solution was added into this hydrazine–surfactant mixed solution. No stirring was necessary after initial 10 min. All experiments were performed in a thermostatted water bath maintained at 27°C and silver colloids obtained were also kept under the same conditions. The schematic process for the formation of silver colloids in a surfactant solution is shown in Fig. 1.

C. Morphology of Particles

The morphology and size of the particles formed in surfactant solutions were investigated by transmission electron microscopy (TEM, JEOL model JEM-2000EX II). A drop of the Ag colloidal solution was placed on Ni grids (200 mesh) covered with a carbon film. TEM micrographs were obtained at a magnification of 100,000 at an operating voltage of 200 kV.

D. UV–Visible Spectroscopy Measurements

UV–visible spectra of the Ag colloids prepared in different surfactant solutions were taken with a SCINCO S-2150 spectrophotometer. The measurements were performed after 12 h because at this time reduction of all silver ions by hydrazine was completed.

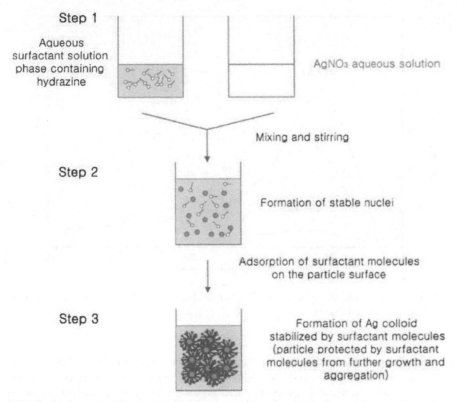

FIG. 1 Schematic diagram of the size control and stabilization of Ag particles by adsorption of a surfactant on the surface of particles.

E. Particle Size Measurement by Image Analysis

The size distribution was calculated from the measurement of particle size for more than 100 particles in the TEM pictures of colloidal particles.

III. RESULTS AND DISCUSSION

A. Nature of the Silver Plasmon Band in the Presence of a Surfactant

The UV absorbance spectra of silver colloids are shown in Fig. 2 and the absorbance values are given in Table 2. Among the surfactants tested, the SDS showed the highest yield. Depending on the nature of the surfactant, different absorbance patterns were observed. The measurement of the ex-

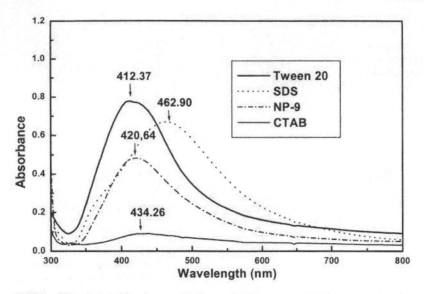

FIG. 2 Ultraviolet absorbance of silver colloids prepared in the presence of various surfactants as stabilizers.

tinction spectra of colloidal solutions provides the preliminary information about the particle size and size distribution. The shape of the plasmon resonance peak of colloidal solutions correlates with the size and size distribution of the particles. A narrow peak is a sign of a narrow particle size distribution, and the shift to a longer wavelength usually means that the particle size is larger. The aggregation of colloidal silver particles causes a decrease in the intensity of the peak. Thus, it can be inferred from Fig. 2 that the size of Ag particles formed in Tween 20 solution would be minimum, which was corroborated from the TEM pictures of colloidal particles. The particles prepared in CTAB solution were mostly aggregated.

Metal nanoparticles have been studied mainly because of their unique optical properties; especially nanoparticles of the noble metals copper, silver, and gold have a broad absorption band in the visible region of the electromagnetic spectrum. Solutions of these metal nanoparticles show a very intense color, which is absent in the bulk material and atoms. The origin of the intense color of noble metal nanoparticles is attributed to the collective oscillation of the free conductive electrons induced by an interacting electromagnetic field. These resonances are also denoted as surface plasmons [15].

Mie [16] was the first to explain this phenomenon by applying classical electrodynamics to spherical particles and solving Maxwell's equations for the appropriate boundary conditions. The cross-sectional area of total ex-

BLE 2 Properties of Silver Colloids Prepared in the Absence of a Surfactant or in the Presence

	UV absorbance (nm)	Particle size (nm)	Stability (days)	Adsorption on the silica wall of glass vials	
ueous phase	—	—	Unstable	No	
S	462.9	28.2	7	Yes	
AB	434.26	50–300	<1	No	
·9	420.64	68.9	<1	Yes	
een 20	412.37	11.6	>7	Yes	

tinction composed of absorption and scattering is given as a summation over all electric and magnetic multipole oscillations. The Mie theory has the advantage of being conceptually simple and has found wide applicability in explaining experimental results. However, all of the material properties are represented by a complex dielectric function of the absorbing metal nanoparticles, thus obscuring the underlying microscopic events, such as the possible decay mechanisms of the coherent motion of the free electrons.

It is well known that the adsorption of a nucleophile onto the particle surface increases the Fermi level of the silver particle due to its donation of electron density to the particle. Similarly, withdrawal of electron density from the particle surface by an electrophile lowers the Fermi level [14,17].

This phenomenon was also reported by Liz-Marzan and Lado-Tourino [18], and the effects of the adsorption of different species onto the surfaces on the optical properties of metal particles have been reviewed by Mulvaney [19]. This author points out that "Sols prepared with different stabilizers often have quite different absorption spectra even though the particle size distributions appear similar." This statement could readily be applied to our sols. The adsorption of iodide or sulfide ions [20] or of stabilizers such as gelatin or poly(vinyl pyrrolidone (PVP) [21–23] onto silver colloids led to a marked red shift, with damping of the plasmon band. The same phenomena were observed in this study experimentally as blue and red shifts of the plasmon maximum peak. The observed red shift, associated with the addition of surfactant stabilizers, is due to the displacement of nucleophiles (i.e., anions adsorbed on the particle surface) by the surfactants from the surface.

B. Stability of Ag Colloids

In the case of the reduction in a simple aqueous phase, the prepared Ag particles were immediately precipitated and then formed very large agglomerates with a gray color because of their hydrophobic nature.

Actually, this sample was completely separated into water and Ag agglomerates after about 3 h. On the contrary, silver colloids prepared in the presence of surfactants showed relatively good stability compared with the sample prepared in a simple aqueous solution without a surfactant. But depending on the surfactant species, quite different stabilities were observed. Especially in the case of Tween 20 and SDS, the colloids remained basically stable for weeks. The silver nanoparticles in the colloid solution were adsorbed on the silica wall of glass vials. This effect was observed for Tween 20 just after 3 min from the start of the reduction in our experimental conditions, but in the case of NP-9 and SDS, this phenomenon was observed after 1 day. Such adsorption phenomenon was also reported by Liz-Marzan and Lado-Tourino [18] after their experiment on silver reduction through oxidation of oxyethylene groups of the surfactant to hydroperoxide. All experiments were performed in plastic vials. But in the case of the NP-9 system, the adsorption of silver nanoparticles onto the walls of glass vials was observed very weakly.

C. Stabilization Mechanism for Silver Colloids

1. Stabilization Effect of Ionic Surfactants

The best known theory to explain the stabilization of lyophobic colloids is the DLVO (Derjaguin, Landau, Verwey, and Overbeek) theory. This theory takes into account the van der Waals attractive force (which covers all attractive forces, the main component being the dispersion forces) and the electrostatic repulsion of similarly charged particles. The net effect is that there is an energy barrier to overcome the agglomeration. The particles will be stable and will not coagulate if there is a net repulsion. The net repulsion will obviously be larger if the electrostatic repulsion is larger than the attractive interaction. The influence of the surfactant is that the adsorption of the hydrophobic tail onto the solid causes the solid to acquire a charge that will repel similarly charged particles, thus increasing the electrostatic force. When the surfactant molecules are adsorbed onto the surface of particles, the electrostatic barrier is produced to prevent the aggregation of the particles. The type of surfactant that is suitable for the preparation of a stable dispersion will depend on the nature of the solid to be dispersed. The adsorption is due to van der Waals interactions between the hydrophobic group and the hydrophobic particle surface [24].

It is well known that the Ag nanoparticles are positively charged in an aqueous solution. In the case of an anionic surfactant, SDS, we observed very good stability in spite of the positive charge of the Ag particles and did not observe any precipitation. This result might be caused by the hydrophobic bonding of the anionic surfactant on the positively charged hydrophobic particle surface. This phenomenon occurs when the combination of mutual attraction between the hydrophobic groups of the surfactant molecules and their tendency to escape from an aqueous environment becomes large enough to permit them to adsorb onto the solid adsorbent by aggregating their chains. The adsorption of surfactant molecules from a liquid phase onto or adjacent to other surfactant molecules already adsorbed on the solid adsorbent can also occur by this mechanism [25].

But in the case of CTAB, because of its long chain length (about 1.8 nm) and electrical repulsion between the particle surface and hydrophilic head groups of the surfactant, only a few surfactant molecules can adsorb at the particle surface. So the surfactant molecules cannot stabilize the silver nanoparticles in colloid solution.

2. Stabilization Effect of Nonionic Surfactants

Among the surfactants tested, Tween 20 also showed good stability by the steric effect because of the adsorption of hydrophobic groups onto the surface and strong hydration of large hydrophilic groups as shown in Fig. 3.

In the case of a nonionic surfactant, there is no electrostatic interaction between the particle surface and the hydrophilic groups of the surfactant. Furthermore, because of the hydrophobic property of Ag particles, adsorption of hydrophobic groups of the surfactant onto the particle surface takes place and the polyoxyethylene chains form the steric barrier. The highly hydrated polyoxyethylene chain in Tween 20 is extended into the aqueous phase in the form of a coil that acts as an effective barrier against aggregation.

But, in the case of the NP-9 system, poor colloidal stability was observed. We assume that this phenomenon was caused by the difference in the surfactant structure compared with that of Tween 20. Tween 20 has a sufficiently long oxyethylene chain to provide the steric hindrance (20 oxyeth-

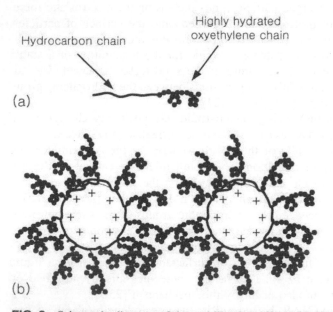

FIG. 3 Schematic diagram of the stabilization effect of hydrophobic particles by a surfactant with a polyoxyethylene chain in an aqueous solution. (a) Surfactant with polyoxyethylene in an aqueous solution. (b) Particle stabilization mechanism by steric hindrance due to surfactant with hydrated polyoxyethylene in an aqueous solution.

ylene units). NP-9 does not provide an effective steric barrier to disperse the Ag particles because of the short oxyethylene chain length (nine oxyethylene units).

The EO/PO copolymers are good at stabilizing hydrophobic particles. The polyoxypropylene chain will adsorb onto the hydrophobic particle surface and allow the polyoxyethylene chain to form a steric barrier because of its highly hydrated nature. The polyoxyethylene chain should be more than 20 units long to provide efficient stabilization.

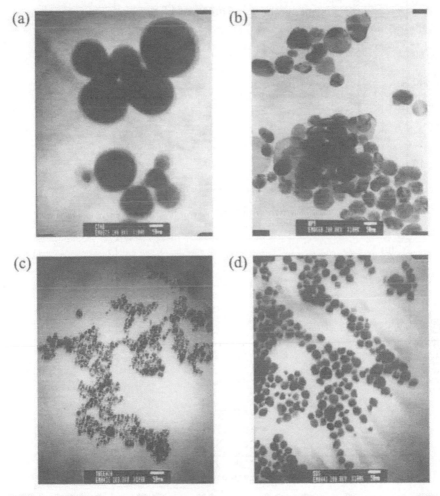

FIG. 4 TEM pictures of silver particles prepared in various surfactants as stabilizers. (a) CTAB; (b) NP-9; (c) Tween 20; (d) SDS.

D. Size Distribution and Morphology of Ag Particles

Particle size and size distribution were obtained using the image analysis technique. The Ag particles prepared in the presence of Tween 20 and SDS were small in size and had a narrow size distribution as shown in Figs. 4 and 5 and Table 2. The Ag particles prepared in the presence of NP-9 had

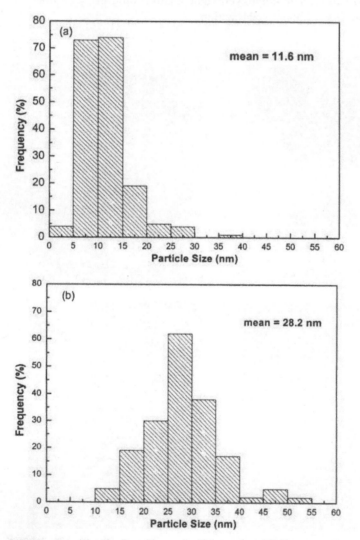

FIG. 5 Size distributions determined from the TEM images of surfactant-stabilized silver nanoparticles. (a) Tween 20; (b) SDS.

a large particle size. In the case of Tween 20 and SDS solutions, the average particle sizes were around 11.6 and 28.2 nm, respectively, and showed narrow size distributions. It is apparent that the instability of the colloid was caused by lack of a hindering effect of surfactant against aggregation and continuous growth of particles in solution. The size of the Ag particles prepared in the presence of surfactants was determined using TEM pictures as shown in Fig. 4. The shape of particles formed in surfactant solutions was spherical in all cases.

IV. CONCLUSIONS

Silver colloids were prepared in both the presence and absence of surfactants by the reduction of silver nitrate with hydrazine. In the case of silver reduction in an aqueous phase without a surfactant, the particles immediately precipitated and the size of particles was very large. But the Ag particles prepared in the presence of surfactants showed different UV absorption patterns and different colloidal stabilities. The SDS provides good Ag colloidal stability because of the electrical repulsion between the particles created by SDS molecules on the surfaces of particles. The Tween 20 also showed good colloidal stability due to the steric hindrance effect caused by its long and highly hydrated polyoxyethylene groups. But the NP-9 does not provide good colloidal stability because of insufficient number of oxyethylene units. The degree of colloidal stability was influenced by the particle size distribution. A wide distribution of particles size yielded poor stability of the colloidal solution.

V. FUTURE RESEARCH

In this study, the preparation of a single component colloid (silver) was investigated and characterized. In general, the nanoscale particles have a very high surface-to-volume ratio, and consequently their properties are quite different from those of bulk materials or atoms. Thus, the mixed nanoparticles of metals such as silver–gold, silver–copper, and silver–palladium would also show very different characteristics from those of bulk mixtures. In the future, we are planning to investigate the morphology, size, size distribution, and physical properties of mixed metal nanoparticles prepared in the presence of various kinds of surfactants.

ACKNOWLEDGMENT

This research was funded by the Center for Ultramicrochemical Process Systems, sponsored by KOSEF.

REFERENCES

1. PV Kamat. J Phys Chem 93:859–864, 1989.
2. K Kobayakawa, Y Nakazawa, M Ikeda, Y Sato. Ber Bunsenges Phys Chem 94:1439–1445, 1990.
3. JL Deiss, P Anizan, S El Hadigui, C Wecker. Colloids Surf A 106:59–62, 1996.
4. H Hirai, Y Nakao, N Toshima. Chem Lett 5:545–548, 1978.
5. KL Tsai, JL Dye. J Am Chem Soc 113:1650–1652, 1991.
6. Y Yonezawa, T Sato, S Kuroda, K Kuge. J Chem Soc Faraday Trans I 87: 1905–1910, 1991.
7. A Henglein, T Linnert, P Mulvaney. Ber Bunsenges Phys Chem 94:1449–1457, 1990.
8. M Koudelka, J Sanchez, J Augustyhsky. J Phys Chem 86:4277–4280, 1982.
9. JH Fendler. Chem Rev 87:877–879, 1987.
10. SG Dixit, AR Mahadeshwar, SK Haram. Colloids Surf A 133:69–75, 1998.
11. P Somasundaran, ED Snell, E Fu, Q Xu. Colloids Surf 63:49–54, 1992.
12. K Wong, B Cabane, R Duplessix, P Somasundaran. Langmuir 5:1346–1352, 1989.
13. VL Alexeev, P Ilekti, J Lambard, T Gulik, B Cabane. Langmuir 12:2392–2401, 1996.
14. T Pal, TK Sau, NR Jana. Langmuir 13:1481–1485, 1997.
15. CF Bohren, DR Huffman. Absorption and Scattering of Light by Small Particles. New York: Wiley, 1983, pp 82–104.
16. G Mie. Ann Phys 25:377–445, 1908.
17. A Henglein. J Phys Chem 97:5457–5471, 1993.
18. LM Liz-Marzan, I Lado-Tourino. Langmuir 12:3585–3589, 1996.
19. P Mulvaney. Langmuir 12:788–800, 1996.
20. F Strelow, A Henglein. J Phys Chem 99:11834–11838, 1995.
21. CR Berry, DC Skillman. J Appl Phys 42:2818–2822, 1971.
22. DC Skillman, CR Berry. J Chem Phys 48:3297–3304, 1968.
23. HH Haung, XP Ni, GL Loy, CH Chew, KL Tan, FC Loh, JF Deng, GQ Xu. Langmuir 12:909–912, 1996.
24. MR Porter. Handbook of Surfactants. 2nd ed. Glasgow: Chapman & Hall, 1994, pp 81–83.
25. MJ Rosen. Surfactants and Interfacial Phenomena. 2nd ed. New York: Wiley, 1989, pp 39–64.

15

Silver and Palladium Nanoparticles Incorporated in Layer Structured Materials

RITA PATAKFALVI, SZILVIA PAPP, and
IMRE DÉKÁNY University of Szeged, Szeged, Hungary

ABSTRACT

Lamellar liquid crystalline systems composed of hexadecyltrimethylammonium bromide/pentanol/water and hexadecylpyridinium chloride/pentanol/water were investigated. The influence of water and pentanol contents on the swelling properties of lamellar liquid crystalline structure was studied. We have investigated the possibility of inserting negatively charged layer silicates and silver nanoparticles in these cationic surfactant/pentanol/water systems. After addition of sodium montmorillonite to the liquid crystalline system, the silicate lamellae were built up into the ordered liquid crystalline structure. The effect of nanoparticle incorporation on structural ordering in liquid crystals was studied by X-ray diffraction measurements and it was established that lamellar distance (d_L) was only slightly altered.

Palladium nanoparticles have been generated by reduction of adsorbed Pd^{2+} ions in the interlamellar spaces of montmorillonite and kaolinite in aqueous media. Expansion of the interlayer space in kaolinite was effected by intercalation of dimethyl sulfoxide (DMSO) at 65°C. Neutral, cationic macromolecules and octylammonium ions were adsorbed on the clay minerals from an aqueous solution, which was followed by adsorption and reduction of Pd^{2+} ions. The size and the size distribution function of the palladium particles formed were determined by transmission electron microscopy (TEM) measurements, which showed that nearly spherical, nearly monodisperse particles were generated.

I. INTRODUCTION

Liquid crystalline systems formed by cationic surfactants have been described in numerous publications. The structural characteristics of liquid crystalline mesophases, the arrangement of the molecules, and the globular, cylindrical, or lamellar structure of the associations have been revealed by analysis of the phase diagrams of surfactant/cosurfactant/water systems by Ekwall et al. [1] and Friberg and coworkers [2,3]. Hoffmann [4] has analyzed the micelles formed by cationic and anionic surfactants in solutions of various concentrations and has studied the properties of the different phases formed by rheological and X-ray diffraction methods. Friberg and Venable [3] studied tetradecyltrimethylammonium bromide/pentanol/water systems and established that the interlamellar distance was dependent on the water/surfactant ratio but the presence of the cosurfactant also affected (reduced) the extent of swelling of the liquid crystalline system. As for cationic surfactants, hexadecylpyridinium chloride/alcohol/benzene/water quaternary systems were studied by Dékány et al. [5] using X-ray diffraction methods. It was shown that water was incorporated into the organic (benzene, cyclohexane, alcohol) inverse liquid crystalline mesophase of the lamellar structure and caused interlamellar swelling. The interlamellar distance can be manipulated not only by adjustment of the water/surfactant ratio but also by the addition of alcohols acting as cosurfactants [5]. Lamellar liquid crystalline systems are also suitable for the preparation and stabilization of nanoparticles. Nanoparticles are incorporated between the bilayer membranes and are thereby stabilized, as verified by small-angle X-ray scattering (SAXS) measurements [6,7]. Wang et al. [8] studied systems composed of silver hydrosol or organosol and lamellar sodium dodecyl sulfate (SDS)/hexanol/dodecane/water. Depending on the hydrophilic or hydrophobic character of the silver nanoparticle, it was incorporated into the aqueous or organic portion, respectively, of the lamellar bilayer phase and a stable dispersion was formed [8].

In the present work, the liquid crystalline characteristics of cationic surfactants were studied. Negatively charged lamellar silicate particles of nanometer thickness were also added to these systems. Our aim was to investigate the swelling properties of liquid crystalline systems with water and pentanol and the incorporation of nanoparticles as layer silicate and hydrophobic silver particles in cationic surfactant/pentanol/water liquid crystalline system. This allowed us to use X-ray diffraction methods for the characterization of the incorporation of cationic surfactants and to investigate whether layer silicates and silver nanoparticles could be inserted between the bilayer membranes.

Several procedures have been devised for the preparation of the nanoparticles in the size range 1–50 nm. In these procedures an important role is assigned to stabilizing agents that protect the nonparticles formed against aggregation, making possible the preparation of nanoparticles of a few nanometers diameter [9–11]. The stabilizing agents most often used are surfactants, polymers, and layer structured materials because particles of controlled size can be prepared within the internal spaces of micelles and microemulsins [12,13]. Silicate minerals of layered structure are excellent supports for the preparation of semiconductor and precious metal particles a few nanometers in diameter on the external surface as well as in the interlamellar space of the layered materials. Clay minerals (montmorillonite, hectorite, etc.) are especially suitable for this purpose because they swell readily in aqueous colloidal suspensions and, therefore, have large internal surfaces [14–16].

In our earlier works, the adsorption layer at a solid–liquid interface was employed as a "nanophase reactor" for the generation of nanocrystalline semiconductor particles (CdS, ZnS, TiO_2) and for their stabilization in the presence of a clay mineral [14–19]. Király et al. [20,21] have reported the synthesis of Pd nanoparticles in organic suspensions on hydrophobized montmorillonite by alcohol reduction of Pd acetate. Teranishi, Lin, and their coworkers [22–24] prepared precious metal sols by alcohol reduction (methanol, ethanol, propanol) using a neutral polymer [poly(vinyl alcohol), poly(vinyl pyrrolidone)] as a stabilizing agent. Chen and Akashi [25] obtained polymer-stabilized Pt sols from H_2PtCl_6 in water–ethanol mixtures. Nanosized palladium particles can also be prepared in the interlamellar space of kaolinite, a mineral that does not swell in water. The large specific surface area necessary for nanoparticle growth can be created by breaking up the hydrogen bonds between the kaolinite lamellae, i.e., by disaggregation of the kaolinite particles. Gábor et al. [26] obtained an intercalation complex of kaolinite with hydrazine and potassium acetate. The values of basal spacing were 0.72 nm for kaolinite, 1.05 nm for the kaolin–hydrazine complex, and 1.41 nm for the kaolin–potassium–acetate complex.

In the present work, novel methods are also described for the generation of palladium nanoparticles on the layered silicate montmorillonite and kaolinite using various polymers. The nonionic poly(N-vinyl-2-pyrrolidone) (PVP) and the cationic poly(diallyldimethylammonium chloride) (PDDA) were used for the synthesis.

II. MATERIALS

Liquid crystalline systems were prepared using the following chemicals: hexadecylpyridinium chloride (HDPCl, 98%, Fluka), hexadecyltrimethylammonium bromide (HDTABr, 98%, Reanal, Hungary), n-pentanol (99%, Reanal, Hungary), and milli-Q water. For sample preparation, the components were combined in a ratio to give the desired composition (see Table 1) and the mixture was vigorously stirred. Before measurements the samples were allowed to stand for 5 days because in our experience such time is necessary for the development of liquid crystalline structures.

Sodium montmorillonite (Wyoming, prepared at Süd-Chemie Ag, Moosburg, Germany, diameter $\leq$ 2 μm) was used for the preparation of the liquid crystal/layered silicate organoclay. Prior to the addition of the liquid crystalline mixture, the mineral was stored under saturated water vapor. After 4 days the adsorbed water amount was 22 wt%. This water content makes possible the incorporation of liquid crystalline components between the silicate lamellae.

The silver sol was prepared by reduction of silver ions ($AgNO_3$, 99.9%, Reanal) using $NaBH_4$ (98%, Sigma) stabilized by sodium oleate (technical grade, BDH Chemicals Ltd). An appropriate amount (25 mL) of $AgNO_3$ solution (10^{-3} M) was added drop by drop into an equivolume part of $NaBH_4$ (4×10^{-3} M) containing sodium oleate ($2,5 \times 10^{-4}$ M) [8]. The solution was stirred vigorously at ice cold temperature, and a brown-yellowish silver sol was obtained. The average diameter of the silver particles was 3.25 nm ($\pm$1.4 nm) as determined by transmission electron microscopy (TEM) experiments. After this, we prepared silver-doped liquid crystalline structures in the following way: we prepared the liquid crystalline systems, as before but silver nanosol (mass fraction of silver sol: 0.20–0.45) was added to the HDTABr/n-pentanol system (mass fractions: 0.55–0.80).

Kaolinite (Zettlitz, Germany, diameter 10–20 μm) was used as a support for the preparation of palladium nanoparticles. The specific surface area (a^s) was determined by N_2 adsorption measurements and the results were $a_{BET}^s = 87.7$ m^2/g and $a_{BET}^s = 14$ m^2/g for sodium montmorillonite and kaolinite, respectively. The reagents were used as received without further purification. The metal precursor was $PdCl_2$ (purity 99%, Aldrich). PVP (K-30, average molecular weight 40,000, Fluka) and a 20% aqueous solution of PDDA (average molecular weight 40,000–50,000, Aldrich) and octylamine (purity 99%, Fluka) were used as protective agents for the Pd nanoparticles. Methanol and ethanol (Reanal, Hungary) were of analytical purity. Dimethyl sulfoxide (DMSO, analytical grade, Reanal, Hungary) was used for the disaggregation of the kaolinite sample. The reducing agent hydrazine hydrate was a 55 wt% aqueous solution (Carlo Erba).

III. EXPERIMENTAL METHODS

Interlamellar distances in the liquid crystalline mesophase and basal spacing in montmorillonite and kaolinite were determined by X-ray diffraction measurements in a Philips PW-1830 diffractometer (CuKα radiation, λ = 0.154

TABLE 1 Composition of Liquid Crystalline Cationic Surfactant Systems and Measured Lamellar Distances (d)[a]

W_s	W_{cs}	W_w	Φ_s	Φ_{cs}	Φ_w	d [nm]	r [nm]	a_s [nm^2]	d_H [nm]
\multicolumn{10}{c}{HDTABr/pentanol/water system, $W_s/W_{cs} = 1.5$}
0.48	0.32	0.20	0.447	0.366	0.186	3.03	2.18	0.89	0.85
0.45	0.30	0.25	0.421	0.345	0.234	3.11	2.10	0.93	1.00
0.42	0.28	0.30	0.395	0.323	0.282	3.19	2.02	0.96	1.17
0.39	0.26	0.35	0.368	0.301	0.33	3.43	2.03	0.96	1.40
0.36	0.24	0.40	0.341	0.279	0.379	3.76	2.06	0.95	1.70
0.33	0.22	0.45	0.314	0.257	0.428	3.85	·1.94	1.00	1.91
\multicolumn{10}{c}{HDTABr/pentanol/water system, $W_w/W_s = 0.7$}
0.51	0.13	0.36	0.493	0.155	0.352	3.67	1.99	0.67	1.68
0.48	0.18	0.34	0.460	0.212	0.328	3.40	1.95	0.78	1.45
0.45	0.23	0.32	0.427	0.268	0.305	3.28	1.98	0.87	1.30
0.42	0.28	0.30	0.395	0.323	0.282	3.19	2.02	0.96	1.17
0.39	0.33	0.28	0.363	0.377	0.260	3.10	2.06	1.08	1.04
\multicolumn{10}{c}{HDPCl/pentanol/water system, $W_s/W_{cs} = 1.5$}
0.48	0.32	0.20	0.447	0.366	0.187	3.28	2.38	0.81	0.90
0.45	0.30	0.25	0.421	0.345	0.234	3.19	2.18	0.89	1.01
0.42	0.28	0.30	0.395	0.323	0.282	3.43	2.19	0.88	1.23
0.39	0.26	0.35	0.368	0.301	0.331	3.53	2.11	0.92	1.42
0.36	0.24	0.40	0.341	0.279	0.379	3.87	2.14	0.90	1.73
0.33	0.22	0.45	0.314	0.257	0.429	4.28	2.18	0.89	2.10
\multicolumn{10}{c}{HDPCl/pentanol/water system, $W_w/W_s = 0.7$}
0.48	0.18	0.34	0.460	0.212	0.328	3.78	2.20	0.69	1.58
0.45	0.23	0.32	0.427	0.268	0.305	3.53	2.16	0.79	1.37
0.42	0.28	0.30	0.395	0.323	0.282	3.43	2.19	0.88	1.23
0.39	0.33	0.28	0.363	0.377	0.260	2.97	1.98	1.11	0.99
0.36	0.38	0.26	0.333	0.429	0.238	2.74	1.91	1.31	0.83

[a] W_s, W_{cs}, and W_w are the mass fractions of the surfactant, cosurfactant, and the water, respectively. Φ_s, Φ_{cs}, and Φ_w are the molar fractions of the surfactant, cosurfactant, and the water, respectively. The values of r, a_s, and d_H are calculated from Eqs. (3)–(5); r is the thickness of the apolar layer, a_s is the effective cross-sectional area of the chain, and d_H is the thickness of the polar layer.

nm). During the measurement the samples were covered with a Mylar foil to prevent changes in composition of the liquid crystalline materials due to evaporation.

All liquid crystalline systems were investigated by polarization microscopy (LEICA Q 500 MC Image Analyzer System).

Transmission electron microscopy is suitable for the examination of subcolloids because its range of measurement covers particles 1 to 50 nm in size. Images were made in a Philips CM-10 transmission electron microscope with an accelerating voltage of 100 kV. Aliquots of the ethanol suspensions of the samples were dropped on copper grids (diameter 2 mm) covered with a Formwar® foil, left to stand for 3–40 s, and then transferred into the microscope. Particle size distributions were determined using the UTSHCSA Image Tool version 2.00 program (the program was developed by Dr. C. Donald Wilcox and coworkers, Department of Dental Diagnostic Science, University of Texas Health Science Center, San Antonio).

A. Preparation of Polymer-Stabilized Pd⁰ Nanoparticles in an Aqueous Dispersion on Montmorillonite

We prepaed Pd^0 particles in situ in the interlamellar space of montmorillonite dispersed in an aqueous medium. The support suspended in a polymer solution of adequate concentration (0.41 wt%) was stirred at 60°C for 24 h, and a Pd chloride solution containing ethanol was next added to the dispersion. The palladium chloride solution was freshly prepared (2 mM, adjusted to pH 4 with hydrochloride acid). Macromolecules were adsorbed on the clay minerals from an aqeuous solution, followed by adsorption and reduction of Pd^{2+} ions. Composites of nonionic PVP/montmorillonite and cationic PDDA/montmorillonite (0.02–0.12 g polymer/1 g clay) were prepared in systems containing various concentrations of adsorbed PVP and PDDA from aqueous solution. The added metal content was 2 g/100 g clay and the metal content of the products was in the range 1.34–2.01% as determined by inductively coupled plasma atomic emission spectrosocpy (ICP-AES) analysis. The composition of the resulting systems and their monomer/Pd ratios are listed in Table 2.

B. Preparation of Pd⁰ Nanoparticles in the Interlamellar Space of Kaolinite

Nanoparticles of Pd were prepared in kaolinite suspension containing surfactant and neutral or cationic polymer in methanol medium. Disaggregation of the kaolinite lamellae was brought about by the intercalation of DMSO at $T = 65°C$. The DMSO was removed from between the lamellae by repeated washes with methanol and decanting for 5 days. The MeOH–kaolin-

TABLE 2 Basal Distances (d_L) of Intercalated Clays and Particle Diameters (d_{ave}) of Pd^0 Clays

Clay samples	Pd content, wt%	PVP/PDDA/C8N content, wt%	d_L, nm (XRD)	d_{ave}, nm (TEM)
Montmorillonite	0.00	0.00	1.44	—
PVP/montmorillonite	0.00	12.41	1.58	—
PDDA/montmorillonite	0.00	12.41	1.53	—
Pd-montmorillonite samples	Prepared in water, reduced by ethanol			
PVPPdM1	1.68	12.41	1.54	2.26
PVPPdM2	1.93	4.07	3.24[a]; 1.46[b]	3.82
PVPPdM3	1.95	2.07	5.68[a]; 1.45[b]	3.82
PDDAPdM1	1.34	12.41	4.14[a]; 1.52[b]	1.93
PDDAPdM2	1.83	5.86	2.84[a]; 1.47[b]	2.45
PDDAPdM3	2.01	3.01	1.45	3.57
Kaolinite	0.00	0.00	0.72	—
DMSO/kaolinite	0.00	0.00	1.12	—
MeOH/kaolinite (K)	0.00	0.00	1.12	—
PVP/MeOH/kaolinite	0.00	1.90	1.12	—
PVP/H_2O/kaolinite	0.00	1.90	3.64[a]; 0.72[b]	—
PDDA/H_2O/kaolinite	0.00	1.90	3.92[a]; 0.74[b]	—
C8N/kaolinite	0.00	56.50	2.48[a], 1.28[b]	—
Pd-kaolinite samples	Prepared in methanol, reduced by N_2H_4			
PdK	0.47	0.00	7.53	3.25
PVPPdK1	0.47	3.80	2.21[a]; 1.00[b]	2.24
PVPPdK2	0.95	3.80	4.56[a]; 0.72[b]	2.45
PVPPdK3	0.95	17.00	4.24[a]; 0.71[b]	2.45
PVPPdK4	0.95	28.00	0.84	2.59
PVPPdK5	1.41	3.80	0.74	2.10
PDDAPdK1	0.95	3.80	3.52[a]; 0.82[b]	6.33
Pd-kaolinite samples	Prepared in methanol, reduced by $NaBH_4$			
PVPPdK6	0.95	3.80	3.78[a]; 0.84[b]	4.16
PDDAPdK2	0.95	3.80	5.07[a]; 0.83[b]	2.67
Pd-kaolinite samples	Prepared in water, reduced by N_2H_4			
C8NPdK1	0.95	11.60	0.73	2.17
C8NPdK2	0.95	20.10	4.42[a]; 0.73[b]	2.04
C8NPdK3	0.95	40.00	2.65	1.95
C8NPdK4	0.95	56.50	5.95[a]; 0.72[b]	2.22

[a]Clay/polymer (surfactant) or Pd^0/clay/polymer (surfactant) basal distance calculated from the Bragg equation.
[b]Clay basal distance calculated from the Bragg equation.

ite complex obtained in this way was used as an intermediate for further preparations. The Pd–kaolinite complexes were prepared by reduction of Pd^{2+} ions as follows: 0.7 mM $PdCl_2$ aqueous solution was adsorbed in the MeOH–kaolinite complex or, alternatively, by employing a polymer or surfactant to ensure binding to the lamellae and steric stabilization. Polymers were adsorbed onto the support from a methanol solution, followed by adsorption and reduction of Pd^{2+} ions. Interacalation complexes of nonionic PVP/kaolinite, cationic PDDA/kaolinite (0.02–0.4 g polymer/1 g kaolinte), and octylammonium/kaolinite (0.12–1.2 g/1 g kaolinite at pH 4.0 in aqueous suspension) were prepared by this method in systems containing various concentrations (0.1–2.0%) of methanol/PVP, PDDA by polymer adsorption at room temperature in a 24-hour reaction. Kaolinite samples containing different amounts of palladium were obtained by hydrazine (0.9–3.6 mL 0.1 M) and $NaBH_4$ (1.8 mL 0.1 M) reduction of palladium ions adsorbed from a solution of H_2PdCl_4 (pH 4). The metal contents of the products were in the range 0.45–1.4% by weight. The systems prepared and their compositions are listed in Table 2.

IV. RESULTS AND DISCUSSION

A. The Structure of Liquid Crystalline Lamellar Systems

Because of their highly ordered structure, liquid crystalline mesophases give very sharp X-ray diffraction signals from which interlamellar distance (d) can be calculated using the Bragg equation. The dimensions of the polar and apolar layers were calculated by Kunieda and coworkers [27,28]. The molar fractions of the surfactant (Φ_s) and of its lipophilic chain (Φ_L) in such a system are given by

$$\Phi_s = \frac{1}{1 + \dfrac{\rho_s}{W_s}\left(\dfrac{W_w}{\rho_w} + \dfrac{W_{cs}}{\rho_{cs}}\right)} \tag{1}$$

$$\Phi_L = \frac{V_L}{V_S}\,\Phi_s \tag{2}$$

where W_s, W_w, and W_{cs} are the mass fractions of the surfactant, water, and the cosurfactant; ρ_s, ρ_w, and ρ_{cs} are the densities of the surfactant, water, and the cosurfactant; and V_S and V_L are the molar volumes of the surfactant and its lipophilic chain, respectively.

In the case of a lamellar phase (Fig. 1), the thickness of the apolar layer (r) (comprising the cosurfactant and the apolar chain of the surfactant) is described by the following formula:

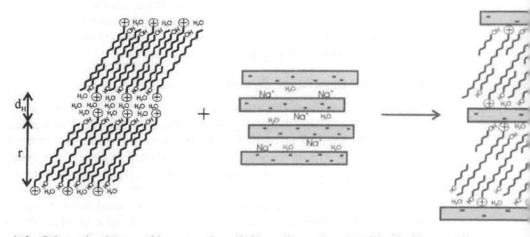

i. 1 Schematic pictures of incorporation of silicate layers into lamellar liquid crystalline systems

$$r = (\Phi_L + \Phi_{cs})d \tag{3}$$

where Φ_{cs} is the volume fraction of the cosurfactant and d is the interlamellar distance measured.

The effective cross-sectional area of the chain, a_s, is obtained from the following equation [27]:

$$a_s = \frac{2V_L}{rN_A}\left(\frac{\Phi_L + \Phi_{cs}}{\Phi_L}\right) \tag{4}$$

(N_A is Avogadro's number)

The thickness of the layer containing the polar head group and water, d_H, is

$$d_H = d - r \tag{5}$$

The length of a single surfactant molecule [29] is $l_{alk} = 0.127n_c + 0.28$ nm, where n_c denotes the number of C—C bonds and 0.28 nm is the size of the terminal CH_3 group, i.e., the so-called van der Waals radius; to this value the length of the polar head group (in our case, pyridinium or trimethylammonium ion) has to be added as well.

Because of the all-*trans* conformation, the surfactant chains include an angle of 55° with the solid surface within the interlamellar space of a hydrophobic layer silicate in organic liquids. If the length of the surfactant chains is calculated and it is assumed that chains within the interlamellar space also include a tilt angle of 55° (see Fig. 1), the thickness of the liquid crystalline lamellae can be calculated.

B. Liquid Crystalline Systems

Because it was necessary to elucidate the role of the alcohol in the formation of the adsorption layer, liquid crystalline systems were studied with pentanol as a cosurfactant. The effect of varying the water and pentanol contents was examined in HDTABr/pentanol/water and HDPCl/pentanol/water systems. Thus, we prepared a series of samples in which W_s/W_{cs} was kept at a constant value of 1.5 and W_w/W_s was varied between 0.4 and 1.4, whereas in another series $W_w/W_s = 0.7$ was constant and W_s/W_{cs} was varied in the range 3.9–0.9. The comparisons are listed in Table 1. The liquid crystalline structure of the samples was verified by polarization microscopy. The lamellar structure is discernible in the photograph (Fig. 2).

The effect of water on lamellar distances was first investigated. In the HDTABr/pentanol/water system (W_w/W_s variable) the Bragg reflections characteristic of d are shifted toward lower wave vectors (angles). In liquid crystalline systems a minimum of 1 mol H_2O/surfactant is necessary for the hydration of the polar head groups. As water content (W_w) increases, d

FIG. 2 Liquid crystalline structure of HDPCl/n-pentanol/water system (51:36:13 wt%) (in polarized light).

changes in the range 3.03–3.85 nm. Two peaks appear in each sample: one sharper peak of higher intensity at a larger angle and the other, less sharp one (characteristic of contamination) of lower intensity at a smaller angle. When lamellar distance is plotted as a function of water content (Fig. 3), it becomes evident that the system is swelling and the distance between the surfactant layers is increasing. The lamellar distance varies in the range $d = 3$ to 3.8 nm.

The values of r, d_H, and a_s calculated using Eqs. (1)–(4) for the systems studied are listed in Table 1. The calculations reveal that the thickness of the apolar layer (r) is nearly constant (≈ 2.2 nm), whereas that of the polar layer (d_H) increases. Assuming an angle of 55°, a model for the arrangement of the surfactant molecules within the lamellae can be constructed. The values of r indicate that the alkyl chains overlap those attached to the neighboring lamellae. Head groups are initially aligned ($d_H = 0.85$–0.9 nm) and poorly hydrated. As W_w increases, they move apart ($d_H = 1.91$–2.10 nm) and become increasingly hydrated: at n_{water}/n_{HDTABr}; = 26, as many as four layers of water molecules are present in the polar layer. In other words, when the ratio of water increases, the system swells.

Similar observations were made in the case of systems containing HDPCl

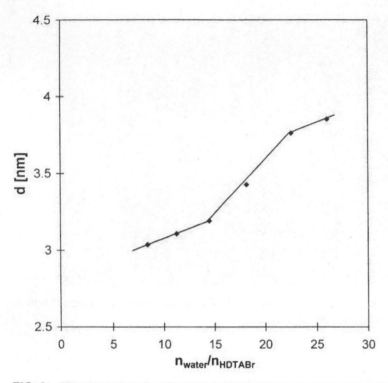

FIG. 3 Changes in lamellar distance d with increasing water contents in HDTABr/ n-pentanol/water liquid crystalline system (composition $W_{HDTABr}/W_{pentanol}$: 1.5%).

(HDPCl/pentanol/water, W_w/W_s variable). Interlamellar distance is slightly larger (3.3–4.3 nm) than in the case of the HDTABr/pentanol/water system. The polar layer increases from $d_H = 0.9$ nm to 2.1 nm and the average thickness of the apolar layer is 2.2 nm. Again, the surfactant chains on neighboring lamellae slide along each other.

The thickness of the layers can also be manipulated by changes in pentanol concentration. When the weight fraction of pentanol is increased and the water/surfactant ratio is kept constant (Table 1), the interlamellar space is reduced. The decrease is $d = 3.7 \rightarrow 3.1$ nm in the HDTABr/pentanol/ water system and $d = 3.8 \rightarrow 2.7$ nm in the HDPCl/pentanol/water system. Because of the reduction of water concentration in the system, however, the aqueous layer becomes thinner (HDTABr/pentanol/water: $d = 1.8 \rightarrow 1.1$ nm; HDPCl/pentanol/water: $d = 1.5 \rightarrow 0.8$ nm).

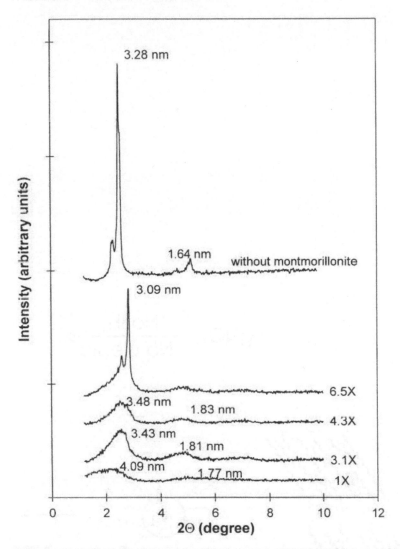

FIG. 4 Lamellar reflections in the XRD spectra of HDTABr/*n*-pentanol/water system before and after addition of Na-montmorillonite. (1X: amount of surfactant corresponding to the cec of the montmorillonite; 3.1X, 4.3X, 6.5X: amount of surfactant corresponding to 3.1, 4.3, 6.5 times the value of cec. The d_L basal distances are denoted in the figure.)

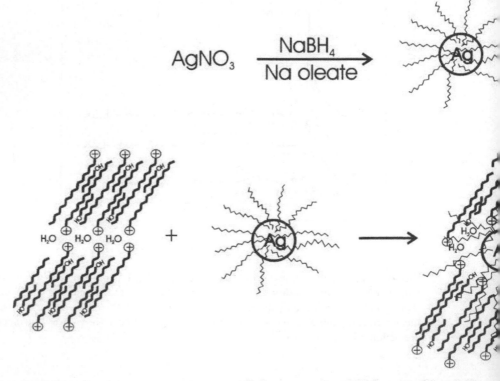

ig. 5 Schematic picture of (a) preparation of silver nanosol and (b) incorporation of silver
stalline systems.

C. Incorporation of Silicate Layers into Liquid Crystalline Systems

The interaction between liquid crystals and clay minerals was studied by adding various amounts of HDTABr/pentanol/water (W_w/W_s or W_s/W_{cs} constant) to sodium montmorillonite (Fig. 1). The effect of the addition of increasing amounts of HDTABr/pentanol/water (42:28:30) system to Na-montmorillonite on basal spacing as revealed by X-ray diffraction (XRD) spectra is presented in Fig. 4. After the addition to montmorillonite of a liquid crystalline mixture containing an amount of surfactant corresponding to the cation exchange capacity (cec) of the clay mineral (0.85 mmol/g), ion exchange on the montmorillonite lamellae is not complete even after 3 days of reaction time. Intercalation is already indicated by the new reflection (d_L = 4.09 nm) as compared with the wet state (d_L = 1.55 nm); the intensity, however, is rather low because the lamellae have not yet attained a sufficiently parallel arrangement. For the 1X sample we can see that due to the

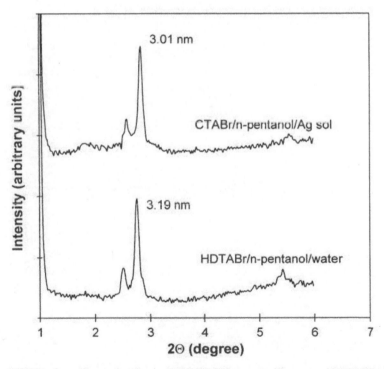

FIG. 6 Lamellar reflections of HDTABr/n-pentanol/water and HDTABr/n-pentanol/ silver nanosol systems. (The d_L basal distances are denoted in the figure.)

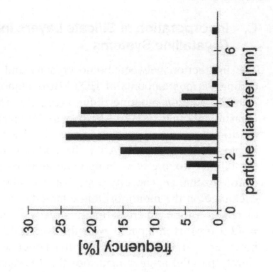

(a)

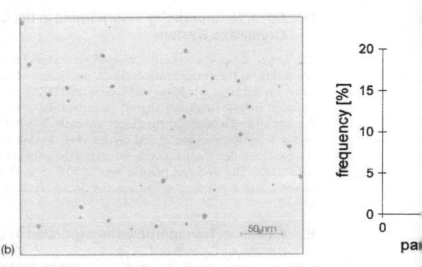

FIG. 7 TEM image and size distribution of (a) 0.5 mM silver nanosol and (b) nanosol liquid crystalline systems.

adsorption of HDTA$^+$ cations on montmorillonite the n_{water}/n_{HDTABr} ratio is very high and this leads to a shift ($d_L = 4.09$ nm) of the interlamellar distance in the HDTABr/pentanol/water system. When the amount of mixture added is increased, the basal spacing $d_L = 3.43$ nm but the second- and third-order reflections also appear, indicating that the silicate lamellae have assumed a more ordered arrangement (Fig. 4). This is also indicated by the increase in intensity. Upon further increasing the amount of liquid crystal mixture, the values of the basal spacing become constant. This supports the formation of liquid crystalline system because the Na-montmorillonite is "loaded" by HDTA$^+$ cations and the n_{water}/n_{HDTABr} ratio is closer to that of the system without montmorillonite. When an amount of surfactant corresponding to six times the value of cec is added to the Na-montmorillonite sample, the spectra obtained display the characteristics of liquid crystalline systems: montmorillonite lamellae are incorporated into the liquid crystalline system.

D. Silver Nanoparticles Incorporated in the Liquid Crystalline System

The lamellar liquid crystalline systems were doped with silver nanoparticles according to the procedure described previously (Fig. 5). The composites were investigated by X-ray diffraction and TEM experiments. The incorporation of silver nanosol slightly disturbed the order of lamellar structure: the lamellar distance decreased, for example, $d = 3.19$ nm in the undoped and $d = 3.01$ nm in the doped phases (Fig. 6). We determined the size of Ag nanoparticles from aqueous sol and in the lamellar phases by TEM measurements. The average particle was $d = 3.35 \pm 1.4$ nm in the aqueous nanosol and $d = 4.9 \pm 1.5$ nm in the liquid crystalline phase (see Fig. 7a and b).

E. Palladium Nanoparticles Incorporated in the Interlayer Space of Clays

When polymers are adsorbed from aqueous solution on montmorillonite, polymer segments are bonded in the interlamellar space and the basal spacing varies between 1.4 and 1.8 nm (Fig. 8). When palladium ions are added to this clay/polymer suspension, they are also adsorbed on the external as well as internal surfaces of the lamellae, and as soon as ethanol is added, they are reduced. Adsorption and reduction are represented schematically in Fig. 9. Average particle sizes determined by TEM are given in Table 2.

The basal spacing in montmorillonite was increased by about 0.15 nm by the incorporation of PVP molecular segments. XRD patterns for a PVPPdM3 sample are presented in Fig. 8 with different basal distances: (1) sodium montmorillinite: $d_L = 1.43$ nm; (2) PVP/montmorillonite (PVPM): $d_L = 1.58$

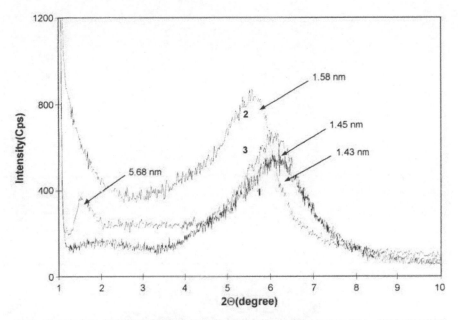

FIG. 8 XRD patterns of Pd⁰/polymer/montmorillonite composites. (1) Na-montmorillonite; (2) PVPM sample; (3) PVPPdM3 sample.

nm; (3) Pd^0/PVP/montmorillonite (PVPPdM3): d_L = 1.45 and 5.68 nm (Table 2). Through interlamellar adsorption, macromolecules are capable of forcing the lamellae apart, opening the possibility of the formation of even larger Pd nanoparticles as indicated by the peak appearing at 2Θ = 1.5°. The existence of these particles was also proved by TEM pictures showing polydispersity. The TEM pictures of the PVPPdM3 sample and its size distribution are shown in Fig. 10.

The Pd^0 nanoparticles of the samples were studied in the angle range $38° < 2\Theta < 42°$ because the basal plane causing the most intense reflection (111) of crystalline metal palladium is at $2\Theta \sim 40°$ (Fig. 11a and b). The more diffuse peak is indicative of amorphous particles with lower degrees of crystallization. On this basis, PVPPdM1 and PDDAPdM1 are identified as the samples containing the smallest nanocrystals.

To produce Pd nanocrystals in kaolinite it is necessary to open the interlamellar space for the synthesis of nanoparticles; i.e., the hydrogen bonds between the kaolinite lamellae must be broken. A schematic diagram of the synthesis is presented in Fig. 12. Disaggregation was nearly 100% complete after the formation of the DMSO/kaolinite intercalation complex. Changes in the basal spacing of kaolinite studied by X-ray diffraction measurements

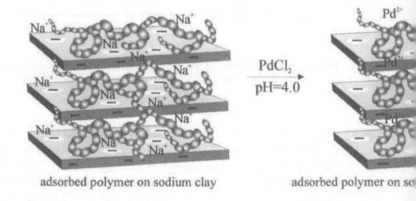

adsorbed polymer on sodium clay

$PdCl_2$
pH=4.0

adsorbed polymer on s

Pd^0 nanoparticles stabilize

FIG. 9 Schematic illustration of preparation of polymer/montmorillonite-s

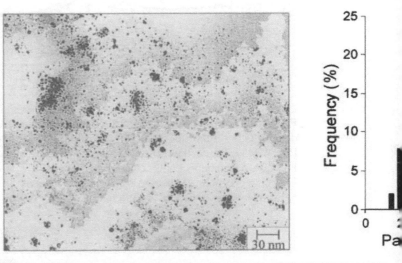

FIG. 10 TEM image and particle size distribution of PVPPdM3 sample.

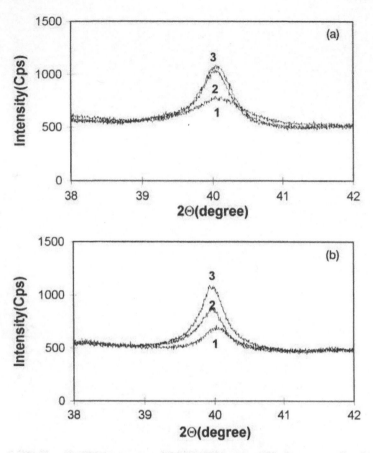

FIG. 11 (a) XRD patterns of Pd⁰/PVP/montmorillonite composites for particle size determination from the Pd(111) reflection. (1) PVPPdM1; (2) PVPPdM2; (3) PVPPdM3. (The d_L basal distances are denoted in the figure.) (b) XRD patterns of Pd⁰/PDDA/montmorillonite composites for particle size determination from the Pd(111) reflection. (1) PDDAPdM1; (2) PDDAPdM2; (3) PDDAPdM3. (The d_L basal distances are denoted in the figure.)

revealed that the basal spacing increased from 0.72 to 1.12 nm (Fig. 13a and b). Interlamellar distance was unchanged after washing of DMSO by methanol (Fig. 13c). The MeOH/kaolinite complex obtained in this way was used as an intermediate in further preparations. The peak at 7.53 nm appeared in the XRD pattern of the Pd/kaolinite complex prepared by reduction of Pd^{2+} ions adsorbed in the MeOH/kaolinite system using hydrazine.

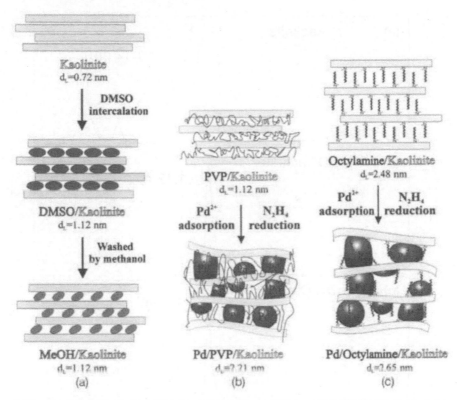

FIG. 12 Schematic illustration of preparation of Pd nanoparticles stabilized by polymer/kaolinite composite. (a) Disaggregation of kaolinite, (b) adsorption of PVP, and (c) octylammonium ions on methanol-treated kaolinite.

The disappearance of the reflection of disaggregated kaolinite also demonstrates the interlamellar particle growth (Fig. 13d). For synthesis employing polymers, macromolecules were first adsorbed onto the support from a methanol or aqueous solution (0.1–2.0% w/v). The incorporation of the amount of polymer necessary for successful particle synthesis (2–20% w/w) does not increase the basal spacing according to the XRD patterns (see Fig. 14 and Table 2). The particle size calculated from the position of the new reflection (2.21 nm − 0.72 nm = 1.49 nm) is in good agreement with the results of TEM measurements (see Fig. 15a and b). When Na-borohydride was used for reduction, the peak appeared at higher lamellar distances (d_L = 5–6 nm).

Kaolinite complexes containing palladium were also prepared using octylamine in an acidic medium (pH 4.0) to separate the lamellae in order to

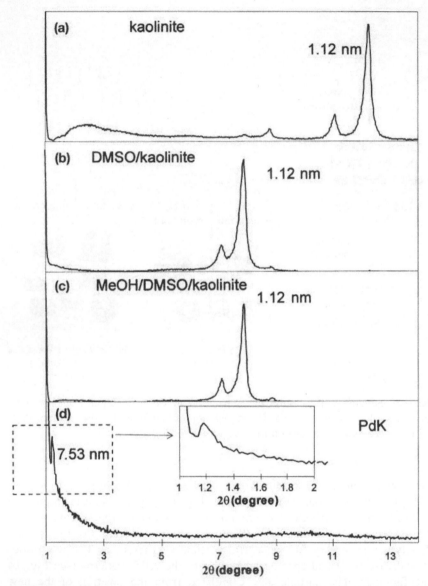

FIG. 13 XRD patterns of kaolinite samples: (a) kaolinite, (b) DMSO-treated kaolinite, (c) methanol-washed DMSO/kaolinite, and (d) palladium nanoparticles between disaggregated methanol-washed kaolinite layers. (The d_L basal distances are denoted in the figure.)

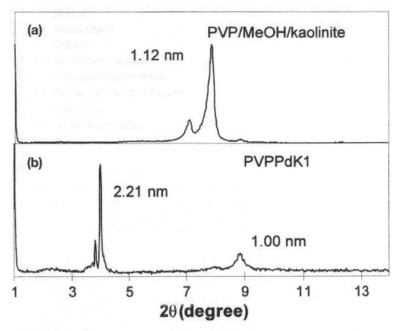

FIG. 14 XRD patterns of kaolinite samples: (a) methanol-treated kaolinite sample with adsorbed PVP, (b) methanol-treated palladium/PVP/kaolinite nanocomposites. (The d_L basal distances are denoted in the figure.)

provide room for the particles as well as to ensure stability. The amount of octylamine was varied between 0.12 and 1.2 g/g kaolinite. The reflection at 2.48 nm appearing in the X-ray diffractogram of the samples indicates the intercalation of octylammonium ions (Fig. 16a). Following the adsorption of palladium ions on the octylammonium kaolinite in the course of the reduction of Pd^{2+} by hydrazine, a further shift of the 001 reflection is observed (2.65 nm, Fig. 16b).

The TEM picture of the PdK sample shows spherical particles 2–4 nm in diameter attached to the kaolinite lamellae without any aggregation (Fig. 15a). Electron micrographs of the complexes containing nonionic PVP display relatively small, nearly monodisperse particles (Fig. 15b) that tend to link up as the polymer concentration is increased. Polymer concentration was found to have no effect on the Pd particle size (see Table 2). In samples containing cationic PDDA as the stabilizer, the size distribution of the particles formed was more polydisperse. Increasing the palladium content not only increases the particle size but also enhances polydispersity.

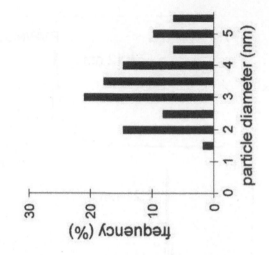

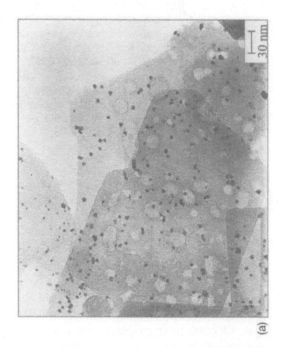

(a)

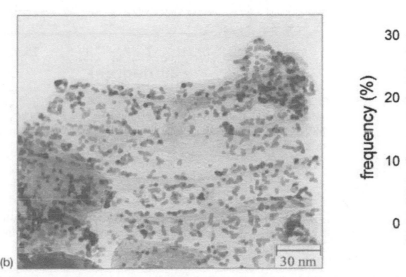

FIG. 15 (a) TEM image and particle size distribution of PdK sample (0.47% solution). (b) TEM image and particle size distribution of PVPPdK6 sample (0.9% treatment, reduced by 0.1 M NaBH₄ solution).

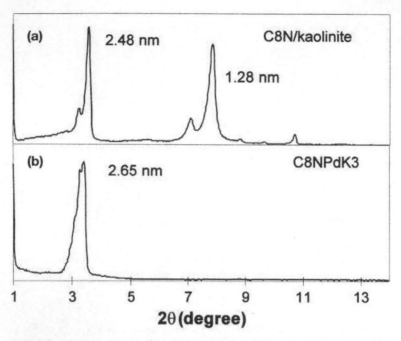

FIG. 16 XRD patterns of kaolinite samples: (a) methanol-treated rehydrated kaolinite sample with adsorbed octylammonium ions, and (b) methanol-treated rehydrated palladium/octylammonium/kaolinite nanocomposite. (The d_L basal distances are denoted in the figure.)

Octylammonium cations as well as nonionic and cationic polymers proved equally suitable for the realization of our aim. In each case, even at high metal contents, perfectly nanocrystalline palladium particles are seen in the pictures (see Table 2).

V. CONCLUSIONS

Liquid crystalline cationic surfactant/pentanol/water systems were investigated. The influence of water and pentanol contents on the swelling properties of lamellar systems was studied. Liquid crystalline systems were mixed with montmorillonite, resulting in the formation of a swollen organocomplex that retained its ordered lamellar structure. It can be assumed that the driving force for maximal swelling and the corresponding adsorption is the development of a liquid crystalline system of ordered structure in the interlamellar space between the silicate lamellae. Silver nanoparticles were also prepared in the liquid crystalline system, and the intercalation was observed by XRD and TEM experiments.

The synthesis of palladium nanoparticles on montmorillonite layer silicates was studied. The Pd^0 particles were prepared in situ in the interlamellar space of montmorillonite dispersed in an aqueous medium. Macromolecules were adsorbed on the support from an aqueous solution, followed by adsorption and reduction of Pd^{2+} ions. The Pd^0 nanoparticles appear and grow in the internal, interlamellar space as well as on the external surfaces of the lamellae. Well-crystallized kaolinite clay can be disaggregated by the intercalation of DMSO to individual lamellae, which may serve as excellent supports for metal nanoparticles. After the adsorption of palladium precursor, metal nanocrystals were reduced by hydrazine or sodium borohydride between the kaolinite lamellae, i.e., in the interfacial layer acting as a nanoreactor. The incorporation of nanoparticles between the lamellae was shown by XRD measurements. This procedure makes possible the steric control and restriction of nanoparticle growth. The stability of nanoparticles can be further enhanced by the addition of polymers (PVP) and surfactants (alkylammonium salts) that are also adsorbed between the kaolinite lamellae. The presence of the particles was also verified and their sizes were quantified by TEM measurements.

ACKNOWLEDGMENT

The authors wish to thank the National Scientific Research Foundation, OTKA T034430, and Ministry of Education, FKFP 0402/1999, for financial support.

REFERENCES

1. P Ekwall, K Fontell, L Mandell. Mol Cryst Liq Cryst 8:157, 1969.
2. SI Ahmad, SE Friberg. J Am Chem Soc 94:5196–5199, 1972.
3. SE Friberg, RL Venable. Colloids Surf 15:285–293, 1985.
4. H Hoffmann, Ber Bunsenges Phys Chem 88:1078–1093, 1984.
5. I Dékány, F Szántó, A Weiss. Colloids Surf 41:107–121, 1989.
6. C Quilliet, P Fabre, V Cabuil. J Phys Chem 97:287–289, 1993.
7. C Quilliet, V Ponsinet, V Cabuil. J Phys Chem 98:3566–3569, 1994.
8. W Wang, S Efrima, O Regev. J Phys Chem B 103:5613–5621, 1999.
9. RW Siegel. In: FE Fujita, ed. Springer Series in Materials Sciences. Vol. 19. Berlin: Springer-Verlag, 1994, p 65.
10. A Henglein, M Guiterrez, Ber Bunsenges Phys Chem 87:474–478, 1983.
11. A Ueno, N Kakuta, KH Park, MF Finlayson, AJ Bard, A Champion, MA Fox, SE Webber, JM White. J Phys Chem 89:3828–3833, 1985.
12. YM Trickot, JH Fendler. J Am Chem Soc 106:7359–7366, 1984.

13. HC Your, S Baral, JH Fendler. J Phys Chem 92:6320–6327, 1988.
14. NA Kotov, K Putyera, JH Fendler, E Tombácz, I Dékány. Colloids Surf A 71: 317–326, 1993.
15. NA Kotov, FC Meldrum, JH Fendler, E Tombácz, I Dékány. Langmuir 10: 3797–3804, 1994.
16. NA Kotov, I Dékány, JH Fendler. J Phys Chem 99:13065–13069, 1995.
17. NA Kotov, T Haraszti, L Túri, G Zavala, RE Geer, I Dékány, JH Fendler. J Am Chem Soc 119:6821–6832, 1997.
18. I Dékány, L Nagy, L Turi, Z Király, NA Kotov, JH Fendler. Langmuir 12: 3709–3715, 1996.
19. I Dékány, L Túri, G Galbács, JH Fendler. J Colloid Interface Sci 195:307–315, 1997.
20. Z Király, I Dékány, Á Mastalir, M Bartók. Magy Kemiai Folyoirat 101:539–545, 1995.
21. Z Király, I. Dékány, Á Mastalir, M. Bartók. J Catal 161:402–408, 1996.
22. T Teranishi, M Miyake. Chem Mater 10:594–600, 1998.
23. H Liu, G Mao, S Meng. J Mol Catal 74:275–284, 1992.
24. W Yu, H Lin. Chem Mater 10:1205–1207, 1998.
25. CW Chen, M Akashi. Langmuir 13:6465–6472, 1997.
26. M Gábor, M Tóth, J Kristóf, G Komáromi-Hiller. Clays Clay Miner 43:223–228, 1995.
27. H Kunieda, K Shigeta, K Ozawa, M Suzuki. J Phys Chem B 101:7952–7957, 1997.
28. H Kunieda, K Shigeta, M Suzuki. Langmuir 15:3118–3122, 1999.
29. I Regdon, Z Király, I Dékány, G Lagaly. Colloid Polym Sci 272:1129–1135, 1994.

16

Water-in-Carbon Dioxide Microemulsions Stabilized by Fluorosurfactants

JULIAN EASTOE and ALISON PAUL   University of Bristol, Bristol, United Kingdom

DAVID STEYTLER and EMILY RUMSEY   University of East Anglia, Norwich, United Kingdom

RICHARD K. HEENAN and JEFFREY PENFOLD   ISIS Facility, Rutherford Appleton Laboratory, Chilton, United Kingdom

ABSTRACT

The uses of fluorosurfactants to stabilize water-in-carbon dioxide (w/c) microemulsions are reviewed. A systematic study with fluorosuccinate surfactants of the effects of the extent of fluorination of hydrophobic chains on the stability and structures of these w/c phases is described. Therefore, it has been possible to delineate a structure–function relationship for these fluorosurfactants with reference to their efficiency of water-in-carbon dioxide microemulsion formation. An important finding is that one of these surfactants, namely sodium bis(1H,1H-perfluoropentyl)-2-sulfosuccinate (di-CF4), is able to stabilize w/c microemulsions close to the bottle pressure of a normal CO_2 cylinder. Such efficient surfactants of this kind have obvious advantages for potential practical applications of CO_2.

I. INTRODUCTION

In its liquid or supercritical state, carbon dioxide (sc-CO_2) represents a cheap ($\sim$\$0.05 kg^{-1}), nontoxic, nonflammable, and environmentally responsible alternative to conventional petrochemical and perchlorinated solvents. At 31.1°C and 73.8 bar, the critical point is readily accessible, and variation of solvent density with temperature and pressure offers a flexibility in solvent power that is not so readily achieved with normal solvents. Many low-molecular-weight and nonpolar materials are soluble in CO_2, and it is already used routinely for various commercial processes including decaffeination of coffee [1], supercritical fluid chromatography [2], and as a reaction medium for synthetic chemistry, e.g., synthesis of fluoropolymers [3]. However, weak intermolecular forces in CO_2 result in very low solubility of polar materials and this has prevented its widespread application. Small molecules such as low-molecular-weight (low-MW) alcohols may be employed as entrainers to enhance solvent polarity and, therefore, solubility of polar entities, but success with this approach has been limited and the mechanism is not well understood.

Formation of a water-in-CO_2 (w/c) microemulsion (or emulsion) is one way of overcoming this problem: polar material may be solubilized in water droplets, while nonpolar material is selectively solubilized in the continuous phase. These systems have potential for a wide range of extraction or reaction processes. Extensive screening studies were carried out in order to identify surfactants capable of forming a w/c microemulsion. Consani and Smith [4] looked at over 130 commercially available surfactants and found a few that were CO_2 soluble and even less of those that could also solubilize water. This qualitative study did, however, highlight the compatibility between a fluorocarbon material and CO_2, a result also observed by Iezzi [5]. Following on from this, Hoefling et al. (1991) [6] synthesized model surfactants containing fluorocarbon chains. These surfactants were expected to interact favorably with CO_2 on the basis of their mutually low solubilization and polarizability parameters. Three fluorinated Aerosol-OT (AOT) analogues were shown to dissolve in dry CO_2 and to extract the dye thymol blue into the CO_2 phase of a Winsor II–type system. In addition, high-MW fluoropolyether hydroxyl aluminum and carboxylate salts were shown to extract the same dye into CO_2 phases under certain conditions. The drawback of the solubilization technique is that thymol blue is soluble in dry reverse micelles, so it was not possible to ascertain whether any water was taken into the CO_2 phase. After this pioneering publication, research into w/c microemulsions branched into two main areas: (1) development of perfluoropolyether (PFPE)-based amphiphiles and modified polymers and (2) the use of fluorinated analogues of hydrocarbon surfactants.

A. Fluorosurfactants

The first water-in-CO_2 microemulsion was reported in 1994 by Harrison et al. [7], who used a dichain anionic surfactant consisting of one hydrocarbon and one fluorocarbon chain attached to the same sulfonate head group $[C_7F_{15}CH(OSO_3^- Na^+)C_7H_{15}]$. When mixed with water and CO_2, this hybrid surfactant (H7F7) exhibited typical cloud point behavior seen previously for microemulsions in supercritical and near-critical alkanes [8–10]. It was possible to stabilize a $w = 32$ system ($w = [H_2O]/[surfactant]$) at 35°C, 262 bar with 1.9 wt% surfactant. The detailed structure of these ternary mixtures was investigated by Eastoe et al. [11], who used small-angle neutron scattering (SANS) to characterize the water domains in single-phase regions. As shown in Fig. 1, scattering data for a $w = 32$ sample are consistent with the presence of polydisperse spherical water droplets with an average D_2O core radius R_c^{av} around 25 Å and polydispersity 0.2, values also typical of water-in-oil microemulsions. At high pressures, and therefore at high CO_2 densities, the SANS data are consistent with weakly interacting droplets, whereas on approaching the low-pressure phase boundary an increase in critical-type scattering indicates interactions. Table 1 reports fit parameters with an Ornstein–Zernicke (O-Z) structure factor model (see Section II) as a function of pressure and density.

This scattering law represents the strength of attractions by $S(Q = 0)$ and the range of interaction by a correlation length ζ. The increases in these parameters at low pressures are consistent with a phase separation driven by interdroplet attractions at the lower densities, a phenomenon seen previously for w/o microemulsions in low-density, short-chain alkanes [9]. This H7F7 compound was later the subject of molecular simulation studies by Salaniwal et al., who looked at aggregation in surfactant/water/CO_2 systems [12]. Droplet sizes obtained were comparable to those from neutron scattering [11] and high diffusivities were predicted, highlighting the potential advantages of using low-density and low-viscosity liquids.

Between 1996 and 1997 most publications on w/c systems focused on PFPE-based surfactants and other polymeric materials, discussed subsequently. The next low-MW surfactant reported to form a w/c phase was a dichain anionic, bis(1H,1H,5H-octafluoro-1-pentyl) sodium sulfosuccinate (di-HCF4). In 1997 Eastoe and coworkers reported phase behavior up to $w = 30$ for 0.1 mol dm^{-3} surfactant at 15°C and up to 600 bar [13]. (Pressure cell constraints rather than microemulsion instability limited the study.) Neutron scattering experiments again showed classical w/o-type structure, with droplet clustering close to phase boundaries.

Water solubility up to $w = 12$ was observed for the nonionic polyether surfactant C_8E_5, although cosurfactant (up to 10 wt% pentanol) was required and no direct observation of solution microstructure was made [14]. Reports of w/c systems based on the hydrocarbon surfactant bis(2-ethylhexyl)sodium sulfosuccinate (AOT) were published by Hutton and coworkers [15]. Again, a significant proportion of pentanol was required as a cosolvent to obtain a single-phase system and solvochromatic absorption studies suggested that no bulk water was present.

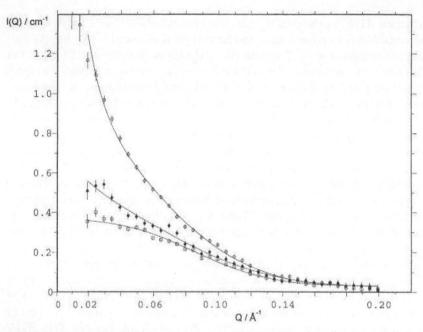

FIG. 1 Small-angle neutron scattering profiles for H7F7-stabilized water-in-CO_2 microemulsion at 25°C. Composition 3wt% H7F7 (0.05 mol dm^{-3}) and 3.5wt% D_2O ($\sim$1.75 mol dm^{-3}) and $w \sim$ 32. ($\square$) 500 bar; ($\bullet$) 260 bar; ($\circ$) 120 bar. The lines are model fits as described in Table 1.

Eastoe and coworkers have used many more structurally related fluoro-succinates (similar to di-HCF4) to determine the effects of surfactant architecture on phase stability. These studies are described in more detail in the following and have illustrated the sensitivity of microemulsion phase boundaries to subtle variations in chain structure [16,17]. (Generic chemical structures are shown in Fig. 4 later.)

B. Fluoropolymers

The CO_2 solubility of high-MV polymeric materials may be enhanced by structural modification [18], e.g., by introduction of fluorinated side groups [19,20], silicone-based segments [21], or fluorocarbon repeat units [22–25]. Few of these materials have been used for stabilizing dispersions of water, but there are several notable exceptions.

Fulton and Pfund [20] used small-angle X-ray scattering (SAXS) to show that a polymer with a CO_2-philic poly(fluorooctyl acrylate) backbone and CO_2-phobic poly(ethylene oxide) grafts formed small aggregates in CO_2.

TABLE 1 Values Obtained from Analysis of SANS Data from a H7F7 Water-in-CO_2 Microemulsion at 25°C as a Function of Pressure and CO_2 Density[a]

P (bar)	ρ (g cm^{-3})	R_c^{av} (Å)	σ/R_c^{av}	$S(Q = 0)$	ζ (Å)
500	1.03	25.0	0.20	—	—
400	1.01	24.6	0.18	—	—
260	0.95	23.0	0.20	0.7	50
120	0.85	20.5	0.19	5.4	96

[a]The model is for polydisperse core–shell spherical particles, and where appropriate an Ornstein–Zernicke $S(Q)$ is included. Uncertainties: water core radius R_c^{av} + 1 Å and correlation length ζ + 5 Å.
Source: Ref. 11.

These micelles were also able to solubilize a small amount of water ([H_2O]/ [PEO groups] = 5.2). Other materials with hydrocarbon and fluorocarbon blocks also formed aggregates but did not solubilize water [20].

PFPE-based surfactants have been studied extensively by Johnston and coworkers [26–29]. Fourier transform infrared (FTIR) measurements on a w = 10 microemulsion based on ammonium carboxylate PFPE [$CF_3O(CF_2CF(CF_3)O)_mCF_2COONH_4$] (MW = 740) indicated the presence of bulk water domains. Water solubility up to w = 14 was reported at 55°C and around 177 bar for 1.4wt% surfactant [26]. Time-resolved fluorescence and electron paramagnetic resonance (EPR) studies suggested the presence of anisotropic or nonspherical micelles. Zielinski et al. [30] studied the phase stability of this system at 35°C, as shown in Fig. 2; SANS was then employed to study droplet structure. Observations with PFPE were similar to the earlier study in H7F7 [11]; discrete water droplets of around 25 Å radius were found to be present, and as indicated in Fig. 3 the droplet size increased with added water.

Data-fitting analyses shown in Fig. 3 gave the droplet radius as 20 Å at 0.8wt% dispersed D_2O, increasing to 35 Å at 2.0wt%. Critical-type scattering was also observed on approaching the phase boundary; hence the O-Z structure factor was used in the data analysis. Despite the onset of interactions, only small variations in structure and droplet radius were seen with variation of T and/or P, a behavior reminiscent of systems stabilized by H7F7 [11] and fluorosuccinates [13,16,17].

A range of PFPE-based amphiphiles are able to stabilize emulsions consisting of 0.01 mol dm^{-3} surfactant and equal quantities of CO_2 and 0.1 mol dm^{-3} NaCl solution [31]. Ammonium carboxylates of MW 672, 940, 2500,

P / bar

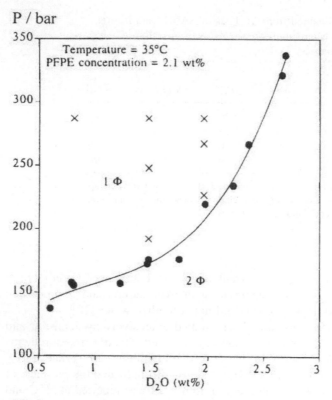

FIG. 2 Temperature–composition phase diagram for $D_2O/CO_2/PFPE$ mixtures at constant PFPE composition of 2.1wt% on a D_2O-free basis. The single-phase region is denoted 1ϕ and the phase-separated region 2ϕ. Crosses show the SANS operating conditions. The line is a guide to the eye. (From Ref. 30.)

and 7500 were studied, and droplet sizes from 3 to 10 μm were observed. The more water-soluble MW 672 material preferentially formed a c/w emulsion and the other three formed w/c phases, with stability ranging from a few seconds for MW 940 to several hours for MW 7500. Emulsion stability was pressure dependent, and systems that existed as w/c phases at high pressure inverted to c/w at low pressure as the surfactants became less CO_2 soluble [31]. Poly(dimethyl siloxane)-based surfactants have also been used to form w/c emulsions, although the stability was lower. The development of poly(ether-carbonate) polymers that are able to stabilize w/c emulsions has demonstrated that, with the right chemistry, hydrocarbon materials may after all have potential for use in CO_2 [32]. Optimization of these materials

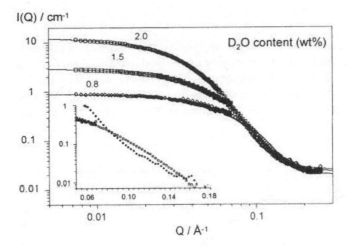

FIG. 3 SANS spectra of 0.89 (◇) 1.5 (□) and 2.0 (○) wt% D_2O in CO_2/PFPE mixtures at 35°C and 287 bar. Lines represent model fits to the data. The inset shows the high-Q portions of the 0.8wt% (filled points) and 2.0wt% D_2O (open points) spectra. The difference is indicative of the increase in droplet radius with increasing D_2O concentration. (From Ref. 30.)

could represent a significant advantage over high-cost fluorocarbon or siloxane-based surfactants.

C. Applications

The slight mutual solubility between water and CO_2 means that an equilibrium exists between the dissolved CO_2 and the carbonic acid formed in the solution. The pH in microemulsion droplets was investigated by Clarke et al. [29] and Niemeyer and Bright [33], both of whom used spectroscopic probe molecules to characterize water domains in PFPE-stabilized microemulsion systems. The pH was found to be between 3.1 and 3.5 and was essentially independent of water content and solvent density. Acidity of the aqueous environment obviously has consequences for potential application of these systems. Holmes and coworkers [34,35], who addressed this issue, showed that it was possible to buffer pH in microemulsion droplets and hence carry out enzyme-catalyzed reactions. The feasibility of carrying out several processes using w/c systems has been demonstrated in the literature, including synthesis of nanoparticles [36] and catalysis of inorganic and organic reactions [37–40]. In an exciting development, a CO_2-based dry cleaning system has been commercialized [41].

D. Aims

In spite of this interest in w/c systems, the issue of how surfactant chemical structure affects properties remains largely unresolved, partly because of lack of suitable, well-characterized compounds. This chapter describes the behavior with various custom-synthesized fluorinated analogues of Aerosol-OT. Example molecular structures are shown in Fig. 4. The main features

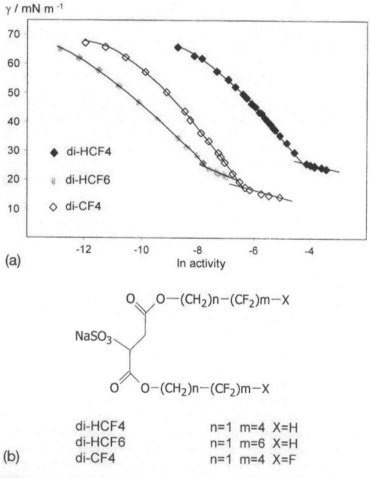

FIG. 4 (a) Surface tensions of aqueous surfactant solutions for di-HCF4, di-HCF6, and di-CF4. The lines fitted to the pre-cmc data are polynomials used for assessing adsorption isotherms. (b) Chemical formulas of the fluorinated surfactants. (From Ref. 17.)

of these succinates are bulky CO_2-compatible fluorocarbon chains and a water-soluble sulfonate group. Hence, a tendency to form reversed curvature micelles and microemulsions is expected, in a similar fashion to AOT in water–oil phases. Three important examples are bis($1H,1H,5H$-octafluoropentyl)-2-sulfosuccinate (di-HCF4), bis($1H,1H,7H$-dodecafluoroheptyl)-2-sulfosuccinate (di-HCF6), and sodium bis($1H,1H$-perfluoropentyl)-2-sulfosuccinate (di-CF4).

There are three reasons for concentrating on these compounds. First, for di-HCF4 the precursor alcohol $1H,1H,5H$-octafluoropentanol is 10 times cheaper than the more fluorinated di-CF4. Clearly, for potential commercial applications cost is important, as is the need to minimize working pressure. Second, there are interesting differences in the hydrophobic groups: a hydrogen atom in the di-HCF4 surfactant replaces a chain tip fluorine atom in di-CF4. This switch introduces a large permanent dipole moment (μ); for example, the μ in CF_3CF_2H is 1.54 D. Therefore, it is of interest to examine how this affects w/c microemulsion formation. Third, comparing the behavior of di-HCF4 and di-HCF6 highlights the effects of chain length and F-H ratio on surfactant performance.

Although it is a minor issue, water partitioning into the CO_2 continuous phase may have an effect on droplet size, but this has not been confirmed directly by scattering methods. However, a previous study using infrared spectroscopy [29] has been made with PFPE phases. We have combined near infra-red (NIR) and small-angle neutron scattering (SANS) experiments using a fluorinated AOT analogue and show that water partitioning is important when the microemulsion is dilute or at elevated temperatures.

II. METHODS AND MATERIALS

A. Chemicals

The procedures for surfactant preparation and purification have been described previously [42,43]. As a final purification step, aqueous solutions made up below the critical micelle concentration (cmc) were foam fractionated. Water (18.2 MΩ cm) was taken from either a RO100HP Purite water system or a Millipore Milli-Q Plus system, D_2O was 99.9% D atom (Fluorochem Ltd., UK), and CO_2 (BOC) was used as received.

B. Surface Tension and Neutron Reflection Measurements

Surface tension measurements were made with a drop-volume (DV) tensiometer (Lauda TVT1) in the presence of trace EDTA (99.5% tetrasodium salt hydrate, Sigma) as described elsewhere [43,44]. The D_2O, which is used in neutron reflection (NR) measurements, can also be a source of ionic contamination [43,44], so it was distilled prior to use. Thermostatting was to $\pm 0.1°$ using a Grant LTD6G bath. To avoid phase separation at high multiples of the cmc with di-CF4 and the Krafft point with di-HCF6, the temperatures were 40°C for di-HCF6, 25°C for di-HCF4, and 30°C for di-CF4.

Neutron reflection experiments were carried out on the CRISP and SURF reflectometers at ISIS, Rutherford Appleton Laboratories (Didcot, UK) [45], using standard cells for liquid surfaces [46]. Selected concentrations were studied in null reflecting water (NRW, 8.0 mol% D_2O in H_2O) at the same temperatures as for tensiometry. A full account of NR theory can be found elsewhere [47], but a relevant summary follows. The momentum transfer $Q(\text{Å}^{-1}) = (4\pi/\lambda)\sin(\theta/2)$, where λ is the neutron wavelength (0.5 → 6.5 Å) and θ the scattering angle (3°). The reflectivity $R(Q)$ from a surfactant layer at the surface of NRW can be modeled in terms of a single, uniform layer using an optical matrix method [47]. This results in fitted values for the layer thickness τ and scattering length density ρ, which are related to an effective area per molecule in the film A by

$$A = \frac{\sum b_i}{\rho \tau} = \frac{1}{\Gamma N_a} \tag{1}$$

In Eq. (1) $\sum b_i$ represents a sum of nuclear scattering lengths over the molecule, Γ is the surface excess in mol m^{-2}, and N_a is the Avogadro number.

C. Water-in-CO_2 Systems: NIR and Small-Angle Neutron Scattering

Pressure–temperature microemulsion phase stability was determined by visual inspection using a stirred high-pressure optical cell as described elsewhere [9,11,13,16,17]. The samples were studied at various surfactant concentrations and water-to-surfactant molar ratios w. Under these T-P conditions the solubility of water in CO_2 is negligible (0.13%) (see later) as compared with the 5% or so of D_2O present and the w values were not corrected for this small effect. To check the validity of this assumption, NIR experiments were carried out to determine the true extent of partitioning in the microemulsions, using a 200 series Bentham spectrometer with a 100 W quartz-halogen lamp source and Peltier-cooled PbS detector. Samples were contained in the high-pressure cell described previously [9] but with a 2.34-cm path length. The spectral features of interest were the water overtone bands close to 1400 nm for (1) water-saturated CO_2 and (2) a w/c microemulsion over a temperature range (25–45°C) at constant pressure (300 bar). For the first experiments CO_2 was equilibrated with excess water, and for the second a $w = 20$ microemulsion stabilized by di-HCF4 was studied. To obtain the pure water signal, backgrounds of (1) CO_2 and (2) di-HCF4 in CO_2 were measured under the same conditions and subtracted from the microemulsion spectrum.

SANS experiments were performed on LOQ at ISIS, UK [48], thereby measuring the scattering cross section $I(Q)$ (cm^{-1}) as a function of momentum transfer, characterized by λ (2.2 → 10 Å) and θ (<7°). Corrections to account for both pressure-induced sample volume changes and the path length were made as before [11,13,16,17]. For liquid CO_2 the scattering length density is given by $\rho_{CO_2} = $ (mass density $\times 2.498 \times 10^{10}$ cm^{-2}) [49]. In the final data analysis, the effects of P and T on ρ_{CO_2} were taken into account. However, given that in these experiments the liquid CO_2 density is around 1.0 g cm^{-3} (typical variation $\pm 10\%$ for the pressures studied here), and for fluorosurfactants $\rho_{surf} \sim 2.0 \times 10^{10}$ cm^{-2}, the interfacial film is essentially contrast matched to the solvent. Hence, because $\rho_{D_2O} = 6.4 \times 10^{10}$ cm^{-2}, the scattering is dominated by contrast at the D_2O–surfactant interface, so the droplets can be sized in terms of a core radius R_c.

1. SANS Data Analysis

The $I(Q)$ data were analyzed using the multimodel FISH program [50]; a Schultz distribution of spherical particles gave the best fits and most physically reasonable parameters. The scattering law can be written as

$$I(Q) = \left(\frac{\phi (\rho_{D_2O} - \rho_{CO_2})^2}{\sum_i V_i X(R_i)} \right) \sum_i [V_i^2 P(Q, R_i) X(R_i)] \qquad (2)$$

The parameters ϕ, R, and V represent the particle volume fraction, radius, and volume, respectively. The function is the particle form factor $P(Q, R_i)$. The contribution $X(R_i)$ is the Schultz function [50,51], which is characterized by an average radius R_c^{av} and root-mean-square (RMS) deviation $\sigma = R_c^{av}/(Z + 1)^{1/2}$, where Z is a width parameter. With samples that were close to the low-pressure phase boundary it was necessary to introduce into Eq. (2) an attractive Ornstein–Zernicke structure factor $S(Q, \zeta)$. This $S(Q)$ is an effective function, accounting for additional scattering at low Q, and has been used before with CO_2 microemulsions [11,13,16,17,30]. The O-Z function describes a decaying particle distribution with ξ the correlation length, and the strength of interactions is related to $S(0)$ via the isothermal compressibility [51]

$$S(Q, \xi) = 1 + \left[\frac{S(0)}{1 + (Q\xi)^2} \right] \qquad (3)$$

Sample compositions and scattering length densities are known quantities, hence the adjustable parameters are R_c^{av} and σ/R_c^{av} and, if used, ζ and $S(0)$. Extensive trial fits showed that $\sigma/R^{av} = 0.20$ gave the best fits, so this was constant in final analyses. Uncertainties in R_c^{av}, σ/R_c^{av}, and ζ may be taken as ± 2 Å, ± 0.02, and ± 10 Å, respectively.

III. RESULTS AND DISCUSSION

A. Surface Tension and Neutron Reflection Measurements with Aqueous Systems

Prior to investigating water–CO_2 systems, the dilute aqueous phase behavior of these surfactants was studied. The importance of this was to establish surface chemical purity and obtain insight into packing at the air–water interface, which is more readily accessible than the CO_2–water interface. Hence, surface tension measurements were used to check the purity and to determine the adsorption isotherms. Figure 4 shows example γ–ln activity curves (activity coefficients were obtained using the Debye–Hückel limiting law) and molecular structures of the compounds.

The plots are characteristic of pure surfactants: clean breaks at the cmc with no minima or shoulders. A detailed description of the experimental procedures necessary to obtain representative adsorption isotherms is given elsewhere [43,44]. Polynomials fitted to pre-cmc surface tension data were used to generate surface excesses Γ using the Gibbs equation.

$$\Gamma = -\frac{1}{mRT}\frac{d\gamma}{d\ln a} \tag{4}$$

In Eq. (4) R is the universal gas constant, T the absolute temperature, γ the surface tension, and a the activity. For a dilute 1:1 ionic surfactant in the absence of swamping electrolyte, the prefactor m should be 2, and the validity of this approach has been thoroughly tested by combining measurements from drop volume tensiometry (DVT) and neutron reflection [44]. Table 2 lists relevant parameters derived from surface tension and neutron reflection measurements. The limiting tension at the cmc clearly depends on chain type; for the terminal-H compound, di-HCF4, it is approximately 9 mN m^{-1} higher than for the equivalent CF_3-terminated compound di-CF4. The presence of a chain tip H atom in di-HCF4 is expected to increase the surface free energy with respect to a perfluoromethyl-ended surfactant such as diCF4. In addition, H—CF_2— dipolar repulsion should increase γ. Increasing the chain length from di-HCF4 to di-HCF6 clearly reduces γ_{cmc}, as found for normal hydrocarbon surfactants; however, the nature of the chain tip dominates over chain length in terms of γ_{cmc}. Neutron reflection experiments described subsequently give evidence for a lower film packing with

TABLE 2 Parameters Derived from Surface Tension and Neutron Reflection (NR)

Surfactant	T (°C)	cmc (mmol dm^{-3})	γ_{cmc} (m Nm^{-1})	Surface tension A_{cmc} ($\pm 3/\text{Å}^2$)	Neutron reflection A_{cmc} ($\pm 2/\text{Å}$)
di-HCF4	25	16.0	26.8	65.0	65.8
di-HCF6	40	0.46	24.1	76.0	—
di-CF4	30	1.57	17.7	56.0	62.7
di-HCF4GLU	30	11.2	25.4	66.0	65.9
AOT	25	2.56	30.8	77.0	78.0

[a]Equation (4) was used to obtain the molecular areas at the cmc from the tensiometric me
were analyzed with Eq. (1). The NR experiments with di-CF4 were at the cmc, but with di
13.4 mmol dm^{-3} ≈ 0.84 × cmc. For AOT the neutron data are from Ref. 53. For di-CF4G

the H-terminated systems compared with a fully fluorinated end group, hence this may also contribute to γ_{cmc}.

The NR curves are not shown for brevity; however, the analyses with the single-layer model resulted in effective film thicknesses τ and scattering length densities ρ given in Table 2. Hence, molecular areas and surfaces excesses Γ were obtained using Eqs. (1) and (4). Figure 5 shows example adsorption isotherms for three different surfactants, comparing results from neutron reflection [Eq. (1)] and tensiometry [Eq. (4), with $m = 2$]. (The chemical structure for di-HCF4GLU is shown in Fig. 7b.)

The agreement between the two methods is an indication of the high surface chemical purity of these surfactants. For tensiometry the uncertainty in A_{cmc} is ± 3 Å^2, and for NR it is $\pm 3-5\%$ (as evidenced by the repeat measurements shown in Fig. 5c for di-CF4GLU). Therefore, at the cmc of di-CF4 the differences between these two results for A_{cmc} are just outside the errors. A key indication of purity is the absolute value of the surface excess Γ as measured by NR at twice the cmc. For both di-HCF4 and di-CF4 these were identical (within errors) to the values at the respective cmc values. Surface-active impurities would adsorb strongly below the cmc, but above it partitioning into micelles would alter both the surface composition and the adsorbed amount, which would show up in the NR experiments.

Because physical cross-sectional area of a fluorocarbon chain is approximately 28 Å^2 [52], the layers are quite densely packed; for di-CF4 the A_{cmc} is 56 ± 3 Å^2 from surface tension data. Although the NR data gave $A_{cmc} = 62.7$ Å^2, the relative packing efficiency of the chains is still high, and fluorocarbon chains account for about 90% of the A_{cmc} alone. With the terminal-H analogue, film packing decreases, consistent with the additional chain dipolar interaction. There is a magnified effect for di-HCF6, which shows $A_{cmc} = 76$ Å^2. The data show that these fluorosuccinates are strongly adsorbed, with the terminal-H surfactant di-HCF4 having a slightly lower surface activity.

It is interesting to compare the interfacial behavior of these straight-chain fluorocarbon succinates with the branched-chain hydrocarbon AOT (Table 2). Tensiometric experiments on AOT were carried out in our laboratory, and the results from neutron reflection experiments by Thomas and coworkers [53] are also given. As can be seen, the F surfactants all exhibit lower limiting surface tension and pack more efficiently than AOT.

B. Phase Stability of Water-in-CO₂ Microemulsions

Figure 6 shows the P-T phase stability for two different w values with di-CF4 and di-HCF4 in water-in-CO_2 microemulsions. Owing to CO_2-hydrate formation at $\leq 10°C$, temperatures below 15°C could not be studied. At el-

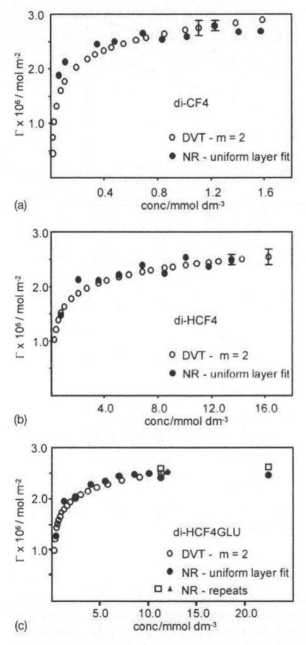

FIG. 5 Adsorption isotherms of fluorinated surfactants obtained by neutron reflectivity and tensiometry measurements: (a) di-CF4, (b) di-HCF4, and (c) di-HCF4GLU. Example error bars are shown for some points. (From Ref. 44.)

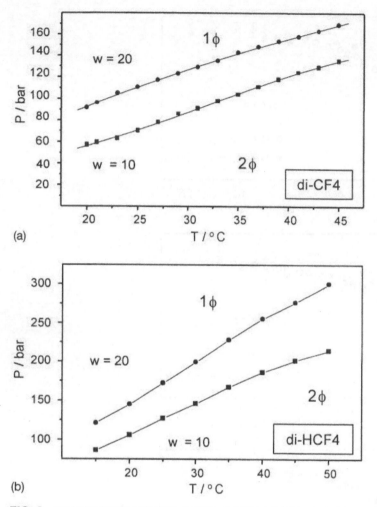

FIG. 6 Phase behavior of water-in-CO_2 microemulsions for two w values with di-CF4 (a) and di-HCF4 (b) at 0.05 mol dm^{-3}. (From Ref. 16.)

evated pressures single-phase (1ϕ) transparent microemulsions are formed, and there is a critical pressure P_c below which the samples rapidly turn opaque (2ϕ). With an efficiently stirred cell these phase transitions are entirely reversible, similar to the behavior seen with AOT-stabilized w/o systems in short-chain alkanes such as ethane and propane, which is documented elsewhere [8–10,54,55].

This *P-T* phase behavior is similar to that seen with the H7F7 and PFPE surfactants in other w/c systems as shown, for example, in Fig. 2 [7,11,30]. Furthermore, for microemulsions in both CO_2 and short-chain alkanes, the phase instability at low pressure (density) is consistent with reduced compatibility between the surfactant chains and the solvent.

Figure 7a shows *P-T* boundaries for seven dichain surfactants under equivalent conditions of concentration and water content. The surfactant molecular structures are shown in Figs. 4b and 7b.

In terms of phase stability, di-CF4 is the best performer because it stabilizes w/c droplets at the lowest pressures. The variation of surfactant chain length has a significant effect; decreasing the fluorination from CF4 to CF3 increases P_c, and increasing the chain length of di-HCF4 to di-HCF6 is beneficial to stability. Di-CF4H is similar to di-CF4 except for an extra —CH_2— close to the ester oxygen; this change represents a reduction in

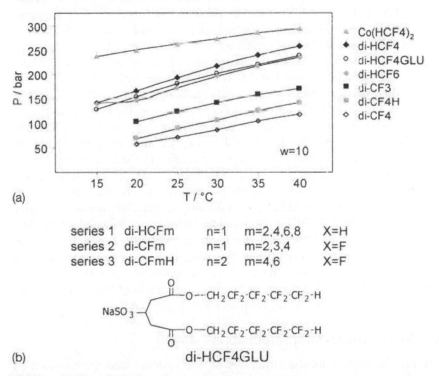

(a)

series 1 di-HCFm   n=1   m=2,4,6,8   X=H
series 2 di-CFm   n=1   m=2,3,4   X=F
series 3 di-CFmH   n=2   m=4,6   X=F

(b)   di-HCF4GLU

FIG. 7 (a) Phase behavior of water-in-CO_2 microemulsions for various fluorinated anionic surfactants with $w = 10$ and [surf] = 0.05 mol dm^{-3}. (b) Chemical formulas of the surfactants used in (a). (From Ref. 17.)

the F/H ratio, which causes an increase in P_c by about 10–20 bar. Hence, there appears to be an optimal hydrophobe chain length, of four fully fluorinated carbons with only one CH_2 group, and subtle deviations from this key structure reduce the phase stability. Indeed, di-CF4 is notable for having very low transition pressures; for example, at 20°C the $w = 10$ microemulsion is stable at close to bottle pressure for CO_2. This is the lowest P_c for any w/c system, suggesting that di-CF4 is currently the most efficient compound available. A direct comparison can be made with the performance of ammonium PFPE: for $w = 20$ at 30°C the P_c is 120 bar for di-CF4, whereas for PFPE the phase is stable only down to 178 bar [35]. Changing from di-CF4 to di-HCF4 causes a reduction in stability; for example, with $w = 20$ at 30°C the critical pressure increases to over 300 bar. With reference to the air–water interface (Table 2), this may be due to a lower packing efficiency in the film and/or a genuine reduced compatibility with the solvent on replacing F with H. If a 0.05 mol dm^{-3} NaCl solution is used in place of pure water P_c is also reduced, consistent with more efficient interfacial packing induced by screening of head group interactions. Therefore, it is important to work with well-defined, pure, compounds if meaningful results are to be obtained from studies of w/c systems.

It has also been possible to explore some effects of changes in the head group. The surfactant di-HCF4GLU is a glutaconate, rather than a succinate, so the fluorocarbon chains are symmetrically balanced about —SO_3^- (Fig. 7b). This extra methylene unit also serves to space out the F chains, although there was no noticeable change in packing at the air–water interface with di-HCF4GLU compared with di-HCF4 [44]. Comparing di-HCF4 and di-HCF4GLU in Fig. 7a shows that a minor change in the head group gives a noticeable improvement in phase stability (P_c 10–15 bar lower). The $Co(HCF4)_2$ is a cobalt salt of di-HCF4, which was prepared by a method described elsewhere for AOT [56]. A divalent counterion increases P_c compared with the Na^+ di-HCF4. With anionic surfactants, switching from M^{1+} to M^{2+} generally reduces the aqueous phase solubility; hence, lower compatibility with water may contribute to change in P_c. To summarize the phase behavior of these fluorosulfonates, it would seem that $C_4F_9CH_2$— is an optimum CO_2-phile, and in terms of head group, switching from the succinate to a glutaconate gives rise to improved w/c phase stability.

C. Water-in-CO_2 Microemulsions—SANS

For certain surfactants the structures of w/c systems were investigated by high-pressure small-angle neutron scattering. Because maximum pressure of the SANS cell is 600 bar, the experimental conditions were $T = 15$°C and $P = 500$ bar. Working at the high-pressure end helps to minimize attractive interactions, which increase on approach to P_c. Figure 8 shows example scattering curves from single-phase microemulsions with di-HCF4 as a function of w and the least-squares fits to the model for polydisperse spheres (see Section II). The curves give clear evidence for D_2O nanodroplets dispersed in CO_2, similar to the scattering seen for H7F7- and PFPE-stabilized systems shown in Figs. 1 and 3, respectively [11,30].

The scaling of experimental absolute intensities for $P(Q)$ was within 10% of expected values given the known compositions and correction factors employed; hence, the model is physically reasonable. The reproducibility was checked by repeat experiments at various pressures, temperatures, and compositions. With di-HCF6 ($w = 30$) at 500 bar the systems are relatively close to P_c, and enhanced scattering was evident at low Q; therefore, it was necessary to add the attractive $S(Q)$ into the model. This is described in more detail elsewhere [13,16,17]. Increased intensities at low Q were also observed with the H7F7 (Fig. 1) and PFPE surfactants close to the respective P_c values [11,30].

The relative packing efficiencies of these different surfactants at the water–CO_2 interface can be examined in terms of the effective area per head group A_h. If it is assumed that the droplets are spherical and all N (m^{-3}) surfactant molecules adsorb then

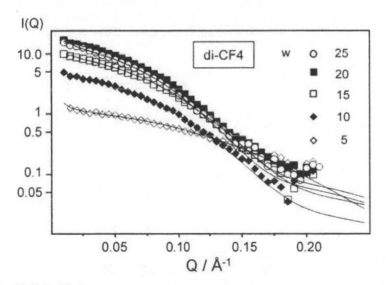

FIG. 8 High-pressure SANS spectra from D_2O droplets in water-in-CO_2 microemulsions with di-CF4 surfactant at 15°C and $P = 500$ bar; [di-CF4] = 0.10 mol dm^{-3}. The lines are model fits as described in the text. (From Ref. 16.)

$$A_h = \frac{3 v_{D_2O} w}{R_c} \tag{5}$$

where v_{D_2O} is the molecular volume of water. Figure 9 shows the dependence of R_c versus w for three principal surfactants under study; characteristic error bars are shown for some example points.

Clearly, at least within the experimental error in R_c (± 2 Å), the surfactants exhibit almost identical behavior. Although phase stability depends on surfactant type (Figs. 6 and 7), there appear to be no obvious effects of chain structure on droplet structure or radius. Hence, for each w value the mean radius for different surfactants was used to generate a linear fit with Eq. (5). This gradient gives $A_h = 115 \pm 5$ Å^2, and an intercept consistent with a "dry" micelle consisting of an 11-Å polar core. The apparent molecular area of di-HCF4 is about 45% larger than that obtained in an air–water film at the cmc (Table 2). At first sight 115 Å^2 seems to be quite large, so an alternative analysis method was employed, based on the Porod equation and assuming all surfactant molecules are bound. More detail can be found elsewhere [17]; however, the result was $A_h = 95 \pm 15$ Å^2. These molecular areas are entirely consistent within the uncertainties. Previous work with H7F7 was also consistent with a relatively large effective head group area, around 140 Å^2 [11]. It is interesting to compare this with AOT in w/o systems, for which a mean value of 72 Å^2 for A_h was found in a range of n-alkanes from propane to decane [55]. Therefore, it would appear that, in general, fluorosurfactants at a water–CO$_2$ interface adopt a noticeably lower packing den-

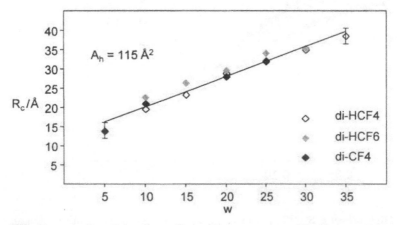

FIG. 9 Variation of droplet radius with water content for three surfactants at $T = 15°C$, $P = 500$ bar, and [surf] = 0.05 mol dm^{-3}. (From Ref. 17.)

sity than a related hydrocarbon surfactant (AOT) at an analogous water–alkane interface.

D.   Water Partitioning in w/c Phases—NIR and SANS

The solubility of water in CO_2 is higher than in oils typically used for w/o microemulsions, and in CO_2 at pressures over 200 bar there is a significant increase in water solubility as a function of temperature [57]. Hence, water partitioning into the CO_2 continuous phase may be important, and a previous study using infrared spectroscopy [29] points to this for PFPE systems. The change in water solubility should have an effect on droplet size, but there are no previous reports to establish this by scattering methods; hence, we performed complementary near-infrared and SANS experiments.

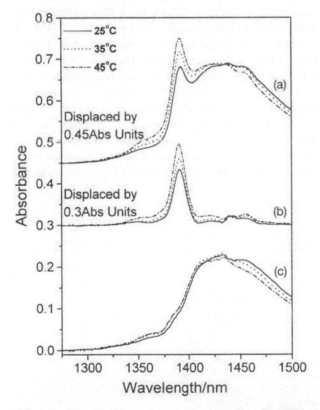

FIG. 10   NIR absorption spectra for water in a di-HCF4, $w = 20$ w/c microemulsion (a) compared with water saturating CO_2 (b). Difference spectra of (a) − (b) showing subtraction (c) of the dissolved water contribution. $P = 300$ bar. (From Ref. 17.)

The NIR spectra of di-CF4 microemulsions are shown in Fig. 10a for various temperatures at 300 bar. Figure 10b shows the sharp peak from water alone in CO_2, which is centered at 1390 nm. This absorption becomes more intense with temperature, reflecting the accompanying increased water solubility. Furthermore, this line characteristic of dissolved water is well resolved from that for water dispersed in w/c droplets (Fig. 10c), which exhibits a much broader band at 1440 nm. In Fig. 10a both contributions are clearly visible, showing the dissolved water component increasing at the expense of the droplet signal as a function of temperature. Subtracting the water-saturated CO_2 spectrum from that of the microemulsion at each temperature gives a broad peak, as shown in Fig. 10c, which is essentially identical to the signal from a w/o microemulsion with AOT in heptane.

For the systems examined by SANS, complete saturation of the CO_2 phase would reduce the w value by 1.5–2.0. Because in this case the effect is small and does not alter the gradient of R_c versus w from which A_h is obtained (Fig. 9), we have chosen to use the total added water w_{add}. However, at lower surfactant concentration the droplet size may be reduced at elevated temperatures and pressures owing to this partitioning. Therefore, SANS measurements were made on a di-HCF6–stabilized w/c microemulsion at w_{add} = 30, i.e., sufficient total water to saturate the CO_2 at the initial condition 15°C and 200 bar. Further SANS measurements were made on this sample at higher pressure (15°C, 450 bar) and temperature (42°C, 500 bar), and the $I(Q)$ curves were analyzed using the polydisperse sphere model. In Fig. 11 the droplet radii under these conditions are compared with calculations based on solubility data (shown as a line), assuming that the CO_2 phase remained

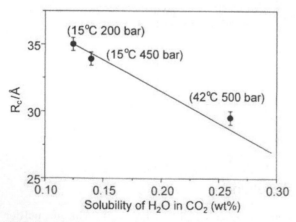

FIG. 11 Radii obtained from SANS showing reduction in droplet size for a di-HCF6 w/c microemulsion with $w = 30$ at three conditions. (From Ref. 17.)

saturated. Overall, the correlation between the observed decrease in R_c and the increase in water solubility is good, confirming that droplet size may be "adjusted" by variation of temperature and pressure.

IV. SUMMARY

A range of fluorinated anionic surfactants, which are structural relatives of Aerosol-OT, were synthesized and characterized, with the aim of investigating the effects of chemistry on the structure and stability of water-in-carbon dioxide microemulsions. The dilute aqueous phase behavior was studied to check for chemical purity and fully characterize the compounds. Once appropriate measures were taken to achieve sufficient purity, the surface excesses measured by both tensiometry and neutron reflection measurements agreed well, and the surface tensions were consistent with a prefactor of 2 in the Gibbs equation.

Phase behavior studies showed that various fluorosurfactants stabilized water-in-carbon dioxide microemulsions; detailed work was carried out on three of the sulfosuccinates. As expected, the added electrolyte affected the phase diagram, highlighting the need for working with pure surfactants. In this series, di-CF4 was the most effective surfactant because it gave microemulsions at significantly lower pressures. Remarkably, with di-CF4, microemulsions can be formed close to the bottle pressure of a normal CO_2 cylinder, and this may have advantages in terms of associated costs in any future practical application. The effect of the variation of the fluorocarbon chain length was studied, and a shorter chain length was found to be detrimental to phase stability. The addition of extra —CH_2— units either in the chains or in the head group or the replacement of the terminal CF_3— by H—CF_2— also reduced the phase stability window. Therefore, for the series with a sodium sulfosuccinate head group $C_4F_9CH_2$— appears to be the optimal CO_2-philic group. Overall, the phase behavior of these sulfosuccinates mirrors that for the H7F7 and PFPE systems that have also been examined [7,11,30]: at low pressures and low CO_2 densities there is a phase transition driven by attractive droplet interactions.

It is interesting to compare adsorption at the two interfaces under investigation here, water–CO_2 interface and air–water interface (i.e., at the cmc). For example, with di-HCF4 the area per molecule is 60% larger in the CO_2 system as compared with the normal air–water interface (105 $Å^2$ versus 65 $Å^2$). On the other hand, with branched-chain AOT at air–water (cmc) and water-in-oil microemulsion interfaces, the molecular area is essentially constant at about 76 $Å^2$ (Table 2 and Refs. 45 and 54). The SANS results for various sulfosuccinates were consistent with an effective head group area of around 110 $Å^2$, which is more than 30 $Å^2$ greater than for AOT in equivalent hydrocarbon–water phases [55]. Other SANS experiments on H7F7 were also consistent with a relatively high surface area per molecule for fluorosurfactants (~140 Å^2 in that case [11]), suggesting that this is perhaps a genuine effect (and pointing to real differences compared with the curved water–oil surface).

Based on the available data, it can be stated that a fluorosurfactant film at a water–CO_2 microemulsion interface appears to be less densely packed than the equivalent hydrocarbon surfactant in water-in-oil phases. The value of 110 Å^2 represents approximately twice ($\times 1.87$) the physical cross section of two fluorocarbon chains (56 Å^2 [52]). For n-alkyl sulfosuccinates the equivalent factor is around $1.55\times$ a close-packed chain cross section [44]. The origin of this may be the finite, although small, solubility of water in CO_2, which is not the case in water–oil systems. This comparison of molecular areas is valid only for strongly adsorbed surfactants, i.e., a negligible amount of free monomer in the bulk phases. For the CO_2 systems, if there were to be a significant loss of surfactant from the interface, then the effective w value would increase because it would become $[D_2O]/([\text{surfactant}]_{\text{total}} - [\text{surfactant}]_{\text{free}})$ [58]. The upshot would be to reduce the gradient of the R_c versus w plot (Fig. 9), thereby further increasing A_h above 115 Å^2. Therefore, a low (negligible) background concentration of free monomer in CO_2 can be assumed.

In summary, this chapter reviews water-in-CO_2 microemulsion formation with three different types of anionic surfactants, hybrids such as H7F7 [7,11], polysurfactants such as PFPE [26–30], and the sulfosuccinates [13,16,17]. There are common patterns in the phase and structural behavior of all these systems and striking similarities with the properties of related AOT-stabilized water-in-oil microemulsions with short-chain low-density alkane solvents. In particular, the fluorosuccinates represent well-characterized surfactants, suitable for w/c systems.

V. FUTURE RESEARCH

The main challenge for any potential application is to minimize costs, and these specialized fluorosurfactants are obviously relatively expensive. Therefore, efforts should be directed toward improving compatibility of hydrocarbon surfactants with CO_2. This may be achieved by employing principles demonstrated by Beckman and coworkers with poly(ether-carbonates) [32] or exploring extremes in chain structures by incorporating the most CO_2-philic hydrocarbon groups. Another important concern is the need to maintain low operating pressures so as to minimize plant and pressure vessel costs. Therefore, simply obtaining hydrocarbon surfactants that aggregate in

CO_2 will be the first step, but the real advance will come with systems that can be used close to CO_2 bottle pressure.

ACKNOWLEDGMENTS

We acknowledge support from EPSRC via grants GR/L05532 and GR/L25653 and studentships to AP and ER. We also thank CLRC for allocation of beam time at ISIS and a grant toward consumables and travel. Sandrine Nave (Bristol) carried out the tensiometric study of Aerosol-OT.

REFERENCES

1. CA Eckert, BL Knutson, PG Debendetti. Nature 383:313–318, 1996.
2. RM Smith. J Chromatogr A 856:83–115, 1999.
3. JM DeSimone, Z Guan, CS Elsbernd. Science 257:945–947, 1992.
4. KA Consani, RD Smith, J Supercrit Fluids 3:51–65, 1990.
5. A Iezzi. In: KP Johnston, JML Penninger, eds. Supercritical Fluid Science and Technology. ACS Symposium Series 406. Washington, DC: American Chemical Society, 1989, pp 122–139.
6. TA Hoefling, RM Enick, EJ Beckman. J Phys Chem 95:7127–7129, 1991.
7. K Harrison, J Goveas, KP Johnston. Langmuir 10:3536–3541, 1994.
8. RW Gale, JL Fulton, RD Smith. J Am Chem Soc 109:920–921, 1987.
9. J Eastoe, BH Robinson, DC Steytler. J Chem Soc Faraday Trans 86:511–517, 1990.
10. EJ Beckman, RD Smith. J Phys Chem 94:345–350, 1990.
11. J Eastoe, Z Bayazit, S Martel, DC Steytler, RK Heenan. Langmuir 12:1423–1424, 1996.
12. S Salaniwal, ST Cui, PT Cummings, HD Cochran. Langmuir 15:1018–1026, 1999.
13. J Eastoe, BMH Cazelles, DC Steytler, JD Holmes, AR Pitt, TJ Wear, RK Heenan. Langmuir 13:6980–6984, 1997.
14. GJ McFann, KP Johnston, SM Howdle. AIChE J 40:543–555, 1994.
15. BH Hutton, JM Perera, F Grieser, GW Stevens. Colloids Surf A 146:227–241, 1999.
16. J Eastoe, A Downer, A Paul, DC Steytler, E Rumsey. Prog Colloid Polym Sci 115:214–221, 2000.
17. J Eastoe, A Downer, A Paul, DC Steytler, E Rumsey, RK Heenan. Phys Chem Chem Phys 2:5235–5242, 2000.
18. F Rindfleisch, TP DiNoia, MA McHugh. J Phys Chem 100:15581–15587, 1996.
19. R Fink, D Hanchu, R Valentine, EJ Beckman. J Phys Chem B 103:6441–6444, 1999.
20. JL Fulton, DM Pfund. Langmuir 11:4241–4249, 1995.
21. G Li, MZ Yates, KP Johnston. Macromolecules 33:4008–4014, 2000.

22. JB McClain, DA Canelas, ET Samulski, JM DeSimone, HD Cochran, GD Wignall, R Triolo. Science 274:2049–2052, 1996.

23. F Triolo, A Triolo, JD Londono, GD Wignall, DE Betts, S Wells, ET Samulski, JM DeSimone. Langmuir 16:416–421, 2000.

24. EG Ghencin, AJ Russell, EJ Beckman, L Steele, NT Becher. Biotechnol Bioeng 58:572–580, 1998.

25. JM DeSimone, EE Maury, YZ Menceloglu, JB McClain, TJ Romack, JR Combes. Science 265:356–359, 1994.

26. KP Johnston, KL Harrison, MJ Clarke, SM Howdle, MP Heitz, FV Bright, C Carlier, TW Randolph. Science 271:624–626, 1996.

27. MP Heitz, C Carlier, J DeGrazia, KL Harrison, KP Johnston, TW Randolph, FV Bright. J Phys Chem B 101:15581–15587, 1997.

28. SRP DaRocha, KL Harrison, KP Johnston. Langmuir 15:419–428, 1999.

29. MJ Clarke, KL Harrison, KP Johnston, SM Howdle. J Am Chem Soc 119: 6399–6406, 1997.

30. RG Zielinski, SR Kline, EW Kaler, N Rosov. Langmuir 13:3934–3937, 1997.

31. CT Lee, PA Psathas, KP Johnston. Langmuir 15:6781–6791, 1999; SRP DaRocha, KL Harrison, KP Johnston. Langmuir 15:419–428, 1999.

32. T Sarbu, T Styranec, EJ Beckman. Nature 405:165–167, 2000.

33. ED Niemeyer, FV Bright. J Phys Chem B 102:1474–1478, 1998.

34. JD Holmes, DC Steytler, GD Rees, BH Robinson. Langmuir 14:6371–6376, 1998.

35. JD Holmes, KJ Ziegler, M Audriani, CT Lee Jr, PA Bhargava, DC Steytler, KP Johnston. J Phys Chem B 103:5703–5711, 1999.

36. M Ji, X Chen, C Xiaoyan, CM Wai, JL Fulton. J Am Chem Soc 121:2631–2632, 1999.

37. CB Jacobson, CT Lee, KP Johnston, W Tumas. J Am Chem Soc 121:11902–11903, 1999.

38. JD Holmes, PA Bhargava, BA Korgel, KP Johnston. Langmuir 15:6613, 1999.

39. CB Jacobson, CT Lee, KP Johnston. J Org Chem 64:1201–1206, 1999.

40. CB Jacobson, CT Lee, SRP DaRocha, KP Johnston. J Org Chem 64:1207–1210, 1999.

41. http://www.micell.com

42. N Yoshino, N Komine, J-I Suzuki, Y Arima, H Hirai. Bull Chem Soc Jpn 64: 3262–3266, 1991.

43. A Downer, J Eastoe, AR Pitt, J Penfold, RK Heenan. Colloids Surf A 156:33–48, 1999.

44. J Eastoe, S Nave, A Downer, A Paul, A Rankin, K Tribe, J Penfold. Langmuir 16:4511–4518, 2000.

45. http://www.isis.ac.uk

46. J Penfold, RK Thomas. J Phys Condens Matter 2:1369, 1990.

47. J Lekner. Theory of Reflection. Dordrecht: Martinus Nijhoff, 1987.

48. RK Heenan, SM King, J Penfold. J Appl Crystallogr 30:1140, 1997.

49. JB McClain, D Londono, JR Combes, TJ Romack, DA Canelas, DE Betts, GD Wignall, ET Samulski, JM DeSimone. J Am Chem Soc 118:917, 1996.

50. RK Heenan. FISH Data Analysis Program. Didcot, UK: Rutherford Appleton Laboratory Report RAL-89-129, CCLRC, 1989.
51. M Kotlarchyk, S-H Chen, JS Huang, MW Kim. Phys Rev A 29:2054, 1984.
52. L Mengyang, A Acero, Z Huang, S Rice. Nature 367:151, 1994.
53. ZX Li, JR Lu, RK Thomas. Langmuir 13:3681, 1997; ZX Li, JR Lu, RK Thomas, J Penfold. Prog Colloid Polym Sci 98:243, 1995; ZX Li, JR Lu, RK Thomas, J Penfold. J Phys Chem B 101:1615, 1997.
54. EW Kaler, JF Bilman, JL Fulton, RD Smith. J Phys Chem 95:458, 1991.
55. J Eastoe, BH Robinson, WK Young, DC Steytler. J Chem Soc Faraday Trans 86:2883–2889, 1990.
56. J Eastoe, BH Robinson, G Fragneto, TF Towey, RK Heenan, FJ Leng. J Chem Soc Faraday Trans 88:461–471, 1992.
57. R Weibe, V Gaddy. Chem Rev 63:475, 1941.
58. PDI Fletcher, BP Binks, R Aveyard. Langmuir 5:1210, 1989.

17

Organic Synthesis in Microemulsions: An Alternative or a Complement to Phase Transfer Catalysis

KRISTER HOLMBERG Chalmers University of Technology, Göteborg, Sweden

MARIA HÄGER Institute for Surface Chemistry, Stockholm, Sweden

ABSTRACT

Microemulsions are excellent media for chemical reactions involving reagents of different polarities. Water-soluble reactants dissolve in the water domain, lipophilic reactants dissolve in the hydrocarbon domain, and the reaction occurs at the oil–water interface. Because the interfacial area of microemulsions is very large, the reaction rate is often satisfactory even with reactants with complete incompatibility. Traditionally, organic reactions involving reactants with a high degree of incompatibility have been performed in a two-phase system with an added phase transfer catalyst. Hence, the use of a microemulsion as a reaction medium can be seen as an alternative to phase transfer catalysis. The two approaches can also be combined. In this chapter we demonstrate that addition of a phase transfer catalyst to a microemulsion may give a reaction rate higher than that of either of the two approaches alone.

I. BACKGROUND

A common practical problem in synthetic organic chemistry is to attain proper phase contact between nonpolar organic compounds and inorganic salts. There are many examples of important reactions where this is a potential problem: alkaline hydrolysis of esters, oxidative cleavage of olefins with permanganate-periodate, addition of hydrogen sulfite to aldehydes and to terminal olefins, and preparation of alkyl sulfonates by treatment of alkyl chloride by sulfite or by addition of hydrogen sulfite to α-olefin oxides. The list can be extended further. In all examples given, there is a compatibility problem to be solved if the organic component is a large nonpolar molecule.

There are various ways to solve the problem of poor phase contact in organic synthesis. One way is to use a solvent or a solvent combination capable of dissolving both the organic compound and the inorganic salt. Polar, aprotic solvents are sometimes useful for this purpose, but many of these are unsuitable for large-scale production due to toxicity and/or difficulties in removing them by low vacuum evaporation.

Alternatively, the reaction may be carried out in a mixture of two immiscible solvents. The contact area between the phases may be increased by agitation. Phase transfer agents, in particular quaternary ammonium compounds, are useful aids in many two-phase reactions. Also, crown ethers are very effective in overcoming phase contact problems; however, their usefulness is limited by their high price. (Open chain polyoxyethylene compounds often give a "crown ether effect" and may constitute practically interesting alternative phase transfer agents.)

Microemulsions are excellent solvents both for hydrophobic organic compounds and for inorganic salts. Being macroscopically homogeneous yet microscopically dispersed, they can be regarded as something between a solvent-based one-phase system and a true two-phase system. In this context, microemulsions should be seen as an alternative to two-phase systems with phase transfer agents added.

II. OVERCOMING REAGENT INCOMPATIBILITY BY THE USE OF MICROEMULSIONS

Schomäcker [1,2] has described the use of microemulsions as a medium for a large number of reactions including nucleophilic substitutions, alkylations, Knoevenagel condensations, ester hydrolyses, oxidations, and reductions. In all reactions there is a potential compatibility problem because a hydrophobic organic compound is reacted with an inorganic salt. By formulating microemulsions based on water, hydrocarbon, and a nonionic surfactant, most often hexa(ethylene glycol)monooctyl ether (C_8E_6), and by adjusting the temperature during the course of the reaction so that the mixture remained clear and isotropic, most reactions were completed in less than 2 h.

Lif and Holmberg [3] have demonstrated the efficiency of microemulsions as a medium for both organic and bioorganic hydrolyses of a 4-nitrophenyl ester; see Fig. 1. The reactions were performed in a Winsor I type of microemulsion and took place in the lower phase oil-in-water microemulsion.

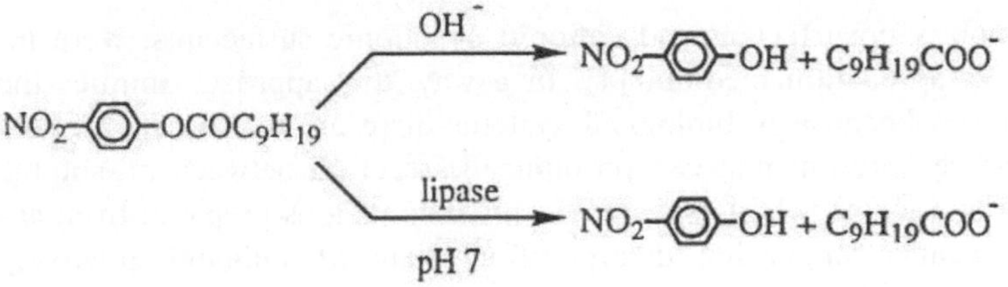

OH⁻
NO₂-⟨O⟩-OH + C₉H₁₉COO⁻
NO₂-⟨O⟩-OCOC₉H₁₉
lipase
pH 7
NO₂-⟨O⟩-OH + C₉H₁₉COO⁻

FIG. 1 Alkaline and lipase-catalyzed hydrolyses of 4-nitrophenyldecanoate.

After the reaction was complete, a Winsor I → III transition was induced by a rise in temperature. The products formed, 4-nitrophenol and decanoic acid, partitioned into the upper oil phase and could easily be isolated by separation of this phase and evaporation of the solvent. The principle is outlined in Fig. 2. The surfactant and the enzyme (in the case of the lipase-catalyzed reaction) resided in the middle-phase microemulsion and could be reused.

Another interesting example of the use of microemulsions to overcome solubility problems is related to metal–ligand substitution. The synthesis of metalloporphyrins has received considerable attention because of their biological importance. Because most naturally occurring porphyrins are water insoluble, reaction with a metal salt requires some measure to overcome reagent incompatibility. To this end, water-in-benzene microemulsions, with

RCOOR'
OH⁻
Na⁺
OH⁻
Na⁺
heat, HCl
RCOOH + R' OH
Na⁺ Cl⁻

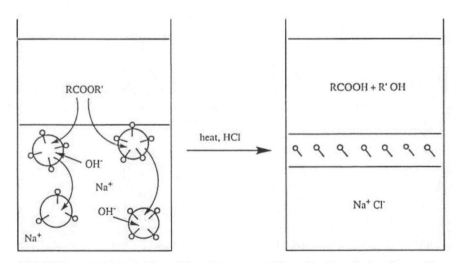

FIG. 2 Ester hydrolysis in a Winsor I system followed by heat-induced transition into a Winsor III system.

cyclohexanol as cosurfactant and cationic or anionic surfactants, were investigated as a reaction medium [4]. In a way, this approach mimics the natural process because in biological systems there are microscopic lipid–water interfaces present that can promote the reaction between oil-soluble porphyrins and water-soluble species. In microemulsions prepared from anionic surfactants, copper ion incorporation occurred, although relatively slowly. The rate of metallation was increased considerably by addition of oil-soluble Lewis bases, e.g., quinoline. On the other hand, water-soluble bases, such as pyridine, slowed down or completely inhibited the reaction. The reaction in the presence of triphenylphosphine (Ph$_3$P) was investigated in detail. It was concluded that Ph$_3$P stabilized the formation of a copper–porphyrin complex in the microdroplet interfacial region. The use of a microemulsion as a reaction medium in this case not only is interesting from a preparative point of view but also provides general insight into the mechanism of metalloporphyrin formation. Several other examples of use of microemulsions for studying biologically significant reactions are given in Ref. 5.

Figure 3 shows alkylation of 4-nitrobenzylpyridine (NBP) with alkyl halide (RX) followed by reaction with hydroxide ion to form a violet product. The reaction, which is used for the detection of alkylating agents, is a good example of the advantage of using a microemulsion as a reaction medium [6,7]. In the conventional procedure a nonaqueous solvent is employed to dissolve the oil-soluble NBP and the RX. This solution is then treated with

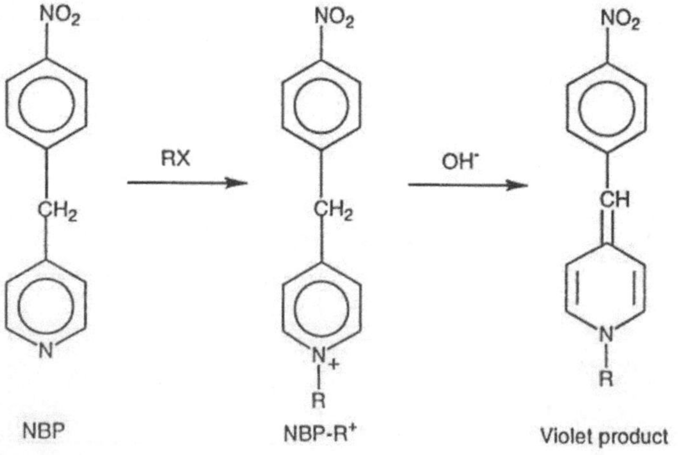

FIG. 3 Reaction of 4-nitrobenzylpyridine (NBP) with an alkyl halide (RX) followed by alkaline hydrolysis of the N-alkyl derivative (NBP-R$^+$) formed.

an aqueous base in a two-step process because NBP and base, when combined, react slowly to form an interfering blue color. However, the use of an anionic microemulsion permits a combination of the two reagents in the same solution. The addition of Ag^+ is required to effect alkylation by methyl iodide. The suggested mechanistic explanation is as follows: NBP is solubilized in the oil phase and is exposed to only a very low hydroxide ion concentration because OH^- is electrostatically repelled by the anionic interface. Added methyl iodide, aided by Ag^+, which is concentrated at the negatively charged interface, reacts with the NBP to form 4-nitrobenzyl-1-methylpyridinium ion ($NBP-Me^+$). This more water-soluble ion is accessible to attack by OH^- to form the final violet product. Due to accumulation of Ag^+ (from $AgNO_3$) and OH^- (from NaOH) in different domains of the microemulsion, the solubility product of AgOH is never exceeded.

Oh et al. [8] have demonstrated that a microemulsion based on a nonionic surfactant is an efficient reaction system for the synthesis of decyl sulfonate from decyl bromide and sodium sulfite. Whereas at room temperature almost no reaction occurred in a two-phase system without surfactant added, the reaction proceeded smoothly in a microemulsion. A range of microemulsions were tested with the oil-to-water ratio varying between 9:1 and 1:1. Nuclear magnetic resonance (NMR) self-diffusion measurements showed that the 9:1 ratio gave a water-in-oil microemulsion and the 1:1 ratio a bicontinuous structure. No substantial difference in reaction rate could be seen between the different types of microemulsions, indicating that the curvature of the oil–water interface was not decisive for the reaction kinetics. More recent studies of the kinetics of hydrolysis reactions in different types of microemulsions showed a considerable dependence of the reaction rate on the oil–water curvature of the microemulsion [9]. This was interpreted as being due to differences in hydrolysis mechanisms for different types of microemulsions.

The effect of microemulsion structure on reaction rate has also been studied in relation to oxidation and reduction of cysteine residues in keratin [10]. The system sodium dodecyl sulfate (SDS)/n-pentanol/water/dodecane was chosen as a microemulsion because in this system the realm of the existence of the isotropic region in the pseudoternary phase diagram is a continuous domain, extending from the water apex to the close vicinity of the hydrocarbon–surfactant edge. It was shown experimentally that the microemulsion structure varied smoothly with composition within the isotropic region.

The reactions studied included oxidation (using hydrogen peroxide) and reduction (using thioglycolic acid) of keratin cystine. Traditionally, reactions of keratin are performed in water–alcohol mixtures. These reactions were carried out along two lines in the pseudoternary phase diagram, one representing constant (35%) surfactant concentration and the other constant (15%)

water content. Both lines are within the microemulsion domain. For both reactions there was only a weak correlation between composition and reaction yield; the higher the hydrocarbon content, the higher the yield. Evidently, in this case the reaction rate was not markedly affected by the microstructure of the system.

Amphiphilic block copolymers are a class of compounds for which synthesis is often a major problem. Such polymers contain one or more hydrophilic blocks with high water solubility and one or more hydrophobic blocks, which are usually not water soluble. The polymers are usually made by linking the blocks by a condensation or addition reaction, and the incompatibility of the reactants often results in very long reaction times or incomplete yields. [One exception is the well-known family of block copolymers made from alkylene oxides, i.e., various types of ABA polymers based on poly(ethylene oxide)–poly(propylene oxide) and poly(ethylene oxide)–poly(butylene oxide). The synthesis of these polymers from the monomeric alkylene oxides is straightforward.]

We have recently demonstrated the use of a microemulsion based on the nonionic surfactant penta(ethylene glycol)monododecyl ether ($C_{12}E_5$) as a medium for the synthesis of a surface-active poly(ethylene glycol) (PEG) derivative [11]. The synthesized compound consisted of a central block of PEG 3400 with dodecyl chains at both ends. The reactants, PEG dihydrazide and dodecylaldehyde, were dissolved in water and aliphatic hydrocarbon, respectively, and the reaction was performed in an oil-in-water microemulsion in which these solutions constituted the aqueous and oil components. The reaction was also performed in a lamellar liquid crystalline phase obtained at an elevated temperature. As a reference, the synthesis was carried out in a hydrocarbon–water two-phase system. As shown in Fig. 4, the reaction was fast in the microemulsion and very sluggish in the two-phase system. The reaction rate in the liquid crystalline system was of the same order as that in the microemulsion. A liquid crystalline phase lacks the dynamics, i.e., the constant formation, disintegration, and re-formation of the aggregates, that characterizes a microemulsion. Both liquid crystals and microemulsions possess a very large oil–water interfacial area, however. The fact that the reaction rate was similar in the liquid crystal and in a microemulsion seems to indicate that the usefulness of microemulsions as a medium to overcome reactant incompatibility is mainly due to the large interface and not to the highly dynamic state of the system.

An interesting new type of reaction system is that of water-in-carbon dioxide microemulsions. Several nucleophilic substitution reactions have been performed in such a system using an anionic perfluoropolyether ammonium carboxylate surfactant. The reactions studied were considerably fas-

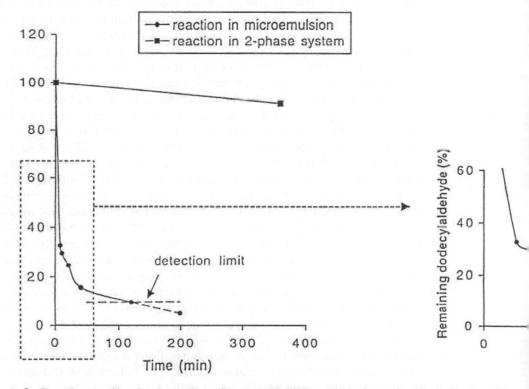

i. 4 Reaction profiles for formation of an amphiphilic poly(ethylene glycol) derivative. The r ·roemulsion and in a water–hydrocarbon two-phase system.

ter in the water-in-carbon dioxide microemulsion than in conventional water-in-hydrocarbon microemulsions [12].

III. MICROEMULSION CATALYSIS

Microemulsion droplets and micellar aggregates can catalyze or inhibit chemical reactions by compartmentalization and by concentration of reactants and products. The catalytic effect in micelles has been widely studied, a typical reaction being base-catalyzed hydrolysis of lipophilic esters. This rate enhancement is normally referred to as micellar catalysis. The analogous effect occurring in microemulsions may be called microemulsion catalysis.

The groups of Menger and Bunton have made early significant contributions to the understanding of the effect the organized media had on catalysis of organic reactions [13,14]. A number of kinetic studies, on hydrolysis and other reactions, have been made in micellar systems, i.e., aqueous solutions of surfactants at concentrations above the critical micelle concentration. The rate enhancement in such systems is often very substantial. The effect of micelle formation on reaction rates is primarily a consequence of reactant compartmentalization. Inclusion or exclusion of reactants from the Stern layer has either a catalytic or an inhibitory effect on the reaction rate, depending on the reaction type and the nature of the micelle. The most widely used model to describe these phenomena is the pseudophase kinetic model. This approach assumes that micelles act as a phase (pseudophase) apart from water. Further, it is assumed that effects on reaction rates are due primarily to the distribution of reactants between the micellar and aqueous pseudophases [15,16]. The rate of a chemical reaction is then the sum of adjusted rates in the aqueous and micellar pseudophases. Although it has been clearly demonstrated that hydrolysis reactions of hydrophobic substrates can be strongly accelerated when carried out in micelles, true micellar catalysis is of limited preparative value (except when the surfactant itself is the reacting species) because the reactant concentration is normally too low. A micellar solution has limited solubilization capacity. By introducing a nonpolar solvent into the system, i.e., by creating a microemulsion, a much higher solubilization capacity can be achieved. The pseudophase model has been successfully applied to microemulsion systems [17,18]. In the following, a few illustrative examples of microemulsion catalysis are given.

Several studies have dealt with alkaline hydrolysis of 4-nitrophenyl diphenyl phosphate, and it was demonstrated that the reaction was very fast in microemulsions based on cationic surfactants and much slower in a system based on a nonionic surfactant [19,20].

It has been shown that the addition of a small amount of the anionic surfactant SDS to a nonionic-based microemulsion increased the rate of decyl sulfonate formation from decyl bromide and sodium sulfite [21,22]. Addition of minor amounts of the cationic surfactant tetradecyltrimethylammonium gave either a rate increase or a rate decrease depending on the surfactant counterion. A poorly polarizable counterion, such as acetate, accelerated the reaction, whereas a large polarizable counterion, such as bromide, gave a slight decrease in reaction rate. The reaction profiles for the different systems are shown in Fig. 5.

The effect of surfactant charge on reaction rate was investigated for a related reaction, ring opening of 1,2-epoxyoctane with sodium hydrogen sulfite. The reaction, which was performed in a Winsor III microemulsion, was fast when a nonionic surfactant was used as the sole surfactant and was considerably more sluggish when a small amount of SDS was added to the formulation [23].

IV. COMPARING MICROEMULSIONS WITH PHASE TRANSFER CATALYSIS AS A MEANS TO OVERCOME REAGENT INCOMPATIBILITY

One of the first examples of a comparison between the microemulsion approach and the more conventional process of phase transfer catalysis was a study by Menger et al. [24] on the hydrolysis of trichlorotoluene to sodium

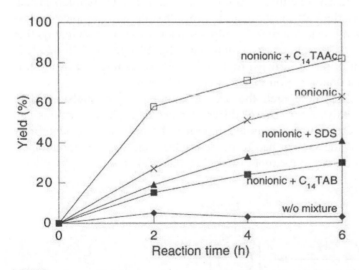

FIG. 5 Effect of addition of ionic surfactant to a microemulsion based on a nonionic surfactant on the rate of the reaction decyl bromide + sodium sulfite → decyl sulfonate. (From Ref. 28.)

benzoate; see Fig. 6. As can be seen from Table 1, the hydrolysis in the presence of the cationic surfactant cetyltrimethylammonium bromide (CTAB) required 1.5 h. The reaction without the surfactant took 60 h. The nonionic surfactant $C_{12}E_{23}$ also accelerated the reaction but to a lesser extent than did CTAB. The commonly used phase transfer agent tetrabutylammonium bromide gave a rate enhancement comparable to that of the nonionic surfactant. Evidently, the solubilizing power of a properly formulated microemulsion was at least as effective as phase transfer catalysis for overcoming the compatibility problems common to hydrolyses of lipophilic organic compounds.

In another comparison between the microemulsion approach and the use of phase transfer agents, Menger and Elrington [25] investigated the decontamination of chemical agents, in particular bis(2-chloroethyl)sulfide, commonly known as mustard. Mustard is a well-known chemical warfare agent. Although it is susceptible to rapid hydrolytic deactivation in laboratory experiments where rates are measured at low substrate concentrations, its deactivation in practice is not easy. Due to its extremely low solubility in water, it remains for months on a water surface. The addition of strong alkali does not increase the rate of reaction. Microemulsions were explored as media for both hydrolysis and oxidation of "half-mustard," $CH_3CH_2SCH_2CH_2Cl$, a less toxic mustard model (Fig. 7). Oxidation with hypochlorite turned out to be extremely rapid in both oil-in-water and water-in-oil microemulsions. In formulations based on either an anionic, a nonionic, or a cationic surfactant, oxidation of the half-mustard sulfide to sulfoxide was complete in less than 15 s. The same reaction took 20 min when a two-phase system, together with a phase transfer agent, was employed [26]. Menger and Rourk [27] have more recently made further progress in optimizing microemulsion formulations for decontamination of chemical warfare agents.

Schomäcker [1,2] compared the use of nonionic microemulsions with phase transfer catalysis for several different types of organic reactions and concluded that the former was more laborious because the pseudoternary phase diagram of the system had to be determined and the reaction temperature needed to be carefully monitored. The main advantage of the micro-

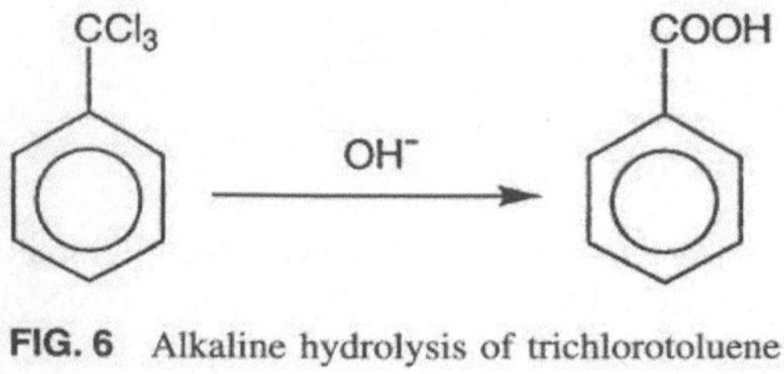

FIG. 6 Alkaline hydrolysis of trichlorotoluene to benzoate.

TABLE 1 Hydrolysis of Trichlorotoluene to Sodium Benzoate Using 20% NaOH at 80°C

Additive[a]	Reaction time (h)	Yield (%)
CTAB (0.01 M)	1.5[b]	98
None	1.5	0
Dioxane (20%)	1.5	0
$C_{12}E_{23}$ (0.006 M)	11[b]	97
TBAB (0.02 M)	15[b]	98
None	60[b]	97

[a]CTAB and TBAB stand for cetyltrimethylammonium bromide and tetrabutylammonium bromide, respectively.
[b]This is roughly the minimum time required for completion of the reaction.
Source: Ref. 24.

emulsion route for industrial use is related to the ecotoxicity of the effluent. Whereas nonionic surfactants are considered relatively harmless, quaternary ammonium compounds exhibit considerable fish toxicity.

Different types of microemulsions were evaluated as reaction media for synthesis of the surface-active compound sodium decyl sulfonate from decyl bromide and sodium sulfite:

$$C_{10}H_{21}Br + Na_2SO_3 \rightarrow C_{10}H_{21}SO_3Na + NaBr$$

The reaction rates of the nucleophilic substitution reaction were compared with the rates obtained by phase transfer catalysis using either a quaternary ammonium salt (tetrabutylammonium hydrogen sulfate) or a crown ether (18-crown-6) as catalyst [28]. The microemulsions were based on the non-

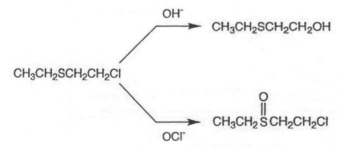

FIG. 7 Transformation of 2-chloroethylethyl sulfide (half-mustard) into 2-hydroxyethylethyl sulfide by alkali or into 2-chloroethylethyl sulfoxide by hypochlorite.

ionic surfactant $C_{12}E_5$ and two compositions were used, one in the water-in-oil region (the L_2 phase) and the other in the bicontinuous region. The reaction profiles are shown in Fig. 8. As can be seen, the reaction was much faster in the microemulsions than in the two-phase systems with added phase transfer agent. In the reference system, hydrocarbon–water with neither a surfactant nor a phase transfer agent, there was virtually no reaction at all.

As can be seen from Fig. 8, the addition of a phase transfer agent to the two-phase system did not affect the reaction rate much. We believe that the reason why phase transfer agents are not effective as catalysts in this reaction is that the product formed, decyl sulfonate, is a much more lipophilic anion than the inorganic ions. Hence, Q^+ will prefer to form an ion pair with $R\text{—}SO_3^-$ rather than with HSO_3^-, which means that transfer of the latter ion across the oil–water interface will decline as more and more decyl sulfonate is produced during the course of the reaction. Figure 9 illustrates the competing processes.

In order to support the statement that the formation of a lipophilic anion is the reason behind the poor yield of the phase transfer reactions as compared with the reactions in microemulsions, the same organic reagent, decyl bromide, was reacted with two other nucleophilic ions, cyanide and azide:

$$C_{10}H_{21}Br + NaCN \rightarrow C_{10}H_{21}CN + NaBr$$

$$C_{10}H_{21}Br + NaN_3 \rightarrow C_{10}H_{21}N_3 + NaBr$$

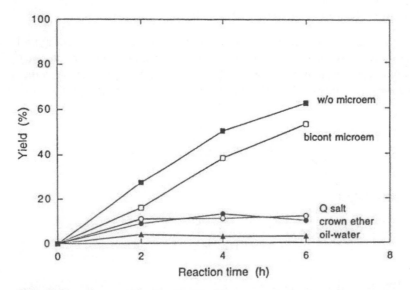

FIG. 8 Reaction profiles for decyl sulfonate synthesis in various reaction media.

Aqueous phase:
$$Na_2SO_3 + QHSO_4 \longrightarrow NaHSO_3 + NaQSO_4$$

Interface:
$$Q^+ + HSO_3^- \longrightarrow Q^+HSO_3^- \text{ (org)}$$

Organic phase:
$$R - Br + Q^+HSO_3^- \longrightarrow R - SO_3H + Q^+Br^-$$

Side reaction

Interface:
$$R - SO_3H \text{ (org)} \longrightarrow R - SO_3^- + H^+$$
$$R - SO_3^- + Q^+ \longrightarrow R - SO_3^- + Q^+ \text{ (org)}$$

FIG. 9 Reaction steps involved in synthesis of decyl sulfonate using phase transfer catalysis.

Both reactions yielded products that were not ionizable; hence, no lipophilic anions were being formed that could compete with the reacting nucleophile in ion pair formation with Q^+. As can be seen from Fig. 10, the formation of both decyl nitrile and decyl azide proceeded at approximately the same rate in the Q^+-assisted reaction in a hydrocarbon–water two-phase system

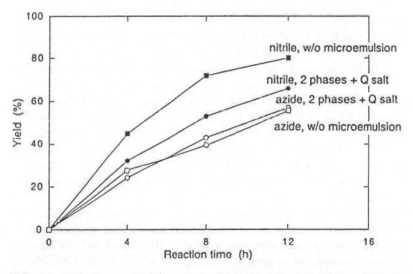

FIG. 10 Reaction profiles for decyl nitrile and decyl azide formation in microemulsion and in a two-phase system with added tetrabutylammonium hydrogen sulfate (Q salt) as phase transfer agent.

as in a microemulsion. One conclusion from this work is that the advantages with microemulsions as reaction media are particularly apparent in the synthesis of lipophilic ionizable species, such as long-chain alkyl sulfonates.

V. USE OF PHASE TRANSFER CATALYSTS IN MICROEMULSIONS

The use of a two-phase system with added phase transfer catalyst and the use of a microemulsion are two alternative approaches to overcoming reagent incompatibility problems in organic synthesis. Both routes have proved useful but on entirely different accounts. In phase transfer catalysis the nucleophilic reagent is carried into the organic phase, where it becomes highly reactive. In the microemulsion approach there is no transfer of reagent from one environment to another; the success of the method relies on the very large oil–water interface at which the reaction occurs.

We have carried out a nucleophilic substitution reaction in a microemulsion in the presence of a phase transfer agent [29]. The aim of the work was to investigate whether a combination of the two approaches would give a reaction rate higher than that obtained in either the microemulsion approach or a two-phase system using phase transfer catalysis.

The reaction chosen was the previously mentioned ring opening of 1,2-epoxyoctane by sodium hydrogen sulfite in an oil-in-water microemulsion based on a chlorinated hydrocarbon, water, and surfactant with tetrabutylammonium hydrogen sulfate added to the formulation. Attempts to formulate a microemulsion with standard nonionic surfactants containing polyoxyethylene chains as the polar head group failed, probably due to too high solubility of these surfactants in the oil domain. Polyol surfactants are less soluble in chlorinated hydrocarbons, and a microemulsion with a relatively broad range of existence could be obtained by using a combination of two alkylglucoside surfactants. Figure 11 shows the reaction profiles for the synthesis carried out in a two-phase system with and without the phase transfer catalyst and in a microemulsion containing the catalyst. As can be seen, the combined approach leads to a very fast reaction. Obviously, a substitution reaction in a microemulsion can be greatly accelerated by addition of a phase transfer catalyst.

An attempt was also made to accelerate the same reaction performed in a microemulsion based on water, nonionic surfactant, and hydrocarbon oil [23]. The reaction was performed in a Winsor III system and the same Q salt, tetrabutylammonium hydrogen sulfate, was added to the formulation. In this case the addition of the phase transfer catalyst gave only a marginal increase in reaction rate. Similar results have been reported for an alkylation reaction performed in different types of micellar media [30]. The addition

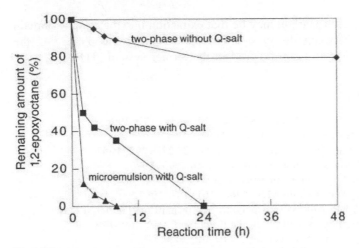

FIG. 11 Reaction profiles for the ring opening of 1,2-epoxyoctane as determined from disappearance of the epoxide.

of a Q salt gave no effect for a system based on cationic surfactant, a marginal increase in rate for a system based on nonionic surfactant, and a substantial effect when an anionic surfactant was used. The last system, also with Q salt added, gave a lower yield than the first two, however, most likely due to electrostatic repulsion of the negatively charged nucleophile by the anionic micelles, as discussed earlier for microemulsion-based reaction media.

VI. CONCLUSION

Microemulsions are useful as reaction media to overcome problems of reactant incompatibility. They should be seen as an alternative to phase transfer catalysis. In some instances, such as in the synthesis of relatively lipophilic molecules that are anionic or that can be deprotonated to become anionic, the microemulsion approach is definitely superior to phase transfer catalysis. In other instances the two synthesis methods are approximately equal in terms of reaction rate.

The combined use of phase transfer catalysis and a microemulsion gave excellent results when the microemulsion was based on chlorinated hydrocarbon as the oil component. When the reaction was carried out in a microemulsion based on hydrocarbon oil, the addition of a Q salt only marginally increased the reaction rate. The reason for the combined approach being more efficient in a microemulsion based on chlorinated hydrocarbon than on aliphatic hydrocarbon may be that extraction into the organic domain of the ion pair between the Q salt and the anionic nucleophile, i.e., the hydrogen sulfite ion, is more favored in the former system [31].

REFERENCES

1. R Schomäcker. Nachr Chem Tech Lab 40:1344–1351, 1992.
2. R Schomäcker. J Chem Res 92–93, 1991.
3. A Lif, K Holmberg. Colloids Surf A 129–130:273–277, 1997.
4. L Lefts, RA Mackay. Inorg Chem 14:2990–2993, 1975.
5. RA Mackay. Adv Colloid Interface Sci 15:131–156, 1981.
6. RA Mackay. Actual Chim May–June: 161–167, 1991.
7. RA Mackay, RP Seiders. J Dispersion Sci Technol 6:193–207, 1985.
8. S-G Oh, J Kizling, K Holmberg. Colloids Surf A 97:169–179, 1995.
9. J Hao. J Dispersion Sci Technol 21:19–30, 2000.
10. P Erra, C Solans, N Azemar, JL Parra, M Clausse, D Touraud. Prog Colloid Polym Sci 73:150–155, 1987.
11. M Häger, K Holmberg, U Olsson. Colloids Surf A 189:9–19, 2001.
12. GB Jacobson CT Lee, KP Johnston. J Org Chem 64:1201–1206, 1999.
13. FM Menger, CE Portnoy. J Am Chem Soc 89:4698–4703, 1967.
14. CA Bunton, L Robinson. J Am Chem Soc 90:5972–5979, 1968.
15. CA Bunton, G Savelli. Adv Phys Org Chem 22:213–309. 1986.
16. CA Bunton. In: KL Mittal, DO Shah, eds. Surfactants in Solution. Vol 11. New York: Plenum, 1991, pp 17–40.
17. RA Mackay. J Phys Chem 86:4756–4758, 1982.
18. CA Bunton, R de Buzzaccarini. J Phys Chem 86:5010–5014, 1982.
19. RA Mackay, BA Burnside, SM Garlick, BL Knier, HD Durst, PM Nolan, FR Longo. J Dispersion Sci Technol 9:493–510, 1988–1989.
20. SM Garlick, HD Durst, RA Mackay, KG Haddaway, FR Longo. J Colloid Interface Sci 135:508–519, 1990.
21. S-G Oh, J Kizling K Holmberg. Colloids Surf A 104:217–222, 1995.
22. K Holmberg, S-G Oh, J Kizling. Prog Colloid Polym Sci 100:281–285, 1996.
23. K Andersson, J Kizling, K Holmberg, S Byström. Colloids Surf 144:259–266, 1988.
24. FM Menger, JU Rhee, HK Rhee. J Org Chem 40:3803–3805, 1975.
25. FM Menger, AR Elrington. J Am Chem Soc 113:9621–9624, 1991.
26. JH Ramsden, RS Drago, R Riley. J Am Chem Soc 111:3958–3961, 1989.
27. FM Menger, MJ Rourk. Langmuir 15:309–313, 1999.
28. S Gutfelt, J Kizling, K Holmberg. Colloids Surf A 128:265–271, 1997.
29. M Häger, K Holmberg. Tetrahedron Lett 41:1245–1248, 2000.
30. C Siswanto, T Battal, OE Schuss, JF Rathman. Langmuir 13:6047–6052, 1997.
31. CM Starks, CL Liotta. Phase Transfer Catalysis. New York: Academic Press, 1978.

18

Physicochemical Characterization of Nanoparticles Synthesized in Microemulsions

J. B. NAGY, L. JEUNIEAU, F. DEBUIGNE, and I. RAVET-BODART Facultés Universitaires Notre-Dame de la Paix, Namur, Belgium

ABSTRACT

This chapter essentially deals with the preparation of nanoparticles using microemulsions. The preparation of inorganic nanoparticles—Ni_2B, Co_2B, Ni-Co-B, Pt, Au, Pt-Au, ReO_2, Pt/ReO_2, AgX—and the synthesis of organic nanoparticles—cholesterol, rhovanil, rhodiarome—are systematically studied as a function of the concentration of the precursor molecules, the size of the inner water cores, and the manner of mixing the various solutions. Two different behaviors are observed in the various systems. The first case shows a dependence of the nanoparticle size on the various physicochemical parameters. Either a monotonous increase of the size or the presence of a minimum is observed as a function of the concentration of the precursor molecules. This case can be easily explained following the La Mer diagram, where the nucleation of the nanoparticles is separated from the particle growth. The second case does not show any dependence of the nanoparticle size on the physicochemical parameters. The size remains constant in all experimental conditions. The constant character of the size can be explained only by thermodynamic stabilization, where particles with a certain size are better stabilized. It should be emphasized that the size distribution is small in all the cases studied. Finally, the aging of the nanoparticles was also checked, especially for the organic nanoparticles. It is concluded that these particles remain stable for months in the microemulsion.

I. INTRODUCTION

The synthesis of monodisperse nanoparticles is of great technological and scientific interest. The quantum size effects of these particles are particularly studied because they lead to interesting mechanical, chemical, electrical, optical, magnetic, electro-optical, and magneto-optical properties that are quite different from those reported for bulk materials [1–4]. The nanoparticles not only are of basic scientific interest but also have resulted in important technological applications, such as catalysts, high-performance ceramic materials, microelectronic devices, and high-density magnetic recordings [5–7].

The synthesis of nanoparticles in microemulsions allows one to obtain monodisperse size of the particles and in some cases to control the size of the particles by variation of the size of the microemulsion droplet radius and of the precursor concentrations.

Although the synthesis of inorganic particles in microemulsions is already widespread, only polymer nanoparticles have been synthesized in microemulsion media as far as the organic particles are concerned. In this chapter, it will be shown that it is also possible to synthesize organic particles by a direct precipitation reaction in the microemulsions.

We emphasize some of the fundamental aspects of monodisperse nanoparticle formation. Two models are proposed for the formation of the particles: the first is based on the La Mer diagram, and the second is based on the thermodynamic stabilization of the particles. In the first case, the particle size varies as a function of either the size of the inner water cores or the precursor concentration; in the second case, the particle size is independent of these parameters.

The monodisperse nanoparticles will be characterized directly in the microemulsion or after transferring them in another medium. First, the size of the nanoparticles will be determined as a function of various parameters. Their composition will be analyzed by X-ray photoelectron spectroscopy (XPS) or energy dispersive X-ray analysis (EDX). The specific surface area is determined by the BET technique. The direct solvation is analyzed by multinuclear magnetic resonance spectroscopy.

II. PREPARATION OF NANOPARTICLES
USING MICROEMULSIONS

A. Description of Microemulsions

A water-in-oil microemulsion is a thermodynamically stable, optically transparent dispersion of two immiscible liquids stabilized by a surfactant. The important properties are governed mainly by the water–surfactant molar ratio (R = [H₂O]/[surfactant]). This factor is linearly correlated with the size of the water droplets.

The nanoparticles have been synthesized in different microemulsion systems. Some of them are shown in Fig. 1. The anionic Aerosol-OT (AOT)/heptane/water system is one of the best characterized microemulsions [8,9]. The system AOT/*p*-xylene/water [10] has also been used. The cationic cetyltrimethylammonium bromide (CTAB)/hexanol/water system contains hexanol, which forms the organic phase and plays the role of cosurfactant [11]. The nonionic penta(ethylene glycol)-dodecylether (PEGDE)/hexane/water was studied by Friberg and Lapczynska [12]. The reverse micellar droplets have a cylindrical shape in which the surfactant molecules are parallel to each other, forming a bilayer impregnated with water. Triton X-100

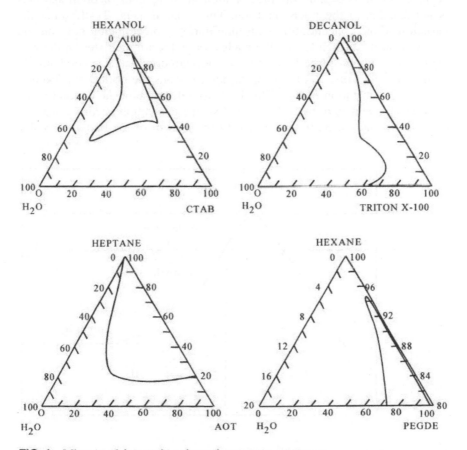

FIG. 1 Microemulsion regions in various ternary systems.

$[p\text{-}(1,1,3,3\text{-tetramethylbutyl})$phenyl-polyethoxyethanol)/decanol/water has been characterized by Ekwall and coworkers [10,13].

B.   Mechanism of Synthesis of Nanoparticles in Microemulsions

The aqueous droplets continuously collide, coalesce, and break apart, resulting in a continuous exchange of solution content. In fact, the half-life of the exchange reaction between the droplets is of the order of 10^{-3}–10^{-2} s [14,15].

Two models have been proposed to explain the variation of the size of the particles with the precursor concentration and with the size of the aqueous droplets. The first is based on the La Mer diagram [16,17], which has been proposed to explain the precipitation in an aqueous medium and thus is not specific to the microemulsion. This diagram (Fig. 2) illustrates the variation of the concentration with time during a precipitation reaction and is based on the principle that the nucleation is the limiting step in the precipitation reaction. In the first step, the concentration increases continuously with increasing time. As the concentration reaches the critical supersaturation value, nucleation occurs. This leads to a decrease of the concentration. Between the concentrations C^*_{max} and C^*_{min} the nucleation occurs. Later, the decrease of the concentration is due to the growth of the particles by dif-

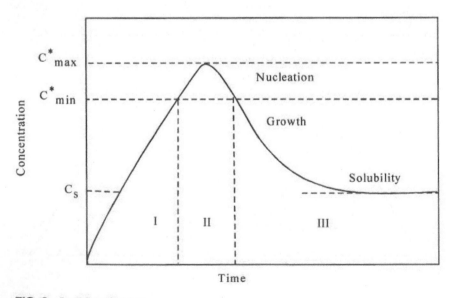

FIG. 2   La Mer diagram.

fusion. This growth occurs until the concentration reaches the solubility value.

This model has been applied to the microemulsion medium, i.e., that nucleation occurs in the first part of the reaction and later only growth of the particles occurs. If this model is followed, the size of the particles will increase continuously with the concentration of the precursor or a minimum in the variation of the size with the concentration can also be expected. This stems from the fact that the number of nuclei is constant and the increase of concentration leads to an increase of the size of the particles.

The second model is based on the thermodynamic stabilization of the particles. In this model the particles are thermodynamically stabilized by the surfactant. The size of the particles stays constant when the precursor concentration and the size of the aqueous droplets vary.

These two models are limiting models; the La Mer diagram does not take into account the stabilization of the particles by the surfactant, and the thermodynamic stabilization model does not take into account that the nucleation of the particles is more difficult than the growth by diffusion.

C. Preparation of Monodisperse Colloidal Particles

The different monodisperse nanoparticles were prepared following either Scheme I or Scheme II of Fig. 3. We first discuss the mechanism of formation of particles following Scheme I, where small amounts of aqueous solutions are added to the initial microemulsion.

1. Size of Metal Boride Particles

Monodisperse colloidal nickel boride and cobalt boride particles were synthesized by reducing, with $NaBH_4$, the metallic ions solubilized in the water cores of the microemulsions. The $NaBH_4/MCl_2$ ratio was held equal to 3 because larger particles were obtained for a lower value, and the particle size remained constant above this ratio [18–20].

The composition of the particles was determined by XPS to be, respectively, Ni_2B and Co_2B. In each case, the size of particles (2.5–7.0 nm) was much smaller than that obtained by reduction of Ni(II) or Co(II) in water (300–400 nm) or in ethanol (250–300 nm), and the size distribution was quite narrow (±0.5 nm).

Figure 4 shows the dependence of the nickel boride particle size on the water content in the microemulsion as well as on the Ni(II) ion concentration. The average size of the particles decreases with decreasing size of the inner water core (decreasing water content), and a complex behavior is observed as a function of the Ni(II) ion concentration; a minimum is detected at approximately 5×10^{-2} molal concentration. These observations can be

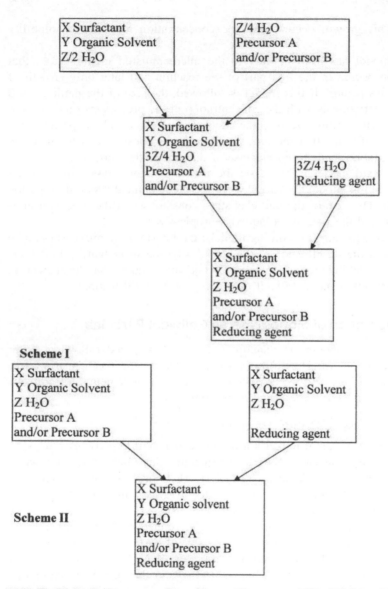

FIG. 3 Methods for preparation of monodisperse particles (X, Y, and Z are in weight % of the various components).

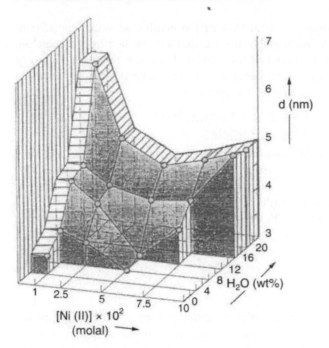

FIG. 4 Variation of the average diameter (in mm) of the nickel boride particles as a function of water content and Ni(II) ion molal concentration.

understood if one analyzes the nucleation and the growth processes of the particles.

To form a stable nucleus, a minimum number of atoms are required [21]. Thus, for nucleation several atoms must collide at the same time, and the probability of this phenomenon is much lower than the probability of collision between one atom and an already formed nucleus. The latter phenomenon is called the growth process. At the very beginning of the reduction, nucleation occurs only in water cores that contain enough ions to form a nucleus. At this moment, the micellar aggregates act as "reaction cages" where the nuclei are formed. On the other hand, the microemulsion being dynamic, the water cores rapidly rearrange. The other ions brought into contact with the existing nuclei essentially participate in their growth process. The latter being faster than nucleation, no new nuclei are formed at this moment. As all the nuclei are formed at the same time and grow at the same rate, monodisperse particles are obtained. In summary, the particle size depends on the number of nuclei formed at the very beginning of the reduction, and this number is a function of the number of water cores, containing enough ions to form stable nuclei, that are reached by the reducing agent before the rearrangement of the system. However, the stabilization of the nuclei by surfactants is probably one of the most important factors in explaining the monodispersity of the particles.

2. Quantitative Aspects of the Formation of Monodisperse Colloidal Particles

The first step in the determination of the essential parameters that control particle size is a study of the distribution of the ions in the microemulsion water cores.

By knowing the average radii of the microemulsion water cores (r_M) and the total volume of water (V_T) per kilogram of microemulsion, one can calculate the number of water cores per kilogram of reverse micelles (N_M), neglecting the solubility of water in the hexanol organic phase:

$$N_M = \frac{V_T}{(4/3)\pi r_M^3} \tag{1}$$

The parameter N_M and the initial concentration of metal ions expressed in molality allow one to determine the average number of ions per water core (n_{ions}):

$$n_{ions} = \frac{[\text{ions}] \times 6.023 \times 10^{23}}{N_M} \tag{2}$$

The ions are statistically distributed in the aggregates. To calculate this distribution, Poisson statistics is quite adequate [22]. This gives the probability (p_k) of having k_{ions} per water core (k is an integer taking the values 0, 1, 2, 3, ...), provided the average number of ions per water core ($\lambda = n_{ions}$) is known:

$$p^k = \frac{\lambda^k e^{-\lambda}}{k!} \tag{3}$$

The number of nuclei formed (N_n), when the ions solubilized in 1 kg of solution are reduced, is proportional to the number of aggregates containing enough ions for nucleation. If the minimum number of ions required to obtain a nucleus is i, then N_n can be calculated from the relation

$$N_n = FN_M \sum_{k=i}^{\infty} p_k \tag{4}$$

where $\Sigma_{k=i}^{\infty} p_k$ is the probability of having i or more ions per aggregate; hence $N_M \Sigma_{k=i}^{\infty} p_k$ is the number of water cores containing i or more ions. The F is a proportionality factor taking into account the proportion of aggregates reached by the reducing agent before rearrangement of the system can occur.

In Eq. (4) we do not know the values of i and F but we can calculate all the other parameters. Indeed, the number of nuclei (N_n) is the number of particles prepared, and it is given by

$$N_n = W_t/W \tag{5}$$

where W_t is the total weight of the particles prepared per kilogram of micellar solution, W is the weight of one particle, and

$$W_t = \frac{[\text{ions}] \times M_{\text{particle}}}{x} \tag{6}$$

where M_{particle} is the molecular weight of the particles and x is the number of metal atoms per particle. The weight of one particle is given by

$$W = \frac{4}{3} \pi \left(\frac{d}{2}\right)^3 M_{v,\,\text{particle}} \tag{7}$$

where d is the diameter of the particles measured by electron microscopy and $M_{v,\,\text{particle}}$ is the volumetric mass of the particle.

All the experimental and computed data are reported in Tables 1a and 1b for Ni_2B and Co_2B particles, respectively.

The diameter of the particles is systematically higher than the diameter of the inner water cores. For all the particles synthesized, we calculated the proportionality factor F by systematically varying the value of the minimum number of ions required to form a nucleus (i). If $i = 1$ or $i > 2$, the values of factor F vary considerably (not shown). However, if $i = 2$, its values are reasonably constant (see Tables 1a and 1b). The order of magnitude of the factor F is always 10^{-3}. This means that at the very beginning of the reduction, i.e., when the nuclei are formed, only one aggregate per thousand leads to the formation of metal boride particles.

There is another indication that the nucleation occurs at the very beginning of the reduction. Indeed, the average radii of the water cores used for the calculation of the formation parameters of colloidal particles are measured for the system containing only three fourths of the total amount of water, which is the composition of the solution before the addition of the reducing agent. If the final composition is used, however, no coherent results based on the preceding analysis can be obtained.

The order of magnitude of the factor F is constant, but its value decreases with increasing water content in the microemulsion (see Tables 1a and 1b). This phenomenon can be easily understood because the rearrangement rate of the microemulsion decreases with the water amount and hence the number of aggregates reached by the reducing agent before rearrangement decreases.

BLE 1a Important Parameters for the Formation of Ni_2B Colloidal Particles

(II)] 0^{-2} ɔlal)	$r_M{}^a$ (nm)	$N_M{}^{a,b}$ ×10^{-22}	$n_{Ni(II)}{}^a$	d (nm)	$W_t{}^c$ (g)	W^d ×10^{19} (g)	$N_n{}^b$ ×10^{-18}	
				CTAB 24%–hexanol 60%–water 16%				
00	1.17	1.86	0.32	4.5	0.64	3.77	1.70	9.
50	1.32	1.29	1.17	4.2	1.60	3.06	5.23	4.
50	1.54	0.81	5.58	4.0	4.81	2.65	18.15	2.
00	1.57	0.77	7.82	5.1	6.41	4.87	11.67	1.
				CTAB 30%–hexanol 50%–water 20%				
00	1.34	1.54	0.39	6.7	0.64	12.44	0.51	3.
50	1.48	1.16	1.30	4.9	1.60	4.87	3.28	2.
50	1.68	0.79	5.72	4.6	4.81	4.03	11.93	1.
00	1.72	0.74	8.14	4.9	6.41	4.87	13.16	1.

BLE 1b Important Parameters for the Formation of Co_2B Colloidal Particles

)(II)] 0⁻²)lal)	r_M^a (nm)	$N_M^{a,b}$ ×10⁻²²	$n_{Co(II)}^a$	d (nm)	W_t^c (g)	W^d ×10¹⁹ (g)	N_n^b ×10⁻¹⁸	N_n/N
				CTAB 38%–hexanol 47%–water 15%				
50	1.04	2.48	0.12	5.9	0.32	8.71	0.37	1.49 ×
50	1.15	1.83	0.82	4.9	1.61	4.99	3.23	1.76 ×
00	1.23	1.50	4.01	3.4	6.43	1.67	38.50	2.57 ×
00	1.24	1.46	6.19	3.8	9.65	2.33	41.42	2.84 ×
				CTAB 37%–hexanol 45%–water 18%				
50	1.21	1.90	0.16	6.7	0.32	12.76	0.25	1.32 ×
50	1.42	1.18	1.28	5.2	1.61	5.96	2.70	2.29 ×
00	1.52	0.96	6.27	4.1	6.43	2.92	22.02	2.29 ×
00	1.54	0.92	9.82	4.6	9.65	4.13	23.37	2.54 ×

lues given for the system containing three fourths of the total amount of water.
lues given for 1 kg of solution.
is calculated with $M(Ni_2B) = 128.23$ g/mol and $M(Co_2B) = 128.68$ g/mol.
is calculated with $M_v(Ni_2B) = 7.9$ g/cm³ and $M_v(Co_2B) = 8.1$ g/cm³.
rrection factor from $N_n = FN_M \sum_{k=2}^{\infty} p_k$ (see text).

As the number of nuclei formed decreases at a constant concentration of precursor ions, the particle size increases with the water content in the system.

The diameter of the particles is plotted as a function of micellar droplet concentration in Fig. 5. The values of the particle size are those obtained by interpolation of previous results in the presence of 0.161 molal aqueous metal ion for the CTAB–hexanol–water systems. Particles prepared in the AOT–heptane–water system at a much lower droplet concentration are included for comparison.

The size of the particles decreases linearly with the micellar droplet concentration. This is a strong indication that the final size obtained for the particles is governed by the presence of reverse micellar aggregates. Indeed, if initial nucleation takes place in the water cores, then nucleation should be related to the micellar droplet concentration of the system. Further, the greater the number of micellar droplets, the greater the number of nucleation sites possible (the aqueous metal ion concentration being obviously maintained constant). The results of Tables 1a and 1b also allow us to explain the minimum in the particle size as a function of the concentration of ions (see Fig. 4).

For a constant microemulsion composition, at low ion concentration, only a few water cores contain the minimum number of ions (two) required to form a nucleus; hence, only a few nuclei are formed at the very beginning of the reduction, and the metal boride particles are relatively large. When the ion concentration increases, the distribution of precursor ions in the microemulsion is very different (Fig. 6), and the number of nuclei obtained by reduction increases faster than the total number of ions (Fig. 7). This results in a decrease in the particle size. When more than 80% of the water cores contain two or more ions, the number of nuclei formed remains quasi-constant with increasing ion concentration. Hence the size of the particles increases again.

Figure 4 also shows the particle size as a function of water content in the microemulsion for different Ni(II) concentrations. An increase in the average diameter is observed with increasing proportion of water. The decrease in the number of micellar aggregates (N_M) with water (Table 1a) is accompanied by an increase in their size. For the same Ni(II) concentration with respect to water (i.e., for the same probability of collision between the ions in the same water core), the total number of nuclei formed in the early stage of the reduction decreases with increasing water concentration, and more ions can participate in the growth process. This results in an increase in the particle size. One should keep in mind that the total number of Ni(II) ions also increases with increasing water content. This is shown if the size of the particles is plotted as a function of micellar droplet concentration (Fig. 5).

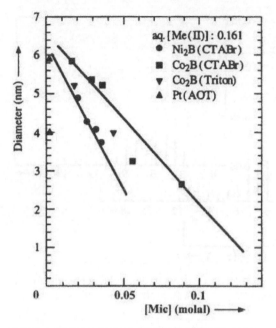

FIG. 5 Size of nanoparticles prepared in various microemulsions as a function of micellar droplet concentration.

For most of the systems studied, a monotonous decrease in the size with increasing N_M is observed. These results reinforce the hypothesis leading to the computation of the number of nuclei and underline the importance of the water cores as reaction cages. From the CTAB 18%–hexanol 70%–water 12% microemulsion and M(II), 5.00×10^{-2} molal bimetallic particles of Ni-Co-B were also prepared.

The F values for nickel boride and cobalt boride particles are quite different. For the former the value obtained for F is equal to 3.2×10^{-3} and for the latter, 17.4×10^{-3}. Because for these experiments the rearrangement rate of the microemulsion system is constant in the first approximation, the difference between the F values is probably due to the different solvations of the two types of ions at the interface. The Co(II) ions contain, on average, one hexanol molecule in their first coordination shell, while the Ni(II) ions are multiply coordinated with hexanol at the interface. The mobility of the latter is hence lower, and the probability of collision between the two reduced Ni atoms required to form a nucleus is also lower. In other words, the rate of nucleation is higher for cobalt boride than for nickel boride particles [20].

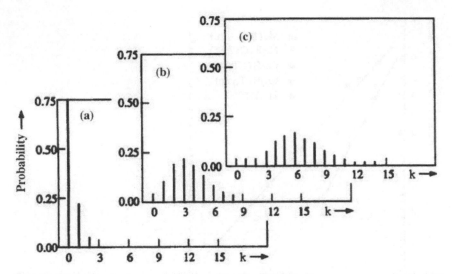

FIG. 6 Variation in the probability of having k Ni(II) ions per aggregate for the microemulsion CTAB 18%–hexanol 70%–H$_2$O 12%. [Ni(II)] (molal): a, 1×10^{-2}; b, 5×10^{-2}; c 7.5 $\times$ 10^{-2}.

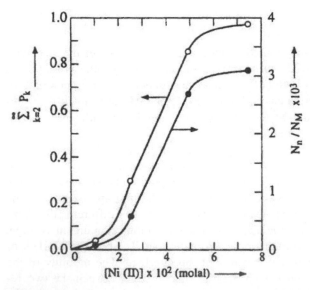

FIG. 7 Variation in (●) the number of nuclei formed per aggregate and (○) the probability of having two or more ions per aggregate as a function of Ni(II) concentration in the microemulsion CTAB 18%–hexanol 70%–water 12%.

The average particle size and the width of the size distribution were measured by electron microscopy (Fig. 8). No consistent values are obtained for the factor F if the particles are considered to be homogeneous bimetallic particles. On the other hand, knowing the values for F for Ni_2B and Co_2B for this microemulsion, we calculated the expected sizes for the case where a mechanical mixture of separate particles of monometallic borides was formed. These values are shown in Fig. 8 as well as the weighted average sizes for these two types of particles. Only the latter can be compared with the experimental results, which are average sizes. The average sizes so calculated are close to those measured experimentally. In most of the cases, the experimental size distributions are narrow, whereas a mechanical mixture of monometallic particles would result in a broad bimodal distribution. Hence, the particles are probably bimetallic but not completely homogeneous. The nucleation rate is higher for Co(II) ions than for Ni(II) ions (see earlier), the nuclei are formed preferentially from Co(II) ions, and the particles contain more nickel at the surface.

The essential parameters for the formation of monodisperse colloidal particles are thus quantified. We have shown that two metal atoms are required to form a stable nucleus and that nucleation occurs only in the aggregates that are reached by the reducing agent before rearrangement of the system can occur (1 per 1000 aggregates). Figure 9 illustrates quite well the mechanism of reduction in a water-in-oil microemulsion.

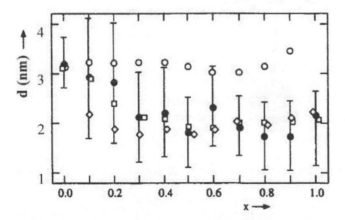

FIG. 8 Variation of particle size as a function of the molar fraction of cobalt in the catalysts (x). (●) Experimental values. Hypothesis of a mechanical mixture of pure Ni_2B and Co_2B particles: (○) Sizes calculated for Ni_2B; (◇) sizes calculated for Co_2B; (□) weighted average sizes for Ni_2B + Co_2B.

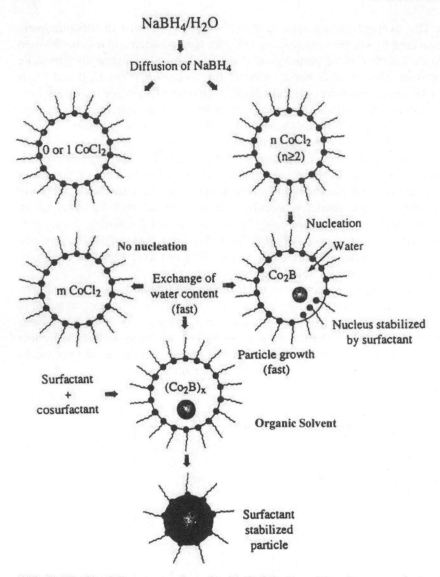

FIG. 9 Model of the preparation of colloidal Co₂B particles from a water-in-oil microemulsion.

After fast diffusion of the reducing agent, nucleation occurs in the water droplets where the preceding conditions are satisfied. The nucleus is stabilized by the adsorbed surfactant molecules. The growth of the particles requires an exchange between different water cores. Finally, surfactant-protected monodisperse particles are formed that can be used directly or by being deposited on a support. This study also allowed us to gain some information about the composition of the bimetallic Ni-Co-B particles.

In the treatment just discussed, the nucleation step could not yet be clearly described. The model is based on the presence of discrete water pools in the microemulsion, whereas conductivity measurements showed that percolation already occurred in these systems, favoring the exchange between water pool contents [23]. More experiments are needed to determine the formation of the first nuclei using fast kinetics measurements.

Nevertheless, the stabilization of the nuclei by the surfactant molecules at the interface could play a definite role in controlling their number formed at the very beginning of the reduction. The method of addition of the reducing agent in the aqueous solution is indeed very important, because if higher amounts of microemulsion systems are used, larger particles are obtained.

Finally, we should mention the formation of monodisperse colloidal Co_2B particles in the Triton X-100–decanol–water microemulsion. The diameter of Co_2B particles is shown in Table 2 as a function of micellar composition. The diameter of Co_2B increases with increasing water content.

3. Characterization of the Ni_2B and Co_2B Nanoparticles

The nature of the boride particles and the composition of the mixed Ni-Co-B particles were determined by XPS and EDX, respectively. Table 3 shows the initial composition of the nickel and cobalt chlorides dissolved in the microemulsion (X). The composition of the nanoparticles (x) was determined by EDX measurements. It can be seen that the compositions X and x are equal within 3%, 5% being the largest difference.

The as-prepared boride nanoparticles adsorb large amounts of BO_2^- and $CTAB^+$ ions. The amount of adsorbed BO_2^- was measured using ^{11}B nuclear magnetic resonance (NMR). It was determined as the difference between the total amount of $NaBH_4$ added and the final concentration of boron in the microemulsion after precipitation of the nanoparticles.

The amount of boron adsorbed on the Ni_2B, Co_2B, and Ni-Co-B nanoparticles is about 85–90% of the total boron present in the system as BO_2^- ions.

The borate ions are eliminated by two successive washings with an aqueous solution of HCl and three successive washings with distilled water. After washing, the remaining adsorbed boron still varies from 15 to 46%.

BLE 2 Size of Colloidal Co_2B Particles Obtained in Triton X-100–Decanol–Water Microer

$Cl_2]$) vs. H_2O	[Triton X-100] (molal)	[NaBH$_4$] (M) vs. H_2O	R^a	r_w^b nm	[Aggregate]^c × (molal)
5	0.43	0.5	4.2	0.62	4.12
	3.59	0.5	13.1	1.42	2.67

= $[H_2O]/[Triton]$.
= radius of inner water cores.
ncentration of micellar droplets.
mber of micellar droplets per kilogram of solution.
ameter of Co_2B particles.

TABLE 3 Composition of Co-Ni-B Particles, % H_2O Adsorbed on the Particles, and % Boron Adsorbed on the Particles before and after Washing with an HCl Solution and Distilled Water

X^a	x^b	% H_2O^c hyd	d^d (nm)	S_{part}^e (m^2)	% B_{ads}^f	% B_{ads}^g
0.00	—	15	5.5	44	85	15
0.20	0.21	28	5.0	49	86	39
0.40	0.40	35	4.0	60	90	46
0.50	0.50	14	3.1	78	88	45
0.60	0.59	25	3.3	73	85	31
0.80	0.78	32	3.2	75	84	34
1.00	—	21	3.7	65	86	28

^{a}Mole fracton of Co(II) in the initial microemulsion.
^{b}Mole fraction of cobalt in the nanoparticles (EDX values).
^{c}% H_2O hydrating the Co-Ni-B nanoparticles.
^{d}Diameter of the Co-Ni-B nanoparticles.
^{e}Total surface of nanoparticles synthesized in 100 g of micellar solution determined from TEM (transmission electron microscopy) measurements [Co(II) and/or Ni(II)] = 5×10^{-2} molal.
^{f}Percent boron adsorbed on the particles before washing.
^{g}Percent boron adsorbed on the particles after washing.

The precipitated particles adsorb a nonnegligible amount of water from the microemulsion inner water core. This amount can rise to about 15–35% of the initial water present in the microemulsion (Table 3).

The specific surface area of the particles was determined on both the as-prepared and the washed nanoparticles. In Table 4 are compared the S_{BET} values, the total surface of the nanoparticles obtained in 100 g of micro-emulsion and determined from the diameter of the particles (S'_{part}), and the total surface of the particles (S_{part}) obtained from BET measurements. It is seen that the unwashed particles present a surface 2.3–5.9 times smaller than the theoretical surfaces (S_{part}/S'_{part}). After washing, the surface is almost clean for the Ni_2B particles and the amount of BO_2^- (and also of Cl^- and Br^- ions) increases with increasing Co content in the mixed Ni-Co-B nano-particles.

4. Sizes of Platinum, Rhenium Dioxide, and Gold Particles

(a) CTAB–Hexanol–Water Microemulsion. Colloidal particles of Pt, ReO_2, and Au were prepared following Scheme I of Fig. 3. The monodis-perse Pt particles prepared from H_2PtCl_6 dissolved in the CTAB–hexanol–water microemulsion had an average diameter of 4.0 ± 0.5 nm, and their

TABLE 4 Specific Surface Area (S_{BET}) of the Co-Ni-B Nanoparticles Prepared in the 18.0% CTAB–70.0% Hexanol–12.0% H_2O^a Microemulsion

X	Before washing the particles			After washing the particles		
	S_{BET} (m²/g)	S'_{part} [b] (m²)	$S_{part}{}^c/S'_{part}$	S_{BET} (m²/g)	S'_{part} (m²)	S_{part}/S'_{part}
0.0	60.6	19.4	2.3	124.2	39.9	1.1
0.2	26.9	8.6	5.7	126.8	40.7	1.2
0.4	31.8	10.2	5.9	161.1	51.7	1.2
0.5	44.4	14.2	5.5	141.3	45.4	1.7
0.6	42.3	13.6	5.4	104.9	33.7	2.2
0.8	42.2	13.6	5.5	157.3	50.7	1.5
1.0	33.6	10.8	5.9	112.5	36.2	1.8

[a] [Co(II) and/or Ni(II)] = 5×10^{-2} molal.
[b] $S'_{part} = W^t S_{BET}$ where W_t is the total weight of the particles synthesized in 100 g of microemulsion.
[c] S_{part} is determined by TEM (see Table 3).

size was not dependent on the H_2PtCl_6 concentration (5×10^{-3}–2×10^{-2} molal with respect to water) [24]. The aqueous solution of hydrazine containing a 10-fold molar excess of hydrazine with respect to H_2PtCl_6 had an initial pH of 10. The metal particle precursor is soluble in both the dispersed inner water core and the continuous (or hexanol) phases. If it is assumed that the nucleation occurs in both phases, the particle size is dependent only on its stabilization by the adsorbed surfactant molecules [20,25,26].

It is interesting to note that particles of a similar size were obtained, independently of water and H_2PtCl_6 concentrations, from the AOT–heptane–water microemulsion [27].

If K_2PtCl_4 is used instead as the particle precursor (for the same hydrazine-to-K_2PtCl_4 ratio), a complex behavior is observed as a function of pH. At low pH values (1 < pH < 4), no Pt particles could be obtained. At 5 < pH < 8, dispersed Pt particles were formed, but the reduction was not complete even after 24 h of reaction. For high pH values (pH < 9), complete reduction of the Pt salt occurred, but the particles thus obtained were aggregated.

It is thus clear that the surface charge does influence the aggregation of the metal particles. In addition, the adsorption of the surfactant molecules, also pH dependent, can greatly influence the particle aggregation.

(b) PEGDE–Hexane–Water Microemulsion. Colloidal Pt particles were prepared following both Schemes I and II of Fig. 3. To avoid particle ag-

TABLE 5 Analysis of Pt Particle Size as a Function of the Number of Nuclei N_n

[K$_2$PtCl$_4$] (molal vs. H$_2$O)	$n_{PtCl_4^{2-}}$[a,b]	d (nm)	$W_t \times 10^3$ (g)[c]	$W \times 10^{19}$ (g)[c,d]	$N_n \times 10^{-16}$
0.001	0.54	1.5 ± 0.3	0.98	0.38	2.6
0.01	5.4	2.5 ± 0.3	9.8	1.75	5.6
0.05	27.0	5.0 ± 0.3	49.0	14	3.5
0.1	54.0	9.0 ± 1.0	98.0	81.9	1.2
0.3	162.0	13.0 ± 1.5	294.0	247	1.2

[a]Number of PtCl$_4^{2-}$ ions per inner water core.
[b]N_M (number of inner water cores) = 5.56 × 10^{18} per kg solution; r_M = 6.0 nm.
[c]Values given for 1 kg of solution.
[d]Assuming volumetric mass of Pt = 21.45 g/cm^3.

gregation, a neutral surfactant, PEGDE, was used to form a microemulsion of composition PEGDE 9.5%–hexane 90%–water 0.5%. Only K$_2$PtCl$_4$ was tested as a precursor salt, however, because it is insoluble in the organic medium. Table 5 and Fig. 10 show the variation in the size of the Pt particles obtained following Scheme I as a function of initial K$_2$PtCl$_4$ concentration.

The particle diameter increases monotonously with increasing K$_2$PtCl$_4$ concentration and approaches a plateau at high concentration. This behavior seems to be different from those previously observed for the Pt particles using H$_2$PtCl$_6$ [27,28] and for the Ni$_2$B or Co$_2$B particles. In the first case, a constant particle size was obtained irrespective of the initial H$_2$PtCl$_6$ con-

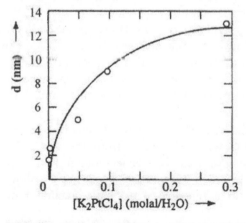

FIG. 10 Variation of the average Pt particle diameter as a function of K$_2$PtCl$_4$ concentration with respect to water prepared according to Scheme I of Fig. 3.

centration, and in the second cases a minimum was observed in the particle size (Ni$_2$B or Co$_2$B) versus NiCl$_2$ or CoCl$_2$ concentration curve.

For low initial K$_2$PtCl$_4$ concentration (up to 0.01 molal with respect to water), N_n increases as a function of Pt concentration. This behavior was observed earlier for the case of Ni$_2$B and Co$_2$B particles. However, for higher K$_2$PtCl$_4$ concentrations, the N_n value decreases, leading to larger particles. The probable nucleus is a surfactant-stabilized Pt atom that is able to form the final Pt particle [29].

If the particles are prepared following Scheme II, where the two micro-emulsions containing the precursor K$_2$PtCl$_4$ and the reducing agent N$_2$H$_4$, are mixed together, smaller sizes are obtained. Indeed, the Pt particles prepared from the microemulsion with [K$_2$PtCl$_4$] = 0.1 molal with respect to water have a diameter of 3.5 $\pm$ 0.5 nm, whereas the diameter is much greater (9.0 $\pm$ 1.0 nm) if Scheme I is used (Table 5). Figure 10 illustrates the variation of the average diameter of the Pt particles as a function of the concentration of K$_2$PtCl$_4$ prepared by Scheme I.

The larger size of the Pt particles obtained by the method of Scheme I can be explained in a first approximation by the diffusion of the aqueous solution through the organic phase being slower than the exchange between the water cores. Although in the PEGDE–hexane–water microemulsion no separate spherical droplets are present, the water is probably the dispersed phase in the microemulsion. The structure of the microemulsion is better represented as a lamellar aggregate where the surfactant molecules are associated head to head along a cylinder.

(c) Preparation of Monodisperse ReO$_2$ Particles. Monodisperse ReO$_2$ particles were obtained by reducing NaReO$_4$ with hydrazine in the system PEGDE 9.5%–hexane 90%–water 0.5% following Scheme I of Fig. 3. The presence of ReO$_2$ was confirmed by XPS experiments. However, the NaReO$_4$ was only partially reduced under these conditions.

Table 6 illustrates the variation of particle size as a function of NaReO$_4$ concentration. Once again, the size of monodisperse particles approaches a plateau for high ReO$_4$ concentrations, and this behavior is quite similar to that of the Pt particles.

(d) Preparation of Monodisperse Pt-ReO$_2$ Particles. Monodisperse Pt-ReO$_2$ particles were prepared following Scheme I from the PEGDE–hexane–water microemulsion using a total ion concentration [K$_2$PtCl$_4$] + [NaReO$_4$] = 0.10 molal with respect to water. Table 7 and Fig. 11 show the variation of the particle size as a function of the mole fraction x of K$_2$PtCl$_4$.

It is surprising that up to $x = 0.7$, the diameter of the particles remains quasi-constant and is close to that of the pure ReO$_2$ particles. For higher

TABLE 6 Variation of the Monodisperse ReO$_2$ Particles Size as a Function of NaReO$_4$ Concentration

[NaReO$_4$] (molal vs. H$_2$O)	d (nm)
0.01	1.8 ± 0.2
0.05	2.7 ± 0.3
0.1	3.1 ± 0.4
0.3	4.2 ± 0.5
1.0	5.5 ± 0.6

initial [K$_2$PtCl$_4$], the diameter of the particles increases monotonously to reach that of the pure Pt particles. The electrochemical potentials of PtCl$_4^{2-}$ and ReO$_4^-$ are, respectively, 0.73 and 0.51 V; furthermore, two electrons are needed for the reduction of K$_2$PtCl$_4$ and three electrons for the reduction for NaReO$_4$. The nucleation should thus be easier for the platinum particles than for the ReO$_2$ particles. It can thus be concluded that the ReO$_2$ is dispersed on the Pt particles. As the particle size is constant for low values of the K$_2$PtCl$_4$ molar fraction, it can be concluded that the size of the particles is controlled not by the nucleation as in the case of the (Ni,Co)$_2$B particles but by the interaction between the ReO$_2$ and the surfactant. This shows the importance of the thermodynamic stabilization of the particles in the case where the particle size seems to be determined by the La Mer

TABLE 7 Variation in Monodisperse Pt-ReO$_2$ Particle Size as a Function of the Mole Fraction (x) of K$_2$PtCl$_4$[a,b]

Mole fraction x of K$_2$PtCl$_4$	d Pt (nm)[c]	d ReO$_2$ (nm)[c]	Pt-ReO$_2$
0	—	3.0	3.1 ± 0.3
0.16	3.0	≈2.9	2.5 ± 0.3
0.33	≈3.5	≈2.8	2.4 ± 0.3
0.5	5.0	≈2.7	2.7 ± 0.3
0.66	7.0	≈2.2	2.5 ± 0.2
0.8	≈8.0	≈2.0	3.8 ± 0.4
0.9	≈8.5	≈1.8	7.0 ± 0.5
1	9.0	—	9.0 ± 1.0

[a]PEGDE 9.5%–hexane 90%–H$_2$O 0.5%.
[b][K$_2$PtCl$_4$] + [NaReO$_4$] = 0.10 molal with respect to water.
[c]Hypothetical particle size estimated for the cases where the systems would contain pure Pt or ReO$_2$ particles.

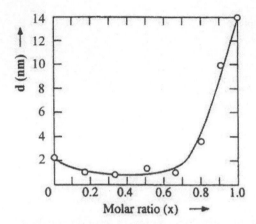

FIG. 11 Variation of Pt-ReO₂ particle size as a function of ratio x of K_2PtCl_4 ($[K_2PtCl_4]$ + $[NaReO_4]$ = 0.10 molal with respect to water).

diagram. It can be noted that a similar variation has been observed in the case of Ag(ClBr) particles.

All these results are different from those one would expect on the basis of a mechanical mixture. Indeed, in that case a bimodal distribution would be expected at least for $x \geq 0.5$, based on the different sizes of the separate Pt and ReO₂ particles. Table 7 also includes the particle sizes estimated for the hypothetical case of a system containing pure Pt and ReO₂ particles (see Tables 5 and 6).

(e)   DOBANOL–Hexane–Water Microemulsion. Particles of Pt, Au, and Pt-Au were prepared in a DOBANOL–hexanol–water microemulsion following Scheme II of Fig. 3. DOBANOL is a mixture of penta(ethylene glycol) undecyl (<1 wt%), dodecyl (41 wt%), tridecyl (58 wt%), and tetradecyl (<1 wt%) ethers. The microemulsion region is smaller than for the PEGDE system [30].

Table 8 illustrates the influence of DOBANOL and PEGDE surfactants. A difference is noted only for low K_2PtCl_4 concentrations, where larger Pt particles are formed with DOBANOL.

The Au particles were obtained from the precursor $AuCl_3$. In the PEGDE–hexane–water microemulsion at low precursor concentrations (less than 0.05 molal with respect to water), only small particles (about 3.0 nm diameter) were formed (Table 9a and Fig. 12), whereas both small and large (about 10 nm diameter) particles were formed at higher precursor concentrations.

TABLE 8 Average Diameter of Pt Particles Synthesized
from PEGDE 9.5%–Hexane 90%–Water 0.5%
and DOBANOL 9.5%–Hexane 90%–Water
0.5% Microemulsions

[K₂PtCl₄] (molal vs. H₂O)	d (nm)	
	PEGDE	DOBANOL
0.001	1.9 ± 0.2	2.3 ± 0.3
0.01	2.2 ± 0.3	2.7 ± 0.4
0.05	2.6 ± 0.3	2.7 ± 0.4
0.1	2.8 ± 0.3	2.8 ± 0.4
0.3	3.8 ± 0.4	3.9 ± 0.4

In the DOBANOL–hexane–water microemulsion only one type of particle is obtained, the size of which increases with increasing precursor concentration (Table 9b).

The Pt-Au particles were prepared in both microemulsion systems (Table 10). Both small (about 3.0 nm diameter) and large (about 12 nm diameter)

TABLE 9 Average Diameter of Au Particles Showing the
Bidispersion in PEGDE–Hexane–Water Microemulsions

(a) PEGDE 9.5%–hexane 90%–water 0.5%		
[AuCl₃] (molal vs. H₂O)	d (nm)	d (nm)
0.001	—	2.9 ± 0.4
0.01	—	3.0 ± 0.3
0.05	8.2 ± 1.2	3.6 ± 0.4
0.1	11.0 ± 1.9	3.8 ± 0.5
0.3	13.9 ± 2.5	4.4 ± 0.5

(b) DOBANOL 9.5%–hexane 90%–water 0.5%	
[AuCl₃] (molal vs. H₂O)	d (nm)
0.001	3.3 ± 0.5
0.01	7.1 ± 1.1
0.05	9.7 ± 1.2
0.1	11.5 ± 1.4
0.3	13.2 ± 1.6

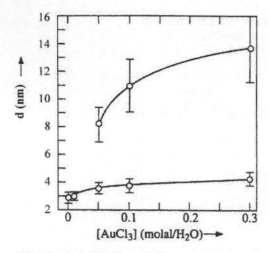

FIG. 12 Variation in gold diameter as a function of precursor AuCl$_3$ concentration versus water synthesized in PEGDE 9.5%–hexane 90%–water 0.5% microemulsion. The presence of larger particles shows particle aggregation.

TABLE 10 Average Diameter of Pt-Au Particles as a Function of Pt Mole Fraction x Showing the Bidispersion in Both Systems[a]

x	d (nm)	d (nm)
(a) PEGDE 9.5%–hexane 90%–water 0.5%		
0.16	9.5 ± 1.8	3.4 ± 0.4
0.33	11.3 ± 2.0	3.2 ± 0.4
0.50	10.7 ± 1.9	3.7 ± 0.5
0.66	13.5 ± 2.4	2.9 ± 0.3
0.80	9.2 ± 1.3	3.4 ± 0.4
(b) DOBANOL 9.5%–hexane 90%–water 0.5%		
0.16	11.2 ± 2.2	2.6 ± 0.4
0.33	12.5 ± 2.5	2.9 ± 0.4
0.50	14.6 ± 3.2	2.6 ± 0.4
0.66	12.7 ± 2.5	2.7 ± 0.5
0.80	12.1 ± 2.6	2.8 ± 0.4

[a][AuCl$_3$] + [K$_2$PtCl$_4$] = 0.1 molal vs. H$_2$O.

particles were obtained in both systems. The size of the particles is not dependent on the composition of the precursor salts. The large particles are clearly formed by aggregation of the small particles. The nanoparticles are true mixed Pt-Au particles, as was shown by scanning transmission electron microscopy (STEM)/EDX measurements [30].

5. Characterization of the Silver Halide Nanoparticles in Microemulsions

In the numerous studies concerning the synthesis of nanoparticles in microemulsion media, the location of water after the nanoparticle synthesis has never been determined. Two models can be proposed (Fig. 13). In the first one the particles are surrounded by a layer of water, and in the second the surfactant molecules (the AOT) are directly adsorbed onto the particles and only a small amount of water molecules is present.

In order to discriminate between these two models, ^{2}H NMR measurements of deuterated water in microemulsions have been carried out. Two NMR lines were observed in the ^{2}H NMR spectra (Fig. 14) for the various microemulsions without particles of silver bromide.

If the same spectrum is taken for a very low R value, such as $R = 0.5$ (Fig. 15), three NMR lines are observed. These lines are not due to the presence of impurities—in fact, their intensity does not decrease as the amount of water decreases—so these lines stem from different types of water molecules. This is illustrated by the measurements of their relaxation times T_1. In fact, for $R = 1$ the following three relaxation times T_1 were

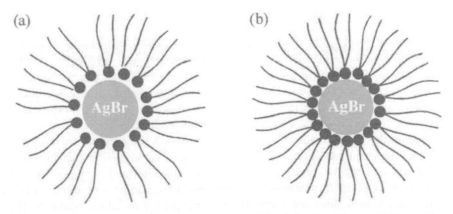

FIG. 13 Two models of the nanoparticles stabilized in the microemulsion media. (a) The particle is surrounded by a layer of water. (b) AOT is directly adsorbed onto the particle.

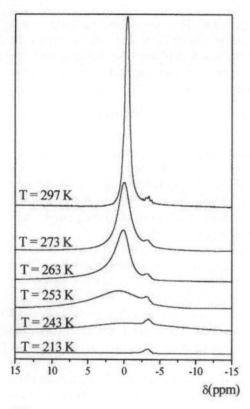

FIG. 14 NMR spectra of the deuterated water in the microemulsion for $R = 3.1$.

obtained at 273 K: 321 ms for the broader line, 804 ms for the line situated at -3.50 ppm, and 1087 ms for the line situated at -3.95 ppm. As the variation of the relaxation time with temperature indicates that we are in a region where the relaxation time increases with the decrease of temperature, these two lines correspond to water molecules that are less mobile and, therefore, more in contact with the surfactant molecules.

Generally, three kinds of water may exist in a microemulsion medium: "bulk" water in the center of the water core, "bound" water that interacts with the hydrophilic part of the surfactant molecule, and "trapped" water that is trapped in the interface in the form of monomers or dimers [31]. Bulk water molecules are normally not present for R values below 6–10, where all the water molecules are structured due to their interaction with Na^+ counterions and the strong dipole of the AOT polar group [32]. As in this case the ratio $R = [H_2O]/[AOT]$ is 3.1, only two kinds of water mole-

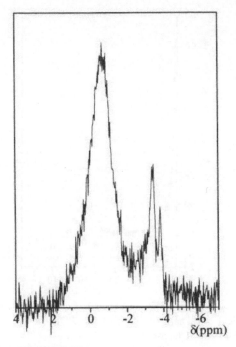

FIG. 15 NMR spectrum of deuterated water in the microemulsion for $R = 0.5$ at $T = 297$ K.

cules would be expected. Therefore, it is assumed that the two NMR lines observed here correspond to bound water and trapped water. In order to check this assumption, the same experiment was carried out for higher R values. The chemical shift increases with the R value until it reached approximately that of the pure deuterated water (used as reference) while the line width at half-height decreases with R (Fig. 16).

Such variation has already been observed [32] and is the result of a fast exchange (faster than 2×10^{10} s^{-1}) between the bulk water and the bound water. At low R values, the observed chemical shift comes from the variation of the number of hydrogen bonds in which the water molecules are involved. In fact, the water molecules adsorbed at the interface (or solvating the Na$^+$ ions) form fewer hydrogen bonds, provoking a high-field chemical shift. The smaller number of hydrogen bonds has previously been shown by Wong et al. [33] using ^{1}H NMR experiments.

Furthermore, if the NMR spectra are recorded at lower temperatures, the NMR line corresponding to the bound water decreases due to the freezing of this kind of water (the bandwidth becomes too large to be detectable)

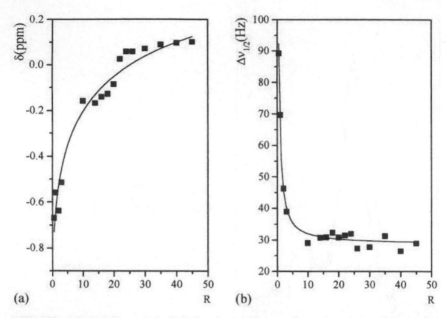

FIG. 16 (a) Variation of the ^{2}H chemical shift as a function of the R factor. (b) Variation of the line width as a function of the R factor.

(Fig. 14). In fact, the freezing point of bound water seems to be about 243 K inside the inverted micelles. This corresponds to the decrease of the freezing point of water with the size of the droplet; for example, the freezing point of water in a droplet corresponding to $R = 4.5$ in AOT/water/2,2,4-trimethylpentane is at around 241 K [34]. On the other hand, the line corresponding to the trapped water shows no freezing and its intensity remains quasi-constant.

In order to distinguish between the two models of AgBr stabilization (see earlier), the NMR experiments mentioned have also been carried out in presence of silver bromide nanoparticles. As the only difference between the two experiments is the presence of silver bromide particles, all observed differences must be due to the particles. In the presence of these particles, the quantity of trapped water is larger, as shown by a comparison of spectra in the presence and in the absence of nanoparticles (Fig. 17). It could be hypothesized that the particles repel the bound water into the interface and, as a consequence, the amount of trapped water increases. The total intensity is also higher in presence of silver bromide particles, also stemming from the greater importance of the trapped water. In fact, this water freezes at a lower temperature. Furthermore, not all the water cores of the microemulsion

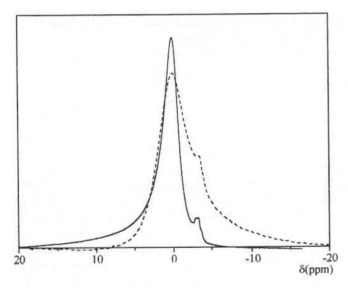

FIG. 17 NMR spectra of the deuterated water in the microemulsion (full line) and in presence of AgBr particles (dotted line) at 263 K.

are occupied by a particle, only 1 water core out of 1.3×10^4 is occupied by a particle. Hence, if the microemulsion structure stayed the same, with the same number of water molecules in each water core, no influence on the NMR spectra could be observed upon addition of AgBr.

The higher amount of trapped water is in favor of model (b), where the particles are in closer contact with the interfacial layer. However, the NMR line of the adsorbed water could overlap that of the trapped water. In order to check this hypothesis, the number of water molecules per AOT was calculated. The spectra in Fig. 17 have been decomposed in two bands corresponding, respectively, to the bound water and to the trapped water. The difference in intensities of the two NMR lines corresponding to the trapped water in the spectra without and with AgBr particles gives the amount of water trapped or adsorbed on the particles. The number of AOT molecules per particle has been calculated using a spherical surface of 4.6 nm diameter and a surface area of 0.41 nm^2 for the polar part of the AOT molecule [35]. It has been computed that if the whole line intensity corresponded to the trapped water there would be 2000 water molecules per AOT molecule. As the trapped water is considered to be in the form of a monomer or a dimer, this value is too high to correspond only to water molecules trapped in the interface. Hence, it has to be assumed that the additional water molecules so computed are adsorbed on the AgBr particles and the NMR lines of the trapped water and the adsorbed water overlap.

If it is assumed that all these additional water molecules are adsorbed on the particles, the number of water monolayers can be calculated by using the van der Waals radius of a water molecule. Approximately 1000 mono-layers of water can be formed around the nanoparticles. (This number must be overestimated because it does not take into account that the number of trapped molecules increases by repelling the water molecules in the inter-face.) These two arguments, the observation of an NMR line corresponding to the adsorbed water molecules and the estimation of the number of water monolayers, are in favor of model (a). Hence, this model will be adopted.

In order to quantify by another method the amount of water adsorbed on the nanoparticles, a microemulsion in which the particles had sedimented was also examined. This microemulsion was obtained by adsorption of pseu-doisocyanine on the particles. This dye causes a rapid sedimentation of the particles [36], and an ^{2}H NMR spectrum was taken after sedimentation of all the particles. From this spectrum, it was established that 68% of the water was adsorbed on the particles. The number of water monolayers formed around the particles was calculated and a value of about 4600 monolayers of water was obtained. This value is too large and physically impossible; in fact, the radius of the corresponding water core should be 2.6 μm. These water cores should scatter the light, and as the colloidal suspen-sion is limpid, the number of water molecules bound to the silver halide particles must be overestimated in this approach. Such a large amount of water in the precipitate can be explained only if the sedimented particles form a sort of gel where a large amount of water is required. This gelation was previously shown in the case of Co$_2$B nanoparticles prepared from CTAB/n-hexanol/water [37] microemulsion. This high amount of adsorbed water molecules is also in favor of model (a).

6. Synthesis of Silver Bromide Particles in the AOT/p-Xylene/Water Microemulsion

Astonishingly, the average diameters of the AgBr nanoparticles prepared in the AOT/p-xylene/water microemulsion remain quite constant, irrespective of the concentrations of precursor salts or the size of the water nanodroplets. Figure 18 shows the average diameters of the nanoparticles as a function of the salt concentrations and the R values. All the average diameters seem to lie between 3.3 and 4.3 nm. The reproducibility of the size measurements was estimated to be approximately 0.9 nm, which corresponds to the range of variation. As we do not observe any correlation between the average diameter of the nanoparticles and these synthesis parameters, we draw the conclusion that in this microemulsion system, the nucleation and growth of

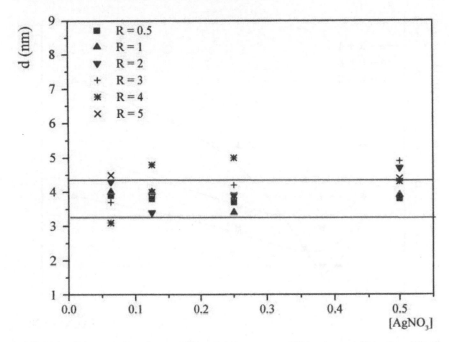

FIG. 18 Average diameters of the AgBr nanoparticles prepared in the AOT/*p*-xylene/water microemulsion.

the nanoparticles do not follow the La Mer diagram. The size of the particles seems to be thermodynamically stabilized by the adsorbed surfactant.

7. Synthesis of Silver Chloride Particles

The preparation of silver chloride nanoparticles has also been studied in the AOT/*n*-heptane/water microemulsion. A size dependence on the synthesis parameters has been observed (Fig. 19). The diameter seems to pass through a minimum value around 0.125 M concentration of $AgNO_3$.

The synthesis of AgCl particles seems to follow a model between the La Mer diagram and the thermodynamic stabilization of the particles. The variation of the particles size seems to correspond to the La Mer diagram but other factors are in favor of the thermodynamic stabilization of the particles. In fact, the size of the particles (and the number of nuclei) does not vary with the contact surface area between the two microemulsions during the precipitation reaction. If the La Mer diagram were followed, the number of nuclei should increase with the contact surface area between the microemulsions and the size of the particles should decrease. It is not astonishing that the synthesis of AgCl particles lies between the La Mer diagram and

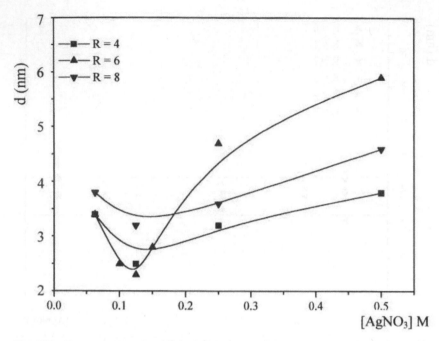

FIG. 19 Average diameter of the AgCl nanoparticles prepared in the system AOT/ *n*-heptane/water as a function of the concentration of precursor salt and the value of *R*.

the thermodynamic stabilization of the particles because these two models are only limiting models.

III. SYNTHESIS OF ORGANIC PARTICLES

A. General Considerations

Different organic nanoparticles have been synthesized in certain micro-emulsions. The active compounds are cholesterol, rhodiarome, and rhovanil (aromas) (Fig. 20). The microemulsions used are AOT/heptane/water, Triton/ decanol/water, and CTABr/hexanol/water.

The general preparation of these organic nanoparticles has already been described in previous chapters. It consists of the direct precipitation of the active compound in the aqueous cores of the microemulsion. After their preparation, nanoparticles are revealed with iodine vapor and observed with a transmission electron microscope (Phillips EM301) [38,39]. A TEM micrograph of nanoparticles is presented in Fig. 21.

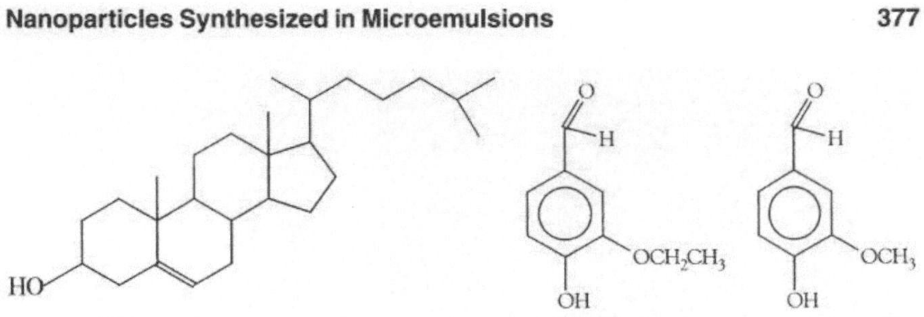

Cholesterol   Rhodiarome   Rhovanil

FIG. 20 Structures of the active compounds.

The mechanism of the formation of nanoparticles has been proposed previously [20,25,26]. This consists of several stages. The solution of the active compound in an appropriate solvent penetrates inside the aqueous cores by crossing the interfacial film. The solvent certainly plays a role in the transport of the active compound inside the aqueous cores. The active compound precipitates in the aqueous cores because of its insolubility in water, and the nuclei are thus formed. The so-formed nuclei can grow because of the exchange of the active compound between the aqueous cores. Finally, the nanoparticles are stabilized by the surfactants.

B. Nanoparticles of Cholesterol Prepared in Different Microemulsions

Figure 22 represents the evolution of nanoparticle size as a function of R at different concentrations of the cholesterol solution in chloroform in the AOT/ heptane/water microemulsion.

It should be noted that the total amount of cholesterol added increases with increasing R, as the volume of chloroform solution is equal to that of the water in the microemulsion. The mean particle size is 3–6 nm and a minimum is observed for a certain R value. A hypothesis is the participation of water as a reaction medium in the precipitation reaction. In this case, for low values of R, the amount of water is not enough to enable the formation of an optimal number of nuclei. As the concentration of water increases, the number of nuclei increases and the size of the particles decreases. For a large amount of water, the number of nuclei is already optimal, and the size of the particles remains constant. In fact, the concentration of cholesterol has increased and the amounts of both water and cholesterol are sufficient and finally the thermodynamic stabilization of the particles plays a role. In this hypothesis, the La Mer diagram is followed for small R values in the synthesis of the cholesterol particles. Another hypothesis to explain the pres-

FIG. 21 TEM micrograph of rhodiarome nanoparticles synthesized with a solution of rhodiarome in acetone (50 g/L) in AOT/heptane/water microemulsion (scale 96,000×).

ence of a minimum stems from the direct participation of chloroform in the stabilization of the cholesterol nanoparticles. Indeed, the amount of chloroform increases with R value and the relative amount of chloroform in the solvation sphere could depend on the size of the particles. In order to check the veracity of this hypothesis, another series of experiments were carried out: the same amount of chloroform solution of cholesterol (0.3 mL) was added to the various microemulsions with different R values (Fig. 23).

In this case, the particle size is constant as a function of R. The size of the particles is hence controlled by the thermodynamic stabilization of the particles. But in this experiment, the amount of cholesterol also stays constant. Hence, the statement that the variation of the chloroform concentration is responsible for the minimum observed in Fig. 22 cannot be accepted as a proof.

Figure 24 shows the variation of nanoparticle size as a function of the concentration of cholesterol in the same microemulsion system. Contrary to the previous graph (Fig. 22), no minimum appears. The size of the particles is thus controlled by a thermodynamic stabilization with the molecules of the surfactant.

Nanoparticles of cholesterol have also been synthesized in two other microemulsion systems: Triton/decanol/water and CTABr/hexanol/water. Similar experiments have been carried out. In these two cases, the nanoparticle size was independent of both the factor R and the concentration of the cholesterol solution. The particles are thus thermodynamically stabilized by the surfactants at certain favored sizes.

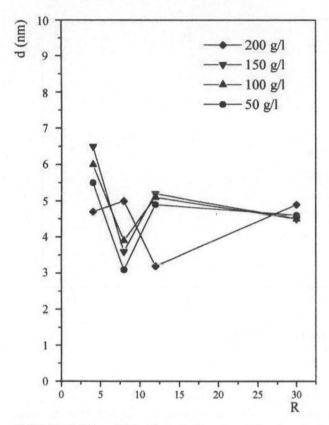

FIG. 22 Variation of the cholesterol nanoparticle size as a function of R at different concentrations.

The nanoparticles were stable for months, no precipitate appeared, and the final solutions were still limpid.

C. Nanoparticles of Rhodiarome (or Rhovanil) Prepared in the AOT/Heptane/Water Microemulsion

1. Influence of the Factor R and the Concentration of the Active Principle on the Nanoparticle Size

An example is presented of the formation of nanoparticles of rhovanil. A solution of rhovanil in acetone (50 g/L) was used. Figure 25 presents the variation of the mean diameter as a function of R.

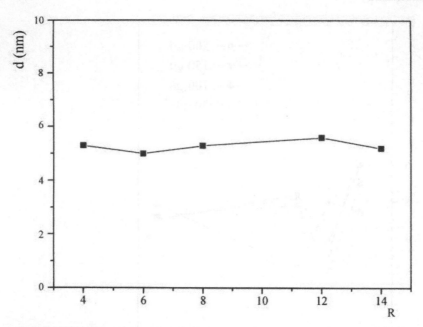

FIG. 23 Variation of the cholesterol nanoparticle size as a function of R at a fixed concentration (50 g/L).

The nanoparticle size is relatively constant as a function of R and is between 4.5 and 6.2 nm for the four concentrations studied. It is the same for rhodiarome, where the nanoparticle size is independent of the factor R. The second parameter studied is the concentration of the active principle in the solvent. Figure 26 shows a constant size between 4.5 and 7.0 nm.

In the two cases, a hypothesis can be made: the nanoparticle size is essentially determined by thermodynamic stabilization by the surfactant molecules at a certain size as it is dependent neither on R nor on the concentration.

2. Recovery of the Nanoparticles Stabilized by Surfactants in a Microemulsion and Their Transfer into an Aqueous Medium

Some potential pharmaceutical applications can be considered if less toxic solvents are used. Thus, the residual solvents (heptane, for example) are evaporated under vacuum and the nanoparticles stabilized by surfactants are recovered. These particles are suspended in distilled water under ultrasound energy in order to obtain a limpid and stable dispersion. Figure 27 shows the variation of the nanoparticle size as a function of R.

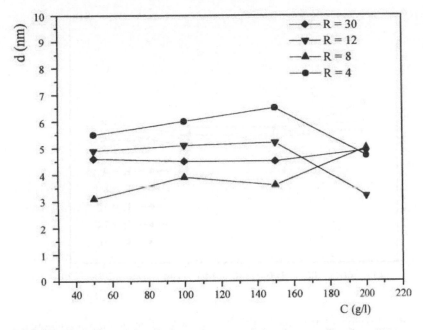

FIG. 24 Variation of the cholesterol nanoparticle size as a function of the concentration.

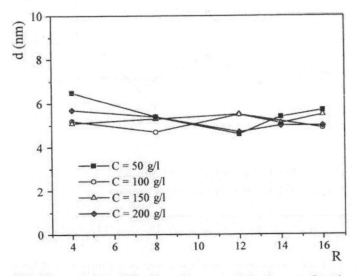

FIG. 25 Variation of the rhovanil nanoparticle size as a function of R.

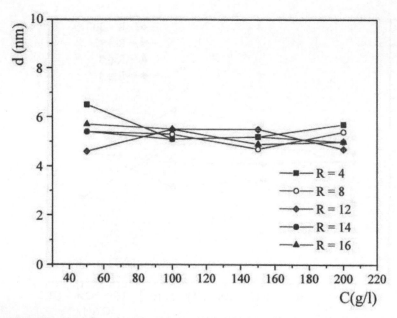

FIG. 26 Variation of the nanoparticle size as a function of the concentration of rhovanil in acetone.

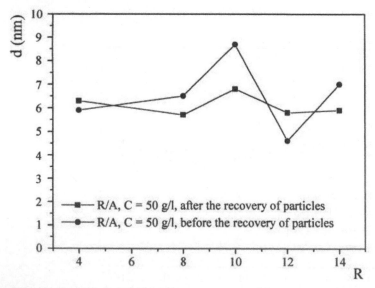

FIG. 27 Variation of the rhodiarome nanoparticle size as a function of R before and after the recovery of the nanoparticles.

The size lies between 5.7 and 6.3 nm and does not change after the recovery. The nanoparticles are thus thermodynamically stabilized by the surfactants. The change of the medium does not influence the nanoparticle size.

Biocompatible microemulsions have also been employed in order to allow their use in drug delivery [39].

IV. CONCLUSIONS

This chapter has placed emphasis on the mechanism of formation of the particles. Two models have been proposed: the La Mer diagram and the thermodynamic stabilization of the particles. These two models are relatively simplistic. The La Mer diagram is based on the separation between the nucleation and the growth of the particles. It is consistent with the mechanism proposed by López-Quintela and Rivas [40] for Fe nanoparticles obtained in AOT microemulsions using a stopped-flow technique and measuring the time-resolved small-angle X-ray scattering (SAXS) with synchrotron radiation. Nucleation implies an increase in the number of scattering centers (number of particles) for a given observation window, and, therefore, it gives an increase in the scattered intensity. On the contrary, the growth of particles is associated with a decrease of the scattered intensity because the observation window corresponds to the diffraction of smaller particles, which are disappearing during the growth process. The presence of this maximum (although not well defined) has also been spectrophotometrically detected by Towey et al. [41] for the formation of CdS in AOT microemulsion. This is an illustration of the La Mer diagram, as according to this diagram the nucleation occurs only in the beginning of the reaction. Theoretical calculation has been carried out by Tojo et al. [42] involving the study of the influence of the concentration and the film flexibility and of the kinetic exchange constant between the droplets using the difference between the nucleation and the growth of the particles. The thermodynamic stabilization is less documented in the literature, but an example shows the formation of secondary monodisperse spherical particles by coagulation of the primary particles [43].

Whether the reaction follows the La Mer diagram or the thermodynamic stabilization of the particles depends on the microemulsion diagram used and on the nature of the particles synthesized. As an example, the synthesis of AgBr particles follows the La Mer diagram in the AOT/heptane/water microemulsion system, but it follows the thermodynamic stabilization of the particles for the AOT/p-xylene/water system. The difference between the two systems can arise from the adsorption of the p-xylene molecule on the particles of AgBr. In fact, the adsorption of p-xylene on the AgBr particles has been shown in the study of the adsorption of pseudoisocyanine on these particles [44]. In the CTAB/hexanol/water microemulsion, the formation of Ni_2B particles follows the La Mer diagram, but the formation of Pt follows the thermodynamic stabilization of the particles. The difference could stem from the different adsorption of the surfactant on the particles. Mixed particles have also been synthesized. The particles are not homogeneous in composition, and the size of these particles does not vary linearly with their composition.

In the case of the organic nanoparticles, all the particles seem to follow the thermodynamic stabilization (with the exception of the cholesterol synthesized in the AOT/heptane/microemulsion). This can be due to a specific interaction of the surfactant with the particles.

REFERENCES

1. JH Fendler, FC Meldrum. Adv Mater 7:607, 1995.
2. GA Ozin, A Kuperman, A Stein. Angew Chem Int Ed Engl 28:359, 1989.
3. J Belloni, M Mostafavi, J-L Marignier, J Amblrad. J Imaging Sci 35:68, 1991.
4. A Henglein. J Phys Chem 97:5457, 1993.
5. RP Andres, RS Averback, WL Brown, LE Brus, WA Goddard III, A Kaldor, SG Louie, M Moscovits, PS Peercy, SJ Riley, RW Siegel, F Spaepen, Y Wang. J Mater Res 4:704, 1989.
6. RW Siegel. MRS Bull 15:60, October 1990.
7. YT Tan. MRS Bull 14:13, May 1989.
8. J Rouvière, J-M Couret, M Lindheimer, J-L Dejardin, R Marrony. J Chem Phys 76:289, 1979.
9. C Cabos, P Delord. J Appl Crystallogr 12:502, 1979.
10. P Ekwall, L Mandell, K Fontell. J Colloid Interface Sci 33:215, 1970.
11. SI Ahmad, S Friberg. J Am Chem Soc 94:5196, 1972.
12. S Friberg, I Lapczynska. Prog Colloid Polym Sci 56:16, 1975.
13. P Ekwall, L Mandell, K Fontell. Mol Cryst Liq Cryst 8:157, 1969.
14. PDI Fletcher, AM Howe, BH Robinson. J Chem Soc Faraday Trans 1 83:985, 1987.
15. SS Atik, JK Thomas. Chem Phys Lett 79:351, 1981.
16. VK La Mer, RH Dinegarn. J Am Chem Soc 72:4847, 1950.
17. T Sugimato. Adv Colloid Interface Sci 28:65, 1987.
18. J B.Nagy, A Gourgue, EG Derouane. Stud Surf Sci Catal 16:193, 1983.
19. J B.Nagy, A Claerbout. In: KL Mittal, DO Shah, eds. Surfactants in Solution. Vol 11. New York: Plenum, 1991, p 363.
20. J B.Nagy, EG Derouane, A Gourgue, N Lufimpadio, I Ravet, J-P Verfaillie. In: KL Mittal, ed. Surfactants in Solution. Vol. 10. New York: Plenum, 1989, p 1.
21. PC Hiemenz. Principles of Colloid and Surface Chemistry. New York: Marcel Dekker, 1977, p 234.

22. RD Vold, MJ Vold. Colloid and Interface Chemistry. London: Addison-Wesley, 1983, p 181.
23. J-P Verfaillie, PhD thesis, FUNDP, Namur, Belgium, 1991.
24. A Wathelet. Mémoire de Licence, FUNDP, Namur, Belgium, 1984.
25. J B.Nagy, D Barette, A Fonseca, L Jeunieau, Ph Monnoyer, P Piedigrosso, I Ravet-Bodart, J-P Verfaillie, A Wathelet. In: JH Fendler, I Dekany, eds. Nanoparticles in Solids and Solutions. NATO ASI Series 3. High Technology. Vol 18. Dordrecht: Kluwer, 1996, p 71.
26. J B.Nagy. In: P Kumar, KL Mittal, eds. Handbook of Microemulsion Science and Technology. New York: Marcel Dekker, 1999, p 499.
27. A Khan-Lodhi, BH Robinson, T Towey, C Herrmann, W Knoche, U Thesing. In: DM Bloor, E Wyn-Jones, eds. The Structure, Dynamics and Equilibrium Properties of Colloidal Systems. NATO ASI Series C324. Dordrecht: Kluwer, 1990, p 373.
28. B Lindman, P Stilbs. In: KL Mittal, B Lindman, eds. Surfactants in Solution. Vol 3. New York: Plenum, 1984, p 1654.
29. A Claerbout, J B.Nagy. Stud Surf Sci Catal 63:705, 1991.
30. D Barette. Mémoire de Licence, FUNDP, Namur, Belgium, 1992.
31. TK Jain, M Varshney, A Maitra. J Phys Chem 93:7409, 1989.
32. H Hauser, G Haering, A Pande, PL Luisi. J Phys Chem 93:7869, 1989.
33. M Wong, JK Thomas, T Nowak. J Am Chem Soc 99:4730, 1977.
34. P-O Quist, B Halle. J Chem Soc Faraday Trans I 84:1033, 1988.
35. MAJ Rodgers, M Da Silva. Chem Phys Lett 78:256, 1974.
36. L Jeunieau, J B.Nagy. Colloids Surf A 151:419, 1999.
37. I Ravet. PhD thesis, FUNDP, Namur, Belgium, 1988.
38. F Debuigne, L Jeunieau, M Wiame, J B.Nagy. Langmuir 16:7605, 2000.
39. F Debuigne. Bachelar thesis DEA, FUNDP, Namur, Belgium, 1999.
40. MA López-Quintela, J Rivas. J Colloid Interface Sci 158:446, 1993.
41. FF Towey, A Khan-Lodhi, BH Robinson. J Chem Soc Faraday Trans 86:3757, 1990.
42. C Tojo, MC Blanco, F Ricadulla, MA Lopez-Quintela. Langmuir 13:1970, 1997.
43. L Lerot, F Lefrand, P De Bruycker. J Mater Sci 26:2353, 1991.
44. L Jeunieau, J B.Nagy. Appl Organomet Chem 12:341, 1998.

19

Phase Behavior of Microemulsion Systems Based on Optimized Nonionic Surfactants

WOLFGANG VON RYBINSKI and MATTHIAS WEGENER
Henkel KGaA, Düsseldorf, Germany

ABSTRACT

Microemulsions can be obtained in different ways. For ethoxylated nonionic emulsifiers the temperature is the decisive parameter; therefore, the microemulsion phase is stable only in a certain limited temperature range. If other specific types of surfactants are used, e.g., alkyl polyglycosides, the temperature dependence is less pronounced or even negligible. In this case, the formation of microemulsions is enabled by adjusting a specific mixing ratio of different emulsifiers, i.e., balancing the hydrophilic to lipophilic components in the surfactant system.

These effects have been studied for a variety of systems using mainly purified model raw materials. As microemulsions are gaining increasing importance for different applications, there is a need to study surfactants and surfactant mixtures that can be used in commercial products and to extend the knowledge to multicomponent emulsifier systems.

In this study, the phase behavior of microemulsions consisting of alkyl polyglycosides and ethoxylates as hydrophilic emulsifiers, a lipophilic coemulsifier, an oily component, and water is evaluated in terms of microemulsion formation and stability. Parameters such as temperature, oil polarity, and composition of the surfactant mixture are discussed. It was shown that both the concentration range and the temperature stability could be extended by using suitable mixtures of emulsifiers and coemulsifiers.

I. INTRODUCTION

Microemulsions have been the subject of intensive research efforts for many years. This research has been concentrated above all on microemulsions with ethoxylated nonionic surfactants [1–7]. Emulsions consisting of oil, water, and ethoxylated nonionic surfactants undergo a temperature-induced phase inversion, during which a microemulsion is formed [1,8]. A system consisting of tetradecane, water, and dodecyl pentaethyleneglycol ether ($C_{12}E_5$) inverts from an oil-in-water to a water-in-oil emulsion in the temperature range between 45 and 55°C (Fig. 1) [8]. A microemulsion is formed in the phase inversion zone, resulting in the so-called Kahlweit fish in the phase diagram. At low emulsifier concentrations (below 15%) the microemulsion phase is in equilibrium with an oil phase and a water phase and is therefore called a three-phase microemulsion (w+D+o). Emulsifier concentrations of more than 15% are sufficient to solubilize the whole volume of water and oil in the form of a single-phase microemulsion (D) or a lamellar phase L_a. The characteristic feature of microemulsion systems containing ethoxylated nonionic surfactants is the limited temperature range within the microemulsion is stable. This is one reason why these microemulsions have not yet been able to establish themselves on the market.

The phase behavior of alkyl polyglycoside (APG)/water mixtures differs in certain aspects from other nonionic surfactants, especially with regard to the influence of the temperature. Whereas the hydrate shell of the ethoxylate head group depends largely on temperature, the interaction of the sugar unit

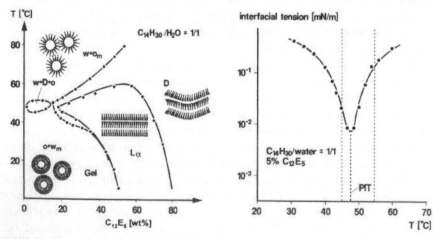

FIG. 1 Phase behavior (left) and interfacial tension (right) of water/tetradecane/$C_{12}E_5$ mixtures [8].

of alkyl polyglycoside with water is only slightly influenced by temperature. The surface activity of alkyl glycosides is evaluated by determining the interfacial tension between oil and water in the ternary system of surfactant, water, and oil with different polarities [9]. The influence of alkyl chain length and temperature on the plateau value of the interfacial tension of an aqueous solution of three alkyl mono- and polyglycosides against decane is shown in Fig. 2 [10]. The plateau value drops with increasing alkyl chain length, but a negligible temperature dependence in the experimental temperature range between 20 and 65°C is the most striking feature. The effect of the salt concentration on the interfacial tension is very weak, which is demon-

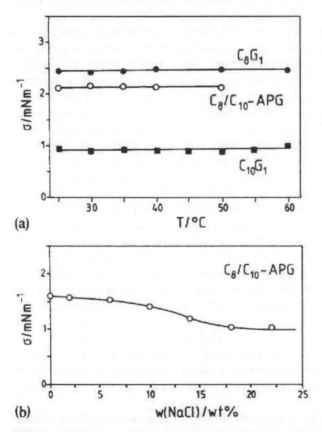

(a)

(b)

FIG. 2 (a) Plateau values of the interfacial tension for the decane/water interface with C_8G_1, $C_{10}G_1$, and $C_{8/10}$ alkyl polyglycoside (APG) as a function of temperature. (b) Interfacial tension as a function of added NaCl concentration for an aqueous solution of $C_{8/10}$ alkyl polyglycoside (0.5 wt%) against decane at 40°C [10].

strated for addition of sodium chloride in Fig. 2b. The partition coefficient of alkyl glycosides is nearly zero in the decane/water system, indicating a negligible oil solubility of alkyl glycosides. In contrast to these characteristics, a strong temperature dependence of the interfacial tension and, depending on the temperature, a significant oil solubility can be observed for fatty alcohol ethoxylates. Thus, the ability of alkyl glycosides to lower the interfacial tension and therefore the emulsifying capacity of the respective surfactant solution is more pronounced for alkyl glycosides than for the analogous fatty alcohol ethoxylates [9]. The effect is also reflected in the fact that the phase behavior of simple binary APG/water mixtures shows comparatively weak temperature effects [11–13]. Accordingly, no temperature-dependent phase inversion can be expected to occur in APG-containing emulsions.

Similarly to anionic surfactants, alkyl polyglycosides react to the addition of cosolvents, which increases the solubility of the surfactant in the oil phase. In the decane/water/APG system, the addition of the cosolvent i-butanol results in a drastic reduction in the interfacial tension between oil and aqueous phase and, hence, in the formation of a third phase, the microemulsion [14]. As expected, the range in which this three-phase microemulsion exists is only slightly dependent on temperature and in contrast to anionic surfactants is also hardly affected by electrolytes [14]. Systematic investigations of the phase behavior confirm these initial results for a number of simple hydrocarbons from hexane to hexadecane and aromatics [15,16].

Figure 3 shows the well-known Kahlweit fish [1,15] in a pseudoternary phase diagram. The ratio of dodecane to water was kept constant at 1:1. The ratio by weight of APG to APG+oil+water is shown in the lower half, and the percentage content by weight of the cosolvent pentanol is shown on the left-hand side. With small APG contents of 5 to 25% and small pentanol contents of 3 to 10%, three-phase microemulsions are formed. Of greater interest for practical applications are single-phase microemulsions, which are formed with an APG content of >25% and a pentanol content of 10% in the system. Figure 3 includes the—almost identical—results for two different alkyl polyglycosides, namely a high-purity C_{10} monoglycoside and a $C_{10/12}$ polyglycoside of commercial purity with an average degree of polymerization of 1.3. In the commercial product, the slightly increased degree of polymerization evidently compensates for the somewhat longer alkyl chain length.

Representation of the microemulsion phases as a function of formulation parameters (for example, temperature for systems containing fatty alcohol ethoxylate) and emulsifier concentration, as described in Refs. 1 and 17, has been successful as an aid for practical formulation work. A basically similar picture emerges for emulsions of oil, water, and an emulsifier mixture of

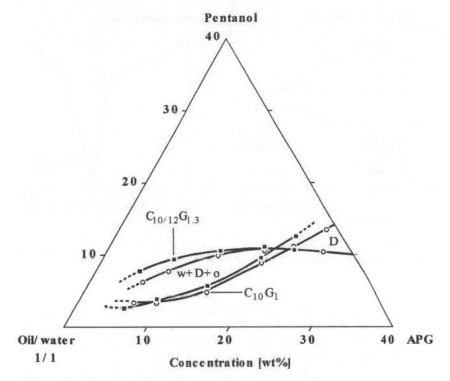

FIG. 3 Pseudoternary phase diagram for the system dodecane/water in a ratio of 1:1, pentanol, C_{10} monoglycoside ($C_{10}G_1$), and $C_{10/12}$ alkyl polyglycoside (APG) ($C_{10/12}G_{13}$) at 40°C [15].

APG and a hydrophobic coemulsifier when, instead of temperature as the formulation parameter, the mixing ratio of APG to hydrophobic coemulsifier is varied [17]. In the case of the specific emulsifier mixing ratio of 1:1, the system of dodecane, water, $C_{12/14}$ APG, and sorbitan monolaurate (SML) as hydrophobic coemulsifier forms microemulsions (Fig. 4) [18]. The emulsions formed with a relatively large SML content are water-in-oil (w/o) emulsions, and the emulsions formed with a relatively large APG content are oil-in-water (o/w) emulsions. On varying the overall emulsifier concentration, a Kahlweit fish again appears in the phase diagram with three-phase microemulsions in its body and a single-phase microemulsion in its tail.

The similarity between alkyl polyglycosides and fatty alcohol ethoxylates is not confined to phase behavior but also applies to the interfacial tension of the emulsifier mixture. With an APG/SML ratio of 4:6, the hydrophilic/lipophilic properties of the emulsifier mixture are balanced and the interfa-

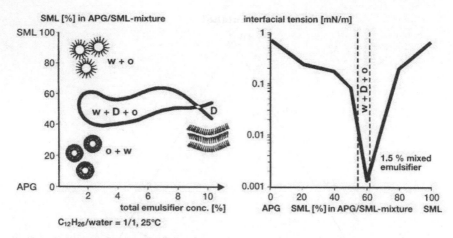

FIG. 4 Phase behavior and interfacial tension of dodecane/water emulsions containing APG/SML mixtures at 25°C [18].

cial tension is minimal. It is remarkable that the APG/SML mixture produces a very low minimum interfacial tension value (around 10^{-3} mN/m), which, once again, is lower by one order of magnitude than that observed in the case of the fatty alcohol ethoxylate system [1,19,20].

In the case of the APG-containing microemulsion, the high interfacial activity is attributable to the fact that the hydrophilic APG with the large polyglycoside head group is present in exactly the right mixing ratio with the hydrophobic coemulsifier SML with its small head group at the oil–water interface. In contrast to ethoxylated nonionic surfactants, hydration and hence the effective size of the head group are hardly dependent on temperature [14,21,22], an attribute that can be utilized for formulating temperature-stable microemulsions [18].

Although for these kinds of emulsifier systems the temperature stability issue is solved, the high concentration of surfactants required to form the microemulsion phase and the sensitivity to the oil structure still exist. Here we describe studies carried out to overcome the difficulties by using specifically designed surfactant mixtures.

II. EXPERIMENTAL

In this study, APGs, glyceryl esters, and fatty alcohol ethoxylates have been used as emulsifiers for microemulsion formation. Emulsifiers were $C_{12/14}$ alkyl polyglycoside (Plantacare 1200 UP) and $C_{8/10}$ alkyl polyglycoside (Plantacare 2000 UP). In addition, the following coemulsifiers were used: glyceryl monooleate (Monomuls 90-O-18), triglyceryl diisostearate (Lameform TGI), glyceryl monohydroxystearate (Rilanit GMHS), isostearyl alcohol ('StOH), and glyceryl monoisostearate (GMI). Furthermore, the emollients dioctyl-cyclohexane (Cetiol S), dicapryl ether (Cetiol OE), and octyldodecanol (Eutanol G) were used. All products were supplied by Henkel. The perfume oil, which was supplied by Dragoco (Holzminden, Germany), was characterized by its dielectric coefficient and the interfacial tension between oil and water. The pH values were adjusted with citric acid.

A. Methods

Phase behavior was identified visually and by polarization microscopy. The emulsion type (o/w, w/o, or microemulsion) was identified by conductivity and turbidity measurements. The emulsion samples—50 mL each—were prepared by mixing the emulsifiers and water for 15 min at 70°C using an Ikavisc MR-D1 stirrer (Jahnke and Kunkel Co, Germany). Afterward, the emollients were added, the pH value was adjusted to 6.0, and the samples were stored for at least 1 day before further characterization. The emulsion type (o/w, w/o) was determined by temperature-dependent conductivity measurements (Radiometer, Denmark). Microemulsions were identified by measuring the turbidity of the sample with a fiber-optic photometer (Metrohm, Germany). The viscosity was determined using a shear rate–controlled rotational rheometer (Rheometrics RFS II) with a plate–plate geometry. The interfacial tension between aqueous and oil phases in emulsions was determined at 25°C by means of a spinning-drop tensiometer (Krüss, Germany) as well as a ring tensiometer (Krüss, Germany). The particle size distributions were studied by performing dynamic light scattering experiments with a Mastersizer 3 (Malvern Co.) The light source was a He-Ne 5-mW laser. The samples were studied at 25°C at an angle of 90°. The dielectric coefficients of the oils at 25°C were determined by means of a Hewlett Packard LCZ Meter.

III. RESULTS AND DISCUSSION

To show the influence of different coemulsifiers, a model emulsion containing $C_{12/14}$ alkyl polyglycoside, a lipophilic coemulsifier, and a certain amount of water and the oil dioctylcyclohexane was chosen. This system can serve as a model emulsion for cosmetic and industrial applications. Thus, the capability of coemulsifiers to facilitate the formation of microemulsions in combination with alkyl polyglycoside [18] could be assessed. The total emulsifier concentration (15%) was high enough to ensure the formation of a one-phase microemulsion and solubilization of both water and emollient if a suitable cosurfactant was used.

The emulsions were characterized by measuring the transparency and electrical conductivity as shown in Fig. 5 [10] for a system containing the coemulsifier glyceryl monooleate (GMO) at 25°C. The GMO-free system is a white o/w emulsion with a conductivity of 400 μS and insufficient storage stability. Addition of GMO reduces the hydrophilicity of the emulsifier system. Consequently, the emulsion undergoes a phase inversion to a w/o emulsion at an alkyl polyglycoside/GMO ratio of roughly 7:3, which is indicated by a drop of the electrical conductivity to zero. A translucent microemulsion is formed for mixing ratios between 6:4 and 7.5:2.5 (Fig. 5). A finely dispersed bluish emulsion is formed in the vicinity of the microemulsion region.

The temperature dependence of the phase behavior of this system is shown in Fig. 6 [10]. The striking features of the phase behavior are the almost negligible influence of temperature and the relatively broad microemulsion region. Phase diagrams such as that shown in Fig. 6 have been determined analogously for systems with various cosurfactants, in which studies of glycerol derivatives were the main focus [10]. The main criteria to distinguish between a microemulsion and a conventional emulsion are the

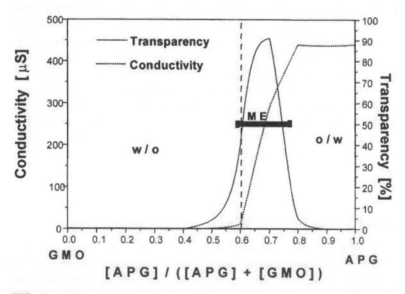

FIG. 5 Electrical conductivity and transparency of a model emulsion comprising $C_{12/14}$ alkyl polyglycoside, glyceryl monooleate (GMO), and a specific amount of water and dioctylcyclohexane at 25°C with varying emulsifier mixture ratios. The dashed line separates the w/o from the o/w region [10].

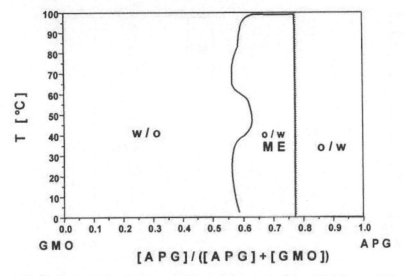

FIG. 6 Phase diagram of a model emulsion comprising 15% $C_{12/14}$ alkyl polygly-coside (APG) and glyceryl monooleate (GMO), 42.5% water, and 42.5% dioctyl-cyclohexane. ME indicates a one-phase microemulsion [10].

transparency and the particle size. A microemulsion is formed if the trans-parency is more than 80%, which is usually combined with an average particle size below 50 nm. Because of the high emulsifier concentration, only one-phase microemulsions are obtained.

Unlike the system with GMO, the formation of a broad microemulsion phase is not observed in the whole temperature range for the phase inversion in the model emulsion containing glyceryl monoisostearate (GMI) (Fig. 7) [10]. Instead, a finely dispersed bluish emulsion is formed at an alkyl poly-glycoside/GMI ratio of about 2:1. A transparent microemulsion can be ob-served only at temperatures above 50°C. The phase inversion concentration ratio does not vary with temperature.

The model emulsion containing glyceryl monohydroxystearate (GMHS) shows a different phase behavior. Apart from a distinct temperature depen-dence similar to the PIT phenomenon at concentration ratios between 2:8 and 7:3, no transparent microemulsion can be observed.

The optimal ratio in the system with triglyceryl diisostearate (TGI) is shifted to the almost balanced value alkyl polyglycoside/GMI = 5.5:4.5. This indicates that TGI is more hydrophilic than GMO and GMI (Fig. 8) [10]. The microemulsion phase extends over the complete accessible temperature range. However, its concentration range is much smaller than for GMO and a small temperature dependence is observed.

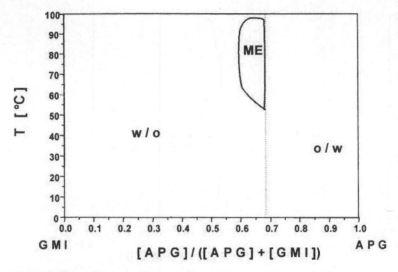

FIG. 7 Phase diagram of a model emulsion comprising 15% $C_{12/14}$ alkyl polygly-coside (APG) and glyceryl monoisostearate (GMI), 42.5% water, and 42.5% dioc-tylcyclohexane. The dotted line separates the o/w from the w/o region. ME indicates a one-phase microemulsion [10].

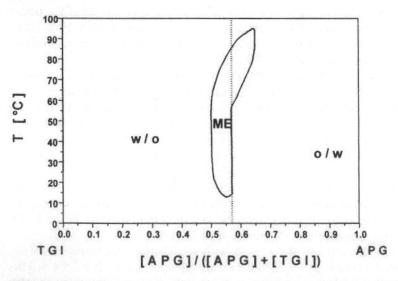

FIG. 8 Phase diagram of a model emulsion comprising 15% $C_{12/14}$ alkyl polygly-coside (APG) and triglyceryl diisostearate (TGI), 42.5% water, and 42.5% dioctyl-cyclohexane. The dotted line separates the o/w from the w/o region. ME indicates a one-phase microemulsion [10].

Finally, isostearyl alcohol is assessed, which is liquid at room temperatures. The phase inversion ratio in the system containing this coemulsifier varies between 7:3 at 25°C and 1:1 at 90°C (Fig. 9) [10]. A bluish emulsion is formed at temperatures below 40°C in the phase inversion region, whereas a translucent microemulsion phase can be identified in the temperature range between 40 and 85°C.

The GMO is by far the most suitable coemulsifier for $C_{12/14}$ alkyl polyglycoside among the candidates investigated. The model emulsions containing this cosurfactant form microemulsions within the broadest concentration range and the temperature dependence can be neglected. Although systems containing the cosurfactants TGI and GMI, in general, have a phase behavior similar to that of GMO with a phase inversion at distinct APG/cosurfactant ratios, a microemulsion is not formed or exists only in a narrow concentration range at room temperature. This can be explained by less efficient incorporation of these cosurfactants into the oil–water interface as compared with GMO because of their branched alkyl chains and therefore bulkier molecules.

A temperature dependence of the phase inversion can be observed for the systems comprising GMHS and ^{i}StOH. A conceivable explanation is that

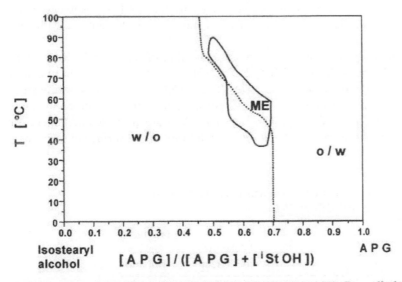

FIG. 9 Phase diagram of a model emulsion comprising 15% $C_{12/14}$ alkyl polyglycoside (APG) and isostearyl alcohol (^{i}StOH), 42.5% water, and 42.5% dioctylcyclohexane. The dotted line separates the o/w from the w/o region. ME indicates a one-phase microemulsion [10].

both materials migrate for the most part not to the surface but are dissolved in the oil phase. They, therefore, behave not as a cosurfactant but more as a cosolvent, altering the polarity of the oil phase. Thus, the oil solubility and, consequently, the interface activity of the alkyl polyglycoside are influenced. A similar effect is reported for systems containing alkyl-β-D-gluco-pyranosides in combination with alkyl ethylene glycol ethers [23].

The study of the phase behavior of a model emulsion is thus extremely helpful in selecting promising emulsifier combinations that can be assigned for further use.

As already mentioned, one outstanding property of microemulsions is their excellent solubilization capacity. This can be explained in terms of very good efficacy of suitable emulsifier combinations resulting in an extremely low interfacial tension between the oil and the aqueous phase. A potential application is, e.g., the solubilization of perfume oils. Perfume oils are rather polar as compared with oils that are usually used in cosmetic formulations, e.g., paraffin or ester oils. This is validated by both the dielectric coefficients and the interfacial tension between oil and aqueous phase (Table 1). In particular, the relatively low interfacial tension without addition of a surfactant indicates that the perfume oil may act, at least partially, as a lipophilic cosurfactant. A similar result was found for geraniol, a doubly unsaturated monoterpene alcohol, which is one of the most used perfume chemicals [24]. It plays a role both as cosurfactant at the interface and as cosolvent in the oil phase in a system consisting of octyl monoglycoside/geraniol/cyclohexane/water.

A model formulation was selected consisting of 20% emulsifier mixture (alkyl polyglycoside and GMO), 20% perfume oil, less than 2% oil (dicapryl ether and octyldodecanol), and water. The GMO was chosen as coemulsifier because it proved to be most suitable for nonpolar oil. No microemulsion is formed if $C_{12/14}$ alkyl polyglycoside and GMO are used as emulsifiers for this system. It was reported that the oil solubility of the surfactant in systems

TABLE 1 Dielectric Coefficient and Interfacial Tension Between Water and Oils at 25°C

Oil	Dielectric coefficient	Interfacial tension (mN/m)
Dioctyl cyclohexane	1.2	25
Dicapryl ether/octyldodecanol	1.5	20
Perfume oil	7.5	6

containing alkyl-β-D-glucopyranosides was a crucial parameter for the formation of microemulsions. The extent of the microemulsion region became larger when the alkyl chain length and, therefore, the hydrophobicity of the alkyl-β-D-glucopyranoside decreased [25]. Consequently, the more hydrophilic $C_{8/10}$ alkyl polyglycoside with shorter alkyl chain length as compared with $C_{12/14}$ alkyl polyglycoside was used as an additional emulsifier in our model system.

The phase behavior of the emulsifier system with fixed oil and water contents is shown in Fig. 10. A microemulsion is formed at a GMO concentration between 15 and 25% of the total emulsifier system with $C_{12/14}$ alkyl polyglycoside/$C_{8/10}$ alkyl polyglycoside in a suitable mixing ratio. The necessary amount of GMO differs only slightly from the optimal alkyl polyglycoside/GMO ratio that was found in the simple model emulsion as discussed earlier.

Figure 11 shows the phase behavior of a decyl glucoside/GMO/dioctyl cyclohexane system and the effect of diluting the system with water. In each

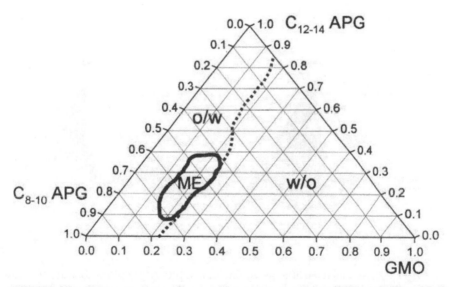

FIG. 10 Pseudoternary phase diagram of a system containing 20% emulsifier (alkyl polyglycoside, GMO), 20% perfume oil, 0.6% oil (dicapryl ether, octyldodecanol), and 59.4% water at 25°C with fixed oil and water contents showing the chain length dependence of the alkyl polyglycosides. The respective compositions of alkyl polyglycosides are obtained by mixing $C_{12/14}$ alkyl polyglycoside and $C_{8/10}$ alkyl polyglycoside in suitable ratios. The dotted line separates the o/w from the w/o region. ME indicates a one-phase microemulsion [10].

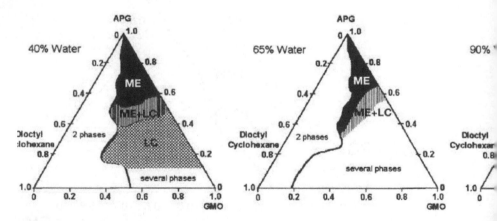

à. 11 Phase behavior of a decyl glucoside (APG)/glyceryl monooleate (GMO)/dioctyl cyclo

1s of 40, 65, and 90% at 25°C.

of the three dilutions in Fig. 11, a one-phase microemulsion is observed at high weight fractions of decyl glucoside. In the most concentrated system, with 40% water present, the microemulsion phase is capable of solubilizing 15% of the oil (dioctyl cyclohexane). Diluting to 90% water, we still have a microemulsion phase containing 3.5% oil.

By decreasing the amount of APG while increasing the GMO weight fraction, a liquid crystalline/microemulsion mixed phase is obtained. In the 40% water case, at even higher GMO weight fractions, a pure liquid crystalline phase is observed. For 65% water no pure liquid crystalline phase is seen, and the system becomes unstable and phase separates. For 90% water a finely dispersed emulsion with a bluish shine is observed; at higher GMO weight fractions the surfactant system becomes too lipophilic to stabilize the system and, consequently, phase separation occurs.

Similar phase diagrams were obtained for isopropyl myristate (IPM) as the oil instead of dioctyl cyclohexane. The IPM has a more polar nature than dioctyl cyclohexane, resulting in a comparable but slightly different phase behavior (Fig. 12). Single microemulsion phases are again observed in Fig. 12 at high decyl glucoside weight fractions. The most concentrated system (40% water) can incorporate 16% of the oil, IPM. AT 90% dilution, 3.8% IPM is present in the microemulsion. At increasing GMO weight fractions, again a microemulsion/liquid crystalline mixed phase is observed for 40 and 65% water. For 40% water a pure liquid crystalline phase is also seen but is much smaller than in the dioctyl cyclohexane system. For the higher water content phase diagrams a finely dispersed emulsion phase is again seen at higher IPM weight fractions.

The addition of a nonionic ethoxylate surfactant, e.g., laureth-4, to a decyl glucoside (APG)/glyceryl monooleate (GMO)/dioctyl cyclohexane/water system was investigated. The nonionic ethoxylate surfactant was used to expand the microemulsion zone. The APG/GMO ratio was held constant at 2:1, and the oil and ethoxylate weight fractions relative to the APG/GMO mixture were varied at 50% water. From the phase diagram (Fig. 13) (which shows actual total weight percent composition, rather than weight fractions) it is clear that 30% oil (dioctyl cyclohexane) is present in the single-phase microemulsion at the optimal surfactant ratio. This is twice as much as for the system without the nonionic ethoxylate present (Fig. 11). Finely dispersed bluish emulsion phases border on the microemulsion region at higher laureth-4 concentrations and at high APG/GMO levels. A liquid crystalline phase is also observed at higher laureth-4 concentrations.

As can be seen from Fig. 14, the microemulsion phase at 25°C is fairly large and shrinks slightly with increasing temperature. The shrinkage effect is much less than in Fig. 1, where the microemulsion phase was stable only within a narrow temperature range. The addition of the APG/GMO mixture

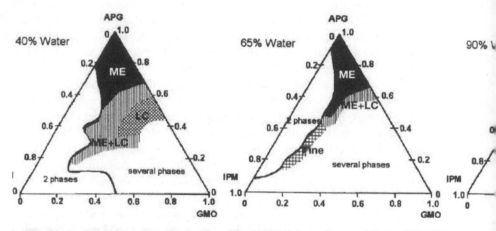

㎓. 12 Phase behavior of a decyl glucoside (APG)/glyceryl monooleate (GMO)/isopropyl
ccentrations of 40, 65, and 90% at 25°C.

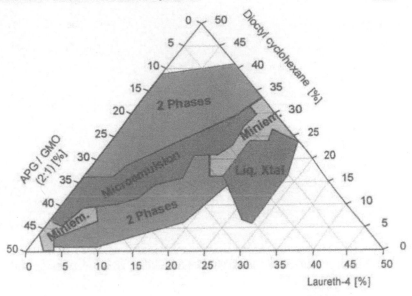

FIG. 13 Phase behavior of a decyl glucoside (APG)/glyceryl monooleate (GMO)/laureth-4/dioctyl cyclohexane system at a water concentration of 50% at 25°C.

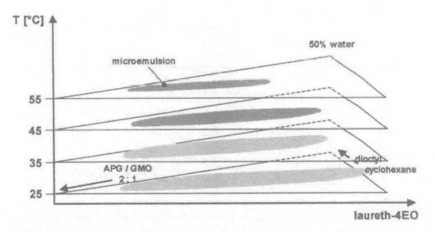

FIG. 14 Temperature effect on the microemulsion phase of a decyl glucoside (APG)/glyceryl monooleate (GMO)/laureth-4/dioctyl cyclohexane system at a water concentration of 50%.

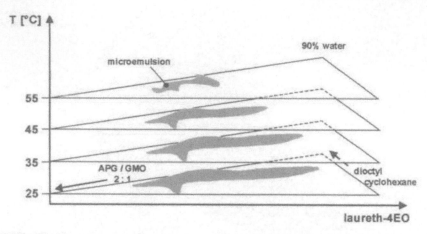

FIG. 15 Temperature effect on the microemulsion phase of a decyl glucoside (APG)/glyceryl monooleate (GMO)/laureth-4/dioctyl cyclohexane system at a water concentration of 90%.

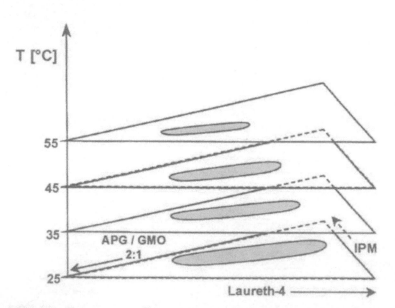

FIG. 16 Temperature effect on the microemulsion phase of a decyl glucoside (APG)/glyceryl monooleate (GMO)/laureth-4/isopropyl myristate (IPM) system at a water concentration of 50%.

renders the microemulsion more temperature stable than an ethoxylated nonionic surfactant without any additives.

Comparing Fig. 15 with Fig. 14, it is observed that on diluting the 50% system to a 90% water system the microemulsion phase is slightly smaller. This is due to the significantly decreased emulsifier concentration. However, even at higher temperatures this microemulsion phase is still observed, although shrinkage with temperature is similar to that observed before.

It is observed that the microemulsion phase is somewhat smaller with IPM as the oil (Fig. 16). The emulsifier system is more suited for the solubilization of less polar oils. Some characteristic shrinkage of the microemulsion phase with increasing temperature occurs as well.

IV. CONCLUSIONS

It has been seen that microemulsions containing APG are much less temperature sensitive than microcmulsions containing just nonionic ethoxylates. The addition of a nonionic ethoxylate to an APG/glyceryl monooleate system dramatically increases the existence range of the single-phase microemulsion and allows the amount of oil solubilized in the microcmulsion system to increase. Diluting the emulsions to 90% water still resulted in a stable single microemulsion phase. The polar oil isopropyl myristate resulted in a smaller microemulsion phase than for nonpolar oils for the selected emulsifier system.

V. FUTURE RESEARCH

With the surfactant mixtures discussed here, microemulsions with high temperature stability can be obtained. In addition, the microemulsions are dilutable up to 90% water and show varying characteristics with different types of oil. The focus of future research should be to decrease the amount of surfactant required still further in order to make these microemulsions suitable and interesting for certain industrial applications.

REFERENCES

1. M Kahlweit, R Strey. Angew Chem 97:665, 1985.
2. K Shinoda, H Kunieda. In: P Becher, ed. Encyclopedia of Emulsion Technology. Vol 1. New York: Marcel Dekker, 1983, p 337.
3. Th Förster, W von Rybinski, A Wadle. Adv Colloid Interface Sci 58:119, 1995.
4. E Nürnberg, W. Pohler. Prog Colloid Polym Sci 69:48, 1984.
5. H Sagitani. J Dispersion Sci Technol 9:115, 1988.
6. F Comelles, V Megias, J Sanchez, JL Parra, J Coll, F Balaguer, C Pelejero. Int J Cosmet Sci 11:5, 1989.

7. T Suzuki, M Nakamura, H Sumida, A Shigeta. J Soc Cosmet Chem 43:21, 1992.
8. M Kahlweit, R Strey. Angew Chem 97:655, 1985 and R Aveyard, B Binks, P Fletcher. Langmuir 5:1210, 1989.
9. EM Kutschmann, GH Findenegg, D Nickel, W von Rybinski. Colloid Polym Sci 273:565, 1995.
10. W von Rybinski, B Guckenbiehl, H Tesmann. Colloids Surf A 142:333–342, 1998.
11. G Platz, C Thunig, J Pölicke, W Kirchhoff, D Nickel. Colloids Surf A 88:113, 1994.
12. GG Warr, CJ Drummond, F Grieser, BW Ninham, DF Evans. J Phys Chem 90:4581, 1986.
13. WD Clemens. Ber Forschungszentr Julich 3028:1, 1994.
14. D Balzer. Tenside Surfactants Deterg 28:419, 1991.
15. M Kahlweit, G Busse, B Faulhaber. Langmuir 11:3382, 1995.
16. H Kahl, K Kirmse, K Quitzsch. Tenside Surfactants Deterg 33:26, 1996.
17. Th Förster, B Guckenbiehl, H Hensen, W von Rybinski. Prog Colloid Polym Sci 101:105, 1996.
18. Th Förster, B Guckenbiehl, A Ansmann, H Hensen. Seife Ole Fette Wachse 122:746, 1996.
19. F Jost, H Leiter, MJ Schwuger. Colloid Polym Sci 266:554, 1988.
20. R Aveyard, BP Binks, PDI Fletcher. Langmuir 5:1210, 1989.
21. K Fukuda, O Söderman, B Lindman, K Shinoda. Langmuir 9:2921, 1993.
22. R Hofmann, D Nickel, W von Rybinski, G Platz, J Pölicke, C Thuming. Prog Colloid Polym Sci 93:320, 1993.
23. LD Ryan, EW Kaler. Langmuir 13:5222, 1997.
24. C Stubenrauch, B Paeplow, GH Findenegg. Langmuir 13:3652, 1997.
25. LD Ryan, K-V Schubert, EW Kaler. Langmuir 13:1510, 1997.

20

Microemulsions in Foods: Challenges and Applications

ANILKUMAR G. GAONKAR Kraft Foods, Inc., Glenview, Illinois, U.S.A.

RAHUL PRABHAKAR BAGWE University of Florida, Gainesville, Florida, U.S.A.

ABSTRACT

The objective of this chapter is to review the literature in the area of food microemulsions. Emphasis is placed on the edible microemulsions and their applications in foods and beverages. In spite of a vast potential for many promising applications in foods and beverages, microemulsions have not been widely used because of the difficulty in their formulation using completely edible food ingredients. In order to apply the microemulsion technology in common foods, more efforts should be directed toward both fundamental and applied research on food-grade microemulsions.

I. INTRODUCTION

The aim of this chapter is to give a brief outline of what the food microemulsions are, their current state of the art, and potential applications in food. Microemulsions are thermodynamically stable, transparent, and homogeneous single-phase solutions of oil, water, and surfactant. This concept also embraces aqueous micellar solutions containing solubilized lipids and reverse micellar solutions containing solubilized water. The characteristics of a microemulsion that find applications in the food industry are (1) transparent quality, which is helpful in clear beverages and in food analysis; (2) small drop size, which provides excellent contact between the lipid and aqueous phases, thereby playing a role in flavor release and perception; (3) enhanced solubilization of vitamins, flavors, and other nutrients whereby the solubilized components are protected from unwanted degradative reactions;

(4) super stability (unaffected by destabilization mechanisms such as creaming, flocculation, and coalescence), which is an added benefit during food processing and storage; and (5) ability to incorporate solutes within the dispersed droplets and action as reaction and/or extraction media. A disadvantage of the thermodynamic stability is that the microemulsions are formed only in specific ranges of temperature, pressure, and composition. Therefore, the food processor does not have complete freedom in designing the food systems.

Microemulsions have distinct advantages over emulsions in providing better contact between the oil and water phases within foods. Principally, microemulsions have a vast potential for many promising applications in foods and beverages. However, they have not been widely used in the food industry so far because of the difficulties in their formulation using completely edible ingredients: (1) the oils used are mostly triglycerides containing long-chain fatty acids, which are semipolar compared with the hydrocarbons and are too bulky to penetrate the interfacial film to assist in the formation of an optimal curvature; (2) the choices of edible surfactants, cosurfactants, and other additives that can be used in foods are very limited; and (3) the concentrations of such surfactant and cosurfactant should be low so as not to impart taste and/or odor to foods.

A literature search (Chemical Abstract Services) covering 33 years (1967 to April 2000) gave the following results:

81842 entries for "Emulsions"
3545 entries for "Food" and "Emulsions"
6387 entries for "Microemulsion"
Only 91 entries for "Food" and "Microemulsion"

It should be emphasized that even among the 91 entries for the food microemulsions, very few of them were made using food-grade, GRAS (generally recognized as safe) ingredients. In addition, out of the vast number of reviews written in the field of emulsions, only three dealt with food microemulsions [1–3]. The literature on food microemulsions is meager, and more work should be done in this area if we wish to increase the application of this innovative technology in foods.

II. SURFACTANTS, COSURFACTANTS, AND PHASE BEHAVIOR

The structure and properties of oil components are fixed because the oils used in foods are triglycerides. Temperature could not be varied to favor microemulsions because the food or beverage has to be stable at the storage temperatures. Hence, the nature and concentration of surfactant become important. Most food surfactants are esters of fatty acids with naturally occurring alcohols. With a few exceptions, food surfactants are mostly nonionic. The surfactants should not impart taste or odor to foods. Some of the surfactants that are used in foods are listed in Table 1 [4].

The geometry of the surfactant determines whether a water-continuous (oil-in-water or o/w) or oil-continuous (water-in-oil or w/o) microemulsion will form at equilibrium. It is known that the surfactants with bulky head groups and nonbulky tail form micelles or o/w microemulsions as it is easier for the short tail to fit within a relatively confined core. The surfactants with small head groups and large tail favor reverse micelles or w/o microemulsions because tails are widely spaced and head groups are constrained. The bulkiness of the tail group can be increased by (1) having two or more tails (e.g., Aerosol-OT, lecithins) in the molecule, (2) adding a second long-chain (or medium chain) molecule known as a cosurfactant, and (3) introducing a "kink" through unsaturation in the tail groups.

The spontaneous curvature, H_0, of the surfactant monolayer at the oil-water interface is depicted in Fig. 1. The H_0 dictates the phase behavior of a microstructure. Hydrophilic surfactants produce o/w microemulsions ($H_0 > 0$), whereas lipophilic surfactants produce w/o microemulsions ($H_0 < 0$). When the hydrophilic–lipophilic tendencies of the surfactant monolayer at the oil–water interface are balanced, a middle-phase microemulsion is formed ($H_0 = 0$). Under these balanced conditions, $H_0 \approx 0$, and maximum solubilization of the oil with a minimum amount of surfactant is achieved. The middle-phase microemulsion may coexist with excess water and oil phases, and ultralow tensions between the coexisting phases are attained. The surfactant packing parameter, defined by Israelachvili [5] as $v/a_0 l_c$ where v is the surfactant molecular chain volume, a_0 is the area per surfactant head group, and l_c is the surfactant alkyl chain length, may be useful in predicting the type of microemulsion more likely to be formed in a given system. This parameter is related to the properties of the surfactant film. For $v/a_0 l_c < 1$, the surfactant prefers curvature toward the oil, whereas for $v/a_0 l_c > 1$ it prefers to curve toward the water. For quantitative analysis it is advantageous to consider the spontaneous curvature, H_0, as the basic property of a film.

The cosurfactant molecules are polar enough to be surface active and reside between the tails of the surfactants, thereby increasing the bulkiness of the tail group region. Cosurfactants have widely different hydrocarbon moiety sizes compared with the surfactants. Typical examples of cosurfactants are medium-chain alcohols, acids, and amines. The role of a cosurfactant is to (1) lower the interfacial tension down to a very small (near zero) value and increase the fluidity, (2) adjust the hydrophile–lipophile balance (HLB) and spontaneous curvature of the interface by controlling surfactant partitioning, (3) destroy the liquid crystalline and/or gel structures, and (4)

TABLE 1 Surfactants Used in Foods

Emulsifier	US 21 CFR[a]	Canadian[b]	EU no.
Mono- and diglycerides (GRAS)[c]	182.4505	M.4, M.5	E471
Succinyl monoglyceride	172.830		
Lactylated monoglyceride	172.852	L.1	E472
Acetylated monoglyceride	172.828	A.2	E472
Monoglyceride citrate	172.832		E472
Monoglyceride phosphate (GRAS)	182.4521	A.94, C.7	
Stearyl monoglyceride citrate	172.755		E472
Diacetyl tartarate ester of monoglyceride (GRAS)	182.4101	A.3	E472
Polyoxyethylene monoglyceride	172.834		
Polyoxyethylene (8) stearate		P.5	
Propylene glycol monoester	172.854	P.14	E477
Lactylated propylene glycol monoester	172.850		
Sorbitan monostearate	172.842	S.18	E491
Sorbitan tristearate		S.18B	
Polysorbate 60	172.836	P.3	E435
Polysorbate 65	172.838	P.4	E436
Polysorbate 80	172.840	P.2	E433
Calcium stearoyl lactylate	172.844		E482
Sodium stearoyl lactylate	172.846	S.15a	E481
Stearoyl lactylic acid	172.848	L.1A	
Stearyl tartarate			E483
Stearyl monoglyceride citrate	172.755	S.19	
Sodium stearoyl fumarate	172.826		
Sodium lauryl sulfate	172.822		
Dioctyl sodium sulfosuccinate	172.810		
Polyglycerol esters	172.854	P.1A	E475
Sucrose esters	172.859	S.20	E473
Sucrose glycerides			E474
Lecithin (GRAS)	184.1400	L.2	E322
Hydroxylated lecithin	172.814	H.1	E322
Triethyl citrate (GRAS)	182.1911		

[a]United States Code of Federal Regulations, Volume 21.
[b]Canadian Food and Drug Regulations, Table IV, Div. 16.
[c]Generally recognized as safe.
Source: Ref. 4.

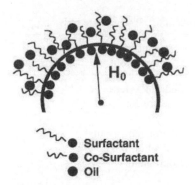

FIG. 1 Optimal curvature, H_0 = radius of spontaneous curvatures of internal phase drops. (From Ref. 3.)

decrease sensitivity to compositional fluctuations. It is known that the use of a cosurfactant increases the microemulsion region in the phase diagram.

III. SOLUBILIZATION OF TRIGLYCERIDES

Very little work has been done so far on the solubilization of triglycerides in water for the reasons mentioned in Section I. Even among the studies that employed triglycerides as an oil phase, only a few used food-grade surfactants and cosurfactants. With the available food-grade surfactants, the areas of solubilized oil phases are restricted to the water corner of the phase diagram, which permits solubilization of only a small amount of oil. The microemulsions containing triglycerides and a nonfood component (surfactant, cosurfactant, or cosolvent) may have some value in understanding the fundamentals of food microemulsions. Some studies related to nonedible microemulsions containing triglyceride oils are listed in Table 2.

Gulik-Krzywidki and Larsson [7], Hernqvist [15], Engstrom [16], and Larsson [17] reported studies that used all food-grade components to formulate w/o microemulsions (L_2 phases). They found that a thermodynamically stable oil-continuous phase could be formed with water, triglyceride, and monoglycerides. A representative phase diagram is depicted in Fig. 2.

Figures 3 and 4 depict the phase behaviors of various triglycerides, ethoxylated mono/diglycerides, and water in conjunction with hydrotropes such as ethanol, propylene glycol, and sucrose [13,18]. It can be seen from Fig. 3 that a w/o microemulsion (L_2 phase) was easily formed at a ratio of 75:25 wt% of ethoxylated mono/diglycerides. In a certain region of the phase diagram, blue phase, droplets separated by lamellar liquid crystals were observed. Ethanol was reported to act synergistically with sucrose to destabilize

TABLE 2 Some Nonedible Microemulsions Containing Triglyceride Oils

Oil	Surfactant(s)	Cosurfactant/cosolvent	Reference
Tricaprylin	Monocaprylin, alkyl aryl polyglycol ether	Sodium xylene sulfonate	6
Soybean oil	Monoglyceride	—	7
Canola oil	Distilled monoglycerides	Isopropanol, t-butyl alcohol, hexanol	8
Canola oil	Acetic acid ester of monoglycerides	Isopropanol	9
Soybean oil	O-Alkyl-3-D-glucose	Ethanol	10
Medium-chain triglycerides, peanut oil	Ethoxylated fatty alcohol, ethoxylated phenol	Lauryl alcohol	11
Medium-chain triglycerides, peanut oil	Sodium oleate	Butanol, pentanol	12
Soybean oil	Polyoxyethylene (40) sorbitol hexaoleate	Ethanol	13
Saffola oil	Triton X-100	Butanol	14

the liquid crystalline mesophases, thus promoting the formation of triglyceride microemulsions, which was confirmed by X-ray diffraction and polarized light microscopy. A short-chain alcohol such as ethanol is believed to create a mixed solvent system with water, thereby making it easier to dissolve the oil compared with pure water. Further, comparing Fig. 4a and c, one observes improved solubilization, as indicated by an increase in the L_1 region (o/w microemulsion), of soybean oil (containing unsaturated fatty acids) in water compared with hydrogenated soybean oil (saturated fatty acids).

In recent work Garti and coworkers [19] concluded that food-grade microemulsions were difficult to formulate from a simple three-component system based on water, oil, and a single surfactant and that short-chain alcohols and polyols were needed. They studied the phase diagrams of five-component systems containing water, oil [such as medium-chain triglycerides or $R(+)$-limonene], short-chain alcohol (such as ethanol), polyols (such as propylene glycol and glycerol), and surfactants (such as ethoxylated sorbitan esters, polyglycerol esters, and sugar esters). The medium-chain alcohol and polyols modify the interfacial spontaneous curvature and the flexibility of

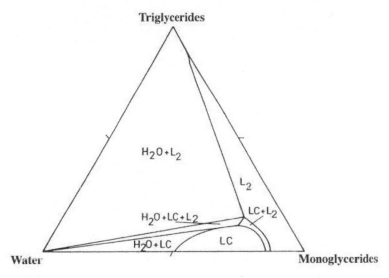

FIG. 2 Phase diagram of a ternary system: sunflower oil-based monoglycerides, triglycerides (soybean oil), and water at 40°C. (From Ref. 17.)

the surfactant film, thus enhancing the oil solubilization capacity of the o/w microemulsions.

The effects of different parameters such as HLB, temperature, and chain length of alcohol on the phase behavior of sucrose ester–containing systems have also been reviewed by Garti et al. [20].

IV. LECITHIN-BASED MICROEMULSIONS

Some studies carried out using lecithin are described in this section as it is one of the important food-grade (GRAS) surfactants. In one of the early studies, Ekman and Lundberg [21] showed that a maximum of 15% (w/w) triolein could be incorporated into the lamellar liquid crystalline phase of hydrated egg lecithin. The solubilization of tributyrin in diheptanoylphosphatidylcholine micelles has been studied by Burns and Roberts [22] and Lin et al. [23] using nuclear magnetic resonance (NMR) and small-angle neutron scattering (SANS) techniques, respectively.

Lecithin is too lipophilic to form spontaneously a mean zero curvature needed for balanced microemulsions. However, by adjusting the polarity of the polar solvent, balanced microemulsions can be obtained. Shinoda et al. [24] showed that when a short-chain alcohol was added as a cosolvent, lecithin formed microemulsions at low amphiphile concentrations. They studied the phase diagram of a lecithin, 1-propanol, water, and n-hexadecane

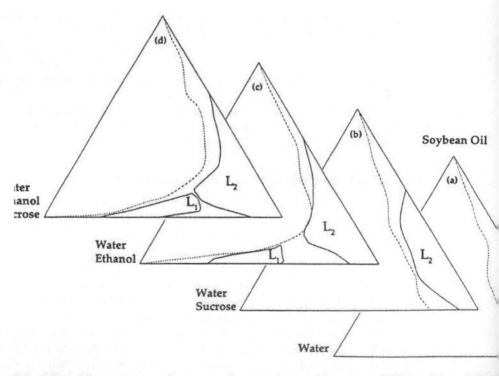

i. 3 Effect of temperature and aqueous phase on phase regions. (———) 35°C and (· · · ·) 40°C
rose, (c) 80/20 wt% water/ethanol, and (d) 70/20/10 wt% water/ethanol/sucrose. (From Ref.

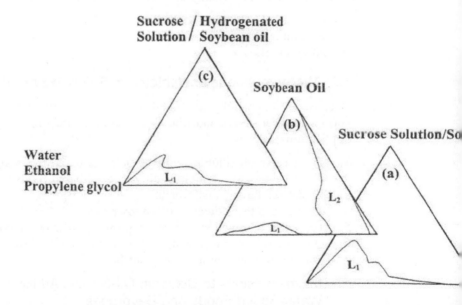

FIG. 4 Effect of soybean oil on isotropic microemulsion regions, L_1 and L_2, at 3[...]taining 75/25 wt% ethoxylated monoglycerides/monoglyceride, 70/25/5 wt% water/e[...] (a) 90/10 wt% mixture of 60 wt% sucrose solution/soybean oil. (b) Same as (a) [...] (c) same as (a) but replacing soybean oil with partially hydrogenated soybean oil.

system and found that the curvature changed progressively with the addition of alcohol. At a low alcohol content, the microemulsion is rich in oil (w/o); at an intermediate alcohol level, it is bicontinuous; and at a high alcohol level, it changes to water rich (o/w). The addition of cholesterol was found to expand the microemulsion region and decrease the gel region in the phase diagram. Gel structure was formed upon increasing the water content.

Aboofazeli et al. [25] and Leser et al. [26] have discussed microemulsions containing water, lecithin, butanol, and long-chain triglycerides. The presence of butanol, propanol, and hexanol makes it unsuitable for food application, and the absence of these alcohols provides a mesophase along with small areas of oil solubilization within the water–surfactant phase. Certain systems could not be diluted with water and hence these microemulsions were not suitable for food applications.

V.   APPLICATION OF MICROEMULSIONS IN FOODS

Microemulsions have many promising applications in the food industry:

As a vehicle to solubilize additives
As reaction media
As extraction media

A.   Microemulsions as Vehicles to Solubilize Additives

Because of their ability to solubilize additives in the core and/or palisade surfactant layer, microemulsions find use as a solubilizing agent in various applications such as

1.   To disperse oil-soluble additives in water-based foods and beverages
2.   To disperse water-soluble additives in oil-based foods
3.   To deliver flavors and aromas
4.   To increase the efficiency of antioxidants
5.   To enhance browning and crisping during microwave cooking
6.   Quick thawing of a frozen food in a microwave oven
7.   Wax microemulsion as a moisture barrier

1.   Microemulsions to Disperse Oil-Soluble Additives in Water-Based Foods and Beverages

The essential oils are a source of many flavors for numerous foods and beverages. Many of these flavors are insoluble in water. In order to disperse them homogeneously in water-based foods or beverages, they have to be mixed in certain solvents such as alcohols. The alternative way to ensure dispersion is to form an o/w emulsion. However, emulsions are inherently unstable and hence ringing may occur over time when added to a beverage.

In addition, because of their milky appearance, emulsions are not suitable if clarity is a desired product attribute. The o/w microemulsions or micelles can greatly increase the solubility of flavor compounds, vitamins, and other additives that are insoluble or marginally soluble in water by solubilizing them in the hydrophobic core.

Table 3 provides some examples of studies dealing with solubilization of oil-soluble flavors in o/w microemulsions or micelles [27,33]. Slocum et al. [33] studied the solubilization capacity of oil-soluble flavors in micelles. The solubilization capacity of various flavors was found to depend on the chemical nature of the flavor molecules. An increase in the chain length and/ or unsaturation in the hydrocarbon chain of the flavor decreases its solubilization capacity. The polarity of the flavor affects the location of its solubilization. Nonpolar flavor molecules go deep in the core, while the polar

TABLE 3 Some Studies Dealing with Solubilization of Oil-Soluble Flavors and Aromas in Oil-in-Water Microemulsions or Micelles

Oil	Surfactant(s)	Cosurfactant/cosolvent	Reference
Oils of peppermint, clove, and lavender	Tweens (polysorbates)	—	27
Flavor oils	Fatty acid ester of alkoxylated phenol	—	28
Edible oils and fats, flavor oils (orange, cinnamon, lemon, etc.)	Edible emulsifiers (HLB = 13–16)	Ethylene glycol, propylene glycol, glycerol, sugar alcohols, etc.	29
Aromatized coffee oil, egg flavor	Tweens (polysorbates)	Ethanol/medium-chain alcohol	30
Ethyl n-butyrate Ethyl benzoate	Sodium dodecyl sulfate (SDS)	Hexanol	31
Spearmint oil	Polysorbates and ethoxy hydrogenated castor oil	—	32
Model flavor compounds (aldehydes, ketones, alcohols and ethyl esters), orange oil	SDS, polysorbate 20, and sodium laurate	—	33

flavor molecules locate in the palisade layer of the surfactant. Figure 5 summarizes the effect of hydrocarbon chain length on the solubilization capacity of different classes of flavor compounds in 0.1 M Tween 80 (polysorbate 80) solution. Flavor solubilization also depends on the hydrocarbon chain length and the nature of the polar head groups of the surfactant because these parameters influence both the size and shape of the aggregates formed. Nonionic surfactants usually form bigger aggregates than the anionic surfactants.

Loss of vitamin A occurs if a conventional vitamin A concentrate is added to whole milk prior to fat separation. This leads to underfortification of vitamin A in low-fat and skim milks. Duxbury [34] reported the use of a vitamin-solubilized o/w microemulsion to disperse vitamin A in milk and maintain its fortification level. Chiu and Jiang [35] reported the formation of a w/o microemulsion in which oil-soluble vitamin E was solubilized in the aqueous phase. However, the surfactants used were not food grade.

2. Microemulsions to Disperse Water-Soluble Additives in Oil-Based Foods

Some functional substances such as aromas, flavors, flavor precursors, salts, minerals, vitamins, antioxidants, enzymes, proteins, and amino acids that are

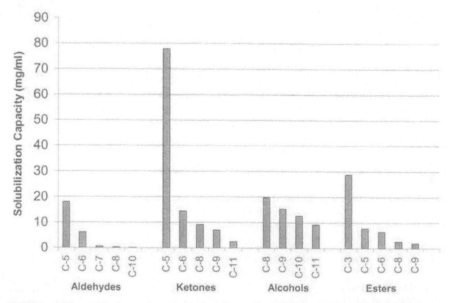

FIG. 5 Effect of hydrocarbon chain length on the solubilization capacity of different classes of flavor compounds in 0.1 M Tween 80 solution. (From Ref. 33.)

required to be homogeneously dispersed in oil-based foods are soluble in water and not in oil. Hence, it is not possible simply to mix these water-soluble substances into the oil phase. Attempts to incorporate these water-soluble substances into oil in the form of a w/o emulsion have not proved very successful. A w/o microemulsion or a reverse micelle seems to be an excellent system to overcome this problem because the water-soluble additives can be solubilized in these systems.

Because of the presence of extremely small aqueous droplets, a w/o microemulsion is an excellent medium to minimize spattering during open pan frying. El-Nokaly et al. [3,36] formulated edible Crisco™ oil–based w/o microemulsions containing solubilized water-soluble additives such as nutrients, vitamins, flavors, or flavor precursors in the oil using a well-selected combination of oils, polar liquids, and surfactants. Water-soluble additives that can be dispersed in the aqueous core of microemulsions can be the following:

1. Natural flavors: derived from leaves, seeds, fruits, or animal materials
2. Artificial flavors: prepared by chemical synthesis
3. Salts: NaCl, KCl, sodium aspartame, and monosodium glutamate (MSG)
4. Flavor precursors that react with heat to form a flavor, e.g., furanone, cysteine, and methionine
5. Browning aids: amino acids and reducing sugars, e.g., fructose and dextrose (Maillard reaction)
6. Vitamins and minerals: vitamin C and calcium
7. Antioxidants: (protect oil from turning rancid), e.g., ascorbic acid and α-tocopherol
8. Water-soluble enzymes, proteins, and amino acids

Food-grade w/o microemulsions containing functional components that can be used in confectionery, margarine, dressings, shortenings, and solid fats were disclosed by Kirby and Needs [37]. Using food-grade emulsifiers such as poly (tri or tetra) glycerol esters, they formulated w/o microemulsions that were resistant to oxidation. Antioxidants, namely ascorbic acid and α-tocopherol, were incorporated in the microemulsion to prevent oxidation of the lipid medium. The antioxidants acted synergistically.

Lester [38] disclosed an edible w/o microemulsion formulation comprising up to 33% water compared with only 5% water solubilization reported by El-Nokaly et al. [3]. Phospholipids and organic acid esters of monoglycerides were used as the surfactants.

3. Microemulsions for Delivery of Flavors and Aromas

Consumer perception of a food product is often significantly improved if the product gives off a pleasant aroma during cooking or on the table. Hence, flavor delivery is an important attribute of a food or beverage. In situ formation of fresh aromas and flavors can be accomplished by rapid delivery of flavor moieties derived from enzymes and precursors just prior to eating.

Taylor et al. [39] formulated flavor-releasing compositions comprising a w/o microemulsion and/or hydrated reverse micelles. These formulations are suitable not only for use as flavor or aroma delivery agents in foods but also for the enzymatic synthesis of various flavors and flavor precursors in vitro. A typical composition had 80% vegetable oil, 15% surfactant (phosphatidylcholine, phosphatidylethanolamine, monoglyceride, and sorbitan ester), 3% ethanol, <3% water, 1% precursor, and 0.5% enzyme. A flavor precursor and an enzyme [inactive, low water activity (a_w)] were present in the core of the microemulsion. When the microemulsion is eaten, higher water activity causes swelling and disruption of microdroplets and the enzyme acts on the precursor to produce the desired flavor upon hydration of the core. Thus, a microemulsion can be exploited to maintain the flavor precursors stably prior to eating while permitting rapid delivery of flavor moieties derived from precursors in the mouth or shortly before eating.

An edible o/w emulsion preconcentrate formulation containing a hydrolyzed fat (melting point = 30–40°C), an aroma or flavor, and a surfactant (polyglycerol mono/diester) was disclosed by Chmiel and coworkers [40,41]. Upon heating the food product to above the melting point of the hydrolyzed fat, the emulsion preconcentrate mixes with the aqueous phase in the food and spontaneously forms an emulsion with microemulsion characteristics that rapidly release the aroma. The utility of this invention was shown for frozen dinners and frozen pizza. For chilled or frozen foods, hydrolyzed fat with a lower melting point can be used.

Chmiel et al. [42] also formulated an emulsion preconcentrate consisting of hydrolyzed coffee oil and coffee aroma and mixed it with a soluble coffee powder to provide a soluble coffee product with an enhanced "preparation aroma." Upon dissolution of the soluble coffee product in hot water, the emulsion preconcentrate spontaneously forms an o/w emulsion including droplets in the microemulsion range. These dispersed oil phase droplets containing aroma provide a burst of coffee aroma.

4. Microemulsion to Increase Efficiency of Antioxidants

Lipid oxidation is a major cause of deterioration of quality in products containing unsaturated fats and oils, as they are highly sensitive to oxidation. Even a small amount of lipid oxidation products can cause a significant reduction in sensory quality and can have negative physiological effects. This poses restrictions on the formulation of foods containing nutritionally important lipids such as polyunsaturated fatty acids (PUFAs), which are known to prevent cardiovascular diseases. Antioxidants are, therefore, added to the food to retard the development of rancid off-flavors.

Cort [43] reported that ascorbic acid at 0.02% was more active than butylated hydroxyanisole (BHA) or butylated hydroxytoluene (BHT) at the same concentration in soybean oil. Unfortunately, ascorbic acid is insoluble in fats and oils and cannot be used as an antioxidant in an oil phase. However, it can be made water dispersible by solubilizing in a w/o microemulsion or reverse micelles. Ruben and Larsson [44] showed that the efficiency of antioxidants could be enhanced by their localization at the interface. Hence, a w/o microemulsion is an ideal system for enhancing the antioxidation behavior of water-soluble antioxidants. Both water- and oil-soluble antioxidants can be incorporated in one stable phase. Chiu and Yang [45] claim that solubilization within a nonionic surfactant structure protects vitamin E from oxidation.

Moberger et al. [46] studied the effect of ascorbic acid alone and with α-tocopherol in an edible microemulsion having three different compositions. Figure 6 shows the antioxidation efficiency measured as conjugable oxidation products (COP) values (measured as an absorbance in a 1-cm cell

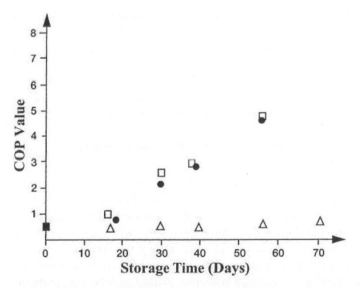

FIG. 6 COP value versus the time of storage of a microemulsion (1% water, 19.8% monoglycerides, 79.2% oil). ($\triangle$) Microemulsions of ascorbic acid (5% w/w) solution, ($\square$) a microemulsion of water only (no ascorbic acid in the water), ($\bullet$) a mixture of soybean oil and monoglycerides in the same ratio, and ($\blacksquare$) the soybean oil batch used at zero time. (From Ref. 46.)

by a 1% w/v lipid solution) versus time of storage for a microemulsion with and without ascorbic acid. It can be seen that the microemulsion with ascorbic acid was stable even after 70 days, whereas the microemulsion without ascorbic acid was stable for only 20 days as the COP values started increasing after 20 days. A synergistic effect was observed between the ascorbic acid solubilized in the water phase and α-tocopherol dissolved in the oil phase of the w/o microemulsion.

Jakobsson and Sivik [47] also used ascorbic acid and α-tocopherol as antioxidants to prevent oxidation of fish oil. The α-tocopherol (0.06%) was incorporated in the oil phase, whereas the ascorbic acid was solubilized in the aqueous phase of the w/o microemulsion, whose composition was 77% fish oil, 20% distilled monoglyceride, and 3% aqueous phase. Oxidation was monitored by measuring the peroxide and anisidine values. A drastic reduction in oxidation was observed when antioxidants were incorporated in the microemulsion system. Oxidation was found to decrease further when ascorbic acid was dissolved in glycerol instead of water. A possible reason for this decrease in oxidation is that there is high internal mobility among the molecules and very short distances between the two phases. This promotes the internal mixing and the ability of the antioxidants to meet and scavenge the free radicals.

Han et al. [48] also studied the effects of various antioxidants such as ascorbic acid, α-tocopherol, and rosemary extract on the oxidation of fish oil and soybean oil using phosphatidylcholine reverse micelles. Peroxide values of the oils indicated that only ascorbic acid (0.02%) solubilized in the reverse micelles was an effective antioxidant in both fish oil and soybean oil. Also, as in the other studies, a combination of ascorbic acid and α-tocopherol was shown to act synergistically in the fish oil.

5. Microemulsions to Enhance Browning and Crisping during Microwave Cooking

One of the drawbacks of microwave cooking is its inability to produce sufficient browning and crispiness in food products. This is related to both temperatures that can be obtained at the product surface and moisture transfer during microwaving. When the product heats up, water vapor will condense on the surface and moisten it. In order to dry the surface, more microwave energy is necessary and this will result in overheating and burning of the core of the product. Usually, a susceptor is used to solve this problem, but it adds to cost and only surfaces adjacent to the susceptor are crisped or browned. Another alternative is to use emulsions containing Maillard reagents or agents that change color upon heating. However, these methods produce only browning without crisping.

Merabet [49] used microemulsion technology to increase browning and crisping taking into consideration the high microwave heating rate and low penetration depth property of the w/o microemulsion. It has been shown that coating the food product with a w/o microemulsion results in more microwave energy absorbed at the surface of the food product and a decreased microwave penetration depth. This allows crisping and even browning of a food product.

6. Microemulsions for Quick Thawing of a Frozen Food in a Microwave Oven

Thawing of a frozen food is traditionally done by leaving it at room temperature. Another way to thaw a frozen food is to use microwave energy. However, these methods are relatively slow and lead to nonuniform thawing.

Merabet [50] found that supercooled water in a frozen w/o microemulsion acted as a microwave energy absorber at temperatures below 0°C and thawed the frozen food rapidly and uniformly. A typical composition of an edible w/o microemulsion used was 84% oil (medium-chain saturated triglycerides, C_8–C_{10}), 11% emulsifiers (sorbitan ester, polyglycerol esters, polysorbates), and 5% water. The microemulsion was applied to the surface of the food product before freezing. Dielectric absorption (ε'') at the microwave frequency (2.45 GHz) at which the food is cooked was 0.4 for frozen food compared with 1.7 for the w/o microemulsion at −20°C. Thawing time of the product containing the microemulsion coating was 6 min compared with 15 min for the product without the microemulsion coating.

7. Wax Microemulsion as a Moisture Barrier

Historically, wax emulsions have been used to prevent moisture loss from fruits and vegetables. However, it is not easy to obtain a thin layer of emulsion because of its relatively high viscosity. This difficulty can be overcome by using a microemulsion because its viscosity is much lower than that of an emulsion.

A wax coating made by drying an o/w microemulsion was found to reduce water loss when applied to fresh fruits and vegetables [51–54]. Baker and Hagenmaier [53] studied the effect of different waxes and fatty acids in microemulsions containing 77% water, 16% waxes, 4% fatty acids, and 3% ammonia solution or morpholine. Candelilla wax exhibited the lowest water loss, lowest gloss, and highest permeability to both O_2 and CO_2, whereas polyethylene wax and carnauba wax coatings provided the best gloss but were most brittle and least effective as a water barrier. Stearic acid produced a brittle coating, and palmitic acid was most effective in suppressing water loss.

B. Microemulsion as Reaction Media

Microemulsions are ideal as reaction media because of their large interfacial area and ability to mix polar and nonpolar reactants in a single homogeneous phase. They are especially useful when reactants have pronounced differences in water and oil solubilities and when the products of two water-soluble reactants are oil soluble or vice versa. In addition, the products of the reaction can be separated physically and the equilibrium can be affected by varying the water activity. They have been used as reaction media for enzymatic synthesis of mono/diglyceride emulsifiers, tailored lipids, and triglycerides and production of flavors.

1. Enzymatic Synthesis of Mono/Diglyceride Emulsifiers

Long-chain monoacyl glycerols (monoglycerides) are widely used in the food industry as emulsifiers. Monoglycerides are typically synthesized by alcoholysis of triglycerides with glycerol. The reaction has low yield and needs high temperature and a catalyst (typically tin and lead compounds).

Microemulsions seem to constitute an ideal medium for in vitro reactions because they possess an enormous interface between the oil and water domains. Holmberg and coworkers [55,56] used w/o microemulsions as reaction media to synthesize monoglycerides. This eliminated the problems associated with higher temperature and insolubility of triglycerides and other lipophilic substrates. The enzymes retained high intrinsic activity compared with the aqueous phase. In their studies, they used 1,3-specific lipase, which leaves the *sn*-2 position of the triglyceride intact. The 2-monoglyceride, thus formed, undergoes an acyl migration to form 1-monoglyceride.

2. Enzymatic Synthesis of Tailored Lipids
or Triglycerides

In certain food applications, plastic or solid fats are desirable. Such fats are conventionally obtained by hydrogenation of liquid triglyceride oil. However, during the hydrogenation of fatty acids in triglycerides, *trans* isomers are formed and essential fatty acids (EFAs), having beneficial nutritional and physiological values, are lost. It is possible to prepare tailored lipids by transesterification of triglycerides or other lipids by chemical means or by using an enzyme such as lipase. In chemical interesterification, physical properties of fats and oils are improved but the reaction occurs randomly. Hence, tailoring is difficult and reaction times are longer.

Lipase solubilized in reverse micelles or w/o microemulsions was found to catalyze the interesterification reaction. This process was found to be very effective, and it was possible to substitute any fatty acid group in any selected position on glycerol [57,58]. Interesterification of liquid oils with a highly saturated fat yields fats with little or no *trans* isomers and high levels of essential fatty acids. Other enzymes such as phosphatases, various lipases, ribonuclease A, phospholipases, and chymotrypsin could also be solubilized to provide other lipids.

The role of (ω-3) fatty acids in health promotion and disease prevention is well known. Phospholipids containing eicosa pentaenoic acid (EPA) (C20:5) and docosahexaenoic acid (DHA) (C22:6) at the 2-position can be easily digested and have nutritional and medical value. Na et al. [59] have incorporated these PUFAs into the 2-position of a phosphotidylcholine (PC) by a phospholipase A_2–catalyzed process in a w/o microemulsion. In another study, Osterberg et al. [58] incorporated γ-linoleic acid into an unhindered triglyceride through lipase-catalyzed transesterification.

As examples of tailored triglycerides, the synthesis of cocoa butter substitute and interesterification of butterfat with desired unsaturated fatty acids are discussed next.

(a) Synthesis of Cocoa Butter Substitute. Cocoa butter, which is an important ingredient in chocolate manufacturing, is in short supply and is quite expensive. Hence, cheaper cocoa butter substitutes are desired. Holmberg and Osterberg [57,60] synthesized cocoa butter from an inexpensive palm oil fraction. Typical triacylglycerol compositions of cocoa butter and palm oil fraction are given in Table 4.

The conversion of inexpensive palm oil fraction into a cocoa butter substitute requires partial replacement of palmitic acid by stearic acid in the 1(3)-position while leaving the 2-position unaffected. This was accomplished by transesterification of a mixture of stearic acid and a palm oil fraction using a 1(3)-specific lipase as a catalyst. The reaction was found to take place at the interface and the interfacial area was rate limiting.

TABLE 4 Typical Triacylglycerol Contents (%) of Cocoa Butter and Palm Oil Midfraction

Triacylglycerol[a]	Cocoa butter (%)	Palm oil midfraction (%)
Sat Sat Sat	1	2
POP	16	65
POS	41	15
SOS	27	2
SatLSat	8	5
Others	7	11

[a]S, P, O, L, and Sat represent stearoyl (C18:0), palmitoyl (C16:0), oleoyl (C18:1), linoleoyl (C18:2), and saturated acyl, respectively.
Source: Ref. 57.

(b)   Production of Butterfat with a Healthy Fatty Acid Profile.   It is generally accepted that saturated fats raise the serum cholesterol level and mono- and polyunsaturated fats lower the cholesterol level. In addition, it has been shown that the position of the fatty acid also plays an important role. Hayes et al. [61] showed the interchange of palmitic acid (C16:0) with oleic acid (C18:1) at the *sn*-2 position suppressed the cholesterol-raising potential of milk fat. Safari and coworkers [62–64] accomplished the conversion (replacement of palmitic acid with oleic acid) by interesterification of butterfat by a specific lipase in the microemulsion system. The total amount of hypercholesterolemic fatty acids (C12:0, C14:0, and C16:0) at the *sn*-2 position decreased by 15%, with a concomitant increase of these fatty acids by 11% at the *sn*-1 position. Also, hypocholesterolemic fatty acids (C18:0 and C18:1) at the *sn*-2 position increased by 39% with a concomitant decrease in these fatty acids at the *sn*-1 and *sn*-3 positions by 25% and 14%, respectively.

3.   Production of Flavors

Chen and Hong [65] showed that butterlike or cheeselike flavors could be obtained by the hydrolysis of butter fat by *Candida cylidracea* lipase solubilized in reverse micelles of soybean lecithin in a continuous butter oil phase. The enzyme exhibited less sensitivity to pH and temperature compared with an emulsified system and its activity increased with increasing surfactant or enzyme concentration. Due to a large interfacial area, a much higher enzyme concentration could be used before the system reached saturation.

C.   Microemulsion as Extraction Media

Because of the ability of o/w microemulsions (or micelles) to solubilize hydrophobic substances and that of w/o microemulsions (or reverse micelles) to solubilize hydrophilic substances, these systems can be used to extract hydrophobic and hydrophilic substances. The ability of o/w and w/o microemulsions to extract various hydrophobic and hydrophilic substances has been used to remove undesirable flavors and nutrients (e.g., cholesterol) from foods and extract valuable food components, e.g., aromas, flavors, proteins, and hydrophobic solutes.

1.   Extraction and Separation of Proteins

Proteins can be extracted from an aqueous phase using a w/o microemulsion without significantly altering their enzymatic or functional properties [66–69]. Extraction is achieved by contacting an aqueous solution of protein with a w/o microemulsion. Proteins are preferentially solubilized within the microemulsion phase and are finally recovered by contacting with a fresh aqueous solution under conditions favoring protein transfer out of the aqueous phase. The extraction process can be readily scaled up using a conventional liquid–liquid extraction technology.

The pH and ionic strength influence the rate of transfer of proteins [2,70]. Surfactant concentration and amount of water in the microemulsion system also affect the protein transfer [71,72]. Because of the dependence of solubilization on the protein properties, it is possible to separate proteins selectively. Goklen and Hatton [67] separated cytochrome c from lysozyme using an Aerosol-OT (AOT)-based microemulsion depending on the solubilizing properties of proteins. One promising application of this ability to separate and concentrate proteins from their mixtures is in the separation of whey proteins to obtain individual proteins with higher value. Other biomolecules such as amino acids, peptides, and nucleic acids have been separated using w/o microemulsions [73,74].

2. Extraction of Cholesterol

The growing awareness of the health risk (arteriosclerosis and heart disease) posed by cholesterol intake has prompted food manufacturers to develop low-cholesterol foods. Sundfeld and coworkers [75,76] demonstrated that cholesterol could be selectively extracted from butter oil using a micellar solution of saponin (GRAS amphiphilic compounds in which sugars are linked to polar groups). A saponin micellar solution, when mixed with butter oil, solubilizes cholesterol because saponins interact selectively with cholesterol. Cholesterol separation can be achieved by discarding the aqueous phase and washing the extracted butter oil with water. The binding ability of a saponin to cholesterol depends on the structure and concentration of saponin and the temperature.

Jimenez-Carmona and Luque de Castro [77] showed that the addition of a microemulsion- or reverse micelle–forming surfactant accelerated the supercritical fluid extraction of cholesterol from food samples containing both low and high levels of cholesterol. The preferred method involves the addition of a microemulsion of a nonionic surfactant (Triton X-100) to the sample followed by a dynamic supercritical CO_2 extraction at 3.83×10^7 Pa and 40°C for 20–40 min. This method allows quantitative extraction of cholesterol from natural food samples including bread, biscuit, milk and egg, with substantial time savings relative to supercritical fluid extraction and conventional extraction techniques.

VI. CONCLUSIONS

Even 55 years after the discovery of microemulsions, their application in the food industry is very limited. Although many promising applications of microemulsions in foods are found in the literature, only a few have been commercialized. The potential, however, exists for applying microemulsion technology innovatively in foods and beverages. In order to realize the possibilities of the microemission technology in foods, further exploration of microemulsions containing triglyceride oils and edible food-grade surfactants, cosurfactants, or cosolvents is desirable. Hence, more efforts in both fundamental and applied research on food-grade microemulsions have to be initiated in both academia and industry.

REFERENCES

1. S Engstrom, K Larsson. In: P Kumar, KL Mittal, eds. Handbook of Microemulsion Science and Technology. New York: Marcel Dekker, 1999, pp 789–796.
2. SR Dungan. In: C Solans, H Kuneida, eds. Industrial Applications of Microemulsions. Surfactant Science Series. Vol 66. New York: Marcel Dekker, 1997, pp 147–174.
3. M El-Nokaly, GD Hiler Sr, J McGrady. In: M El-Nokaly, D Cornell, eds. Microemulsions and Emulsions in Foods. Washington, DC: American Chemical Society, 1991, pp 26–43.
4. CE Stauffer. Fats and Oils. St. Paul, MN: Eagan Press, 1996, p 44.
5. JN Israelachvili. Intermolecular and Surface Forces. London: Academic Press, 1985, p 368.
6. S Friberg, L Rydhag. J Am Oil Chem Soc 48:113–115, 1971.
7. T Gulik-Krzywicki, K Larsson. Chem Phys Lipids 35:127–132, 1984.
8. AM Vesala, JB Rosenholm, SJ Laiho. J Am Oil Chem Soc 62:1379–1385, 1985.
9. JB Rosenholm, AM Vesala, S Laiho. Tenside Surfactants Deterg 25:230–235, 1988.
10. F Chelle, GL Ronco, PJ Villa. World patent WO88/08000, 1988.
11. J Alander, T Warnhelm. J Am Oil Chem Soc 66:1656–1660, 1989.
12. J Alander, T Warnhelm. J Am Oil Chem Soc 66:1661–1665, 1989.
13. R Jourban, N Parris, D Lu, S Trevino. J Dispersion Sci Technol 15:687–704, 1994.
14. BK Paul, ML Das, DC Mukherjee, SP Moulik. Ind J Chem 30A:328–334, 1991.
15. L Hernqvist. In: E Dickinson, ed. Food Emulsions and Foams. London: Royal Society of Chemistry, Special Publication 58, 1987, pp 158–169.
16. L Engstrom. J Dispersion Sci Technol 11:479–489, 1990.
17. K Larsson. In: M El-Nokaly, D Cornell, eds. Microemulsions and Emulsions in Foods. Washington, DC: American Chemical Society, 1991, pp 44–50.
18. N Parris, RF Joubran, DP Lu. J Agric Food Chem 42:1295–1299, 1994.
19. N Garti, A Yaghmur, ME Leser, V Clement, HJ Watzke. J Agric Food Chem 49:2552–2562, 2001.

20. N Garti, A Aserin, S Ezrahi, E Wachtel. J Colloid Interface Sci 169:428–436, 1995.
21. S Ekman, B Lundberg. Acta Chem Scand B32:197–202, 1978.
22. RA Burns, MF Roberts. J Biol Chem 256:2716–2722, 1981.
23. TL Lin, SH Chen, NE Gabriel, MF Roberts. J Phys Chem 94:855–862, 1990.
24. K Shinoda, M Araki, A Sadaghiani, A Khan, B Lindman. J Phys Chem 95: 989–993, 1991.
25. R Aboofazeli, N Patel, M Thomas, MJ Lawrence. Int J Pharm 125:107–116, 1995.
26. ME Leser, WC van Evert, WGM Agterof. Colloids Surf A 116:293–308, 1996.
27. K Thomas, G Pfaff. Perfum Flavorist 2:27, 1978.
28. B Thir, EM Dexheimer. US patent 4,568,480, 1986.
29. PA Wolf, MJ Havekotte. US patent 4,835,002, 1989.
30. AG Gaonkar. US patent 5,376,397, 1994.
31. S Hamdan, BHA Faujan, CR Laili, WBW Ahmad, KJ Dzulkefly. J Agric Food Chem 44:962–963, 1996.
32. SL Chung, CT Tan, IM Tuhill, LG Scharpf. US patent 5,283,056, 1994.
33. SA Slocum, A Kilara, R Nagarajan. In: G Charalambous, ed. Flavors and Off-Flavors. Amsterdam: Elsevier Science Publishers, 1989, pp 233–247.
34. DD Duxbury. Food Process 49:62–64, 1988.
35. YC Chiu, FC Jiang. J Dispersion Sci Technol 20:449–465, 1999.
36. M El-Nokaly, GD Hiler Sr, J McGrady. US patent 5,045,337, 1991.
37. CJ Kirby, EC Needs. UK patent GB 2 297 759A, 1996.
38. ME Leser. European patent 0 657 104A1, 1995.
39. AJ Taylor, MJ Alston, KM Hemingway, CG Chappel, JA Mlotkiewicz. World patent WO:99/62357, 1999.
40. O Chmiel, H Traitler, K Voepel. US patent 5,674,549, 1997.
41. O Chmiel, H Traitler, K Voepel. World patent WO: 96/23425, 1996.
42. O Chmiel, H Traitler, H Watzke, SA Westfall. US patent 5,576,044, 1996.
43. WM Cort. J Am Oil Chem Soc 51:321–325, 1974.
44. C Ruben, K Larsson. J Dispersion Sci Technol 6:213–221, 1985.
45. YC Chui, WL Yang. Colloids Surf 63:311–322, 1992.
46. L Moberger, K Larsson, W Buchheim, H Timmen. J Dispersion Sci Technol 8:207–215, 1987.
47. M Jakobsson, B Sivik. J Dispersion Sci Technol 15:611–619, 1994.
48. D Han, OS Yi, HK Shin. J Food Sci 55:247–249, 1990.
49. M Merabet. European patent EP 0 829 206, 1998.
50. M Merabet. European patent EP 0 848 912, 1998.
51. RD Hagenmaier, RA Baker. J Agric Food Chem 42:899–902, 1994.
52. RD Hagenmaier, RA Baker. J Food Sci 61:562–565, 1996.
53. RA Baker, RD Hagenmaier. J Food Sci 62:789–792, 1997.
54. RA Baker, RD Hagenmaier. J Agric Food Chem 45:349–352, 1997.
55. K Holmberg, E Osterberg. J Am Oil Chem Soc 65:1544–1548, 1988.
56. K Holmberg, B Lassen, MB Stark. J Am Oil Chem Soc 66:1796–1800, 1989.
57. K Holmberg, E Osterberg. Prog Colloid Polym Sci 74:98–102, 1987.

58. E Osterberg, AC Blomstrom, K Holmberg. J Am Oil Chem Soc 66:1330–1333, 1989.

59. A Na, C Eriksson, SG Eriksson, E Osterberg, K Holmberg. J Am Oil Chem Soc 67:766–770, 1990.

60. K Holmberg, E Osterberg. US patent 4,839,287, 1989.

61. KC Hayes, A Pronczuk, S Lindsey, D Dierson-Schade. Am J Clin Nutr 53: 491–498, 1991.

62. M Safari, S Kermasha. J Am Oil Chem Soc 71:969–973, 1994.

63. M Safari, S Kermasha, F Pabai, JD Sheppard. J Food Lipids 1:247–263, 1994.

64. S Kermasha, M Safari, M Goetghebeur. Appl Biochem Biotechnol 53:229–244, 1995.

65. JP Chen, P Hong. J Food Sci 56:234–237, 1991.

66. KE Goklen, TA Hatton. Biotechnol Prog 1:69–74, 1985.

67. KE Goklen, TA Hatton. Sep Sci Technol 22:831–841, 1987.

68. M Dekker, KV Riet, SR Weijers, JWA Baltussen, C Laane, BH Bijsterbosch. Chem Eng J 33:B27–B33, 1986.

69. EM Leser, G Wei, PL Luisi, M Maestro. Biochem Biophys Res Commun 135: 629–635, 1986.

70. SR Dungan, TA Hatton. J Colloid Interface Sci 164:200–214, 1993.

71. PDI Fletcher, D Parrott. J Chem Soc Faraday Trans I 84:1131–1144, 1988.

72. B Kelley, RS Rahaman, TA Hatton. In: WL Hinze, ed. Analytical Chemistry in Organized Media: Reversed Micelles. Greenwich, CT: JAI Press, 1991, pp 123–142.

73. TA Hatton. In: WL Hinze, DW Armstrong, eds. Ordered Media in Chemical Separations. Washington, DC: American Chemical Society, 1987, pp 170–183.

74. PL Luisi, VE Imre, H Kaeckle, H Pande. In: DD Breimer, P Speiser, eds. Topics in Pharmaceutical Sciences. Amsterdam: Elsevier, 1983, pp 243–254.

75. E Sundfeld, S Yun, JM Krochta, T Richardson. J Food Process Eng 16:191–205, 1993.

76. E Sundfeld, JM Krochta, T Richardson. J Food Process Eng 16:207–226, 1993.

77. MM Jimenez-Carmona, MD Luque de Castro. Anal Chem 70:2100–2103, 1998.

21

Microemulsion-Based Viscosity Index Improvers

SUREKHA DEVI and NAVEEN KUMAR POKHRIYAL
M. S. University of Baroda, Baroda, Gujarat, India

ABSTRACT

Viscosity index improvers used in base oils are mainly of three types: olefin copolymers, diene copolymers, and polymethacrylates. Among these, the polymethacrylates have shown better performance at higher temperatures. The microenvironment for polymerization influences the polymer properties to a large extent. In addition, the monomer polarity, the interfacial area between the two phases, and the initiator type play an important role in governing the polymer properties. The microemulsion medium has been shown to give more stereoregular products with high molecular weights. Hence, conventionally synthesized higher polyalkyl(acrylates and methacrylates) have been synthesized in a microemulsion medium and have been characterized by spectral, thermal, and light scattering techniques. Their potential as viscosity index improvers was tested in a paraffin SN 500 base oil (viscosity index 90). The thickening power ($Q = \eta_{\text{spec.}}100°C/\eta_{\text{spec.}}40°C$) of the products in the base oil was estimated. The Q values indicated that the products were more effective at high temperatures. An increase of nearly 40 units in viscosity index of the base oil was observed with poly(dodecyl methacrylate) (viscosity index 126) with just a 1% dose. The effectiveness of the microemulsion-based products was compared with that of emulsion-based products and a commercial viscosity modifier. Microemulsion-based products performed better. The results obtained are explained on the basis of higher molecular weight and more stereoregular structure of the polymers.

I. INTRODUCTION

The use of oils of animal or vegetable origin as lubricants in vehicles or machinery dates back almost to the birth of civilization [1]. However, mineral oil–based products produced commercially rapidly became the essential lubricants of the 20th century. Mineral oil–based lubricants were more effective in reducing the friction, wear, and tear at bearing contacts and hence enabled the machinery to operate smoothly even under severe operational conditions such as high pressures. However, the mineral oils have two main drawbacks:

1. They readily oxidize at temperatures above 100°C.
2. They have low flowability at temperatures below −20°C.

With the development of the aerospace industry it became necessary to overcome these drawbacks of mineral oil–based lubricants. This was achieved in one of the following ways:

1. The use of synthetic base oils. However, the synthesis of base oils requires sophisticated techniques and high investment, hence synthetic base oils still represent a small portion of the base oils used today.
2. The addition of polymeric substances to these lubricants to restrict their oxidation and to modify their flow properties. The synthesis of these polymeric materials is a comparatively easier and cheaper approach. Hence, the use of random or tailored additives in the lubricants has progressively increased.

II. VISCOSITY–TEMPERATURE RELATIONSHIP OF A BASE OIL

The absolute viscosity, which was defined by Newton as the ratio between the applied shear stress and the resulting rate of shear, is the most important parameter used to monitor the performance of a base oil (Fig. 1).

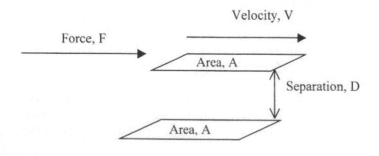

$$\text{Absolute viscosity} = (\,F\,/\,A\,)\,/\,(\,V\,/\,D\,)$$

FIG. 1 Definition of absolute viscosity.

The method most frequently used to study the variation of the viscosity of a liquid with temperature is to calculate a dimensionless number known as the viscosity index, VI. This can be done by measuring the kinematic viscosity of the sample oil at two different temperatures (40 and 100°C) and comparing the viscosity changes with an empirical reference scale. The original reference scale devised by Dean and Davis [2] was based on two sets of base oils derived from two separate crude oils. The Pennsylvania crude was arbitrarily assigned a VI value of 100 and the Texas Gulf crude was assigned a VI of zero. The higher the VI number, the less the effect of the temperature on the viscosity of the sample. According to the ASTM D2270 and Indian Petroleum 226 manual, the viscosity index can be calculated from the following equation:

$$\text{VI} = \frac{(L - U)}{(L - H)} \times 100$$

where L is the viscosity at 40°C of an oil with a zero VI that has the same viscosity at 100°C as the sample under test, H is the viscosity at 40°C of an oil with VI 100 and the same viscosity at 100°C as the oil under test, and U is the viscosity at 40°C of the sample.

The viscosity–temperature relationship of unprocessed base oil regardless of its chemical nature can be improved by the addition of polymeric substances. These additives are generally called viscosity index improvers, VIIs, or thickeners (Fig. 2).

A. Viscosity Index Improvers (VIs) or Thickeners

The early workers in the field of lubricants observed that small amounts of rubber dissolved in a mineral oil raised the VI substantially. However, high unsaturation in the polymer led to oxidation and sludge formation. Otto et al. [3] discovered that this could be overcome through the use of a synthetic polymer prepared from the light ends of the gasoline. Similar observations were later made for polymethacrylates by Rohm and Haas Co. [4,5] and for polyisobutylenes by Farbenindustrie AG [6,7]. Because these materials were initially used mainly to increase the VI, they became known as the viscosity index improvers.

Since the early developments in the use of polymethacrylates and polyisobutylenes, a wide variety of the polymers have been explored to examine their potential as viscosity index improvers. Table 1 lists the representative classes useful as VI improvers.

Out of these, four classes are in commercial use as viscosity index improvers (Scheme 1). Each of these important commercial VI improver families represents one of the most important commercial polymerization tech-

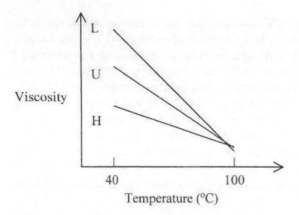

FIG. 2 Viscosity versus temperature relation of mineral oils. The *L*, *U*, and *H* are defined in the text in Section IV.E.

niques for the manufacture of high-molecular-weight polymers: olefin copolymers by Ziegler–Natta polymerization, styrene–isoprene copolymers by anionic polymerization, styrene–polyester copolymers by modification of polystyrene or condensation polymerization, and polymethacrylates by free radical polymerization.

Currently, the VI improvers of the polymethacrylate family are more in use as additives in industrial and motor oils because of their low glass tran-

TABLE 1 Representative Classes of
Viscosity Index Improvers

Polymerized light mixed olefins
Polyisobutylenes
Polyacrylates
Polymethacrylates
Poly(acrylate-co-methacrylate)
Poly(methacrylate-co-styrene)
Polybutadiene
Poly(alkyl fumarate-co-vinyl acetate)
Poly(*n*-butyl vinyl ether)
Esterified poly(styrene-co-maleic anhydride)
Poly(ethylene-co-propylene)
Hydrogenated (styrene-co-butadiene/isoprene)
Hydrogenated polybutadiene/isoprene

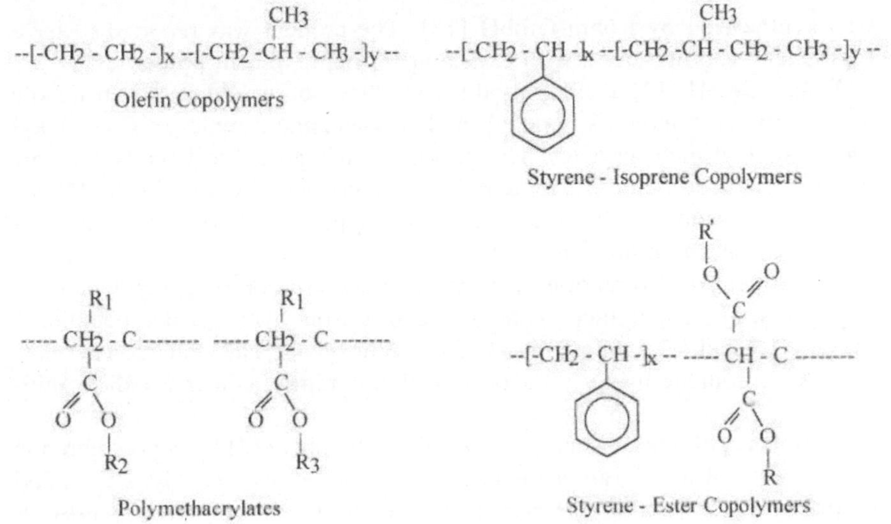

$$--[CH_2 - CH_2 -]_x --[CH_2 -\overset{\overset{\displaystyle CH_3}{|}}{CH} - CH_3 -]_y --$$

Olefin Copolymers

$$--[CH_2 - \overset{\overset{\displaystyle \bigcirc}{|}}{CH} -]_x --[CH_2 -\overset{\overset{\displaystyle CH_3}{|}}{CH} - CH_2 - CH_3 -]_y --$$

Styrene - Isoprene Copolymers

Polymethacrylates:

$$----- CH_2 - \overset{\overset{\displaystyle R_1}{|}}{\underset{\underset{\displaystyle R_2}{|}}{\underset{\displaystyle O}{C}}}{C} --------- \quad ----- CH_2 - \overset{\overset{\displaystyle R_1}{|}}{C} --------$$

with R_1, ester groups C, O, O, and R_2 / R_3

Polymethacrylates

Styrene - Ester Copolymers

$$--[CH_2 - CH -]_x -- ----- CH - C --------$$

SCHEME 1

sition temperatures, high thermal and shear stabilities, and minimum sludge formation properties.

B. Polymethacrylates as VI Improvers

Useful overviews of VI improvers based on polymethacrylates are provided by Arlie et al. [8,9] and Neudoerful [10]. The later development after 1990s in the polymethacrylate family as VI improvers is briefly presented here.

Röhm GmbH [11] have disclosed the synthesis of methacrylate-based VI improvers. The majority of the products are reported to be multifunctional in nature. Pennewiss et al. at Röhm GmbH [11] reported the copolymerization of C_{12-18} alkyl methacrylates in the presence of ethylene–propylene copolymer, using C_7H_{15}-(CO)-OOBu-*tert.* as an initiator. After treatment with N-vinyl-pyrrolidone and N-vinyl-imidazone, the resulting product was reported to be a good dispersant and VI improver.

Röhm GmbH [12] further reported the synthesis of a VI improver that was prepared from isodecyl methacrylate, C_{16-18} alkyl methacrylates, and methyl methacrylate. The VI-improving property of the product was further enhanced by the addition of polyolefins and hydrogenated styrene–diene copolymer and/or by grafting hydrogenated polyisoprene onto the product.

An additive that was a mixture of esters of methacrylic acid and 60% C_{12-15} branched alcohols, methyl methacrylate, methacrylic esters derived from 23% C_{12-15} branched alcohols, and C_{16-18} linear alkyl methacrylates was synthesized by Röhm GmbH [13]. The product was reported to have high shear stability along with excellent VI improvement properties.

Röhm GmbH [14] also reported the synthesis of an additive from 1 mole of 1-decene, 0.5 mole of isodecyl methacrylate, and 2 moles of C_{12-15} alkyl (90% isoalkyl methacrylates. The product was reported to have better flow improvement properties than a poly(α-olefin) additive in trimethylolpropane–adipic acid ester base oil when both types of additives were used at the same concentration (20 wt%).

Röhm GmbH also reported [15] the synthesis of comb copolymers as VI improvers. The copolymers were synthesized from hydrogenated butadienes (having 1,4 and 1,2 configurations), 2-ethyl hexyl acrylate, and butyl acrylate at 77°C in toluene using 2,2-azo-bis-isobutyronitrile as a free radical initiator.

A graft copolymer was prepared by Röhm GmbH [16] by copolymerizing 10–90 wt% of a poly(alkyl methacrylate) macromonomer with C_{6-30} alkyl methacrylates, C_{1-5} alkyl methacrylates, styrene, C_{1-4} alkyl styrene, <60 wt% C_{2-12} fatty acid vinyl esters, and <40 wt% functionalized comonomers from a group of vinyl heterocycles and functionalized methacrylates and amides. The lubricant additive was reported to have pronounced VI-improving and dispersing efficiency.

Self-stabilizing polymer emulsions containing a bimodal polymethacrylate solution as a continuous phase and a block or graft copolymer derived from ethylene–polypropylene copolymer and hydroxyethyl methacrylate, C_{16-18} alkyl methacrylate, and isodecyl methacrylate were prepared by Röhm GmbH [17] and the emulsions were found to be useful for improving the VI of lubricants.

By reacting a mixture of 60 g of 70:30 C_2H_4–C_3H_6 copolymer (molecular weight 2×10^6 as a continuous phase), 40 g of C_{10-18} alkyl methacrylates, 2 g of bis(tert-butyl peroxy)-butane, and 1 g of tert-butyl peroxy ethyl hexanoate at 145°C for 20 min, a polymer dispersion, which was tack free at $-50°C$, was obtained by Fengler et al. [18] at Röhm GmbH and showed 11.4 and 78.1 centistokes (mm^2/s) solution viscosities, respectively, at 40 and 100°C when used at 1.7 wt% in mineral oil.

Polymethacrylate polymers containing (0–25 wt%) C_{16-24} alkyl methacrylates of high molecular weight and (25–70 wt%) C_{16-24} alkyl methacrylates of low molecular weight were found especially effective in improving the low-temperature fluidity of a broad range of base oils by Kinker et al. [19] at Röhm GmbH.

Shell [20] has reported the synthesis of a VI improver by carrying out polymerization of C_{9-18} alkyl methacrylates or a blend of C_{12-15} alkyl methacrylates and 4-vinyl-pyridine in the presence of a hydrogenated divinyl benzene cross-linked star-shaped polyisoprene. Shell also reported [21] the synthesis of block copolymers of dienes, vinyl arenes, and alkyl methacrylates. The products were found to give VI-improving properties.

Lubrizol, India has reported [22] a process for the synthesis of an additive for mineral oils. The additive was snythesized by copolymerization of an alkyl methacrylate monomer and a C_{8-12} branched-chain methacrylate such as isodecyl methacrylate using an azo initiation in a hydrocarbon solvent at 50–100°C. This concentrate was further reacted with diethylenetriamine to obtain a multifunctional nitrogen-containing lubricant. Similar nitrogen-containing dispersant VI improvers based on alkyl acrylates or methacrylates for lubricating oils were also prepared by Lubrizol, USA [23].

Lubrizon, India [24] compared the stability of the three main types (polymethacrylates, olefin copolymers, styrene–isoprene copolymers) of VI improvers in a base oil at high temperatures and shear rates. The viscosity loss in the olefin copolymer type of VI improvers was minimal. Lubrizol, India [25] also explored the possibility of using isodecyl methacrylate and 1-decene copolymers as VI improvers. These copolymers were prepared by a free radical polymerization procedure in toluene using 2,2-azobisisobutyronitrile. The performance of the copolymers was compared with that of standard polymethacrylates. The results showed a moderate improvement in the performance of the synthesized copolymer as compared with standard polymethacrylates.

Sanyo [26] has reported a polymer composition that was found to be useful as a VI improver in mineral oils. This polymer composition consisted of an olefin copolymer, an olefin–methacrylate copolymer, and an oxyalkylated alcohol or amine or amide-based surfactant. Sanyo [27] has also reported the use of polyolefin-graft-dimethylaminoethyl methacrylate copolymers as VI improvers. Copolymers of N-vinyl pyrrolidone and N,N'-dialkylamino alkyl methacrylates were also reported as useful VI improvers by Sanyo [28].

Sanyo developed VI-improving agents with excellent low-temperature performance [29]. These agents were a mixture of polymers of C_{8-12} alkyl methacrylates with glass transition temperature (T_g) < 40°C, polymers of 0–90 mol% C_{1-13} alkyl methacrylates, and polymers of 13–97 mol% C_{14-24} alkyl methacrylates in weight ratios, 80–90:0.1–19.9:1–19.9. The agents were found to be useful for isoparaffin oils.

Sanyo [30] has disclosed a novel VI improver blend for lubricating oils. The blend was prepared by mixing > 70 wt% $C_{n<10}$ poly(alkyl methacrylate) with a mixture of base oils. When compared with conventional polymethacrylate-based additives this blend showed better low-temperature performance and oxidation resistance.

The copolymers of 98% polymethacrylates (8% methyl methacrylate, 92% dodecyl methacrylate), 1% polymethacrylates (52% Dobanol 23 methacrylate, 48% Dobanol 45 methacrylate), and 1% polymethacrylates (90% Dobanol 45 methacrylate, 7% cetyl methacrylate, and 3% octyl methacrylate) were found to be useful as VI improvers for isomerized paraffin–containing mineral oils by Takigawa [31] at Sanyo.

Neveu and Huby [32] at Rohm and Haas studied the applicability of Kraemer and Huggins equations to determine the thickening power of polymethacrylate-based VI improvers in engine base oils. After determining the intrinsic viscosity, the amount of polymethacrylates needed to bring about a given viscosity to any paraffin base oil at 37.7–176.6°C was predicted. Good agreement was reported between the predicted and experimentally determined values.

Rohm and Haas also reported [33] the synthesis of copolymers of 40–100 wt% C_{1-10} alkyl methacrylates, 0 to 60 wt% C_{11-20} alkyl methacrylates, and 30–60 wt% C_{11-15} alkyl methacrylates in their products. These copolymers were found to be useful as VI improvers for phosphate-ester–containing hydraulic fluids used in aircraft.

Bataille and Shariffi-Sanjani [34] reported the synthesis of a copolymer of 2-ethylhexyl acrylate and vinyl 2-ethyl hexanoate by solution polymerization using benzene as a solvent and benzoyl peroxide as an initiator at 65°C. The viscosity index improvement property of the purified copolymer was checked in two ISO 20 base oils. The polydispersity of the copolymers was observed to influence greatly the viscosity index improvement property.

Lath et al. [35] examined the role of poly(n-alkyl methacrylates) as VI improvers for lubricants by performing calculations for poly(butyl methacrylate)–hexadecane model systems.

Tercopolymers of C_{1-4}, $C_{12,13}$, and $C_{14,15}$ methacrylates have also been reported as viscosity modifiers for base oils by Sakai and Takigawa [36].

Kapur et al. [37] studied the structure–performance relationship of olefinic copolymers by nuclear magnetic resonance (NMR) spectroscopy. Apart from high molecular weight of the additives, the spatial arrangement of components and groups in the polymer chains also influenced the additive performance.

The work carried out in the last 25 years in the synthesis of the poly(alkyl methacrylates) as VI improvers is in the area of solution polymerization and is available mainly in patents. A microemulsion [38], which is a thermodynamically stable, isotropic system of oil, water, surfactant, and/or cosurfactant, was selected as the medium for polymerization as it gives rise to high-molecular-weight and more stereoregular polymers with narrow polydispersity. The products were also synthesized in an emulsion medium, and the performance of the additives from the two media was compared.

Although emulsion polymerization has been studied extensively, the polymerization of monomers in microemulsions originated only in 1979 when Bone and Stoffer [39] carried out polymerization of methyl methacrylate in a water-in-oil (w/o) microemulsion using a persulfate initiator. It was found that the addition of potassium persulfate to the microemulsion system did not influence the one-phase region. The particle nucleation in microemulsion polymerization in o/w or w/o systems was believed to be a continuous process, the locus of nucleation was shown to be the microemulsion droplet, and the products produced were of high molecular weight [40,41]. Since then, the number of reports on the polymerization in o/w, bicontinuous or w/o, microemulsion systems, exploring polymerization kinetics, synthesis of porous materials and nanoparticles, and synthesis of more ordered polymers, has continuously increased [42,43].

The performance of an additive depends strongly on its molecular weight, stereoregularity, and compatibility with other additives. Of these, the first two parameters can be controlled, to some extent, by selecting a suitable polymerization medium. Hence, a microemulsion was selected as a medium for the polymerization as it has shown the potential to produce additives with desired properties.

III. EXPERIMENTAL

A. Materials

2-Ethylhexyl acrylate and 2-ethylhexyl methacrylate from Fluka, Switzerland and the synthesized C_{6-12} alkyl acrylates and methacrylates were distilled under reduced pressure to remove the inhibitor. Potassium persulfate (KPS) from Sisco Chemicals, Mumbai, India and sodium dodecyl sulfate (SDS) from Qualigens, Mumbai, India were used as received. The paraffin base oil SN 500 with a viscosity index of 90 was received from Savita Chemicals, Mumbai, India. Commercial VI improver ORTHOLEUM, which is a product of Associated Octel Ltd., U.K. was obtained from Prashant Hydrocarbon Pvt. Ltd., Baroda, India. All solvents were distilled before use and double-distilled deionized water was used throughout the work.

B. Synthesis of Monomers

The C_6, C_8, C_{10}, and C_{12} alkyl acrylates and methacrylates were synthesized by a method reported in 1944 by Reheberg and Fisher [44] by the following reaction scheme:

$$CH_2{=}C{-}R_1 + nR_2OH \xrightarrow[\text{Hydroquinone}]{\underset{\text{reflux}}{\text{Conc. } H_2SO_4}} CH_2{=}C{-}R_1 + CH_3OH$$

with pendant groups $COOCH_3$ (left) and $COOR_2$ (right).

where R_1 = H or CH_3 and $R_2 = C_6H_{13}{-}C_{12}H_{25}$. The products were washed with 5% sodium hydroxide solution before purification by fractional distillation.

C. Polymerization Procedure

The single-phase microemulsion region at 30°C was determined visually by titrating aqueous micellar solutions of SDS with monomer mixtures. Batch polymerization was carried out at different temperatures under a nitrogen atmosphere with constant stirring conditions. The compositions shown in Table 2 were charged in a 250-mL five-neck reaction kettle equipped with a mechanical stirrer, nitrogen inlet, water condenser, and a thermometer. Initiator concentrations used were in moles per liter of water. In the emulsion polymerization, the system was continuously stirred at 400 rpm. The conversion of monomer to polymer was determined gravimetrically.

The product was isolated by filtration after precipitation with a fourfold excess of methanol. The products were washed several times with hot water and methanol to remove the surfactant; reprecipitated from solvents of respective polymers such as toluene for poly C_{6-8} acrylates and methacrylates and hexane for higher polyacrylates, using methanol to remove the traces of unreacted monomers; and dried under vacuum at 60°C.

IV. CHARACTERIZATION

A. Spectroscopic Analysis

The infrared (IR) spectra of the synthesized monomers and their homopolymers were recorded on a Perkin Elmer 16 PC IR spectrophotometer.

TABLE 2 Compositions Used for Polymerization of C_{6-12} Alkyl Acrylates and Methacrylates

| | (wt/wt) % composition of | | | | | |
| | Microemulsion | | | Emulsion | | |
Monomer	Water	SDS	Monomer	Water	SDS	Monomer
C_6	85	10	5	93.0	1.15	5.85
C_8 (linear)	80	15	5	93.0	1.25	5.75
C_8 (branched)	80	15	5	93.0	1.25	5.75
C_{10}	80	15	5	93.5	1.50	5.00
C_{12}	75	20	5	93.5	1.50	5.00

B. Gas Chromatographic Analysis

The purity of the synthesized monomers was checked by gas chromatography using a Hewlett Packard 5980 Series II gas chromatograph and nitrogen as a carrier gas (flow rate 30 mL/min) and a FAL-M (10%, 540 × 0.4 cm) pack column containing 30–60 mesh size SHIMALITE TPA support. Injector and flame ionization detector temperatures were maintained at 175°C.

C. Thermal Analysis

The thermograms of the purified products were recorded on a Shimadzu DT-30 thermal analyzer at a heating rate of 10°C/min under a nitrogen atmosphere.

D. Particle Size Distribution

The latex particle size was determined using a Brookhaven BI 90 particle size analyzer with a 5 mW helium–neon laser of 623.8 nm wavelength at room temperature. Prior to the analysis, the latexes were diluted with double-distilled deionized water to minimize the particle–particle interactions until the volume fractions of the particles were in the range of 0.01 to 0.1. An average hydrodynamic radius of latex particles (R_h) was calculated from the intrinsic diffusion coefficient (D_0) as

$$R_h = KT/6\pi\eta D_0$$

where η is the viscosity of the dispersing medium, i.e., water; K is the Boltzmann constant; and T is the absolute temperature. The polydispersity index (PI), which is an indicator of the size distribution, was obtained using the compute software provided with the instrument.

E. Viscometric Studies

The intrinsic viscosities of the purified products were determined in respective solvents at 30°C using an AVS 350 Schott Geratte autoviscometer. For testing the potential of the products as viscosity index improvers, their viscosities and viscosity index in the paraffin base oil SN 500 were determined at 40 and 100°C according to the ASTM D2270 and IP226/84 methods using the following equations.

(a) *For Oils of Viscosity Index 0 to 100:*

$$\text{Viscosity index} = \frac{L - U}{L - H} \times 100$$

where U is the kinematic viscosity at 40°C of the oil whose viscosity index is to be determined, L is the kinematic viscosity at 40°C of an oil of zero viscosity index having the same viscosity at 100°C as the oil whose viscosity index is to be calculated, and H is the kinematic viscosity at 40°C of an oil of viscosity index 100 having the same viscosity at 100°C as the oil whose viscosity index is to be calculated.

(b) *For Oils of Viscosity Index 100 and Higher:*

$$\text{Viscosity index} = \frac{(\text{antilog } N)}{0.00715} + 100$$

and N is calculated as

$Y^N = H/U$

$N = (\log H - \log U)/(\log Y)$

where Y is the kinematic viscosity in centistokes (cSt or mm²/s) of the unknown oil at 100°C, H is the kinematic viscosity in cSt at 40°C of an oil having a viscosity index of 100 using procedure (a) and having the same kinematic viscosity at 100°C as the oil whose viscosity index is to be calculated, and U is the kinematic viscosity in cSt at 40°C of the oil whose viscosity index is to be calculated ($H > U$).

The Q factor is the ratio of the specific viscosity (η_{spec}) of oils with additive at two different temperatures, 40 and 100°C, which monitors the thickening effect of the additive and was calculated according to the following equation:

$$Q = \frac{\eta_{spec} \ 100°C}{\eta_{spec} \ 40°C}$$

To determine the intrinsic viscosity, the concentration of the polymers in the base oil was varied from 1 to 4 wt% and the intrinsic viscosities were calculated on the basis of the Huggins and Kraemer equations:

$\eta_{spec}/C = |\eta| + K'|\eta|^2 C$

$\ln \eta_r/C = |\eta| - K''|\eta|^2 C$

where η_{spec}/C and η_r are, respectively, the reduced and relative viscosities of the polymer solutions; K' and K'' are constants that depend on the solvent, polymer nature and, temperature; $|\eta|$ is the intrinsic viscosity in dL/g; and C is the concentration of polymer in the solvent in g/L.

V. RESULTS AND DISCUSSION

The physical properties and the purity index of the synthesized and purified monomers are given in Table 3. These values are in good agreement with

TABLE 3 Physical Properties and Purity Index of the Synthesized Monomers

Side chain length in monomer	Density at 30°C (g/mL)		Boiling point (°C)		% Purity by GC	
	(a)	(b)	(a)	(b)	(a)	(b)
C_6	0.889	0.868	186 (40/1.1 mm Hg)	190 (40/1.1 mm Hg)	94.9	98.9
C_8	0.861	0.874	198 (57/0.05 mm Hg)	220 (60/0.05 mm Hg)	82.7	90.7
C_{10}	0.884	0.880	232 (118/5 mm Hg)	240 (120/5 mm Hg)	98.1	95.9
C_{12}	0.896	0.897	262 (120/0.8 mm Hg)	271 (122/0.8 mm Hg)	67.5	99.8

(a) Acrylate; (b) methacrylate.
GC, gas chromatography.
Values given in parentheses are the literature values.

the literature values [44,45]. The IR spectra of all the monomers exhibited strong bands at 1728 cm^{-1} for $>$C$=$O stretching, at 3436 cm^{-1} for —OH stretching, at 2876 cm^{-1} for —C—H stretching, at 1642 cm^{-1} for $>$C$=$C$=$ stretching, at 1168 cm^{-1} for —C—O—C— stretching, and at 725 cm^{-1} due to the long alkyl group side chain.

From the gas chromatographic (GC) analysis, the synthesized higher acrylates and methacrylates were found to be contaminated with the corresponding alcohols and ethers formed by the condensation of two alcohol molecules. The corresponding homopolymer spectra did not show bands due to —C$=$C— and —OH or —C—O—C— groups.

A. Phase Diagrams

Partial phase diagrams for C_{6-12} alkyl acrylate or methacrylate/SDS/water systems are shown in Figs. 3 and 4. The one-phase microemulsion region was very small due to the highly hydrophobic character of the monomers. Initially, a decrease in one-phase boundary was observed with an increase in alkyl chain length. However, the one-phase region was larger for C_{12} monomers than for C_{6-10} monomers. This can be explained by considering the following two points:

The cosurfactant character of the monomer
Chain length compatibility between the monomers and sodium dodecyl sulfate

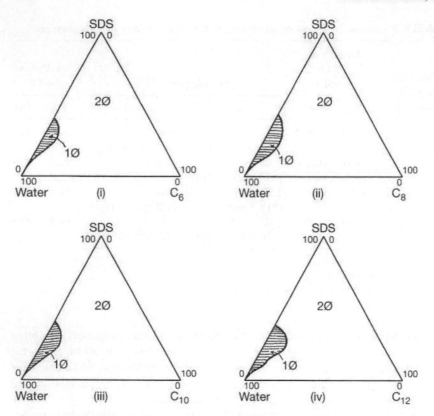

FIG. 3 Ternary phase diagrams of alkyl acrylate/SDS/water systems at 30°C. (i) C_6-hexyl acrylate, (ii) C_8-octyl acrylate, (iii) C_{10}-decyl acrylate, (iv) C_{12}-dodecyl acrylate, (1Ø) one-phase region. (2Ø) two-phase region.

Sharma and Shah [46] reported that the surface tension of the SDS/alkyl alcohol aqueous solutions was minimum when the chain lengths of the surfactant and that of the alcohol were equal, e.g., for the SDS/C_{12}OH solutions. This was attributed to the tight packing of the molecules at the air–water interface. Patist et al. [47] reported similar results for SDS/alkyl trimethylammonium bromide/water systems. By the same token, C_{12} alkyl acrylates and methacrylate/SDS/water systems are expected to have a minimum surface tension and more solubilization of the monomer resulting in an increase in the one-phase region. The role of acrylates as cosurfactants has already been established in the case of hydroxy alkyl methacrylates [48].

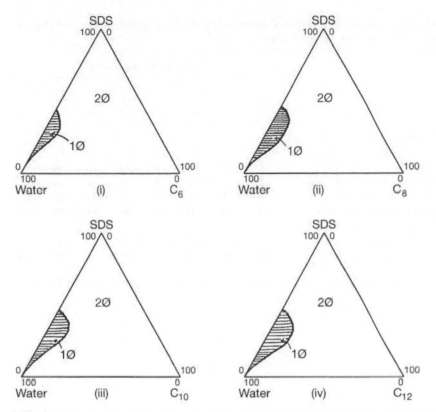

FIG. 4 Ternary phase diagrams of alkyl methacrylate/SDS/water systems at 30°C. (i) C_6-hexyl methacrylate, (ii) C_8-octyl methacrylate, (iii) C_{10}-decyl methacrylate, (iv) C_{12}-dodecyl methacrylate, (1Ø) one-phase region, (2Ø) two-phase region.

B. Polymerization

The optimized reaction conditions for the polymerization of higher acrylates are given in Table 4. All microemulsions were transparent and fluid before polymerization but were slightly turbid after completion of the reaction due to increased particle size. The final latex particle sizes and intrinsic viscosity values of the polymers are compiled in Table 5.

The particle sizes of all synthesized higher polyacrylate/methacrylate microemulsion latexes were in the range of 40–70 nm. An increase in alkyl chain length had little effect on the final particle sizes (Table 5). From the intrinsic viscosity values of the products it can be noted that the microemulsion products were of higher molecular weight than the emulsion products. This can be attributed to the continuous nucleation mechanism opera-

TABLE 4　Optimized Reaction Conditions for the Homopolymerization of Higher Alkyl Acrylates and Methacrylates

	Linear alkyl chain length							
	C_6		C_8		C_{10}		C_{12}	
Reaction conditions	(a)	(b)	(a)	(b)	(a)	(b)	(a)	(b)
Microemulsion polymerization								
SDS/monomer (wt/wt)	2:1	2:1	3:1	3:1	3:1	3:1	4:1	4:1
Reaction temperature (°C)	70	70	70	70	70	70	70	70
KPS concentration (mM)	2	2	4	4	4	4	4	4
Reaction time (h)	5	5	6	6	10	10	10	10
Percentage conversion	95	90	90	95	90	95	85	90
Emulsion polymerization[a]								
SDS/monomer (wt/wt)	0.15	0.15	0.20	0.20	0.20	0.20	0.25	0.25
Percentage conversion	98	95	95	95	98	95	95	95

[a]Other experimental conditions for emulsion polymerization were the same as for the microemulsion.
(a) Acrylates; (b) methacrylates.

TABLE 5　Final Latex Particle Sizes of Poly(C_{6-12} Alkyl Acrylate and Methacrylate) Latexes and Intrinsic Viscosities of the Corresponding Homopolymers[a]

	Monomer chain length									
	C_6		C_8		C_8[b]		C_{10}		C_{12}	
Parameter	(a)	(b)	(a)	(b)	(a)	(b)	(a)	(b)	(a)	(b)
D(nm)										
A	113	46	142	52	62	66	134	54	119	56
MA	119	54	116	57	96	78	138	49	119	62
$|\eta|$(dL/g)										
A	1.4	1.4	2.8	2.9	2.3	2.5	1.3	1.5	2.7	2.9
MA	1.4	1.5	2.7	3.1	1.3	1.6	1.2	1.7	2.8	2.9

[a]Synthesized in (a) emulsion medium; (b) microemulsion medium; D, particle diameter; $|\eta|$ intrinsic viscosity; A, acrylate; MA, methacrylate monomer.
[b]Branched monomers (2-ethyl hexyl acrylate/methacrylate).

tive in microemulsion polymerization. In microemulsions, because of the very high interfacial area of the microemulsion droplets compared with nucleated particles, droplets preferentially capture the primary radicals formed in the continuous phase. This leads to a continuous nucleation process with each particle formed in a single step and hence a low number of polymer chains per particle (hence very high molecular weight) [40,41], whereas in emulsion polymerization the number of polymer chains per particle can go up to hundreds.

C. Thermal Stability of the Products

As these polymeric additives are generally used in lubricating oils, they are frequently subjected to high local temperatures due to friction at gears, roll contacts, bearings, etc. Hence, the thermal stability of the additive is also one of the important parameters in dictating the additive performance. The thermal decomposition range for all synthesized products is given in Table 6. They were all fairly stable in the working range 30–250°C. The products synthesized in a microemulsion showed some improvement in thermal stability compared with those from an emulsion medium. It may partially be due to the higher molecular weight of the microemulsion products. However, no decisive trend was observed in thermal stability with increase in alkyl chain length and the products appeared to have similar thermal stability (Table 6). Similar results were reported by Sazanov et al. [49] for the thermal degradation of higher polyacrylates. The decomposition range for $C_{10,12}$ poly(alkyl acrylates) and methacrylates could not be determined due to their

TABLE 6 Thermal Decomposition Range of the Synthesized Poly(C_{6-12} Alkyl Acrylates and Methacrylates) in Emulsion and Microemulsion Media

	Degradation temperature (°C)							
	t_1		t_{10}		t_{50}		t_{90}	
Product	E	M	E	M	E	M	E	M
Poly(2-ethyl hexyl acrylate)	135	190	410	420	460	470	—	—
Poly(2-ethyl hexyl methacrylate)	305	305	330	345	376	380	418	425
Poly(hexyl acrylate)	301	310	320	335	340	360	405	430
Poly(octyl acrylate)	315	325	315	340	350	380	410	450
Poly(hexyl methacrylate)	352	368	387	410	465	480	508	530
Poly(octyl methacrylate)	310	315	365	380	415	425	—	—

E, Emulsion medium; M, microemulsion medium.

highly viscous nature, as weighing for the thermogravimetric analysis (TGA) is difficult for sticky materials.

D. Performance of the Products

The intrinsic viscosities of the products are compiled in Table 7. It can be noted that the intrinsic viscosities increase with increase in temperature. This can be attributed to the expansion of the polymeric chains in the base oil with increase in the temperature. Similar results were reported for the poly-methacrylate-based VI improvers by Mueller [50] and Neveu and Huby [32].

The homopolymers of C_4–C_7 acrylates/methacrylates and C_8 acrylate were found to be insoluble in the base oil even at very high temperatures (90°C) irrespective of the synthesis route. The additives increased the vis-cosity index of the untreated base oil from 90 to 135 with 1–4 wt% dose, which is comparable with the literature values for polymethacrylate additives in paraffin base oils (or with the synthetic polymethacrylate mineral oils with viscosities 11.4 and 78.1 cSt at 100 and 40°C, Table 8).

At low temperatures, the polymer remains in the coil form and does not contribute to the viscosity of the oil. At higher temperatures the coil opens up and compensates for the drop in viscosity of the oil due to rise in tem-perature. The microemulsion products performed better than emulsion prod-ucts. Their performance was comparable with commercial VI improver ORTHOLEUM. This was attributed to the high molecular weight of the products and their more controlled stereostructure. The second point is sup-

TABLE 7 Intrinsic Viscosity Values (dL/g) of the Additives in the Base Oil SN 500

Additive	$\|\eta\|$ (40°C) (dL/g)	$\|\eta\|$ (100°C) (dL/g)
2-Ethyl hexyl methacrylate (M)	0.41	0.58
2-Ethyl hexyl methacrylate (E)	0.28	0.43
Octyl methacrylate (M)	0.29	0.48
Octyl methacrylate (E)	0.23	0.32
Decyl methacrylate (E)	0.83	0.86
Decyl methacrylate (E)	0.79	0.57
Dodecyl methacrylate (E)	0.96	1.22
Dodecyl methacrylate (E)	0.91	0.95
Decyl acrylate (E)	0.78	0.81
Decyl acrylate (E)	0.68	0.67

(M) and (E) signify products synthesized in microemulsion and emul-sion media, respectively.

TABLE 8 Performance Data of the Products Synthesized in Emulsion and Microemulsion Media

Sample	Dose[a] (wt%)	Viscosity (cSt) 40°C	100°C	Viscosity index	Q
SN 500 (base oil from IOCL)	0	97.2	11.0	90	—
SN 500 + poly(octyl methacrylate) (M)	4	352.1	50.1	133	1.3
SN 500 + poly(octyl methacrylate) (E)	4	339.5	33.0	120	0.8
SN 500 + poly(2-ethyl hexyl methacrylate) (M)	4	353.3	58.5	135	1.6
SN 500 + poly(2-ethyl hexyl methacrylate) (E)	4	309.4	34.2	126	0.9
SN 500 + poly(decyl methacrylate) (M)	1	278.3	28.0	120	0.8
SN 500 + poly(decyl methacrylate) (E)	1	263.4	21.4	110	0.5
SN 500 + poly(dodecyl methacrylate) (M)	1	298.1	33.0	126	1.1
SN 500 + poly(dodecyl methacrylate) (E)	1	276.5	27.0	117	0.8
SN 500 + poly(decyl acrylate) (M)	1	270.4	23.6	106	0.6
SN 500 + poly(decyl acrylate) (E)	1	196.4	17.4	96	0.6
SN 500 + ORTHOLEUM	1	265.6	26.4	119	0.9

[a]Dose: weight percent addition of the polymeric additive to the base oil AN 500.
(M) and (E) signify products synthesized in microemulsion and emulsion media, respectively.

ported by the report of Pelcher and Ford [51] on the preparation of mainly syndiotactic poly(methyl methacrylate) by microemulsion polymerization.

The more branched polymers are supposed to offer more resistance to the flow of oil. From Table 8 it can be noted that poly(2-ethyl hexyl methacrylate) was more effective in viscosity enhancement than its linear counterpart poly(octyl methacrylate). From Table 8 it is clear that for any additive [except poly(dodecyl acrylate)] synthesized in our laboratory in a microemulsion medium the $Q > 1$, which indicates that additives improve the viscosity index because the thickening effect is greater at 100°C than at 40°C. However, poly(dodecyl methacrylate) was found to be the most effective additive as its Q value ~ 1, indicating that its thickening effect is the same at both high and low temperatures.

VI. CONCLUSIONS

The polymerization of C_{6-12} alkyl acrylates and methacrylates in microemulsion media produced translucent and stable latexes with final particle sizes in the range 30–70 nm. The molecular weights of the microemulsion products were found to be higher than those of the emulsion products synthesized under similar conditions. The microemulsion products were also more effective as viscosity index improvers in the paraffin base oil SN 500 than the emulsion products.

ACKNOWLEDGMENT

The authors gratefully acknowledge the financial support for the work from the Department of Science and Technology, New Delhi, India.

REFERENCES

1. D Dowson. Lubricants and Lubrication in Nineteenth Century. Newcomen Society Lecture. Joint Institution of Mechanical Engineers, 1974, pp 1–8.
2. EW Dean, GHB Davis. Chem Metall Eng 36:618–619, 1929.
3. M Otto, FL Miller, AJ Blackwood, GHB Davis. Oil Gas J 33:98–106, 1934.
4. Rohm and Haas Co. US patent 2,091,627, 1937.
5. Rohm and Haas Co. US patent 2,100,993, 1937.
6. IG Farbenindustrie AG. US patent 2,106,232, 1938.
7. IG Farbenindustrie AG. US patent 2,130,507, 1938.
8. JP Arlie, J Denis, G Parc. Viscosity Index Improvers 1. Mechanical and Thermal Stabilities of Polymethacrylates and Polyolefins. IP Paper 75–005. London: Inst Petr, 1975.
9. JP Arlie, J Denis, G Parc. Viscosity Index Improvers 2. Relations Between the Structure and Viscometric Properties of Polymethacrylate Solutions in Lube Oils. IP Paper 75–006. London: Inst Petr, 1975.
10. P Neudoerful. In: WJ Bartz, ed. Additives for Lubricants and Operational Fluids. Ostfieldern, Germany: Technische Akademie Esslingen, 1986, 11, pp 8.2-1–8.2-15.
11. H Pennewiss, R Benda, H Jost, H Knoell. German patent 2,904,954, 1980.
12. Röhm GmbH. Belgian patent 906,116, 1987.
13. H Pennewiss, R Benda, H Jost, H Knoell. German patent 3,544,061, 1987.
14. C Beyer, R Jelitte, H Pennewiss, H Jost. German patent 4,025,494, 1992.
15. J Omeis, H Pennewiss. European patent 621,293, 1994.
16. C Auschra, H Pennewiss. European patent 744, 457, 1996.
17. H Pennewiss, C Beyer, R Jelitte, B Will, C Auschra, J Omies. European patent 773,260, 1995.
18. S Fengler, H Pennewiss, S Massoth. German patent 19,641,954, 1998.
19. BG Kinker, TA Mcgregor, JM Souchik. World intellectual property organisation patent 9910,454, 1999.

20. RJA Eckert. European patent 113,138, 1984.
21. RJ Sutherland, DA Dubois, US patent 5,272,211, 1993.
22. A Mammen, AS Sarma, K Mallik, PK Rudra. Indian patent 173,176, 1994.
23. CP Bryant, BA Grisso, R Cantiani. European patent 750,031, 1996.
24. P Ghosh, AV Pantar, US Rao, AS Sarma. Indian J Chem Technol 5:309–314, 1998.
25. P Ghosh, AV Pantar, AS Sarma. Indian J Chem Technol 5:371–375, 1998.
26. S Takigawa, K Teranishi, T Nomura, T Suzuki, K Sozo. British patent 2,206,600, 1989.
27. S Takigawa, K Sozo. Japanese patent 04,309,598, 1992.
28. S Takigawa. Japanese patent 06,184,579, 1994.
29. S Takigawa. Japanese patent 10,306,291, 1998.
30. K Sakai, H Matsuya, Y Ohta. World intellectual property organisation patent 95,24,458, 1995.
31. S Takigawa. Japanese patent 10,298,576, 1998.
32. C Neveu, F Huby. Lubr Sci 1:27–49, 1988.
33. BG Kinker, RH Gore, CW Hyndman, BM Stevens. US patent 5,817,606, 1998.
34. P Bataille, N Shariffi-Sanjani. Lubr Eng 51:996–1004, 1995.
35. D Lath, E Lathova, M Bohdanecky. Pet Coal 38:34–36, 1996.
36. K Sakai, S Takigawa. Japanese patent 08,53,687, 1996.
37. GS Kapur, AS Sarpal, SK Mazumdar, SK Jain, SP Srivastava, AK Bhatnagar. Lubr Sci 8:49–60, 1995.
38. TP Hoar, JH Schulman. Nature 152:102–103, 1943.
39. T Bone, JO Stoffer. J Dispersion Sci Technol 1:37–54, 1979.
40. YS Leong, F Candau. J Phys Chem 86:2269–2271, 1982.
41. JS Guo, MS El-Aasser, JW Vanderhoff. J Polym Sci A Polym Chem 27:691–710, 1989.
42. F Candau. Polymeric Dispersions: Principles and Applications. NATO ASI Ser No. 335. Dordrecht: Kluwer, 1997, p 127.
43. M Antonietti, HP Hentz. Chem Ing Technol 69:369–373, 1997.
44. CE Reheberg, CH Fisher. J Am Chem Soc 66:1203–1207, 1944.
45. I Kirk, DF Othmer, eds. Encyclopaedia of Chemical Technology. 3rd ed. Vol 15, New York: Wiley, 1984, pp 353–355.
46. MK Sharma, DO Shah. Ind Eng Chem Fundam 23:213–220, 1984.
47. A Patist, R Chhabra, R Pagidipathi, R Shah, DO Shah. Langmuir 13:432–434, 1997.
48. C Larpent, E Bernard, J Richard, S Vaslin. Macromolecules 30:354–362, 1997.
49. Yu N Sazanov, LA Shibaev, NG Stepanov, NA Sokolosvskava. Int J Polym Mater 14:85–90, 1990.
50. HG Mueller. Tribology Int 11:189–192, 1978.
51. SC Pelcher, WT Ford. Macromolecules 31:3454–3460, 1988.

22
Foams, Foam Films, and Monolayers

DOMINIQUE LANGEVIN   Université Paris Sud, Orsay, France

ABSTRACT

This chapter discusses the relation between foam behavior, properties of the soap films that separate the bubbles, and properties of the surfactant monolayers that cover the film surfaces. Different aspects are taken into account —foaming, ripening, drainage and bubble coalescence—and are illustrated with experimental results. We show, in particular, that the relation between foam properties and foam films and monolayers is far from obvious in mixed surfactant-polymer solutions.

I. INTRODUCTION

Foams have been the subject of a large number of studies for many years [1]. Foams made from aqueous solutions rapidly lose water by gravity drainage when the viscosity of the solution is not too high. When the liquid volume fraction is below about 35%, the bubble surfaces are interconnected in such a way that the total area is minimal in order to minimize the energy that is proportional to this area. The generated complex shapes are "minimal" surfaces and are of interest to mathematicians. The time evolution of foams obeys statistical laws similar to those found in many other domains of physics: crystal growth, flow in granular media, etc. In materials science and biology, many different types of solid foams can be encountered: foams made from polymers, glass and metals, skin, bones, etc. In this chapter we focus on the physicochemical aspects of aqueous foams. This type of foam has many practical applications, e.g., in detergency, the food industry, oil recovery, coating processes, and fire fighting. Despite both fundamental and practical interests, many open questions as simple as "Why does a soap bubble burst?" are still awaiting answers.

In the following we try to relate the foam behavior to the properties of the soap films that separate the bubbles and of the surfactant monolayers that cover the film surfaces. We discuss different aspects: foaming, ripening, drainage, and bubble coalescence. We illustrate these different aspects with experiments done in our laboratory. We show, in particular, that the relation between foam properties and foam films and monolayers is far from obvious in mixed surfactant-polymer solutions.

II. ROLE OF SURFACE PROPERTIES

A. Surface Tension

Most surfactants decrease the surface tension of water by similar amounts, provided the surfactant concentration is large enough (above the critical micelle concentration or cmc, if the surfactant forms micelles) [2]. However, the foam behavior can be very different. An extreme example is found with two nonionic surfactants, $C_{10}E_{10}OH$, decyldecaethylene glycol ether, a classical nonionic surfactant, and $C_{10}E_{10}Cl$, the same molecule in which the terminal OH group has been replaced by a chloride group. Solutions of the first surfactant foam easily, whereas those of the second hardly foam at all [3]. Above the cmc, the solutions have the same surface tension; however, their appearance is different: $C_{10}E_{10}Cl$ solutions are turbid and phase separate after some time. This is because the cloud point of $C_{10}E_{10}Cl$ is below room temperature. It has been shown that the surfactant-rich phase nucleates in the form of droplets of micrometer size that play the role of an antifoam [3]: the droplets trapped in foam films break them when the film thickness is comparable to the droplet size by a dewetting mechanism. Usual antifoams make use of hydrophobic particles, and the antifoam action is effective if the following condition is satisfied:

$$\gamma_{wa}^2 + \gamma_{pw}^2 - \gamma_{pa}^2 > 0 \tag{1}$$

where γ_{wa}, γ_{pw}, and γ_{pa} are, respectively, the interfacial tensions between water and air, particles and water, and particles and air. It was shown that Eq. (1) was also satisfied with the hydrophilic droplets of the surfactant-rich phase [3]. Solutions generating their own antifoam activity can also be found with polymers [4].

B. Surface Coverage and Surface Elasticity

In the preceding example, knowledge of the different surface tensions is sufficient to explain the observed differences in foam properties. However, this is not sufficient in most cases. In general, foam properties depend more on surface coverage than on surface tension. The surfactant concentration in the surface monolayer can be evaluated from the surface tension variation with bulk surfactant concentration [2]:

$$\Gamma_i = \frac{1}{\alpha_i k_B T} \frac{\partial \gamma}{\partial \ln c_i} \tag{2}$$

where Γ_i and c_i are, respectively, the surface and bulk concentrations of surfactant species i, $\alpha_i = 1$ for a nonionic species and 1/2 for an ionic one (in the absence of salt), k_B is the Boltzmann constant, and T is the absolute temperature.

We will see in the following that the ability of the surface layer to oppose the resistance to surface motion is also a very important factor. Surface flow induces surface concentration gradients leading to surface tension gradients responsible for very large surface forces (Marangoni effect). This can be quantified by the surface elasticity ε_G defined as

$$\varepsilon_G = -\Gamma \frac{\partial \gamma}{\partial \Gamma} \tag{3}$$

for a single surfactant species. This elasticity is the so-called *Gibbs elasticity*. When the time scale of the motion is long enough that there are exchanges between the surface and bulk, the *effective* elasticity is smaller [5]. Therefore, the resistance to surface motion depends strongly on the time scale.

Because the Gibbs elasticity ε_G involves a derivative of the surface tension with respect to the surface concentration and the surface concentration is itself a derivative of the surface tension with respect to the bulk concentration, ε_G is proportional to $\partial^2 \gamma / \partial c^2$. The Gibbs elasticity is, therefore, sensitive to very small differences in surface tension variation with surfactant bulk concentration, more sensitive than the surface coverage itself. Both Γ and ε cannot be evaluated by the preceding two equations when the bulk concentration is above the cmc. However, in general, the surface coverage does not change appreciably above the cmc, and the surface properties determined at this concentration can be used as a first approximation for the more concentrated solutions.

III. FOAMING AND DYNAMIC SURFACE TENSION

The question of time scales has already appeared in this discussion. When a foam is generated, the time scale of surfactant adsorption is very important. If the bubble surfaces are not rapidly covered by surfactant monolayers, bubble rupture is easy, and both foam quantity and stability are poor. The adsorption kinetics can be conveniently studied by *dynamic* surface tension devices. In these instruments, a fresh surface is created and the surface tension decrease with time due to adsorption can be monitored [6]. Above the cmc, the micelle lifetime plays an important role; indeed, when new surfaces are created during the foaming process, they are covered by new surfactant molecules coming from the micelles. It has been shown that the amount of foam produced correlates well with the micelle lifetime [7].

The foam height does not depend only on the adsorption kinetics. Indeed, if a foam is very unstable, the bubbles are destroyed just after being formed, and the amount of foam produced is small. Foaming and foam stability cannot, therefore, be considered separately. In the following, we will recall the two main mechanisms leading to foam destruction: ripening and coalescence.

IV. FOAM STABILITY

A. Ripening

The gas pressure inside the smaller bubbles is larger than the pressure in the larger bubbles, so the small bubbles lose their gas content by diffusion across the aqueous phase. The process is faster for gases such as CO_2, which is very soluble in water. This can be slowed down by adding fractions of less soluble gases [8]. The process is also faster for foams with small bubbles, and it is difficult in practice to produce foams with sizes smaller than millimeters when air is used as a blowing agent. Ripening is a rather well-understood process compared with coalescence. Contrary to the case of ripening, stability against coalescence is better for smaller bubbles. The compromise for good stability is, therefore, very difficult to find, much more difficult than with emulsions, where oils with very low solubilities in water can be used to obtain oil-in-water emulsions that are stable against ripening; in the case of water-in-oil emulsions, salt can be added to the water phase to reduce its solubility in the oil.

The foam coalescence process can be separated into two stages: drainage and film rupture.

B. Drainage

Foams drain under the influence of capillary forces and of gravity. Indeed, the zones called *Plateau borders* where bubbles are interconnected are curved; the liquid pressure in the Plateau borders is smaller than in the center of the films, and the liquid is sucked into the borders. Gravity also acts on nonhorizontal films. When the bubbles are large, *dimpling* instabilities frequently occur during the early stages of the drainage. When the surface elasticity is not large enough, this is followed by rapid rupture. This behavior was observed, for instance, with foams made with $C_{10}E_{10}Cl$ solutions below the cmc, where no phase separation occurs (and no antifoam effect is expected). In this concentration range, the stability of foams made with $C_{10}E_{10}Cl$ is also poorer than that of $C_{10}E_{10}OH$ foams. The surface elasticity is larger for $C_{10}E_{10}OH$ monolayers and no film dimpling is observed [3]. If the surface elasticity is large enough, even if a dimple is formed, it remains centrosymmetric; the film drainage is slower because it is limited by the flow through the thinnest regions (drainage velocity varies roughly as $1/h^3$, h being the film thickness [9,10]. The dimpling disappears when h reaches values of the order of 1 μm with films of radius of the order of 1 mm. It should be noted that for films of very small radius, dimpling does not occur [11]. This is probably one of the reasons why small bubbles are more stable against coalescence than large ones. When the film surfaces are flat and parallel, the film thins until it reaches an equilibrium thickness at which the liquid pressure difference between the film center and the Plateau border is equilibrated by the repulsive force per unit area between the film surfaces called the *disjoining pressure* (see Fig. 1).

When there is no early rupture of the foam films, foam drainage is controlled mainly by the flow of the liquid in the channels formed by the Plateau borders. Indeed, when the liquid volume fraction is small, the ratio of the

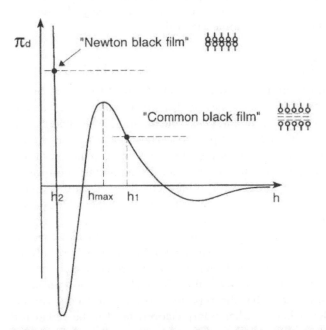

FIG. 1 Schematic representation of the variation of the disjoining pressure Π_d with film thickness h. The dashed lines correspond to different applied pressures ΔP and show the final equilibrium thicknesses of the film.

amount of liquid in the films and in the Plateau borders is also small, so most of the liquid flows through the Plateau borders.

Foam drainage is a very complex hydrodynamic problem, much more difficult to model than film drainage. It has been proposed that *forced drainage* should be easier to interpret [12]. In this type of experiment, a foam column is allowed to drain, and when it has reached its equilibrium liquid fraction, the surfactant solution is poured on the top. It can be shown that the liquid front moves downward at a constant velocity (hydrodynamic soliton). The theory predicts that according to the boundary conditions at the film surface, rigid (surface velocity equal to zero) or fluid (nonzero surface velocity), the front velocity V varies with the flow rate Q as $V \sim Q^{\alpha}$, with $\alpha = 1/2$ in the first case and $1/3$ in the second [12,13]. The transition between the two regimes was observed with mixed sodium dodecyl sulfate (SDS)–dodecanol solutions in which the ratio R = SDS/dodecanol was varied to alter the surface coverage [14]. It is known that both surface elasticity and viscosities (shear and dilational) increase with the addition of dodecanol to SDS solutions [15]. It is not yet clear which of the three parameters is the most important in the preceding experiments; indeed, the transition between the regimes where the surfaces are, respectively, fluid and rigid occurs around $\varepsilon \sim \gamma$ or $\kappa \sim 10\ \eta R$, where κ and η are, respectively, the surface and bulk viscosities and R is the bubble size [5].

V. FILM RUPTURE

Models for film rupture were proposed independently by Vrij and by Scheludko [16]. In these models, it is assumed that the rupture is due to the growth of thermally induced film thickness fluctuations. The models are formally analogous to spinodal decomposition. The thickness growth can occur when short-range surface forces are attractive, for instance, van der Waals force. In practice, when surface coverage is important, short-range repulsive forces, due to hydration, steric hindrance, or other causes, dominate van der Waals forces. These short-range forces are responsible for the stability of the so-called *Newton black films* (Fig. 1).

It is known that foam becomes more stable when the surfactant concentration increases above a value c^* at which Newton black films start to form [17]. This is correlated with the rapid increase in surface coverage and the Gibbs elasticity above a concentration a called the *Szykowski concentration*, $c^* \sim a$. It can be conjectured that the rupture below $c = a$ occurs via the Vrij–Scheludko model, whereas other interpretations need to be found for the rupture of the Newton black films. Exerowa and coworkers [17] proposed a mechanism involving the nucleation of vacancies. In this model, it is assumed that the chemical potential of the surfactant is different in the film and in the solution; because in the solution $\mu = \mu_0 + k_BT \ln c$, the average time of film rupture τ is found to depend ultimately on $\ln c$:

$$\tau = A \exp \left[\frac{B}{\ln(c_0/c)} \right] \tag{4}$$

where c_0, A, and B are constants. When foam films are in equilibrium with Plateau borders, it is, however, not obvious that there is a difference in chemical potential between the film and the solution.

The rupture mechanism can also be due to the amplification of surface concentration fluctuations. The mean square amplitude $\langle \delta\Gamma^2 \rangle$ of these fluctuations scales as [18]

$$\langle \delta\Gamma^2 \rangle \sim \frac{k_BT\Gamma^2}{\varepsilon} \tag{5}$$

The corresponding time for the amplification of a fluctuation $\delta\Gamma$ scales as [19]

$$\tau \sim \exp \left(\frac{\delta\Gamma^2}{\langle \delta\Gamma^2 \rangle} \right) \tag{6}$$

$\delta\Gamma$ should be at least equal to $\Gamma - \Gamma_a$, where Γ_a is the coverage at the Szykowski concentration where the short-range repulsive forces disappear. After sufficient growth of this fluctuation, film rupture can proceed via the Vrij–Scheludko mechanism. Because Γ_a is very small, $\delta\Gamma \sim \Gamma$ and according to Eqs. (5) and (6), the rupture time varies exponentially with the surface elasticity: $\tau \sim \exp(\varepsilon)$.

The measurements of rupture times for Newton black films were found to be in good agreement with Eq. (4) [17]. However, it has been remarked that above the Szykowski concentration $\varepsilon \sim 2\Pi$, where Π is the surface pressure of the monolayer ($\Pi = \gamma_0 - \gamma$, γ_0 being the surface tension of water) [20]. Because Π is usually well described by the Langmuir adsorption equation, $\Pi = k_BT\Gamma \ln(c/a)$, it follows that if the rupture time is an exponential function of ε, the bulk concentration dependence of τ is formally identical to that of Eq. (4). We have simultaneously measured the film rupture time and the surface elasticity and found that $\tau \sim \exp(\varepsilon)$, as predicted by Eqs. (5) and (6) [21].

This mechanism would also explain the general lack of correlation between foam stability and surface forces. For instance, in the alkyltrimethylammonium bromide surfactant series, DTAB (dodecyltrimethylammonium bromide) does not lead to stable foams, whereas TTAB and CTAB, the surfactants with tetradecyl and hexadecyl chains, respectively, give much more stable foams. The measured surface forces (at the cmc) are, however, similar [22]. The Gibbs elasticities are similar too, but the finite frequency elasticities (a few hundreds hertz) are similar for TTAB and CTAB and much smaller for DTAB [23]. The DTAB elasticities are smaller because of the solubilization process mentioned earlier, which is faster than for the other two compounds. This could play a role in the growth process of the surface fluctuations.

VI. RELATION BETWEEN FOAM AND FOAM FILM STABILITY

It is generally admitted that the foam stability is related to foam film stability when coalescence is the governing mechanism for film stability. Foam ripening indeed proceeds without film rupture.

Many experimental devices for the study of foam films have been proposed. The so-called *porous plate* method allows the study of horizontal films under a controlled external applied pressure to study film drainage and to measure surface forces as a function of film thickness h [24]. We have used this device to study films made from mixed polymer–surfactant solutions. In these experiments, the surfactant were cationic, DTAB, TTAB, and CTAB, and the polymers were anionic, polyacrylamidopropane sulfonate (PAMPS) and xanthan, a natural polysaccharide. We have used PAMPS polymers with two different degrees of sulfonation, 10 and 25% (number percentage of charged monomers). In the case of mixed solutions with DTAB, the foams are not very stable, but their stability is not very different from those of pure DTAB solutions (Fig. 2) [25]. The only difference is in the amount of foam obtained for a fixed gas flow rate: much smaller for mixed solutions with xanthan than with the other polymers. When horizontal foam films are formed on holes drilled in porous glass frits, the behavior of PAMPS and xanthan mixed solutions is very different. Let us consider first the case of solutions in which the surfactant concentration is below the cac (critical concentration for the formation of polymer–surfactant aggregates in bulk; cac < cmc) and below the precipitation limit (number concentration of anions > number concentration of cations). In these conditions, the DTAB/ PAMPS films are much more stable than the pure DTAB films. The DTAB/ xanthan films are completely unstable and break immediately. With TTAB and CTAB/xanthan solutions, the films are also completely unstable, although the pure TTAB and CTAB films are much more stable than pure DTAB films. Despite these striking differences in film stability, no differences in foam stability are observed (Fig. 2b).

In the precipitation region, we observed that the mixed DTAB/PAMPS films were still more stable, impossible to rupture. The images of the films showed that microgel precipitates were trapped in the films and were prob-

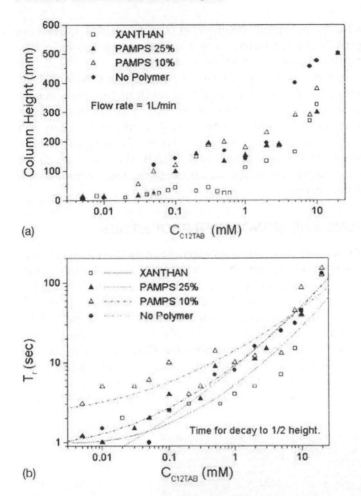

FIG. 2 (a) Foam height as a function of surfactant concentration for the different solutions and a constant flow rate of nitrogen. (b) Foam lifetimes T_r versus surfactant concentration for the different solutions. T_r is the time taken by the foam column to drop to half its original height after the gas flow is stopped. The lines are second-order polynomial fits to the experimental points and are shown as a guide to the eye. (Data from Ref. 25.)

ably responsible for the enhanced stability (Fig. 3) [26]. However, when the precipitation region is crossed (around $c = 1$ mM for the solutions of Fig. 2), no particular feature is observed in the foam behavior.

The explanation of this puzzling behavior is probably the following. During foam production, the foam films are formed much more rapidly than in the porous plate device. The adsorption of the mixed layers at the liquid surface is extremely slow in these systems [25]. It is, therefore, likely that the surface coverages are different in the two types of experiments, much lower in the case of the foams in which only DTAB has time to reach the surface. This would explain the similarities between the foam behavior with and that without the polymers.

VII. FOAM FILMS AND MONOLAYER PROPERTIES

There is generally a good correlation between the monolayer coverage and the foam film drainage and rupture. The case of the mixed polymer–surfactant solutions is again an exception. There is a great difference in the behavior of the films made with PAMPS and xanthan. The surface properties are, however, similar: similar surface tension curves for xanthan and PAMPS 10%, similar monolayer thicknesses (in the range 3–5 nm for the polymer region), and similar surface elasticities, both the Gibbs elasticity and the finite frequency elasticity. The only difference that we could detect is the behavior of the monolayers upon large compressions: the xanthan mixed layers can be compressed to a much larger extent than the PAMPS layers

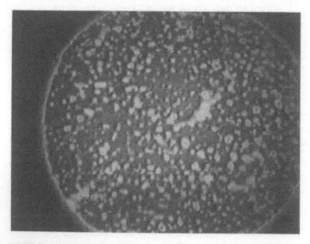

FIG. 3 Image of a film made from a mixed solution of DTAB and PAMPS 25% in the precipitation region. (Data from Ref. 26.)

[25]. However, the relation between this feature and the film rupture is not yet clear.

VIII. CONCLUSION

There are still many unsolved questions posed by aqueous foam behavior. Foam film rupture is one of the most difficult unsolved questions. So far, the admitted correlation between foam and foam film stability does not even hold in polymer–surfactant solutions, however these are currently used in practical applications. In order to understand foam behavior, the time scales of the different surface processes need to be evaluated with care and compared together. The role of surface elasticity in film rupture seems important, but because this quantity is time scale dependent, more work is needed to clarify the relation between elasticity and rupture time.

REFERENCES

1. D Weaire, S Hutzler. The Physics of Foams. New York: Oxford University Press, 1999.
2. A Adamson. Physical Chemistry of Surfaces. New York: Wiley, 1976.
3. A Bonfillon-Colin, D Langevin. Langmuir 13:599, 1997; A Colin, J Giermanska-Kahn, D Langevin, B Desbat. Langmuir 13:2953, 1997.
4. S Ross. Colloids Surf A 118:187, 1996.
5. VG Levich. Physico-Chemical Hydrodynamics. Englewood Cliffs, NJ: Prentice Hall, 1962.
6. SS Dukhin, G Kretzschmar, R Miller. Dynamics of Adsorption at Liquid Interfaces. Amsterdam: Elsevier, 1995.
7. SG Oh, DO Shah. Langmuir 7:1316, 1991.
8. F Gandolfo, H Rosano. J Colloid Interface Sci 194:31, 1997.
9. I Ivanov, DS Dimitrov. In: I Ivanov, ed. Thin Liquid Films. Surfactant Sci Ser 29. New York: Marcel Dekker, 1988, p 379.
10. JL Joye, G Hirasaki, CA Miller. Langmuir 8:3085, 1992; 10:3174, 1994; JL Joye, G Hirasaki, CA Miller. J Colloid Interface Sci 177:542, 1996.
11. OD Velev, GN Constantinides, DG Avraam, AC Payatakes, RP Borwankar. J Colloid Interface Sci 175:68, 1995.
12. D Weaire, S Hutzler, G Verbist, EAJ Peters. Adv Chem Phys 102:315, 1997.
13. S A Koehler, S Hilgenfeldt, HA Stone. Phys Rev Lett 82:4232, 1999.
14. M Durand, G Martinoty, D Langevin. Phys Rev E 60:R6307, 1999.
15. DO Shah, NF Djabarrah, DT Wasan. Colloid Polym Sci 251:1002, 1978; NF Djabarrah, DT Wasan. Chem Eng Sci 37:175, 1982.
16. A Vrij. Discuss Faraday Soc 42:23, 1966; A Scheludko. Adv Colloid Interface Sci 39:1, 1967.
17. D Exerowa, D Kashchiev, D Platikanov. Adv Colloid Interface Sci 40:201, 1992.

18. L Kramer. J Chem Phys 55:2097, 1971.
19. L Landau, E Lifshitz. Statistical Physics. New York: Pergamon Press, 1980.
20. J Lucassen, D Giles. J Chem Soc Faraday I 71:217, 1975.
21. F Bauget, D Langevin, R Lenormand. J Colloid Interface Sci 239:501, 2001.
22. V Bergeron. Langmuir 13:3474, 1997.
23. F Monroy, J Kahn, D Langevin. Colloids Surf A 143:251, 1998.
24. K Mysels, M Jones. Discuss Faraday Soc 42:42, 1966.
25. C Stubenrauch, PA Albouy, RY Klitzing, D Langevin. Langmuir 16:3206,
     2000; A Bhattacharya, F Monroy, D Langevin, JF Argillier. Langmuir 16:8727,
     2000; H Ritacco, PA Albouy, A Bhattacharyya, D Langevin. Phys Chem Chem
     Phys 2:5243, 2000.
26. V Bergeron, A Asnacios, D Langevin. Langmuir 12:1550, 1996.

23

Role of Entry Barriers in Foam Destruction by Oil Drops

ASEN D. HADJIISKI, NIKOLAI D. DENKOV,
SLAVKA S. TCHOLAKOVA, and IVAN B. IVANOV
Sofia University, Sofia, Bulgaria

ABSTRACT

Different oils (mainly hydrocarbons or silicone oils) and their mixtures with hydrophobic solid particles are widely used for destruction of undesirable foam. For a long time, the entry, E, spreading, S, and bridging, B, coefficients (which can be calculated from the oil–water, oil–air, and water–air interfacial tensions) were used to evaluate the activity of such oil-based antifoams (AFs). However, recent studies showed that there was no correlation between the magnitudes of E, S, and B and the antifoam activity—the only requirement for having an active AF, in this aspect, is to have positive E and B. Instead, it was shown that the so-called entry barrier, which characterizes the ease of entry of pre-emulsified oil drops in the solution surface, was of crucial importance; an easy entry (low entry barrier) corresponded to an active AF and vice versa. We developed a new method, the film trapping technique (FTT), which allows one for the first time to measure directly the critical capillary pressure, P_C^{CR}, which induces the entry of micrometer-sized oil drops, identical to those in real AFs. This chapter describes the main results obtained so far by the FTT with various systems.

The results show that P_C^{CR} can be used as a relevant quantitative characteristic of the entry barrier. The value of P_C^{CR} determines the boundary between two rather different classes of AF: fast antifoams (defoaming time < 5 s, P_C^{CR} < 20 Pa) and slow antifoams (defoaming time > 5 min, P_C^{CR} > 20 Pa). These two classes differ in the mechanism by which they destroy foam. The fast AF destroys the foam films in the first several seconds after their formation, whereas the drops of the slow AF destroy the foam only after being compressed by the walls of the shrinking Gibbs–Plateau borders, at a much later stage of foam evolution. Furthermore, a linear relationship between the value of P_C^{CR} and the final height of the foam is established and explained theoretically the slow AFs. The effects of several factors (type of oil, size of the oil drops, surfactant concentration, presence of a spread oil layer over the solution surface) on the entry barrier are studied. It is shown for one specific system that the presence of a prespread oil layer on the solution surface strongly affects the entry barrier—the latter is reduced by the spread layer for decane and dodecane, whereas the effect is the opposite for hexadecane (fivefold increase). The calculations show that there is a big difference between the numerical values of the critical *capillary* pressure, P_C^{CR}, and the critical *disjoining* pressure, Π_{AS}^{CR}, for micrometer-sized oil drops; therefore, one should analyze separately the relation between these two quantities and the antifoam activity. In conclusion, the FTT has provided valuable new information about the role of the entry barrier in the activity of oil-based antifoams.

I. INTRODUCTION

Oily additives are used in various technologies (e.g., paper and pulp production, ore flotation, fermentation) and commercial products (detergents, paints, some pharmaceuticals) to avoid the formation of an excessive foam, which would impede the technological process or the product application [1–3]. In other cases (oil recovery and refinement, shampoos, emulsions for metal processing machines) oil drops are present, without being specially introduced for foam control, and can also affect the foamability of the solutions. Small fractions of hydrophobic solid particles of micrometer size, such as hydrophobized silica or alumina, plastic grains, or stearates of multivalent cations, are often premixed with the oil because the solid–oil "compounds" obtained exhibit a much stronger foam destruction effect than the individual components taken separately [4–8]. Such oily additives are termed *antifoams* in the literature and can be based on hydrocarbons, polydimethylsiloxanes (PDMSs, silicone oil) or their derivatives [1,4].

The mechanisms by which oil drops destroy foams are still a matter of discussion in the literature [2–15]. Two of the possible mechanisms (called bridging–stretching and bridging–dewetting) are illustrated in Fig. 1. Observations by optical microscope and high-speed video camera showed that the bridging–stretching mechanism was operative when compounds of silicone oil and silica were added to some surfactant solutions [11]. Several of the mechanisms proposed in the literature relate the antifoam activity of the oil to its spreading behavior (see the respective discussion in Refs. 4 and 5 and the literature cited therein). The so-called entry, E, spreading, S, and bridging, B, coefficients (which can be calculated from the air–water, air–

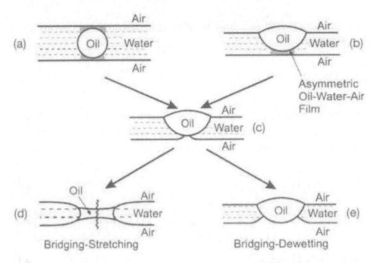

FIG. 1 Formation of asymmetric oil–water–air films (shaded areas) in two of the possible mechanisms of foam destruction by oil drops or lenses: bridging–stretching (a-c-d) and (b-c-d) [11,12]; bridging–dewetting (a-c-e) and (b-c-e) [2–6].

oil, and oil–water interfacial tensions) have often been invoked to characterize the oils with respect to their antifoam properties. However, several studies [4,9,13–17] demonstrated that there was no direct relation between the values of these coefficients and the antifoam activity of the oils. Instead, a correlation between the antifoam activity and the so-called entry barrier, which characterizes the ease of oil drop entry in the surface of the surfactant solution (see later for quantitative definitions), has been established. The primary reason for this correlation is that whatever the mechanism of foam destruction by emulsified oil might be, it must include the stage of formation and rupture of asymmetric oil–water–air films (see Fig. 1). As noted by Kruglyakov [18] and Kulkarni et al. [19], these asymmetric films could be stabilized by various surface forces (electrostatic, van der Waals, etc.), which hinder or suppress the drop entry and thus impede the antifoam action of the oil. Later on, Wasan and coworkers [3,10] studied the importance of the oscillatory structural forces for the stability of the asymmetric films formed from surfactant solutions with concentration well above their critical micelle concentration (cmc). Furthermore, they showed that the introduction of oil into the foaming solution could lead to a more stable foam if the asymmetric film was very stable (due to decelerated water drainage, as a result of the obstruction of the Plateau borders by oil drops) [20].

Several different characteristics have been suggested in the literature to quantify the entry barriers for oil drops. Lobo and Wasan [10] suggested using the energy of interaction per unit area in the asymmetric oil–water–air film, f, as a criterion of its stability:

$$f(h_E) = -\int_{h\to\infty}^{h_E} \Pi_{AS}\, dh \tag{1}$$

where $\Pi_{AS}(h)$ is the disjoining pressure and h_E is the equilibrium thickness of the asymmetric film at a certain capillary pressure. Bergeron et al. [9] suggested the so-called generalized entry coefficient

$$E_g(h_E) = -\int_0^{\Pi_{AS}(h_E)} h\, d\Pi_{AS} \tag{2}$$

where the lower limit of the integral corresponds to $\Pi_{AS}(h\to\infty) = 0$. As seen from their definitions, Eqs. (1) and (2), f and E_g are closely interrelated

$$E_g(h_E) + f(h_E) = -h_E\Pi_{AS}(h_E) \tag{3}$$

The determination of the values of f and E_g and their comparison with the antifoam efficiency of different oils is a difficult task because one needs to know the dependence of the disjoining pressure, Π_{AS}, on the thickness of the asymmetric film (for h varying from infinity to h_E). The most thorough analysis of this type was carried out by Bergeron et al. [9], who measured the disjoining pressure isotherms of planar foam (air–water–air) and asymmetric (oil–water–air) films by the porous plate method [21] and determined E_g for two surfactant–oil couples. Although qualitative agreement between the values of E_g and the foam stability was established, a rigorous quantitative comparison for all systems was impossible because one cannot measure the attractive regions of the $\Pi_{AS}(h)$ isotherms by the porous plate method.

One important result of that study [9] was that the authors convincingly showed that the destabilizing effect of oil was related to much lower stability of the asymmetric films as compared with the foam films. Furthermore, a good correlation was found between the stability of foams that were formed in porous media in the presence of oil (thus resembling the foams in oil reservoirs) and the critical capillary pressure leading to rupture of the asymmetric films, as measured by the porous plate method. Indeed, the critical capillary pressure seems to be an appropriate measure of the film stability in such systems because the capillary pressure is the actual external variable that compresses the film surfaces toward each other, against the repulsive surface forces (disjoining pressure) stabilizing the film. It is worth noting that the authors [9] studied *planar* films, where the imposed capillary pressure in equilibrium was exactly equal to the stabilizing disjoining pressure —that is, the concept of the critical *capillary pressure*, P_C^{CR}, was equivalent to the concept of the critical *disjoining pressure*, Π_{AS}^{CR}, in their studies. As explained subsequently, the asymmetric films formed in typical antifoam systems are strongly curved and there is a large difference between the values of P_C^{CR} and Π_{AS}^{CR}.

Another experimental tool has become available for quantifying the entry barriers of oil drops. Hadjiiski and coworkers [17,22–25] developed a new method, the film trapping technique (FTT), which consists of trapping oil drops in a wetting film, formed from a surfactant solution on a solid substrate, and subsequent measurement of the critical capillary pressure that leads to entry of the oil drop in the fluid surface of the wetting film (see later for details). The FTT has several advantages compared with the porous plate method. First, experiments with real antifoam drops of micrometer size can be carried out, giving a quantitative measure of the entry barrier that can be used to explain the antifoam activity [13–17]. Second, the FTT allows independent variation of the radius of the asymmetric oil–water–air film and of the applied capillary pressure; i.e., the dependence of P_C^{CR} and Π_{AS}^{CR} on the size of the asymmetric film can be investigated. Third, the method can be applied to different types of films (asymmetric oil–water–air, emulsion and foam films) so that a comparison of their stability for a given surfactant–oil system is possible. Last, but not least, the FTT requires relatively simple and inexpensive equipment, and after accumulating some experience one can rapidly obtain a large set of data. These features make the method an interesting complement and/or alternative to the other methods for studying liquid films.

In this chapter we present a brief overview of the results obtained so far by the FTT with various oils and surfactants in relation to antifoaming. As shown here, the critical capillary pressure, determined in the FTT experiments, has a close relation to the actual process of foam destruction by oil drops. Several conclusions about the mechanism of antifoaming and the antifoam activity of the oils have been drawn and presented in quantitative terms by using the concept of the critical capillary pressure, P_C^{CR}, and the FTT results.

II. EXPERIMENTAL DETAILS

A. Materials

Sodium dioctylsulfosuccinate, AOT (Sigma Chemical Co., St. Louis, MO); octylphenol decaethylene glycol ether, Triton X-100 (Merck KGaA, Darmstadt, Germany); alkyl-$C_{12/14}$ (glucopyranoside)$_{1,2}$, also called alkyl-polyglucoside or APG (Henkel KGaA, Germany); sodium dodecyl polyoxyethylene sulfate, SDP3S (Kao Co., Tokyo, Japan); sodium dodecyl benzene sulfonate,

SDDBS (Aldrich, Steinheim, Germany); and sodium dodecyl sulfate, SDS (Sigma) were used as surfactants. In two series of experiments lauryl amide propyl betaine, Betaine (Kao) and n-dodecanol, n-C$_{12}$OH (Sigma) were added to SDP3S and SDS solutions, respectively, as foam boosters. The compositions of the surfactant solutions are summarized in Table 1.

The following oils were studied: n-octane, n-C$_8$; n-decane, n-C$_{10}$; n-dodecane, n-C$_{12}$; n-hexadecane, n-C$_{16}$; n-dodecanol, n-C$_{12}$OH (all products of Sigma), and two silicone oils (polydimethylsiloxane, PDMS) of dynamic viscosities 5 and 1000 mPa·s.

A mixed, solid–oil compound was also used as an antifoam, which consisted of 4.2 wt% hydrophobized silica particles dispersed in PDMS of viscosity 1000 mPa·s. The oil concentration in the working solutions was 0.1 wt% for pure oils and 0.01 wt% for mixed solid–oil compounds.

Hexadecane was purified by passing it through a glass column filled with chromatographic adsorbent (Florisil). The other chemicals were used as received. The solutions were prepared with deionized water from Milli-Q Organex system (Millipore).

B. Methods and Procedures

1. Foam Formation and Foam Stability Evaluation

Because the solid–oil compounds destroy the entire foam column within a few seconds, whereas the pure oils are much slower (minutes or tens of

TABLE 1 Composition of the Surfactant Solutions Studied and Critical Micelle Concentration (cmc) of the Surfactant

Surfactant	Surfactant concentration (mM)	cmc (mM)	Electrolyte concentration (mM)
AOT	10	3	0
APG	0.45	0.15	10
Triton X-100	1	0.18	0
SDP3S	0.5; 20; 100	0.5	0.33; 13.2; 66
SDP3S/betaine (80:20 molar ratio)	100	<0.5	81
SDP3S/n-dodecanol (97:3 molar ratio)	· 100	<0.5	0
SDDBS	0.16–12.8	0.5	12
SDS	20	8	0
SDS/n-dodecanol (97:3 molar ratio)	20	<8	0

minutes), different tests were used to produce foams and to compare their stability in the presence of various antifoams.

(a) *Ross–Miles Test.* This test is appropriate for studying the antifoam action of pure oils [13,14]. Briefly, 0.3 mL of oil was introduced into 300 mL of the surfactant solution. The oil was emulsified in the solution by intensive stirring for 20 min using a magnetic stirrer. The emulsion obtained was additionally homogenized by several hand shakes before placing it into the glass cylinder of the Ross–Miles test. The cylinder had a working volume of 1 dm^3 and diameter of 37 mm. The solution was circulated (pumped) for 20 s at a rate of 125 cm^3/s through an orifice (7 mm in diameter), which was placed at 23 cm above the level of the liquid. The volume of the initially formed foam and the evolution of the foam volume with time were monitored for a period of 15 min (100 min in some of the experiments) after ceasing the liquid circulation. The accuracy in the foam volume determination was ±2 mL, whereas the reproducibility was typically ±5 mL.

(b) *Automated Shake Test (AST).* This test is appropriate for evaluating the antifoam activity of solid–oil compounds [6,11,15]. The tests were performed on an automatic shake machine Agitest (Bioblock). The foaming solution (100 mL) was placed in a standard 250-mL glass bottle, and 0.01 mL of the compound was introduced into this sample. The bottle was then mechanically agitated by the machine, at a frequency of 360 min^{-1} and an amplitude of 2 cm. After 10 s of agitation, the shaking was stopped and the time for complete foam destruction (defoaming time) was monitored for a period that did not exceed 60 s. A micropipette Nichiryo M800 (Nichiryo Co., Tokyo, Japan), specially designed to supply small volumes of viscous substances, was used to load the compound.

2. Methods for Studying Foam Films

Several complementary methods were used to observe the process of foam film thinning and destruction in the presence of antifoam globules.

(a) *Capillary Cell.* Single, millimeter-sized foam film was formed by the method of Scheludko and Exerova [26,27]. A drop of surfactant solution (sometimes containing dispersed antifoam drops) was placed in a short capillary tube with an internal diameter of 2.5 mm. The drop acquired a biconcave shape with the thinnest region being in the center of the capillary. The foam film was formed by sucking out liquid from the drop through a side orifice drilled in the capillary wall. This film was observed in reflected monochromatic light by a microscope (Zeiss Axioplan, Germany; objectives Plan-Neofluar, 10×/0.30; LD Epiplan, 20×/0.40). The image was recorded by a CCD camera (Panasonic WV-CD20, 25 fps). The interference of the light reflected from the upper and lower surfaces of the foam film led to the appearance of dark and bright interference fringes, each corresponding to a given film thickness. The difference, Δh, in the film thickness between two neighboring dark (or two neighboring bright) fringes was equal to

$$\Delta h = \frac{\lambda}{2n} \approx 205 \text{ nm} \tag{4}$$

where $\lambda \approx 546$ nm is the wavelength of the illuminating light and $n = 1.33$ is the refractive index of the surfactant solution.

A major advantage of the capillary cell is that experiments can be performed with surfactant solutions containing micrometer-sized oil globules or lenses, just as in the case of practical antifoams [11,28]. Thus, the films in the capillary cell closely mimic the behavior of relatively small films (of diameter around 1 mm) in real foams.

(b)   Dippenaar Cell [29].   We used this technique to observe the evolution of the oil bridge that formed when an oil drop came in direct contact with the two surfaces of a foam lamella [11]. Similarly to the experiments in the capillary cell, the foam lamella was formed between two concave menisci in a short capillary tube (4 mm internal diameter). The lamella thickness was controlled by sucking liquid through a side orifice. The placement of an antifoam drop ($\approx$2 μL in volume) on the upper meniscus led to the formation of an oil lens floating on the solution surface. When the bottom of this lens touched the lower air–water meniscus, an oil bridge could be formed. The shape of the oil bridge was monitored in transmitted white light. The main advantage of the Dippenaar cell is the possibility for direct optical observation of the bridge shape.

3.   Film Trapping Technique (FTT)

(a)   Experimental Setup and Basic Principles of Operation.   The critical capillary pressure leading to entry of the oil drops was measured by the FTT [25]; see Fig. 2. A vertical glass capillary of radius 4 mm was positioned at a small distance above the flat bottom of a glass vessel. The lower end of the capillary was immersed in the working surfactant solution, which contained dispersed antifoam globules. The capillary was connected to a pressure control system, which allowed one to vary and to measure the difference, ΔP_A, between the air pressure in the capillary, P_A, and the ambient atmospheric pressure, P_A^0. The data acquisition equipment included a pressure transducer (Omega Engineering, Stamford, CT) and a digital multimeter Metex M-4660A (Metex Instruments) connected to a personal computer (PC).

When P_A increased, the air–water meniscus in the capillary was pushed against the glass substrate and a wetting film was formed, which trapped

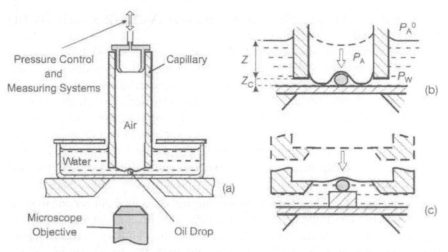

FIG. 2 Scheme of the experimental setup and the basic principles of operation of the film trapping technique (FTT) [25]. (a) Vertical capillary, partially immersed in surfactant solution containing oil drops, is held close to the bottom of the experimental vessel. (b) The air pressure inside the capillary, P_A, is increased and the water–air meniscus in the capillary is pressed against the glass substrate. Some of the oil drops remain trapped in the formed glass–water–air film and are compressed by the meniscus. At a given critical capillary pressure (see Section II.B.3 for details) the asymmetric film formed between the oil drop and the solution surface ruptures and a drop entry event is observed by an optical microscope. (c) Another modification called gentle FTT is used for measuring entry barriers lower than 20 Pa—a flat meniscus is formed, which allows the trapping of drops at virtually zero capillary pressure.

some of the antifoam globules (Fig. 2b). These drops remained sandwiched between the air–water meniscus and the glass substrate. The setup allows one to measure the capillary pressure of the air–water meniscus around the trapped drops, which can be expressed by the following relationship:

$$P_C = \Delta P_A - \rho_w g Z \qquad (5)$$

where the second term on the right-hand side of the equation accounts for the hydrostatic pressure at the bottom of the vessel (ρ_w is the mass density of water, g is the gravity acceleration). The depth of the liquid, Z, was measured by a micrometer translater having an accuracy of ± 5 μm, which corresponded to a precision of ± 0.05 Pa in the determination of the hydrostatic pressure. A Carl Zeiss Jena inverted microscope, equipped with objective LD Epiplan, $20\times/0.40$, digital CCD camera (Kappa CF 8/1 DX) connected to a PC and videocassette recorder (VCR) (Panasonic NV-HD 680), was used for these observations.

The experiments showed that the trapped antifoam globules entered (pierced) the surface of the wetting film at a given, critical capillary pressure, P_C^{CR}. The moment of drop entry, which was accompanied by a significant local change in the shape of the air–water interface, was clearly seen in both reflected and transmitted light. Therefore, the equipment allows one to measure P_C^{CR} as a function of the solution composition and drop radius. As mentioned before, for brevity we refer to P_C^{CR} as the *barrier to drop entry*. A larger P_C^{CR} corresponds to a higher barrier (more difficult drop entry) and vice versa.

(b) Gentle-FTT. The experimental setup just described allows one to measure entry barriers higher than about 20 Pa. This limit is determined by the capillary pressure of the meniscus formed in the capillary before trapping the drops. Because the surfactant solution wets the inner surface of the capillary, a spherical meniscus is formed with capillary pressure $P_C \approx 2\sigma_{AW}/R_{CAP} \sim 15$ Pa ($\sigma_{AW} \approx 30$ mN/m is the surface tension of the solution and $R_{CAP} \approx 4$ mm is the capillary radius). However, the entry barriers are sometimes lower and another modification of the method, called "gentle FTT," has been developed [25] for such cases (Fig. 2c).

The main idea of gentle FTT is to create an almost flat air–water interface in the capillary before trapping the drops so that P_C in the beginning of the experiment is virtually zero. For this purpose, a sapphire disk of special design was attached to the lower end of the capillary. The disk had an opening with a wedgelike shape (Fig. 2c), which ensured stable attachment of the air–water interface to the sapphire upper edge. In addition, a substrate with a small stub, cut out onto a glass plate, was used in these experiments. The plate was placed on the vessel's bottom so that the stub was projected upward into the central hole of the sapphire disk. One could move the capillary precisely in the *x-y-z* directions and juxtapose the flat liquid interface with the glass stub. Thus, one could achieve trapping of drops by a flat interface, followed by a gentle increase of P_C until P_C^{CR} was reached.

(c) Drop Entry Measurements in the Absence of a Spread Oil Layer. The working surfactant solutions, containing oil-based antifoams, usually have a spread layer of oil on their surface as a result of the coalescence of antifoam globules with this surface [11]. Some of the experiments were aimed at revealing how the presence of this prespread oil layer affected the height of the entry barrier. For this purpose, comparative experiments, in both the presence and absence of spread oil, were performed with solutions of the anionic surfactant SDDBS and several oils.

For oils that were not very soluble in the surfactant solution, such as dodecane and hexadecane, a clean surface (free of spread oil) was created by pouring the studied emulsions into the experimental vessel for FTT experiments by the two-tip procedure (TTP) [11]. The latter consisted of gentle injection of the solution, containing emulsified oil drops, through a narrow orifice (syringe needle or pipette tip) into the working vessel—in this way the oil layer, spread on the surface of the "mother" emulsion, was retained and a clean solution surface was created for the FTT experiment. It took some time, which depended very much on the oil and surfactant used, before a new portion of spread oil appeared due to coalescence of oil drops with the solution surface or to molecular transfer of oil [11,17]. Independent surface tension measurements revealed that for dodecane and hexadecane the value of σ_{AW} of the emulsions poured by the TTP was virtually the same as that of the pure SDDBS solution (without oil) and decreased very slowly with time, by less than 0.5 mN/m for a period of 1 h, which was about the time span of the typical FTT experiment. In these FTT tests, the number of the drop entry events observed in a single experiment was restricted to six to avoid the accumulation of a detectable layer of spread oil from the entering drops.

For oils that were soluble in water and/or solubilized in the surfactant micelles (such as octane and dodecanol), it was impossible to create a solution surface free of spread oil for a period of time long enough to carry out the FTT experiments. In these systems, the entry barrier was measured only in the presence of spread oil.

All experiments were carried out at the ambient room temperature ($T = 25 \pm 2°C$). The experiments with dodecanol were performed at a temperature above its melting point (mp = 24°C). The atmosphere above the wetting film was saturated with aqueous vapors so that the water evaporation inside the capillary was suppressed.

4. Surface Tension Measurements

The surface tension of the surfactant solutions was measured by the Wilhelmy plate method, whereas the surface tension of the oils was measured by the Du Nouy ring technique on a Krüss K10T digital tensiometer. The interfacial tension of the oil–solution interfaces was measured by the pendant drop method.

III. EXPERIMENTAL RESULTS AND DISCUSSION

A. Relation Between the Drop Entry Barrier and the Antifoam Activity

1. Drop Entry Barrier and Foam Lifetime—Fast and Slow Antifoams

A careful look at the results published in the literature [1–8,30] and our own foam tests (Refs. 11 and 13–16 and some new results presented later) show a segregation of the antifoam systems into two distinct groups, depending on the time scale of foam destruction. Some of the antifoams (usually these are mixed solid–oil compounds) are very active and destroy the entire foam column in less than 10 s [3,6,11,15]. On the other hand, oils deprived of solid particles typically require a much longer time (at least several minutes) to destroy a significant fraction of the foam [8,13–16,20]. For example, 0.01 wt% silica-PDMS compound destroys completely the foams stabilized by AOT (10 mM) or Triton X-100 (1 mM) in 2–3 s in the automatic shake test; whereas the same silicone oil, when deprived of silica, has very slow action and several minutes elapse between the foam generation and the beginning of the foam destruction process. Moreover, if pure oil is used, the foam destruction is incomplete and a long-standing residual foam often remains stable for hours. To distinguish these two types of antifoams, we call them "fast" and "slow," respectively [15].

To clarify the actual reason for the different activities of the fast and slow antifoams, we performed a large set of experiments aimed at revealing the relation between the entry barrier and the characteristic defoaming time. As an illustration of the results obtained, we will discuss first the FTT experiments with 10 mM AOT solutions. Two antifoam systems were compared: (1) pure silicone oil of viscosity 1000 mPa·s and (2) compound comprising the same oil and hydrophobic silica particles. The gentle-FTT setup was used for these measurements because some of the entry barriers were rather low.

The results for the critical capillary pressure, P_C^{CR}, as a function of the equatorial radius of the antifoam drops, R_E, are shown in Fig. 3. The triangles present the results for P_C^{CR} measured with compound globules, whereas the open circles show data for pure oil drops. The two horizontal lines represent the respective averaged values for a given antifoam. As seen from the figure, the magnitude of the entry barrier does not depend significantly on the drop size in the range studied (typical for real antifoams). On the other hand, the mean entry barrier for pure oil is about 20 Pa, whereas it is much lower for the mixed solid–oil antifoam, about 1 Pa. Therefore, the introduction of hydrophobic silica into the oil leads to a significant decrease of the entry barrier, by a factor of 20. As mentioned before, this reduction is accompanied by a very strong acceleration of the foam destruction process. These results reinforce the hypothesis of Garrett [4], supported by the experiments of Bergeron et al. [6], Koczo et al. [7], and Aveyard and Clint [8], that the main role of the solid particles in mixed solid–oil antifoams was to reduce the entry barrier of the antifoam globules.

Further, we performed similar experiments with a variety of different systems—oils, compounds, and surfactant solutions. The results are summarized in Fig. 4, which shows the relationship between the foam lifetime

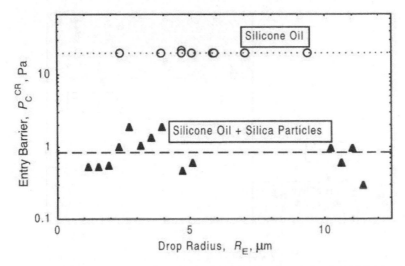

FIG. 3 Drop entry barrier versus drop radius measured by gentle FTT for solution of 10 mM AOT: pure silicone oil (open circles), and silica–PDMS compound (filled triangles). The horizontal lines represent the respective average values.

and the entry barrier. Each experimental point corresponds to a different surfactant antifoam couple (see the figure caption for specification of the systems studied). One sees that the experimental data fall into two distinct regions: (1) systems where the foam is destroyed in less than 5 s (fast antifoams)—for all of these systems the entry barrier is below 15 Pa; and (2) systems where the defoaming time is longer than 8 min (slow antifoams), for which the entry barrier is above 20 Pa. One can conclude that there is a well-defined threshold value of the entry barrier, somewhere between 15 and 20 Pa, which separates the region of the fast antifoams from the region of the slow ones. Therefore, the magnitude of the entry barrier, P_C^{CR}, is of crucial importance for the time scale of foam destruction by a given antifoam.

As shown in the following, the threshold value of P_C^{CR} is related to a transition from one mechanism of foam destruction (rapid rupture of the foam films) to another, much slower mechanism, which includes a compression of the antifoam globules in the Gibbs–Plateau borders of the foam.

2. Observations of Single Foam Films— Position of Drop Entry

One important question for any mechanism of foam destruction by oils is which structural element of the foam (foam film or Gibbs–Plateau border)

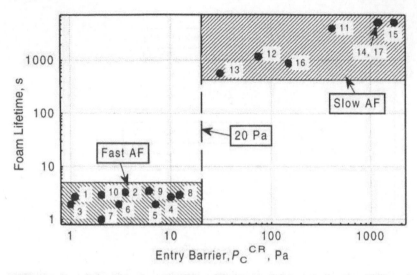

FIG. 4 Correlation between the foam lifetime and the entry barrier, P_C^{CR}, measured for different systems (see also Table 1): (1–6) 10 mM AOT and different silica–PDMS compounds; (7) 0.45 mM APG and silica–PDMS compound; (8–10) 1 mM Triton X-100 and different silica–PDMS compounds: (11–13) 0.1, 0.02, and 5 × 10^{-4} M SDP3S and silicone oil; (14) 0.1 M SDP3S/betaine = 80:20 molar ratio and silicone oil; (15) 0.1 M SDP3S/n-dodecanol = 97:3 molar ratio and silicone oil; (16) 0.02 M SDS and silicone oil; and (17) 0.02 M SDS/n-dodecanol = 97:3 molar ratio and silicone oil. The experimental data fall into two distinct regions: systems in which the foam is destroyed in less than 5 s (fast antifoams) and $P_C^{CR} < 15$ Pa and systems for which the defoaming time is longer than 8 min (slow antifoams) and $P_C^{CR} > 20$ Pa.

is actually destroyed by the antifoam globules. Most of the mechanisms discussed in the literature [2,4,6,11,12,18] imply that the *foam films* are ruptured by the antifoams because these films rapidly thin down to a thickness comparable to the globule diameter (on the order of a micrometer). On the other hand, Koczo et al. [7] suggested that the antifoam globules might first escape from the foam films into the neighboring *Gibbs–Plateau borders* (GPBs) and be trapped there. According to these authors, the antifoam globules are compressed by the walls of the narrowing GPBs, where eventually the drop entry occurs and the foam is destroyed. Our experiments have shown that both scenarios can be realized depending on the particular system [11,13,14]. However, the actual reason for the different locations of the antifoam action has not yet been clarified. Therefore, we performed systematic experiments for investigation of the relationship between the entry barrier and the structural element (foam film or GPB) where the globule entry occurs.

The capillary cell provides the possibility for a direct check of whether the antifoam ruptures the foam films [11,13]. We observed the process of thinning (and sometimes rupture) of foam films of diameter 0.7 ± 0.1 mm in the presence of dispersed globules of fast or slow antifoams. We describe next, as an example, the foam films formed from 10 mM AOT and 0.1 M SDP3S solutions; the results for the other studied systems of low-molecular-mass surfactants were very similar.

(a) Films Without Antifoam Globules. Let us first describe briefly the main stages of foam film thinning in the absence of an antifoam. Immediately after its formation, the foam film has a nonuniform thickness with a lens-shaped thicker region (called "dimple") in its central part—see Fig. 5. The thickness in the dimple center is typically 3 to 4 μm, and the thickness is about 1 μm in the film periphery. The dimple is hydrodynamically unstable, and the liquid captured in it spontaneously leaves the film (by an asymmetric outflow) within a few seconds. Afterward, the film surfaces become approximately plane parallel with several channels (dynamic regions of thickness 200–500 nm larger than the main planar portions of the film) spanning across the film area. The film gradually thins down to 100 nm in less than a minute, and the channels almost disappear. Further, several consecutive stepwise transitions in the film thickness are observed, which occur through formation and expansion of thinner spots. Such multistep film thinning is called "stratification" in the literature and is due to the oscillatory structural forces created by the surfactant micelles [31–36]. About 2–3 min after its formation, the foam film reaches its final thickness (10–20 nm, which corresponds to a common black film) and remains stable for many hours in the absence of an antifoam.

The stages of foam film thinning just described are typical for aqueous solutions of most low-molecular-mass surfactants, and the time scale of the process is approximately the same; the film thickness becomes on the order of 1 μm in a few seconds, about 1 min is needed for thinning of the film down to about 100 nm, and 2 to 4 min are needed until the final equilibrium film thickness is established. The main difference between the various systems is in the number of the stepwise transitions, which depends strongly on the surfactant concentration. Close to the cmc, when the volume fraction of the micelles is low, either there is only one transition from a common black film to a very thin Newton black film or there are no transitions at all because the equilibrium film thickness corresponds to a common black film, stabilized by electrostatic or steric forces. However, when the surfactant concentration is well above the cmc, up to five to seven transitions are

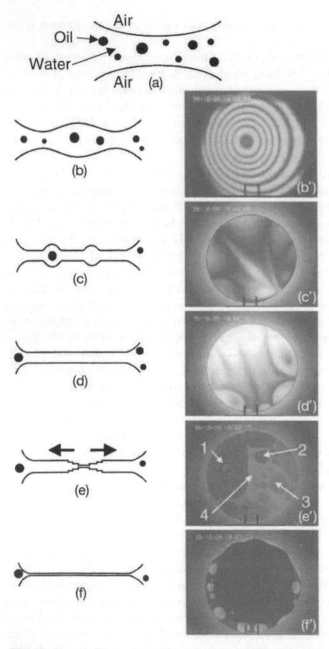

FIG. 5 Stages of foam film thinning: (a)–(f) Schematic presentation; (b')–(f') photographs made in reflected light, which show the typical interference pattern for the corresponding stage, as observed in the capillary cell with 0.1 M SDP3S solution,

observed due to the multiple barriers in the disjoining pressure curve $\Pi(h)$, which are caused by the oscillatory structural forces [34–36].

(b) Films Containing Drops of Pure Silicone Oil. The films formed in the capillary cell from solutions containing drops of pure silicone oil are rather stable, similar to the films formed in the absence of an antifoam. The film-thinning behavior is slightly affected by the presence of the oil drops, and the same stages are observed on approximately the same time scale. Micro-scopically, one can see that several oil drops are usually captured in the dimple, immediately after film formation. Most of these drops leave the foam film together with the dimple. The drops, which remain in the foam film after dimple expulsion, move from the planar film areas toward the channels (where the film thickness is larger) and then leave the film, following the drainage of liquid through the channels. At a film thickness of about 100 nm and below, no oil drops remain in the film because the latter is thinner than the diameter of the smallest drops. Therefore, the foam film is deprived of oil drops during the later stages of its thinning because the drops are expelled into the thicker meniscus region that surrounds the film (Fig. 6).

The fact that the real foams containing oil drops are unstable (although at least several minutes are needed before the foam destruction starts), while the foam films in the capillary cell remain stable for much longer time, indicates that the mechanism of foam destruction by oil drops does not occur through rupture of the foam films. Another mechanism that is in agreement with the results obtained for these systems is discussed in Sections III.A.4 and III.A.5.

(c) Films Containing Mixed Solid–Oil Antifoam Globules. The stages and the time scale of film thinning are almost the same as in the experiments described above, with one important exception—in most cases the antifoam globules induce a rupture of the foam film in the early stages of its thinning (1 to 10 s) at a relatively large thickness (≈ 1 μm). One important feature of the process observed is the formation of a characteristic interference pattern, clearly seen with a high-speed video camera [11], just before the film

---

containing 0.1 wt% silicone oil; the film diameter is 1 mm. (a) Two concave surfaces approach each other. (b) Formation of a film with a thicker central region called a "dimple." (c) An almost planar film crossed by several thicker regions (channels). (d) Film of average thickness around 120 nm; the channels almost disappear. All drops are expelled from the film. (e) A stepwise thinning of the film through formation and expansion of thinner (darker) spots—stratification. The arrows in (e') show the film areas containing different numbers of micelle layers, as indicated by the integers. (f) The film area is occupied by an equilibrium black film.

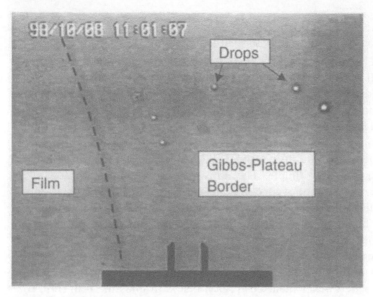

FIG. 6 Photograph showing oil drops that are expelled from the foam film into the surrounding Gibbs–Plateau border. The photograph is made in transmitted light to provide a good contrast for the oil drops; under these conditions the boundary between the film and the Gibbs–Plateau border is not well seen (this boundary is indicated by a dashed arc). The solution is 0.1 M SDP3S and contains 0.1 wt% silicone oil; the diameter of the oil drops seen in this picture is approximately 4 μm.

rupture. This interference pattern, which we call "fish eye" because of its appearance, corresponds to a local reduction of the film thickness by 100 to 300 nm (see Fig. 7). This reduction of the local film thickness is due to the formation of an oil bridge between the two foam film surfaces; see Fig. 8 and Section III.A.3. The film region perturbed by the oil bridge is relatively small (10–50 μm in diameter) and the rest of the foam film thins without being notably affected by the bridge [11,12]. Typically, the foam film ruptures soon (from several milliseconds up to several seconds) after the formation of the first fish eye. In most of the cases, one can specify exactly which antifoam globule became a bridge.

The process of hole formation and expansion inside the foam films was directly observed with larger vertical foam films (2 × 3 cm^2) by using a high-speed video camera [11]. Solutions of 10 mM AOT and silica-PDMS compound were used in these experiments. The results unambiguously demonstrated that the fast antifoams ruptured the foam films (not the GPBs).

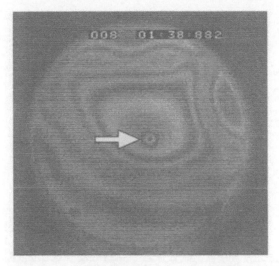

FIG. 7 Interference pattern ("fish eye," indicated by the arrow) indicating the formation of an oil bridge in the foam film just before its rupture in the capillary cell. The film is made from 10 mM AOT solution containing 0.01 wt% silica–PDMS compound.

Most of these large films ruptured almost immediately (within 0.5 s) after their formation at a thickness of several micrometers.

3. Foam Destruction by Fast Antifoams—Bridging–Stretching Mechanism of Foam Film Rupture

The fact that the foam films are destroyed almost immediately after the formation of an oil bridge shows that these bridges are unstable. The microscopic observations [11] and the theoretical analysis [12] showed that once formed, the oil bridge stretched with time due to uncompensated capillary pressures at the oil–water and oil–air interfaces and eventually ruptured, leading to a destruction of the entire foam film. This mechanism of foam film rupture, called bridging–stretching [11,12], is illustrated in Fig. 8 by a schematic drawing and by photographs taken in the Dippenaar cell with 10 mM AOT solutions and silica-PDMS compound. As seen from the figure, the oil bridge acquires a shape with concave oil–water interfaces and the thinnest region is in the bridge center. By changing the amount of the surfactant solution in the Dippenaar cell, we are able to stretch the bridge in the radial direction, which eventually causes its rupture and destruction of the entire foam film. The conditions for stability of oil bridges along with the details of the bridging–stretching mechanism are discussed in Ref. 12.

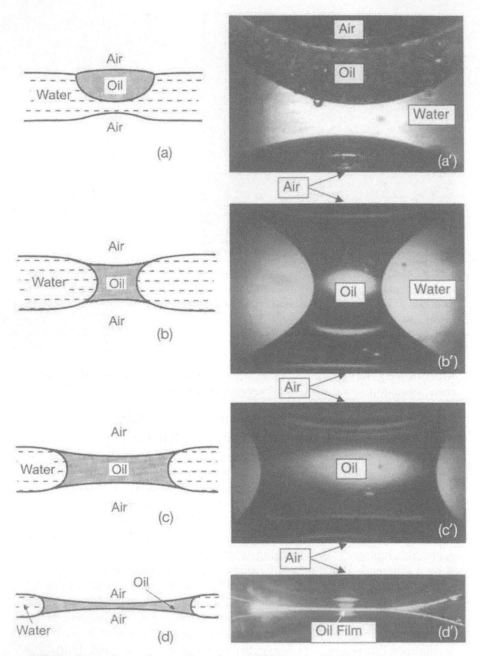

FIG. 8 Formation and stretching of an oil bridge. (a–d) Schematic presentation; (a′–d′) photographs of the corresponding stages made with the Dippenaar cell.

The bridging–stretching process is an alternative to another possible scenario for foam film destruction [3–8], which implies that the antifoam globules are dewetted by the surfactant solution (bridging–dewetting mechanism, Fig. 1). One cannot exclude the possibility that in other systems the antifoam globules destroy the foam films by the bridging–dewetting mechanism.

4. Foam Destruction by Slow Antifoams

See Fig. 9. As already explained, the globules of the slow antifoams escape from the foam films and accumulate in the GPBs, as suggested by Koczo et al. [7]. This process was directly observed [13] by a long-focus optical lens in experiments with vertical foam films suspended on a three-legged glass frame—in this configuration a single GPB was formed between the three films hung on the frame legs [37]. Immediately after the withdrawal of the glass frame from the surfactant solution containing pre-emulsified antifoam, one observes numerous oil drops trapped in the thick foam films just formed [13]. However, these oil drops are expelled out of the film almost instantaneously because the film thickness becomes smaller than the drop diameter due to the liquid drainage. A fraction of these oil drops is trapped in the narrowing GPB—a drop entry and subsequent film rupture are observed when the GPB cross section becomes smaller than the drop diameter and the compressing pressure exceeds the entry barrier of the drops. Numerous oil drops are seen captured in the GPBs of real foams as well; see Fig. 9f. The process of foam destruction in the presence of oil drops (slow antifoams) is described in more detail in Refs. 13–16.

It is worth mentioning here that details of the actual mechanism(s) of foam destruction, occurring through entry of oil drops in the GPBs, are still unclear. One interesting experimental fact is that the foam destruction often occurs through sudden avalanches of multiple bubble bursts in the top layer of the foam, separated by still periods. To explain this observation, a compression–bursting mechanism was suggested in Ref. 13 which implies that the entry of strongly compressed (deformed) droplets in the GPBs creates a mechanical shock (due to the release of energy in the moment of drop entry), which is able to destroy the neighboring foam films. The mechanical energy released upon the foam film rupture creates a propagating mechanical stress that might break the thinnest (least stable) foam films in the upper layer of the foam column.

5. Relation Between the Final Foam Height, Entry Barrier, and Drop Size for Slow Antifoams

The experiments reported in Ref. 14 showed that the foam evolution in the presence of oil drops consisted of several stages (see Fig. 9e). In all cases,

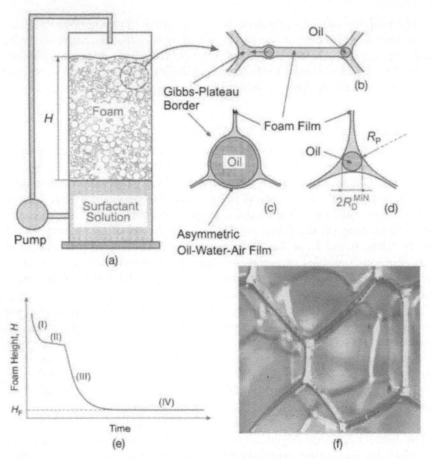

FIG. 9 Foam destruction by oil drops (slow antifoams) [13,14]. (a and b) The oil drops are rapidly expelled from the foam films into the neighboring Gibbs–Plateau borders (GPBs) soon after the foam is formed. (c) The larger drops are strongly compressed in the narrowing GPBs and asymmetric oil–water–air films are formed. (d) Drops of radius smaller than R_D^{MIN} remain noncompressed and cannot induce foam rupture. (e) Schematic presentation of the main stages of foam evolution in the presence of oil drops: (I) drainage of liquid without bubble coalescence; (II) stable foam due to the insufficient compression of the oil drops; (III) foam destruction as a result of drop entry in the GPBs; (IV) long-living residual foam with final height H_F. (f) Photograph of real foam cells with many oil drops trapped in the GPBs.

the foam destruction by slow antifoams was incomplete (on the time scale of interest, ~2 h) and a residual, very stable foam of a certain height was observed. This final foam height, H_F, depended on both the system composition and oil drop size [13,14]. The following consideration explains the magnitude of H_F by analyzing the relations between (1) the drop entry barrier and the compressing capillary pressure imposed by the shrinking walls of the GPBs and (2) the drop size and the cross section of the GPBs.

The capillary pressure that compresses the oil drops in the GPBs gradually increases with the drainage of liquid from the foam. The condition for a hydrostatic equilibrium in the foam column requires the appearance of a vertical gradient of the capillary pressure that opposes the gravity [2,38]. At equilibrium, the capillary pressure at the top of the foam column should be approximately equal to the hydrostatic pressure [13,38]:

$$P_{CF}(H) \approx \Delta\rho g H \tag{6}$$

where $\Delta\rho$ is the mass density difference between the aqueous and gaseous phases, g is the acceleration of gravity, and H is the height of the foam column; see Fig 9a. The capillary pressure determines the radius of curvature, R_P, of the GPB wall (Fig. 9d)

$$R_P \approx \frac{\sigma_{AW}}{P_{CF}} \approx \frac{\sigma_{AW}}{\Delta\rho g H} \tag{7}$$

and, hence, also determines the cross section of the GPB. The radius of a sphere that fits the cross section of a GPB (i.e., touching its walls of radius R_P; Fig. 9d), can be found from geometric considerations [13,16]:

$$R_D^{MIN} = \left(\frac{2\sqrt{3}}{3} - 1\right) R_P \approx 0.15 \frac{\sigma_{AW}}{\Delta\rho g H} \frac{1}{H} \tag{8}$$

The notation R_D^{MIN} is used in Eq. (8) because this presents, in fact, the minimal radius of a drop that can be compressed by the walls of the GPBs in a foam column of height H.

It is seen from Eqs. (6) and (8) that the compressing capillary pressure P_{CF} is higher and R_D^{MIN} is smaller for taller foam columns. If the drops trapped in the GPBs have entry barrier $P_C^{CR} < P_{CF}(H)$ and radius $R_D > R_D^{MIN}$, then the foam destruction will begin after a certain period of liquid drainage because the asymmetric oil–water–air films will be unable to resist the compressing pressure imposed by the shrinking walls (Fig. 9c). The foam destruction will continue until $P_{CF}(H)$ becomes approximately equal to P_C^{CR} (i.e., the asymmetric films become stable) or, alternatively, until the cross section of the GPBs becomes larger than the drop radius, $R_D^{MIN} \geq R_D$ (i.e., when the oils drops are no longer compressed by the GPB walls). Therefore, the final height of the foam column, H_F, reached as a result of the foam destruction by the oil drops (Fig. 9e) must be close to the larger of the two estimates

$$H_F = \max\{H_{FP}, H_{FR}\} \tag{9}$$

$$H_{FP} = \frac{P_C^{CR}}{\Delta \rho g} \tag{10a}$$

$$H_{FR} = 0.15 \frac{\sigma_{AW}}{\Delta \rho g} \frac{1}{R_D} \tag{10b}$$

One can conclude from this analysis that if $H_{FP} > H_{FR}$, the final foam height, H_F, is determined mainly by the entry barrier—the oil drops are still compressed, but the asymmetric films are stable. On the contrary, if $H_{FP} < H_{FR}$, the final foam height is determined by the drop size, and the entry barrier is of secondary importance.

It is worth noting several complications that might be important in some systems. First, the entry barrier might depend strongly on the drop size in some cases (see Section III.B.1) so that the preceding two conditions are not entirely independent, i.e., P_C^{CR} should be regarded as a function of R_D. Second, because most of the antifoam emulsions are rather polydisperse, one must be careful what average drop size is used in these estimates. Also, a coalescence between the trapped oil drops may occur inside the GPBs, which would lead to an increase of the drop size and, possibly, to reduction of the entry barrier and foam stability.

To check whether the preceding estimates, Eqs. (9) and (10), described real foams, we performed a series of parallel experiments with various surfactant–antifoam couples for determination of the entry barrier, P_C^{CR}, by FTT and of the final foam height, H_F, by the Ross–Miles test. The results are summarized in Fig. 10, where H_F is shown as a function of P_C^{CR} (see the figure caption for the specific systems). The results show that the theoretical linear relationship between H_F and P_C^{CR}, Eq. (10a), holds very well for $P_C^{CR} \geq 400$ Pa (see the dashed line in Fig. 10). A further decrease of the entry barrier almost does not affect H_F, which remains around 3–4 cm for $P_C^{CR} < 400$ (see the shaded area in Fig. 10). As already explained, the reason for this result is that the GPBs in short foam columns are too wide to compress the emulsified oil drops. Indeed, one can estimate from Eq. (8) that $R_D^{MIN} = 13$ μm for $H_F = 3.5$ cm ($\sigma_{AW} \approx 30$ mN/m). The size distribution of the drops was determined for one of the systems studied—SDP3S and silicone oil. The main fraction of drops fell in the size range between 2 and 20 μm in radius, although single larger drops were also observed. The *number* av-

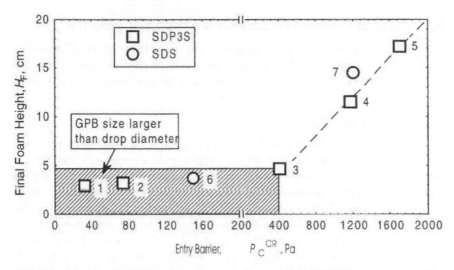

FIG. 10 Final foam height, H_F, versus entry barrier, P_C^{CR}, for different systems: (1–3) 5×10^{-4}, 0.02, and 0.1 M SDP3S; (4) 0.1 M SDP3S/betaine = 80:20 molar ratio; (5) 0.1 M SDP3S/n-dodecanol = 97:3 molar ratio; (6) 0.02 M SDS; (7) 0.1 M SDS/n-dodecanol = 97:3 molar ratio.

eraged, arithmetic mean drop radius was about 5 μm, whereas the geometric mean radius was about 4 μm (the peak width calculated for a log-normal size distribution of the drops was $\sigma_g \approx 2$). These results indicate that R_D in Eq. (10b) should represent the typical radius of the larger drops (which are most active as antifoam entities) in the size distribution curves. The calculations showed that one could use the *volume* averaged, geometric mean drop size (which was $R_D \approx 16$ μm in these experiments) as a reasonable estimate of R_D in Eq. (10b).

Equations (9) and (10) predict that one can use the entry barrier and/or the drop size to control the final height of the foam. Indeed, FTT measurements showed that the addition of different amphiphiles (foam boosters, such as dodecanol, betaines, and others) to the main surfactant (e.g., SDS or SDP3S) led to a significant increase of the entry barrier at a fixed total surfactant concentration [13,14]. The observed increase of the entry barriers was in very good agreement with the enhanced foam stability found with the same systems in the foam tests. On the other hand, the foam stability was found to increase noticeably with the reduction of the oil drop size at fixed composition of the surfactant solution, just as predicted by the preceding consideration—see Figs. 4 and 5 in Ref. 13.

490   Hadjiiski et al.

B. Effect of Different Factors on the Drop Entry Barrier

The foregoing results clearly demonstrate the importance of the entry barrier for the activity of the oil-based antifoams. That is why we carried out systematic experiments to clarify how the entry barrier depended on various factors.

1. Effect of the Drop Size

The FTT experiments with many different systems showed that, in general, P_C^{CR} decreased monotonously with the drop size. In many cases this trend is very weak (e.g., the results shown in Fig. 3) and can be neglected for the size range of typical antifoams (R_D between 1 and 10 μm), whereas in some systems the size effect is very pronounced. Interestingly, we found that the dependence of P_C^{CR} on the drop size was sensitive to the surfactant concentration. As an example, the dependence of P_C^{CR} on the equatorial drop radius, R_E, is shown in Fig. 11 for silicone oil (viscosity 5 mPa·s) at three different concentrations of SDP3S. The radius of the drops studied varied between 1 and 8 μm. To compare quantitatively the effect of the drop size for the different concentrations, we define the ratio $p = P_C^{CR}$ (2 μm)/P_C^{CR} (6 μm),

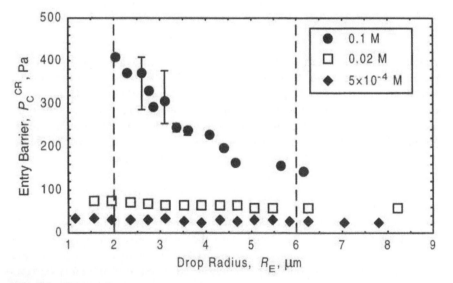

FIG. 11 Entry barrier, P_C^{CR}, for silicone oil as a function of the oil drop radius measured at three different concentrations of SDP3S. As a quantitative measure of the effect of drop size on the entry barrier, we use the ratio $p = P_C^{CR}$ (2 μm)/P_C^{CR} (6 μm). The two vertical lines indicate the drop radii (2 and 6 μm, respectively), which are used to determine p.

which shows how steep the decrease of the entry barrier is with the drop size. At the lowest concentration (5×10^{-4} M, which is around the cmc of this surfactant) the ratio $p \approx 1.1$; for the intermediate concentration (0.02 M), this ratio is slightly larger, $p \approx 1.3$; and for a concentration far above the cmc (0.1 M) a significant increase of p is observed, $p \approx 2.9$, i.e., the entry barrier decreases about three times in the size range studied. As discussed at the end of the previous section, this dependence can be used to control the foam stability by varying the size of the oil drops introduced into the foaming solution. Why the ratio p is larger for more concentrated surfactant solutions is still an open question. The explanation might be in the different types of surface forces (electrostatic and van der Waals at low concentrations; oscillatory structural forces at high concentrations) that stabilize the asymmetric films in the different concentration ranges.

2. Effect of the Surfactant Concentration

A series of experiments were performed with SDDBS solutions of different concentrations to measure the entry barrier for hexadecane drops. The surfactant concentration, C_S, was varied between 0.16 and 12.8 mM, while the salt concentration was fixed at 12 mM NaCl. The cmc of SDDBS at this ionic strength is about 0.5 mM. The working solutions were poured in the experimental cell by using the two-tip procedure (TTP) described in Section II.B.3.c, to avoid the presence of a spread oil on the solution surface. The entry barriers, measured with oil drops of radius $R_E \approx 2.3 \pm 0.3$ μm, are plotted in Fig. 12 as a function of the surfactant concentration. At least three independent experimental runs were carried out at a given concentration, with two or three entry events observed in each run. The reproducibility of the measured value of P_C^{CR} was very good, typically $\pm 5\%$.

The results shown in Fig. 12 indicate a complex dependence of P_C^{CR} on the surfactant concentration. At concentrations below 0.16 mM, the entry barrier is too low to be measured by the experimental procedure used. In this case the drops entered the solution surface without being compressed by the water–air interface. At the lowest concentration where FTT measurements were possible, $C_S = 0.16$ mM, $P_C^{CR} = 10$ Pa was obtained. The entry barrier rapidly increases in the concentration range between 0.16 and 0.5 mM, up to ≈ 40 Pa. At higher concentrations, between 0.5 and 9 mM, the barrier exhibits a slow but steady increase from about 40 to 150 Pa with the surfactant concentration. A much steeper increase of P_C^{CR} is observed at concentrations above 9 mM, and the barrier is above 400 Pa at $C_S = 12.8$ mM.

The observed sharp increase of P_C^{CR} at $C_S > 9$ mM is probably related to the stabilizing effect of the surfactant micelles trapped in the asymmetric oil–water–air film [3,10]. One can estimate that the effective volume frac-

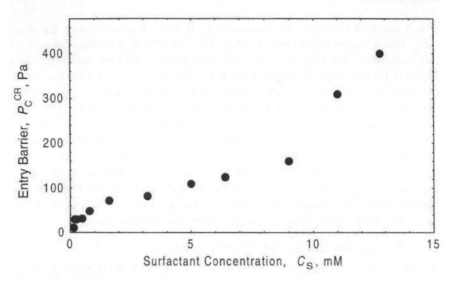

FIG. 12 Dependence of the entry barrier, P_C^{CR}, on the SDDBS concentration, C_S, for hexadecane drops. All solutions contain 12 mM NaCl. The entry barriers are measured with drops of radius $R_E = 2.3 \pm 0.3$ μm. The solution surface is free of oil.

tion of the SDDBS micelles, Φ, including the contribution of the counterion atmosphere, is about 6% at the kink point corresponding to 9 mM [17]. From the micellar concentration one can estimate the height of the last maximum (corresponding to one layer of micelles trapped in the film) in the oscillatory component of the disjoining pressure, created by the micelles, by using the formulas derived in Ref. 36. The estimate shows that this maximum is about 73 Pa, which is not far from the measured values of $P_C^{CR} \approx$ 160 Pa at this concentration (note that the electrostatic and van der Waals forces also contribute to the height of this maximum in the film). Therefore, a detectable contribution of micelles in the stability of the films might be expected in this concentration range and above it. More detailed discussion of these results is presented in Ref. 17.

3. Effect of the Chemical Structure of Oil

All experiments described in Sections III.B.3 and III.B.4 were carried out with solutions containing 2.6 mM SDDBS and 12 mM NaCl. The drop entry barriers for a series of n-alkanes (octane, decane, dodecane, hexadecane), dodecanol, and silicone oil were measured. Drops of diameter between 2 and 12 μm were studied, and no significant dependence of P_C^{CR} on the drop size was observed. The mean values of the drop entry barrier, P_C^{CR}, measured for the different oils are summarized in Table 2. The values given in parentheses correspond to experiments performed in the absence of a spread oil layer over the solution surface; the other values were measured with a prespread, molecularly thin oil layer on the solution surface. The results show that the entry barrier for n-alkanes increases with their molecular mass: for octane $P_C^{CR} = 30 \pm 2$ Pa, for decane 35 ± 5 Pa (>70 Pa with solution surface free of spread oil), for dodecane 48 ± 5 Pa (96 ± 5 Pa), and for hexadecane it is 400 ± 10 Pa (80 ± 5 Pa). Such a significant increase of the entry barrier with the alkane chain length is certainly important for the antifoam action of the alkanes, and systematic foam tests are planned to understand better the relation between the entry barrier and the foam stability for these systems. The significant effect of the spread oil on the entry barrier, found with most of the alkanes, will be discussed in the next section.

The experiments with drops of n-dodecanol and silicone oil revealed very high entry barriers, above 1500 Pa. Not surprisingly, the foam tests showed that emulsified drops of both these oils were inefficient foam breakers, although the E, S, and B coefficients were strongly positive for the silicone oil [16]. More detailed discussion of these results and some possible explanations for the different entry barriers of the studied oils are presented in Ref. 17.

4. Effect of the Prespread Oil Layers

The experiments with dodecane and decane demonstrated a significantly lower drop entry barrier in the presence of a prespread layer of the same oil on the solution surface. On the contrary, the spreading of hexadecane (which makes a mixed adsorption layer with the SDDBS molecules) leads to about

TABLE 2 Drop Entry Barrier, P_C^{CR}, Measured with Different Oils in the Presence or in the Absence (Data in Parentheses) of Spread Oil on the Surface of the Aqueous Solution Containing 2.6 mM SDDBS and 12 mM NaCl

Oil	P_C^{CR} (Pa)
n-Octane	30 ± 2
n-Decane	35 ± 5 (>70)
n-Dodecane	48 ± 5 (96 ± 5)
n-Hexadecane	400 ± 10 (80 ± 5)
n-Dodecanol	>1500
PDMS	>3000

a fivefold increase of the entry barrier compared with that in the absence of spread oil (see Table 2). It is possible that the high entry barrier observed with dodecanol is also related to the formation of a dense mixed adsorption layer on the solution surface [16,17]. Therefore, the presence of a spread oil layer is a significant factor in the magnitude of the entry barriers. Note that this effect has an important implication for the antifoaming action of these oils. For example, one could not explain the poor antifoam activity of hexadecane in SDDBS solutions without taking into account the increase of the entry barrier due to the oil spreading. However, as discussed elsewhere [4,5,13,16,17,39], different factors are often more important and no straightforward correlation between the spreading behavior of the oil and its antifoam activity is observed.

5. Discussion of the Critical Disjoining Pressure for Drop Entry

The results presented in Figs. 3 and 11 show that the critical *capillary* pressure, P_C^{CR}, is usually a weak function of the size of the asymmetrical oil–water–air film. Additional analysis is needed, however, to understand how the critical *disjoining* pressure, Π_{AS}^{CR}, depends on the film size. In this section we investigate this dependence and discuss it from the viewpoint of the mechanism of rupture of the thin asymmetric films.

(a) Disjoining Pressure for Spherical Films. The disjoining pressure, Π_{AS}, accounts for the interactions between the two film surfaces (van der Waals, electrostatic, steric, etc.) and is conventionally defined as the surface force per unit area [40–42]. Positive disjoining pressure corresponds to repulsive surface forces (i.e., to film stabilization) and vice versa. In the case of planar films, the condition for mechanical equilibrium requires that the capillary sucking pressure be exactly counterbalanced by the disjoining pressure. However, the thin films in our experiments are curved and the condition for mechanical equilibrium is more complex because it should account for the capillary pressure jumps across the curved film surfaces. The relevant theoretical approach to this configuration was developed by Ivanov and co-workers [25,42,43], who showed that the disjoining pressure was related to the capillary pressure across the water–air interface, $P_C = P_A - P_W$, by the expression

$$\Pi_{AS} = P_F - P_W = (P_F - P_A) + (P_A - P_W) = \frac{2\sigma_{AW}}{R_F} + P_C \tag{11}$$

where P_F is the pressure in the asymmetric oil–water–air film and R_F is its radius of curvature (Fig. 13a). The aqueous phase (from which the asymmetric film is formed) is chosen as a reference phase for the definition of the disjoining pressure as usual [42,43].

For micrometer-sized drops R_F is on the order of the drop size and $2\sigma_{AW}/R_F > 10^4$ Pa. In most of our systems $P_C \approx 10^2$ Pa and its contribution can be neglected in Eq. (11). Thus, only the radius of film curvature, R_F, would be sufficient to calculate Π_{AS} because σ_{AW} is a known quantity. Note, however, that R_F depends on the drop deformation, which, in turn, is determined by the applied capillary pressure, P_C. For large drops or bubbles, one can measure directly the radius of film curvature, R_F, by using the microscopic method of differential interferometry [44]; however, this method cannot be used for micrometer-sized drops. That is why an indirect method was used in Ref. 17 to estimate the magnitude of Π_{AS} from the accessible experimental data and to study how the critical disjoining pressure for drop entry, Π_{AS}^{CR}, depended on the size of the asymmetric film. We refrain from describing here the exact numerical procedure because it would require too much space. Briefly, it consists of determining the shape of the trapped oil drop and of the contiguous water–air meniscus from the accessible experimental data— the capillary pressure P_C, the equatorial drop radius R_E, and the interfacial tensions, σ_{AW} and σ_{OW}. From the oil drop shape one finds R_F^{CR} and Π_{AS}^{CR} at the moment of film rupture. For a detailed explanation of the numerical procedure, the reader is referred to the original article [17].

(b) *Numerical Results.*   The calculated dependence of Π_{AS}^{CR} on the inverse radius of the asymmetric film is shown in Fig. 13b for the system 3.2 mM SDDBS, 12 mM NaCl, and hexadecane drops (no spread layer of oil). Because the asymmetric film is curved, there are different possible definitions of its size. For this plot we have chosen the "effective" film radius to be equal to the radius of a planar film that has the same area as the real asymmetric film

$$R_{EFF} = \sqrt{\frac{A_F}{\pi}} \tag{12}$$

where A_F is the actual area of the asymmetric film. As seen from Fig. 13b, Π_{AS}^{CR} is a linear function of $1/R_{EFF}$. The calculations performed in Ref. 17 showed that for a given system, the drop entry occurred at approximately the same relative deformation of the drops (independent of the drop size).

The observed dependence of Π_{AS}^{CR} on R_{EFF} is by no means a trivial fact. The isotherm $\Pi_{AS}(h)$ is not expected to depend on either the film size or the film curvature because the film thickness h is much smaller than both R_{EFF} and R_F. Therefore, if the film rupture were accomplished by surmounting the maximum in the isotherm $\Pi_{AS}(h)$, the rupture event for a given system would always be expected to occur at $\Pi_{AS}^{CR} = \cap_{AS}^{MAX}$, independent of the drop size.

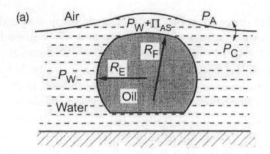

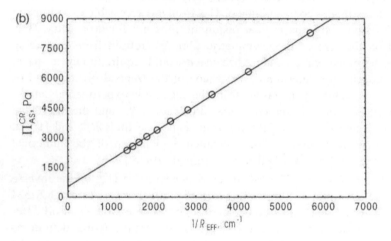

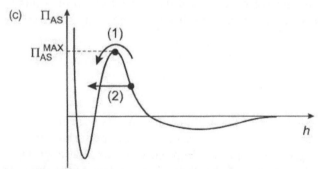

FIG. 13 Determination of the disjoining pressure. Π_{AS}, in the asymmetric oil–water–air film [17]. (a) Schematic presentation of an oil drop trapped in a wetting film; $P_W + \Pi_{AS} = P_F$ is the pressure in the asymmetric film. (b) Calculated critical disjoining pressure Π_{AS}^{CR} as a function of the inverse effective film radius $1/R_{EFF}$ [see Eq. (12)]; the circles show calculated values from experiments with different oil drops and the line represents the respective linear fit; the calculations are made for 1 mM SDDBS solution containing 12 mM NaCl and n-hexadecane drops. (c) Schematic presentation of the disjoining pressure isotherm $\Pi_{AS}(h)$. Two possible ways to

One possible explanation of the observed dependence might be that the film rupture in our systems occurs by passing below the barrier Π_{AS}^{MAX} (Fig. 13c). Indeed, Bergeron [45] showed with large planar foam films (studied by the porous plate method) that in some systems Π_{AS}^{CR} corresponded to an actual maximum of the calculated curve $\Pi_{AS}(h)$, whereas in other systems Π_{AS}^{CR} was well below the maximum of the calculated $\Pi_{AS}(h)$ curves (for a possible explanation see Ref. 45). Such a possibility is offered by different theoretical models of film rupture, in which the formation of unstable spots in large liquid films by various mechanisms is considered [40,45–47]. However, all these models are developed for large planar films and cannot be applied directly to our system without a careful analysis of the role of film curvature in the film rupture process. Further experimental and theoretical work is under way to reveal the actual mechanism of film rupture, to develop an adequate model of this process, and to explain the observed linear dependence of Π_{AS}^{CR} versus $1/R_{FFF}$.

IV. CONCLUSIONS

A systematic experimental study was performed to clarify further the role of the entry barrier in the foam destruction by oil-based antifoams. The critical capillary pressure, P_C^{CR}, which leads to rupture of the asymmetric oil–water–air film (formed between a pre-emulsified oil drop and the solution surface) and to subsequent drop entry, was measured by the film trapping technique (Fig. 2)—for brevity, P_C^{CR} is denoted as the "entry barrier" throughout. The results obtained and the conclusions drawn can be summarized as follows:

1. The experiments reveal that P_C^{CR} determines the boundary between two different classes of antifoam (Figs. 3 and 4).
    a. Fast antifoams (defoaming time < 5 s, no residual foam), which have $P_C^{CR} < 20$ Pa and break the foam films (Figs. 7 and 8).
    b. Slow antifoams (defoaming time > 5 min, stable residual foam),

---

overcome the barrier and possible film rupture are indicated. (1) The film surfaces are compressed toward each other by a capillary pressure that drives the system to surmount the barrier Π_{AS}^{MAX}—in this case, the critical disjoining pressure Π_{AS}^{CR} should be equal to Π_{AS}^{MAX} independently of drop radius. (2) A local fluctuation in the film leads to the formation of unstable spot and local film rupture. In this case, the latter may occur at a critical disjoining pressure $\Pi_{AS}^{CR} < \Pi_{AS}^{MAX}$ and hence Π_{AS}^{CR} could depend on the film size.

which have $P_C^{CR} > 20$ Pa and destroy the foam only after being compressed inside the Gibbs–Plateau borders (Figs. 6 and 9).

2. A relation between P_C^{CR}, the final height of the foam, H_F, and the radius of the antifoam drops, R_D, is found and explained theoretically for the slow antifoams (Figs. 9 and 10).

3. The dependence of the entry barrier on the concentration of the anionic surfactant sodium dodecylbenzene sulfonate (SDDBS) was studied for hexadecane oil drops. A steep increase of the barrier is observed at a concentration above 9 mM (effective volume fraction of the micelles $\approx$ 6%), which implies that the oscillatory structure forces, created by the micelles, play a significant role above this concentration (Fig. 12).

4. The presence of a prespread oil layer on the surface of the surfactant solution is found to affect strongly the entry barrier for alkanes (Table 2). The barrier is reduced by the prespread layer for decane and dodecane, whereas the effect is the opposite for hexadecane (fivefold increase).

5. There is a big difference between the numerical values of the critical *capillary* pressure, P_C^{CR}, and the critical *disjoining* pressure Π_{AS}^{CR}, for micrometer-sized oil drops, such as those in real antifoams. The analysis shows that P_C^{CR} is a more convenient quantity for description of the entry barriers because its magnitude correlates with the final height of the foam, whereas the magnitude of Π_{AS}^{CR} does not.

6. The experiments show that P_C^{CR} usually depends only slightly on the oil drop size and on the radius of the asymmetric film, while Π_{AS}^{CR} scales as (film radius)$^{-1}$ for all of the studied systems (Fig. 13b). The strong dependence of Π_{AS}^{CR} on the film radius shows that the rupture of the asymmetric film does not occur simply by surmounting the barrier in the $\Pi_{AS}(h)$ curve. A possible explanation of this result is discussed (Fig. 13c).

One can conclude that the FTT has provided valuable and nontrivial information about the role of the entry barrier in the antifoam activity of oils and oil-based compounds.

REFERENCES

1. PR Garrett, ed. Defoaming: Theory and Industrial Applications. Surfactant Science Series, Vol 45. New York: Marcel Dekker, 1993.
2. D Exerowa, PM Kruglyakov. *Foams and Foam Films*. Amsterdam: Elsevier, 1998, Chapter 9, pp. 610–652.
3. DT Wasan, SP Christiano. In: KS Birdi, ed. *Handbook of Surface and Colloid Chemistry*. Boca Raton, FL: CRC Press, 1997, Chapter 6, pp. 179–215.

4. PR Garrett. In: PR Garrett, ed. *Defoaming: Theory and Industrial Applications.* Surfactant Science Series, Vol 45. New York: Marcel Dekker, 1993, Chapter 1, pp. 1–118.

5. PR Garrett, J Davis, HM Rendall. *Colloids Surf A* 85:159–197, 1994.

6. V Bergeron, P Cooper, C Fischer, J Giermanska-Kahn, D Langevin, A Pouchelon. *Colloids Surf A* 122:103–120, 1997.

7. K Koczo, JK Koczone, DT Wasan. *J Colloid Interface Sci* 166:225–238, 1994.

8. R Aveyard, JH Clint. *J Chem Soc Faraday Trans* 91:2681–2697, 1995; R Aveyard, BP Binks, PDI Fletcher, TG Peck, PR Garrett. *J Chem Soc Faraday Trans* 89:4313–4321, 1993.

9. V Bergeron, ME Fagan, CJ Radke. *Langmuir* 9:1704–1713, 1993.

10. L Lobo, DT Wasan. *Langmuir* 9:1668–1677, 1993.

11. ND Denkov, P Cooper, J-Y Martin. *Langmuir* 15:8514–8529, 1999.

12. ND Denkov. *Langmuir* 15:8530–8542, 1999.

13. E Basheva, D Ganchev, ND Denkov, K Kasuga, N Satoh, K Tsujii. *Langmuir* 16:1000–1013, 2000.

14. E Basheva, S Stoyanov, ND Denkov, K Kasuga, N Satoh, K Tsujii. *Langmuir* 17:969–979, 2001.

15. ND Denkov, K Marinova. *Proceedings of the 3rd EuroConference on Foams, Emulsions and Applications.* Bremen: Verlag Metal Innovation Technologie MIT: 2000.

16. L Arnaudov, ND Denkov, I Surcheva, P Durbut, G Broze, A Mehreteab. *Langmuir* 17:6999–7010, 2001.

17. A Hadjiiski, S Tcholakova, ND Denkov, P Durbut, G Broze, A Mehreteab. *Langmuir* 17:7011–7021, 2001.

18. PM Kruglyakov, TA Koretskaya. *Kolloid Zh* 36:682–686, 1974; PM Kruglyakov. In: IB Ivanov, ed. *Thin Liquid Films: Fundamentals and Applications.* Surfactant Science Series, Vol 29. New York: Marcel Dekker, 1988, Chapter 11, pp. 767–828.

19. RD Kulkarni, ED Goddard, B Kanner. *J Colloid Interface Sci* 59:468–476, 1977.

20. K Koczo, LA Lobo, DT Wasan. *J Colloid Interface Sci* 150:492–506, 1992.

21. KJ Mysels, MN Jones. *Discuss Faraday Soc* 42:42–50, 1966.

22. A Hadjiiski, R Dimova, ND Denkov, IB Ivanov, R Borwankar. *Langmuir* 12:6665–6675, 1996.

23. SM Patrick, H An, MB Harris, IB Ivanov, NS Braunshtein, EF Leonard. *Ann Biomed Eng* 25:1072–1079, 1997.

24. IB Ivanov, A Hadjiiski, ND Denkov, TD Gurkov, PA Kralchevsky, S Koyasu. *Biophys J* 75:545–556, 1998.

25. A Hadjiiski, S Tcholakova, IB Ivanov, TD Gurkov, E Leonard. *Langmuir* 18:127–138, 2002.

26. A Scheludko, D Exerova. *Kolloid Z* 165:148–151, 1959.

27. A Scheludko. *Adv Colloid Interface Sci* 1:391–464, 1967.

28. AS Aronson. *Langmuir* 2:653–659, 1986.

29. A Dippenaar. *Int J Miner Process* 9:1–14, 1982.

30. P Garrett. *Langmuir* 11:3576–3584, 1995.

31. AD Nikolov, DT Wasan, PA Kralchevsky, IB Ivanov. In: N Ise, I Sogami, eds. *Ordering and Organization in Ionic Solutions*. Singapore: World Scientific, 1988, pp 302–314.

32. AD Nikolov, DT Wasan, PA Kralchevsky, IB Ivanov. *J Colloid Interface Sci* 133:1–12, 13–22, 1989.

33. DT Wasan, AD Nikolov, PA Kralchevsky, IB Ivanov. *Colloids Surf* 67:139–145, 1992.

34. V Bergeron, CJ Radke. *Langmuir* 8:3020–3026, 1992.

35. ML Pollard, CJ Radke. *J Chem Phys* 101:6979–6991, 1994.

36. PA Kralchevsky, ND Denkov. *Chem Phys Lett* 240:385–392, 1995.

37. K Koczo, G Racz. *Colloids Surf* 22:97–110, 1987.

38. G Narsimhan, E Ruckenstein. In: RK Prud'homme, SA Khan, eds. *Foams: Theory, Measurements, and Applications*. Surfactant Science Series, Vol 57. New York: Marcel Dekker, 1996, Chapter 2, pp. 99–188.

39. K Marinova, ND Denkov. *Langmuir* 17:2426–2436, 2001.

40. BV Derjaguin. *Theory of Stability of Colloids and Thin Liquid Films*. New York: Consultants Bureau, 1989, Chapter 12.

41. IB Ivanov, BV Toshev. *Colloid Polym Sci* 253:558–565, 593–599, 1975.

42. IB Ivanov, PA Kralchevsky. In: IB Ivanov, ed. *Thin Liquid Films: Fundamentals and Applications*. Surfactant Science Series, Vol 29. New York: Marcel Dekker, 1988, Chapter 2, pp. 49–130.

43. PA Kralchevsky. *Effect of Film Curvature on the Thermodynamic Properties of Thin Liquid Films*. PhD thesis, Sofia University, Sofia, Bulgaria, 1984 [in Bulgarian].

44. AD Nikolov, PA Kralchevsky, IB Ivanov. *J Colloid Interface Sci* 112:122–131, 1986.

45. V Bergeron. *Langmuir* 13:3474–3482, 1997.

46. D Kaschiev, D Exerowa. *J Colloid Interface Sci* 77:501–511, 1980.

47. PA Kralchevsky, AD Nikolov, DT Wasan, IB Ivanov. *Langmuir* 6:1180–1189, 1990.

24

Principles of Emulsion Formulation Engineering

JEAN-LOUIS SALAGER, LAURA MÁRQUEZ, ISABEL MIRA,*
ALEJANDRO PEÑA,† ERIC TYRODE,‡ and
NOELIA B. ZAMBRANO¶ University of the Andes,
Mérida, Venezuela

ABSTRACT

Emulsion properties depend mainly upon three kinds of variables: physicochemical formulation, composition, and manufacturing protocol. The current state of the art allows the interpretation of the effects of these variables on such properties in the framework of a generalized phenomenology that includes temporal changes, either instantaneous or delayed, as they take place in manufacturing processes. The know-how can be readily translated into guidelines and constraints concerning the process operation and equipment design. This approach is referred to as formulation engineering.

I. INTRODUCTION

Emulsions are encountered both in nature and in many man-made goods. They are used in two-phase products such as foodstuff, paints, pharmaceuticals, cosmetics, and many others. Alternatively, they provide some interfacial or operational property of interest such as a high contact area in liquid–liquid extraction and emulsion polymerization or a controlled mass transfer rate in drug release and pollution remediation. They are increasingly involved in industrial processes, from the small-scale batch preparation of

---

Current affiliation:
*Institute for Surface Chemistry, Stockholm, Sweden.
†Rice University, Houston, Texas, U.S.A.
‡Royal Institute of Technology (KTH), Stockholm, Sweden.
¶M.W. Kellogg Ltd., Middlesex, United Kingdom.

fine-tuned products such as cosmetics to the large-scale manufacturing of millions of tons per year of emulsified fuel [1–4].

In many cases, emulsion specifications are stringent, and meeting them is a challenge that requires not only the guidelines found in industrial chemical recipes but also a full engineering treatment of the involved phenomena. The present chapter deals with the formulation engineering approach to emulsion making.

II. ORGANIZING EMULSION SCIENCE INTO KNOW-HOW

Emulsions are liquid-in-liquid dispersions that can occur as two simple types, namely oil drops in water (O/W) or water drops in oil (W/O), and some more complex morphologies such as double or multiple emulsions in which the drops contain droplets.

The type of emulsion and other properties are known to depend on four kinds of variables: (1) the physicochemical formulation variables, (2) the composition variables, (3) the mixing conditions that prevail during emulsification, and (4) the physical properties of components. This also includes the way in which these variables are manipulated during emulsification.

The effects of the physical properties of the components include, for instance, the role of the external phase viscosity on emulsion stability. These effects are well known or easy to ascertain in most cases and may be handled in a corrective or complementary fashion. They will not be discussed in detail in this chapter, which is dedicated to clarifying the coupled effects of the three other types of variables.

A. Physicochemical Formulation

Physicochemical formulation refers to intensive variables, which are characteristics of the nature of the components, along with temperature and pressure. They determine the affinity or negative of the standard chemical potential of the different species—particularly the surfactant—in all phases at equilibrium. They determine the phase behavior, as well as interfacial properties such as tension or natural curvature.

Although emulsions are systems out of mechanical equilibrium because they would finally end up in a separated two-phase system, the formulation is of paramount importance during the formulation of the emulsion and its useful lifetime. This is probably due to the fact that in many cases, the time scale of the emulsion persistence is large enough to attain physicochemical equilibrium between the phases.

The main practical problem in formulation is the large number of components, often many more than the simplest ternary case that contains only surfactant, oil, and water. In most cases there are other components or additives such as cosurfactants, electrolytes, or polymers. Moreover, the components are not usually pure substances. More often, they are mixtures that could be as complex as a crude oil or that could contain many electrolytes such as seawater. The surfactant is commonly a mixture either because of cost or manufacturing constraints or by choice, in order to adjust some property.

As a consequence, a typical emulsion could contain scores of different chemical species, each of them able to influence the formulation in a way not necessarily proportional to its concentration. Thus, even for a commonplace practical case, a systematic study could require thousands of research hours to be completed. This is why formulation has been considered an art rather than a science. This assessment has been changing with the growth of surface science in the past half-century and the uncovering of an extraordinarily rich variety of phenomena and structures in surfactant, polymer, and colloid chemistry [5].

However, most of the available knowledge is still too specific, simplistic, or naive to be useful to deal with the intricacies of even very simple practical cases. This is why a rational approach based on cause-and-effect trends has been favored by formulators of emulsions and other systems involving surfactants, oils, and water when numerical relationships are not available.

As proposed 50 years ago by Griffin [6], the empirical hydrophilic–lipophilic balance (HLB) method has still its supporters because of its extreme simplicity, although it falls short of taking into account many factors. At the same time, Winsor [7] proposed a theoretical interpretation based on the molecular interactions of the adsorbed surfactant molecules at the interface and the neighboring oil and water molecules. This was an enlightening and pedagogical contribution as far as the physicochemical understanding was concerned, but no numerical value was attainable. In the 1960s Shinoda introduced the phase inversion temperature (PIT), which is an experimentally attainable parameter [8–12]. It took into account all the variables because it could be measured even in extremely complex systems. Since then, it has been used amply to deal with nonionic surfactants that are slightly hydrophilic.

In the 1970s the enhanced oil recovery research effort promoted the development of a more complete description of the formulation effects, from both theoretical and empirical points of view. Empirical correlations involving the effect of the oil type, electrolyte type and concentration in water, surfactant type, alcohol type and concentration, as well as temperature and even pressure were developed [13–21].

More recently, these empirical equations were justified from a physico-chemical point of view as representative of the surfactant affinity difference (SAD), i.e., the free energy of transfer of a surfactant molecule from the oil phase to the water phase [22]. This free energy can be estimated from the measurement of the surfactant partitioning coefficient and from the way it changes with the different formulation variables [23,24]. For simplicity, the relationship has been written as the hydrophilic–lipophilic deviation (HLD), which is the same concept as SAD but is related to a reference state [25].

In the simple case of an ethoxylated nonionic surfactant, an n-alcohol cosurfactant, an n-alkane oil, and a sodium chloride brine, the HLD can be written as

$$HLD = \frac{(SAD - SAD_{ref})}{RT}$$

$$= \alpha - EON - kACN + bS + aC_A + c(T - T_{ref}) \qquad (1)$$

where EON is the average number of ethylene oxide groups per nonionic surfactant molecule, ACN is the alkane carbon number, S is the salinity as wt% NaCl, C_A is the alcohol concentration, T is the temperature, and α is a parameter that is characteristic of the surfactant lipophilic group type and branching. It increases linearly with the number of carbon atoms in the alkyl tail. The k, a, b, and c are numerical coefficients; SAD_{ref} equals $RT \ln K_{ref}$, where K_{ref} is the partition coefficient of the surfactant between oil and water in the reference state at the optimum formulation; K_{ref} is near unity for systems containing ionic surfactants but could be different for those formulated with nonionics.

The HLD is equivalent to the Winsor R ratio in the sense that it gathers all formulation effects in a single generalized formulation variable. When the HLD is equal to, larger, or smaller than zero, R is equal to, larger, or smaller than unity. However, and unlike R, the HLD can be calculated numerically when it is different from zero.

At HLD = 0, the temperature is equal to the PIT of the surfactant in the corresponding state (oil, water, alcohol). Equation (1) at HLD = 0 then makes it possible to determine the variation of the PIT with the system variables, and therefore to evaluate PIT values extrapolated outside the 0–100° C experimental range.

The HLD could be considered as a numerical generalization of previous concepts such as HLB but including the contribution of all variables involved. It is thus a parameter that describes the formulation state in all its generality. Using HLD allows a fine-tuning of formulation by changing the most convenient variable(s) or by combining several changes at once because there are several degrees of freedom.

An expression analogous to Eq. (1) for ionic surfactants, a listing of typical parameter values, and the rules to calculate the parameter corresponding to different mixtures of surfactants, oils, and electrolytes are available in the literature [14,16–22].

B. Phenomenological Approach

1. Changes in Phase Behavior Along a Formulation Scan

The relationship between the formulation variables and the phase behavior is exhibited through a formulation scan. This technique consists of preparing a sequence of SOW systems with identical compositions (a few percent of surfactant and equal amounts of oil and water) and the same formulation with the exception of one formulation variable, which is the selected scanned variable. In most cases the scanned variable is the aqueous phase salinity for ionic surfactant systems and the temperature or the average EON for nonionic ones. However, it may be any variable likely to change the value of HLD in Eq. (1).

The purpose of a formulation scan is to switch from HLD < 0 to HLD > 0, or vice versa, by changing a single formulation variable in a monotonous way. When the HLD < 0 the affinity of the surfactant for the aqueous phase dominates, and a so-called Winsor type I phase behavior is exhibited in which a surfactant-rich aqueous phase (micellar solution or microemulsion) is in equilibrium with an essentially pure oil phase. When the HLD > 0, a Winsor type II phase behavior is exhibited, and this time it is the oil phase that contains most of the surfactant. At the intermediate HLD = 0 formulation, the affinity of the surfactant is the same for both phases, and a very low minimum of interfacial tension is exhibited, which is the reason why the researchers involved in enhanced oil recovery in the 1970s called it the optimum formulation. This label has been conserved ever since even for other applications [13].

An optimum formulation can be characterized in many cases by the occurrence of a three-phase behavior, i.e., the so-called Winsor type III as described elsewhere [7,21]. Figure 1 gathers the results concerning the effect of a formulation scan.

2. Changes in Emulsion Properties Along a Formulation Scan

When a formulation is scanned from HLD < 0 to HLD > 0, i.e., when the surfactant affinity switches from hydrophilic to lipophilic, several transitions in emulsion properties are known to take place. Figure 2 summarizes some of these transitions, as observed in a large number of experimental results from different research groups [26–33].

Salager et al.

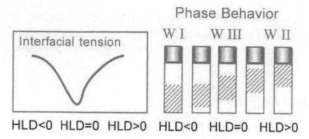

FIG. 1 Change in interfacial tension and phase behavior along a formulation scan.

The emulsion conductivity changes drastically near HLD = 0, indicating that emulsion inversion takes place there, irrespectively of the variable used to alter the HLD. According to the Bancroft rule, the wedge theory, and more modern curvature conceptualizations, HLD < 0 is associated with O/W emulsions and HLD > 0 with W/O emulsions. Near HLD = 0 an emulsion of the microemulsion–oil–water (MOW) three-phase system could be occurring, but there is no clear-cut indication about what constitutes its external phase.

The emulsion stability undergoes a very deep minimum in the vicinity of HLD = 0, regardless of the variable that is scanned to change the formulation. Near HLD = 0, it seems that no surfactant is available to stabilize the emulsion. This phenomenon has been interpreted in different ways [34–37].

Sometimes two maxima are observed on both sides of the optimum formulation, at some HLD distance from HLD = 0, usually within ±3 HLD units according to Eq. (1).

The emulsion viscosity also shows a minimum at the optimum formulation. The value of this minimum is unexpectedly low because the low interfacial tension is likely to result in very small droplets. Actually, it seems that the instability associated with HLD = 0 makes the droplets coalesce at

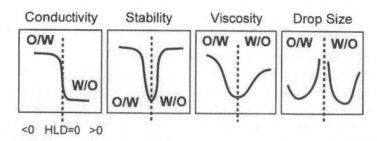

FIG. 2 Summary of emulsion property changes during a formulation scan.

once. On the other hand, the extremely low tension allows easy elongation of the droplets in some threadlike morphology. The real situation at HLD = 0 is not easy to analyze because of the extreme instability of the emulsions formed, but the point is that the resistance to flow is much lower than expected from a common emulsion [38,39].

Emulsion drop size is the result of competing effects that take place during emulsification: the drop breakup and the drop coalescence processes. Many properties and phenomena are likely to influence one or the other effect, sometimes in a complex way. As the formulation approaches HLD = 0 the interfacial tension decreases, thus facilitating the drop breakup and the formation of smaller drops. In a concomitant way, the emulsion stability becomes extremely low, allowing rapid coalescence, which favors the occurrence of larger drops. As a consequence of these opposite effects, the drop size exhibits a minimum for each type of emulsion, i.e., on each side of HLD = 0. For each system, the location of the minimum depends not only on the formulation (HLD value) but also on the stirring energy and efficiency [40].

3. Formulation–Composition Map

The formulation dominates the properties of SOW systems when the surfactant concentration is not too low and when the water-to-oil ratio is close to unity. When this is not satisfied, then the composition, i.e., the relative proportions of different substances, has to be taken into account. Provided that the surfactant concentration is not high enough to produce a single-phase microemulsion, say less than 10 to 20%, the most critical composition variable is the water-to-oil ratio, which is often expressed as the oil or water fraction, because the surfactant amount is small.

The water-to-oil ratio is known to influence, often greatly, the emulsion type, viscosity, and stability, sometimes counteracting the effect of formulation variables. A way to understand the combined or antagonist effects of these variables is to draw a bidimensional map of emulsion properties [41].

Figure 3 shows such formulation–composition schematic maps, which resemble those found experimentally. In these maps, the formulation is indicated in terms of HLD. The composition is expressed as water content in the water–oil mixture, which is essentially the water fraction in the system because the surfactant concentration is low in most practical cases. It is worth noting that because temperature is a formulation variable, formulation–composition maps and temperature–composition maps can be interpreted analogously. This is particularly important for systems containing nonionic surfactants.

The bold line that separates the O/W and W/O regions in Fig. 3 is called the standard inversion line. It has been drawn from the emulsion conductiv-

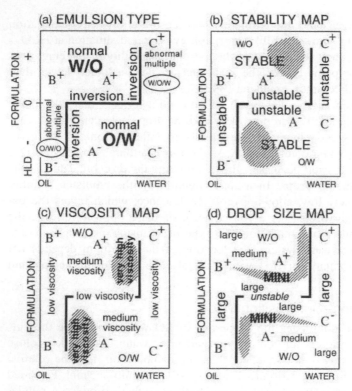

FIG. 3 Emulsion properties on a formulation–composition bidimensional map.

ity data. In most instances the aqueous phase contains some amount of electrolyte and thus conducts electricity, while the oil phase is a nonconductor. Thus, it is straightforward to determine the emulsion types from a conductivity measurement because O/W emulsions are conductors whereas W/O ones are not.

In Fig. 3a, it is seen that the standard inversion line is formed with three branches. First there is a "horizontal" branch, located at the optimum formulation (HLD = 0) in the central part of the map, i.e., when the relative amounts of oil and water are similar. This region is labeled A, with a + or − superscript depending on the sign of HLD. In this A region, which typically spans 30 to 70% water, the emulsion type strictly depends on the formulation, and the discussion presented in Section II.B.2 fully applies.

The other two branches of the standard inversion line are essentially vertical and are typically located at 30% water on the negative HLD side and at 70% water on the positive side. These vertical branches define the low water and low oil content regions, B and C, respectively. At low water content (respectively low oil content), the W/O (respectively O/W) emulsion dominates whatever the formulation. In these extreme water-to-oil ratio regions, the phase that is present in larger volume becomes the external phase of the emulsion, as mentioned by Ostwald [42] almost a century ago. Consequently, in these B and C regions, the composition dominates.

Nevertheless, a closer look at the conductivity value indicates the presence of multiple emulsions in B^- and C^+, the so-called abnormal zones, where there is a conflict between the composition and formulation effects. For instance, in the C^+ region a multiple w/O/W emulsion is found. In this case, the composition determines what is the main or outer (O/W) emulsion, whereas the formulation induces the secondary droplet-in-drop (w/O) inner emulsion. A similar situation, but with o/W/O multiple emulsions, is found in the B^- region. The relative amounts of these two emulsion types depend on the emulsification process, particularly on the way the formulation and composition are varied during the stirring.

The interest in such a combined formulation–composition map is not only because of its generality as far as the emulsion type is concerned but also due to its adequacy for rendering the qualitative variations of emulsion stability, viscosity, and drop size, as indicated in the maps in Fig. 3b–d, which summarize a large amount of experimental data [43–49].

Both A^+ and A^- regions and adjacent B^+ and C^- normal regions are associated with stable emulsions. The maximum emulsion stability is often attained in the corresponding A zone near the vertical branch of the inversion line and at some distance from the optimum formulation, e.g., 3–4 HLD units (shaded zone in map 3b). This is due to the fact that far away from the optimum formulation the emulsion stability tends to decrease because the surfactant is too hydrophilic or too lipophilic. The emulsion stability often decreases as well when the internal phase ratio decreases because the drops are often larger due to inefficient stirring, and thus settling is quicker.

On the other hand, the strip near the optimum formulation, say $|HLD| = 0–1$, exhibits very unstable emulsions, in accordance with Section II.B.2. Unstable emulsions are also found in abnormal B^- and C^+ regions. However, it is worth noting that multiple emulsions are often found in these regions and that the low stability refers to the outer emulsion, e.g., the O/W emulsion in a w/O/W multiple emulsion located in the C^+ region and the W/O emulsion in a multiple o/W/O emulsion found in the B^- region. In both cases, the inner emulsion is stable because it obeys the formulation requirement.

The emulsion relative viscosity increases in the A regions in the direction of higher internal phase ratio (at constant formulation), so that the viscosity maximum is located near the vertical branches of inversion line (see Fig. 3c). This high viscosity, which is due to a high internal phase content, is enhanced by the particularly efficient stirring conditions in these (shaded) regions, which result in small droplets. On the other hand, the viscosity decreases when the formulation approaches HLD = 0 at constant composition, as discussed in Section II.B.2.

In most cases, multiple emulsions located in the B^- and C^+ regions exhibit low relative viscosity because their external phase content is relatively high. There is, however, an exception to this trend, e.g., when most of the external phase has been transferred as droplets inside the drops. Such a situation could happen either at once during the emulsification process or slowly as a consequence of osmotic migration from the most external to the most internal one.

As discussed in Section II.B.2, the emulsion drop size is the result of competing breakup and coalescence processes. As the formulation approaches HLD = 0, the concomitant decrease in interfacial tension and increase in coalescence rate result in a drop size minimum. As a consequence of this effect, there is a minimum drop size region (shaded) on each side of HLD = 0, parallel to the horizontal branch of the inversion line (Fig. 3d).

On the other hand, the slow shear mixing of high internal phase ratio emulsions located in the shaded zones of Fig. 3c has been found to be very efficient in producing extremely small droplets, irrespective of the surfactant concentration and stirring energy. There is thus another minimum drop size (shaded) strip located in each of the A regions, near and parallel to the vertical branch of the inversion line [3,50,51].

C. Shifting the Inversion Boundary

In no case was the horizontal branch of the inversion line displaced by the effect of some variable. Contrariwise, the vertical branches may be shifted in different ways by various means.

1. Effect of Other Variables on the Inversion Line

Although the formulation and oil–water composition are certainly the most important variables as far as the general phenomenology is concerned, it is well known that many other variables are likely to affect the emulsion type and properties. Some of these variables are the surfactant concentration, the phase viscosity, the stirring energy, the nature of the surfactant, and the emulsification protocol. The effect of some of them on the formulation–composition mapping has been identified [52–54]. In general, the enlargement of some regions and the shrinking of others are observed, but the general phenomenology regarding changes in emulsion properties remains unchanged.

An increase in oil viscosity tends to shift the A^+/C^+ vertical branch of the inversion line to the left, thus shrinking the A^+ region where the oil is the external phase [52]. In a similar way, the increase of the water phase viscosity tends to reduce the extension of the A⁻ region.

An increase of surfactant concentration tends to widen the A region, on both the − and + formulation sides. In other terms, it expands the zone where the formulation dictates the emulsion type [53].

An increase in stirring energy seems to produce the opposite trend, i.e., to shrink the central A region and to expand both B and C regions [54]. However, this result is to be taken with caution because the effect of the stirring could depend not only on the energy but also on the duration, and there might be more intricate kinetic issues involved.

Consequently, these effects allow expansion or shrinkage of the regions where some specific emulsion property, such as high stability, is found. Sometimes, the placement of the branches of the inversion line could make a region disappear, with concomitant vanishing of the feasibility to attain some property such as a high-viscosity or small drop size emulsion. In other cases some enhanced or new property could be made to appear instead.

2. Dynamic Inversion and Memory

The emulsions discussed in the previous sections were prepared at fixed formulation and composition in the map. In practice, the formulation and composition of a system can change as time elapses or as emulsification proceeds. For instance, one of the phases could be added little by little, such as oil drops in a homemade mayonnaise preparation. In another case, the formulation or temperature could be changed according to certain programming protocol as in emulsion polymerization.

Such changes may be taken into account by shifting the representative point of the emulsion on the formulation–composition map. In some cases this point could trespass on the standard inversion line and emulsion inversion could take place in a dynamic fashion. Recent studies have shown that there are two kinds of dynamic inversions: (1) the vertical crossing of the horizontal branch, which is produced by changing a formulation variable from A⁻ to A⁺ region or vice versa, and (2) the horizontal crossing of one of the vertical branches, which takes place by changing somehow the water-to-oil ratio. The first type has been called transitional inversion because it happens smoothly in some reversible way. The second one was termed catastrophic dynamic inversion because it develops as a sudden instability and exhibits several characteristics of the cusp catastrophe model, such as hysteresis and metastability [55].

By the way, both the phase behavior at equilibrium and the emulsion dynamic inversion features can be interpreted in a relatively simple way by a sixth-order catastrophe, the so-called butterfly model [56].

This coincidence corroborates the strong relationship between the phase behavior (thus formulation and composition) and the emulsion properties exhibited in the bidimensional map. A recent review describes the state of the art relative to this matter, which is not yet settled [57].

As far as we are concerned here, it is enough to mention that the dynamic transitional inversion along the vertical path in the A region of the map always takes place at the crossing of the HLD = 0 horizontal branch of the standard inversion line, whatever the direction of change, provided that the change is not too quick. On the contrary, there is a delay in the catastrophic inversion produced by the change in composition, i.e., along a horizontal path crossing any of the vertical branches of the inversion line. Figure 4 indicates the typical shift of the inversion line according to the path of change indicated by the arrows. The tip of the arrow is located at the position where the dynamic inversion takes place. The dashed lines indicate the location of the vertical branches of the standard inversion. The triangular shaded zones in Fig. 4 (center) are the hysteresis regions where the emulsion can be one type or the other, depending on the direction of change [29,45,57,58].

The triangular shape of the hysteresis regions is characteristic of the absence of delay at the optimum formulation and of an increasing delay as the formulation departs from HLD = 0. As seen in Fig. 4, left and right, these regions can be made to belong either to the O/W or W/O type depending on the way the dynamic emulsification is carried out. This memory feature thus makes it possible to displace the inversion line to suit applications.

For instance, the home preparation of mayonnaise, a high internal phase content emulsion, consists of adding drops of oil to egg yolk, which is the water phase with a hydrophilic surfactant. The corresponding change follows the horizontal bold arrow located in the lower part of zone A⁻ in Fig. 4, left graph. It is worth noting that adding a teaspoon of mustard to the egg yolk

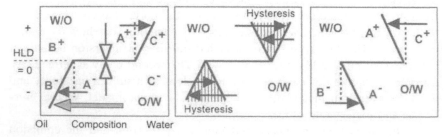

FIG. 4 Transitional and catastrophic dynamic inversions on a formulation–composition map.

shifts the formulation to an even more negative HLD value, thus producing two effects. First, the dynamic inversion takes place more to the left because the arrow is in a lower position. This means that a higher internal phase content O/W mayonnaise can be attained, which is more viscous and has smaller droplets because it is made in the favorable location according to Fig. 3d. Second, the emulsion is more stable because it is in the shaded zone in Fig. 3b, far enough from HLD = 0, and is thus probably less sensitive to a change in temperature, which could move the HLD toward zero.

As with the standard inversion, the catastrophic branches of the dynamic inversion can be shifted in different ways, and almost any situation is feasible provided that a proper path, sometimes very complex, is used. Although some systematic trends have been reported [53,54], such as the effect of the phase viscosity, the surfactant concentration, the stirring energy, the effect of solids, or the inversion protocol, the matter is not yet completely settled and considerable rationalization is required to obtain a clear-cut picture of the optimum way to attain some specific requirement in practice.

The preceding phenomenology describes in a qualitative way how the properties are expected to change in a general framework, where the minima and the maxima are encountered, as well as where little change is likely to take place. This allows the formulators to seek the desired property in the right region of the diagram and to focus their trial on the most probable or most feasible region in the formulation–composition–stirring space. It is worth remarking that this phenomenology only indicates the location of the maxima or minima. Whether the actual value of the maximum of a property is higher or lower, or whether a region is more or less extended, depends on more specific effects that can be seen as quantitative modifiers of the general trends. For instance, any factor that tends to slow down the interdrop film drainage, such as an increase in external phase viscosity or stronger electrostatic or steric repulsion, is likely to increase the emulsion stability [59,60].

Some factors can exhibit a double effect, the first one along the general phenomenology and the other through a qualitative or secondary modification. For instance, an increase in n-pentanol cosurfactant concentration will in most cases drive a formulation transition from HLD < 0 to HLD > 0 because of an increase of the C_A term in Eq. (1). As a consequence of the general phenomenology indicated in Fig. 2, the emulsion stability will pass through a minimum in the neighborhood of HLD = 0 and then will rise again for positive values of HLD. However, such an alcohol concentration increase will result in at least two other effects that could be of importance.

When the concentration of alcohol is augmented in the system, the interfacial adsorption of alcohol tends to increase, and the alcohol molecules compete more and more with the surfactant to occupy the interface. Consequently, the surfactant density at the interface goes down as well as the surfactant stabilizing ability. Thus, the increase in emulsion stability on the positive HLD side will be less than expected from the symmetrical change indicated in Fig. 2. On the other hand, the presence of alcohol suppresses the formation of liquid crystals, which can have an influence on the emulsion viscosity and stability as well.

In addition, the presence of a high concentration of alcohol is likely to change the value of the interfacial tension between the oil and aqueous phases, thus affecting the efficiency of the drop breaking mechanism. Finally, a lipophilic alcohol such as n-pentanol is a good candidate to favor the so-called partitioning phenomenon, which alters the interfacial formulation in a way that depends on the concentration as well as the alcohol/surfactant ratio. All these effects together are likely to modify the values of the properties of the emulsion but not the general phenomenology described previously.

III. CONVERTING KNOW-HOW INTO FORMULATION ENGINEERING

The general phenomenology described in the previous section was independent of the particular system, i.e., whether it was with ionic or nonionic surfactant, whether the oil phase was olive oil or petroleum, and whether the formulation variable to be manipulated in the process was temperature or salinity. This situation reminds us of the genesis of chemical engineering when it evolved from industrial chemistry with the development of the concept of unit operation.

In the second part of this chapter, basic operations of formulation engineering dedicated to emulsion making are proposed, using the formulation–composition framework discussed in the first part.

A. Programming Changes Without Inversion

The maps shown in Fig. 3 indicate that on each side of the standard inversion line, the properties of the emulsion depend on its formulation and composition and also in some way on the stirring energy as discussed previously. If an emulsion made at some formulation–composition point is changed to another location of the map, by modifying either the formulation (including temperature) or the composition or both at the same time, the new representative point could be in a region of the map where the emulsion is expected to have different properties.

Does the emulsion exhibit the properties corresponding to its new location or does it retain all or part of the properties exhibited by the original one?

The answer to this question depends on the way the change is carried out. It seems that there is a general response in two extreme cases, i.e., when the formulation–composition change is applied in a very slow or a very quick fashion, as discussed next.

1. Slow Change

The slow change type corresponds to the situation in which the formulation or composition or both are modified at a rate that allows the system to equilibrate or to attain pseudoequilibrium from two points of view. First, the surfactant partitioning between the phases and at the interface should be at equilibrium or near equilibrium. This implies that the time scale of the change is long enough for diffusional processes and adsorption to take place significantly when the formulation or composition is changed. Second, the dynamic equilibrium between the breakup and coalescence mechanisms that determine the drop size must be reached. This implies that stirring is maintained while the formulation–composition change is taking place.

In these conditions, the properties of the emulsion essentially change as a function of the position on the bidimensional map and thus the characteristic features of each region of the map can be attained by slowly shifting the representative point of the emulsion to this region. This basic operation is referred to as slow formulation–composition programming without inversion.

The time scale depends upon the magnitude and nonequilibrium characteristic of the variation produced by the change. For instance, if an emulsion of the O/W type located in the center of the A^- region is diluted with water, the process reaches equilibrium very quickly because most of the surfactant is already in water, and little diffusion will take place. On the contrary, if a water phase containing some surfactant and alcohol is added to a W/O emulsion located in the A^+ region, some time might be required for the surfactant to migrate to the external oil phase. The location of the change in the map is important as well. It has been found that near HLD = 0 both mass transfer and equilibrium take place much more quickly than far away from it [61].

2. Quick Change with Quench Effect

At the other extreme is the situation in which a change in formulation (including temperature), composition, or stirring is carried out rapidly so that some characteristics have no time to change, such as those related to geometry and structures, in particular drop size and associated properties.

For instance, when an O/W emulsion containing a nonionic surfactant is made in the center of the upper part of the A^- region, not far away from HLD = 0, very small droplets are produced. However, this emulsion is not very stable because it is too near HLD = 0. If it is cooled quickly after being made, the representative point is going to be shifted to a lower position in the A⁻ region, where high stability is found. In such a quench, the emulsion drop size is conserved, so that the quenched emulsion exhibits a drop size smaller than the one attainable directly in the same final position (see path P1 in Fig. 5). The programming thus allows memorizing the small drop size feature.

In this case of rapid cooling, the operation deserves the name "quench." This label could be used in a general fashion to refer to all quick changes that move the representative point of the emulsion from one place of the map to another in a quick way. For instance, a change in HLD equivalent to a rapid cooling can be attained by adding a small amount of concentrated hydrophilic surfactant solution followed by efficient mixing to distribute it throughout the system (see next example).

3. Intermediate Programming Without Inversion

In many cases, the time scale of change could be halfway between slow and quick and some additional action can be taken to move it in one or the other way.

In the crude oil dehydration process, a water in crude oil emulsion coming from the well is treated by adding a very hydrophilic surfactant [62,63]. The original W/O emulsion is located at HLD $\gg$ 0 and at a high content of oil, in the B⁺ region. The final emulsion has essentially the same contents of oil and water, but its formulation, attained by the mixture of the lipophilic natural surfactants and the added chemical demulsifier, is just at HLD = 0, where the coalescence rate is highest (see path P2 in Fig. 5). In crude oil dehydration, the limiting process from the kinetic point of view is the migration of the demulsifier molecules to the water drop interface. This is

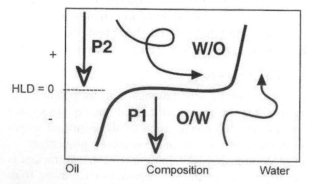

FIG. 5 Programming formulation and composition on the same side of the inversion line.

accelerated by dissolving the emulsifier in gasoil and mixing it with the crude oil as early as possible in the production equipment. Such a case could be considered a formulation quench.

B. Shifting or Pushing Boundaries

It has been seen that while the horizontal branch of the inversion line is essentially immutable, the vertical branches can be displaced or pushed around.

In the case of the standard inversion line, the phase viscosity, the surfactant concentration, and the stirring energy are the most convenient variables for producing a shift. For instance, if a high internal phase ratio O/W emulsion is sought, say with 85% oil, and if the map indicates that in the current conditions the B^-/A^- branch of the inversion line is located at 75% oil (Fig. 6a), then a 10% shift to the left is required to extend the A^- zone to 85% water. Using slower mixing or a higher surfactant concentration or a combination of the two could attain this (Fig. 6b).

If this is not sufficient, a dynamic process could be applied, starting with an emulsion containing a low internal phase ratio, say 50 or 60% oil, and then adding oil little by little so that the hysteresis phenomenon discussed previously pushes the A^-/B^- dynamic inversion line beyond 85% oil (Fig. 6c). This is how homemade mayonnaise is prepared. As discussed before, the dynamics of the change influences the result as well as the formulation deviation from HLD = 0 (see arrows in Fig. 6c).

It is worth noting that such an inversion shifting or pushing process can be combined with formulation–composition–stirring programming. For instance, in manufacturing an O/W emulsion containing 50% of a viscous oil phase, it is often difficult to attain a small drop size by direct stirring. The answer to this problem is to emulsify the system in the A^- region at 70% or more oil with a low-energy stirring device, which results in small drop size (see Fig. 3d), and afterward to dilute it to the final internal phase content

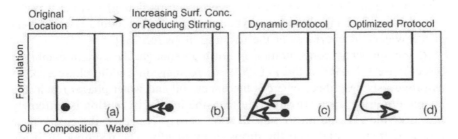

FIG. 6 Shifting and pushing the inversion boundary.

by adding water (Fig. 6d). The dilution can be combined with cooling, for instance, to take advantage of other effects such as the change of viscosity with temperature or the influence of temperature on the formulation.

C.   Trespassing the Inversion Boundary

Crossing the inversion boundary triggers an instability process that is not yet fully understood in all cases, although it is often used in practice to make paints, cosmetics, and other emulsified products. Several cases have been clearly identified as follows.

1.   Transitional Inversion via Formulation or Temperature Change

When a change in formulation (or temperature) shifts the representative point of the emulsion from HLD > 0 to HLD < 0 or conversely, a so-called transitional inversion takes place. The mechanism of this process seems to depend partially upon the surfactant concentration and thus on the number of phases exhibited by the system in the neighborhood of HLD = 0.

If the surfactant concentration is high enough, the system exhibits so-called Winsor IV monophasic behavior in the vicinity of HLD = 0. This means that when the formulation is changed, the emulsified system starts as a two-phase emulsion, then becomes a single-phase microemulsion, and finally ends up in the other type of two-phase emulsion.

Figure 7 depicts the case of a transition by cooling for a nonionic system, the so-called PIT emulsification method [64], because the formulation variable is temperature, and the HLD = 0 optimum formulation is attained at the phase inversion temperature. In this case the emulsion at a temperature above the PIT is W/O; then as temperature decreases the microemulsion oil phase solubilizes more and more water and the water drops vanish.

Below the PIT, the microemulsion starts exuding oil droplets that grow both in number and in size as the temperature keeps descending, to end up in an O/W emulsion that becomes stable at 20 to 30°C below the PIT. The sizes of the final drops depend on the protocol, particularly on the way the temperature is changed, and the eventual deposition of liquid crystal layers at the water–oil interfaces of the forming drops [65–67].

If the surfactant concentration is not high enough, the system exhibits a three-phase behavior at the PIT. The microemulsion middle phase evolves as previously described, whereas the excess oil and water phases result in a coarse emulsion upon stirring. In effect, the interfacial tension is extremely low near the PIT and thus emulsification is easy. However, the coalescence rate is extremely high and the drops grow rapidly. The resulting emulsion is often a bimodal type, with the small droplets exuding from the micro-

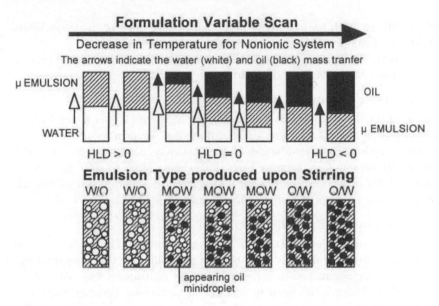

FIG. 7 Phase behavior transition and emulsion transitional inversion due to a change in physicochemical formulation.

emulsion and the coarse emulsion resulting from the stirring of oil and water excess phases.

In any case, a good stirring and temperature programming strategy can change the drop size in a very appreciable way [68–72] (this volume, chapter by Solans et al.).

2. Catastrophic Inversion via Water-to-Oil Ratio Change

Catastrophic inversion takes place when the internal phase is added upon stirring to an emulsion, irrespective of its type. In the direction from a normal emulsion (A region) to an abnormal one (B⁻ or C⁺), an extremely viscous and very high internal phase ratio emulsion is often attained before reaching the inversion. The emulsion sometimes becomes so viscous that the stirring operation has to be interrupted at 95 or 98% internal phase ratio with no evidence of inversion. The way the internal phase is added and the stirring seem to be paramount in triggering the inversion sooner or later for a given system [57,73].

The crossing of the catastrophic inversion line from an abnormal region to the corresponding A region seems to involve even more complex mechanisms. In most cases the original emulsion is, or soon becomes, a multiple emulsion and many intermediate morphologies can happen such as a multiple emulsion with an extremely high inner phase content, which exhibits a foamlike behavior [74].

Considerable research still has to be carried out to achieve a full picture that obviously depends on both thermodynamics and kinetic phenomena.

3. Catastrophic Inversion by Stirring

Increased stirring is known to trigger the inversion of a higher internal phase ratio emulsion against the formulation influence in some practical cases, such as heavy hydrocarbon emulsified fuels. This may be easily related to the general phenomenology because increased stirring tends to shift the vertical branch of the inversion line toward a lower internal phase ratio so that the representative point of the emulsion changes sides [2,73].

4. Spontaneous Emulsification

Spontaneous emulsification refers to the production of an emulsified system in the absence of stirring. It is an instability mechanism in which a substance, generally a surfactant and/or a cosurfactant, is transferred from one phase to the other. There is no need to assume an unrealistic situation such as a negative interfacial tension because the decrease in chemical potential of the transferred substance is the energy source that induces the increase of surface area. How spontaneous emulsification takes place is not fully understood yet, although the diffusion and stranding mechanism seems to offer a good hypothesis [75].

In practice, spontaneous emulsification can be combined with emulsion inversion. For instance, if a water phase is poured little by little into an oil phase containing a dissolved hydrophilic surfactant and/or alcohol, the first dispersion to occur is a W/O emulsion because there is very little water. As the number of water drops increases, the surfactant migrates from the oil to the water phase and the dynamic interfacial tension can be close to zero. A multiple emulsion often occurs as an intermediate situation. Then an O/W emulsion appears after some time when the kinetic phenomena finally prevail. This cannot be interpreted straightforwardly from the bidimensional map unless the formulation is assumed to change as the surfactant migrates from oil to water. In such a view, the trajectory of change moves from B^+ to A^-, crossing the inversion line somewhere.

D. Formulation Engineering Wrap-Up

All the basic operations mentioned in the previous sections that involve changes in formulation, composition, and stirring and other emulsification protocol programming, e.g., heating, mixing, adding substances, and diluting, may be readily translated into process engineering specifications for equipment design.

IV. CONCLUSION

We have shown that the current state of the art in emulsion science, particularly the formulation–composition mapping of emulsion properties, is general know-how that may be segregated into basic operations for the formulation engineering approach to emulsion making.

ACKNOWLEDGMENTS

The authors are grateful to CDCHT-ULA and CONICIT (Agenda Petroleo Program) for sponsoring the Lab. FIRP research effort on emulsion science, particularly inversion and related topics.

REFERENCES

1. H Rivas, ML Chirinos, L Paz, I Layrisse, EL Murray, A Stockwell. Proceedings of the 3rd UNITAR International Conference on Heavy Crude and Tar Sands, Long Beach, CA, 1985, p 1483.
2. LA Pacheco, J Alonso. Proceedings of the 6th UNITAR International Conference on Heavy Crude and Tar Sands, Houston, TX, 1995, vol. 1, p 203.
3. GA Nuñez, MI Briceño, C Mata, H Rivas. J Rheol 40:405, 1996.
4. JL Salager, MI Briceño, CL Bracho. In: J Sjöblom, ed. Encyclopedic Handbook of Emulsion Technology. New York: Marcel Dekker, 2001, p 455.
5. See, for instance, the annual reviews in Current Opinion in Colloid and Interface Science.
6. WC Griffin. J Soc Cosmet Chem 1:311, 1949 and 5:249, 1954.
7. P Winsor. Solvent Properties of Amphiphilic Compounds. London: Butterworth, 1954.
8. K Shinoda, H Arai. J Phys Chem 68:3485, 1964.
9. H Arai, K Shinoda. J Colloid Interface Sci 25:396, 1967.
10. K Shinoda, H Takeda. J Colloid Interface Sci 32:642, 1970.
11. K Shinoda, H Kunieda. In: P Becher, ed. Encyclopedia of Emulsion Technology. Vol. 1. New York: Marcel Dekker, 1983, p 337.
12. T Förster, F Schambil, H Tessman. J Cosmet Sci 12:217, 1996.
13. DO Shah, RS Schechter, eds. Improved Oil Recovery by Surfactant and Polymer Flooding. New York: Academic Press, 1977.
14. JL Salager, J Morgan, RS Schechter, WH Wade, E Vasquez. Soc Petrol Eng J 19:107, 1979.
15. M Bourrel, JL Salager, RS Schechter, WH Wade. J Colloid Interface Sci 75:451, 1980.
16. RE Antón, N Garcés, A Yajure. J Dispersion Sci Technol 18:539–555, 1997.
17. M Baviere, RS Schechter, WH Wade. J Colloid Interface Sci 81:266, 1981.
18. H Kunieda, K Hanno, S Yamaguchi, K Shinoda. J Colloid Interface Sci 107:129, 1985.
19. P Fotland, A Skauge. J Dispersion Sci Technol 7:563–579, 1986.

20. A Skauge, P Fotland. SPE Reservoir Eng November:601, 1990.
21. M Bourrel, RS Schechter. Microemulsions and Related Systems. New York: Marcel Dekker, 1988.
22. JL Salager. In: G Broze, ed. Handbook of Detergents—Part A: Properties. New York: Marcel Dekker, 1999, pp 253–302.
23. A Graciaa, J Lachaise, JG Sayous, P Grenier, S Yiv, RS Schechter, WH Wade. J Colloid Interface Sci 93:474, 1983.
24. N Márquez, RE Antón, A Graciaa, J Lachaise, JL Salager. Colloids Surf A 100: 225, 1995.
25. JL Salager, N Márquez, A Graciaa, J Lachaise. Langmuir 16:5534, 2000.
26. K Shinoda, H Saito. J Colloid Interface Sci 30:258, 1969.
27. M Bourrel, A Graciaa, RS Schechter, WH Wade. J Colloid Interface Sci 72: 161, 1979.
28. JE Viniatieri. Soc Petrol Eng J 20:402, 1980.
29. JL Salager, I Loaiza-Maldonado, M Miñana-Pérez, F Silva. J Dispersion Sci Technol 3:279, 1982.
30. JC Noronha, DO Shah. AIChE Symp Ser 212:42, 1982.
31. LM Baldauf, RS Schechter, WH Wade, A Graciaa. J Colloid Interface Sci 85: 187, 1982.
32. FS Milos, DT Wasan. Colloids Surf 4:91, 1982.
33. S Qutubuddin, CA Miller, T Fort Jr. J Colloid Interface Sci 101:46, 1984.
34. RE Antón, JL Salager. J Colloid Interface Sci 111:54–59, 1986.
35. R Hazzlett, RS Schechter. Colloids Surf 29:53, 1988.
36. A Kalbanov, H Wennerström. Langmuir 12:276, 1996.
37. A Kalbanov, J Weers. Langmuir 12:1931–1935, 1996.
38. S Vijayan, C Ramachandran, H Doshi, DO Shah. Proceedings 3rd International Conference on Surface and Colloid Science, Stockholm, 1979, p 327.
39. JL Salager, M Miñana-Pérez, J Andérez, J Grosso, C Rojas, I Layrisse. J Dispersion Sci Technol 4:161, 1983.
40. JL Salager, M Pérez-Sanchez, Y Garcia. Colloid Polym Sci 274:81, 1996.
41. JL Salager, M Miñana-Perez, M Perez-Sanchez, M Ramirez-Gouveia, CI Rojas. J Dispersion Sci Technol 4:313, 1983.
42. W Ostwald. Kolloid Z 6:103, 1910.
43. M Miñana, P Jarry, M Perez-Sanchez, M Ramirez-Gouveia, JL Salager. J Dispersion Sci Technol 7:331, 1986.
44. BW Brooks, HN Richmond. Colloids Surf 58:131, 1991.
45. BW Brooks, HN Richmond. Chem Eng Sci 49:1053, 1994.
46. HT Davis. Colloids Surf A 91:9, 1994.
47. JL Salager. In: A Chattopadhyay, KL Mittal, eds. Surfactants in Solution. Surfactant Science Series 64. New York: Marcel Dekker, 1996, p 261.
48. BP Binks, SO Lumsdon. Langmuir 16:2539, 2000.
49. BP Binks, SO Lumsdon. Langmuir 16:3748, 2000.
50. TJ Lin. J Soc Cosmet Chem 29:117, 1978.
51. TJ Lin, YF Shen. J Soc Cosmet Chem 35:357, 1984.
52. JL Salager, G Lopez-Castellanos, M Miñana-Perez. J Dispersion Sci Technol 11:397, 1990.

53. F Silva, A Peña, M Miñana-Pérez, JL Salager. Colloids Surf A 132:221, 1998.
54. A Peña, JL Salager. Colloids Surf A 181:319, 2001.
55. E Dickinson. J Colloid Interface Sci 84:284, 1981.
56. JL Salager. In: P Becher, ed. Encyclopedia of Emulsion Technology. Vol 3. New York: Marcel Dekker, 1998, p 79.
57. JL Salager, L Márquez, A Peña, MJ Rondón, F Silva, E Tyrode. Ind Eng Chem Res 39:2665, 2000.
58. F Groeneweg, WGM Agterof, P Jaeger, JJM Janssen, JA Wieringa, JK Klahn. Trans Inst Chem Eng 76(part A):55, 1998.
59. IB Ivanov, ed. Thin Liquid Films. New York: Marcel Dekker, 1988.
60. IB Ivanov, PA Kralchevsky. Colloids Surf A 128:155, 1997.
61. L Fillous, A Cardenas, J Rouvière, JL Salager. J Surfactants Deterg 3:303, 1999.
62. JL Salager. Int Chem Eng 30:103, 1990.
63. A Goldszal, M Bourrel. Ind Eng Chem Res 39:2746, 2000.
64. K Shinoda, H Saito. J Colloid Interface Sci 30:258, 1969.
65. N Pipel, ME Rabbani. J Colloid Interface Sci 119:550, 1987.
66. J Rouvière, JL Razakarison, J Marignan, B Brun. J Colloid Interface Sci 133: 293, 1989.
67. J Rouvière. Inf Chim 325:158, 1991.
68. BW Brooks, HN Richmond. Chem Eng Sci 49:1953, 1994.
69. M Miñana-Perez, C Gutron, C Zundel, JM Andérez, JL Salager. J Dispersion Sci Technol 20:893, 1999.
70. C Solans, R Pons, S Zhu, HT Davis, DF Evans, K Nakamura, H Kunieda. Langmuir 9:1479, 1993.
71. A Forgiarini, J Esquena, C González, C Solans. Langmuir 17:2076, 2001.
72. H Sagitani. J Am Oil Chem Soc 58:738, 1981.
73. A Peña, J Valero, L Castro, L Márquez, M Rondón, JL Salager. Paper presented at the 13th International Symposium on Surfactants in Solution (SIS), Gainesville, FL, 2000.
74. L Márquez. PhD dissertation, University of Pau (France) and University de Los Andes (Mérida Venezuela), 2001.
75. CA Miller. Colloids Surf 29:89, 1988.

25

Nano-Emulsions: Formation, Properties, and Applications

**CONXITA SOLANS, JORDI ESQUENA,
ANA MARIA FORGIARINI, NÚRIA USÓN, DANIEL MORALES,
PAQUI IZQUIERDO, and NÚRIA AZEMAR** Institut
d'Investigacions Químiques i Ambientals de Barcelona,
Barcelona, Spain

MARÍA JOSÉ GARCÍA-CELMA Universitat de Barcelona,
Barcelona, Spain

ABSTRACT

Nano-emulsions are defined as a class of emulsions with uniform and extremely small droplet size (typically in the range 20–500 nm). The formation of kinetically stable liquid/liquid dispersions of such small sizes is of great interest from fundamental and applied viewpoints. In this review, nano-emulsion formation, with special emphasis on low-energy emulsification methods, is first discussed. This is followed by a description of nano-emulsion properties, focusing on their kinetic stability. Finally, relevant industrial applications of nano-emulsions in the preparation of latex particles, in personal-care formulations, and as drug delivery systems are reported.

I. INTRODUCTION

Nano-emulsions are a class of emulsions with uniform and extremely small droplet size (typically in the range 20–500 nm). Because of their characteristic size, some nano-emulsions appear transparent or translucent to the naked eye (resembling microemulsions) and possess stability against sedimentation or creaming. Nano-emulsions of the oil-in-water (O/W) type have been investigated and used in practical applications for a long time [1–5]. However, they have experienced growing interest and very active development in recent years, as reflected by publications [6–16] and patents [17–27] on this subject.

Although nano-emulsions are thermodynamically unstable systems, they may possess high kinetic stability. This property together with their transparent or translucent visual aspect and a viscosity similar to that of water makes them of special interest for practical applications. Nano-emulsions are used in the pharmaceutical field as drug delivery systems [8,17, 18,25,28–33], in cosmetics as personal-care formulations [2,4,6,7,10,19– 21,23,24,27], in agrochemical applications for pesticide delivery [3,34,35], in the chemical industry for the preparation of latex particles [9,22,26,36– 38], etc. In addition, the formation of kinetically stable liquid/liquid dispersions of such small sizes is of great interest from a fundamental viewpoint.

The terminology to designate this type of liquid/liquid dispersions is very varied. They are often referred to in the literature as submicrometer-sized emulsions [8], finely dispersed emulsions [2], ultrafine emulsions [4,39], miniemulsions [1,5,9,11], nano-emulsions [39], etc. The term miniemulsion was introduced in the early 1970s to describe kinetically stable oil-in-water emulsions with average droplet sizes in the range 100–400 nm, containing low concentrations of an emulsifier mixture (1–3 wt% based on the oil phase) and prepared under mechanical shear [1,5,40–43]. These miniemulsions were used to prepare polymer latexes either by polymerization of monomer droplets [1] or by direct emulsification of polymer solutions [44]. Submicrometer emulsion (SEM) is a term usually used to describe parenteral and other types of pharmaceutical emulsions showing nano-emulsion characteristics. In the cosmetic field, these formulations are often designated as fine, ultrafine, and finely dispersed emulsions. The term nano-emulsion has been increasingly adopted because, in addition to being concise, it gives an idea of the nanoscale size range of the droplets and it avoids misinterpretation with the term microemulsion.

Nano-emulsions, being nonequilibrium systems, cannot be formed spontaneously. Consequently, energy input, generally from mechanical devices or from the chemical potential of the components, is required [45]. The methods using mechanical energy are called dispersion or high-energy emulsification methods, and those making use of the chemical energy stored in the components are referred to as condensation, low-energy, or "spontaneous" emulsification methods [46]. In practice, a combination of both methods has proved to be an efficient way to obtain nano-emulsions with small and very uniform droplets. The order of mixing the components is also decisive in nano-emulsion formation and properties, as in conventional emulsions. Although the preparation of nano-emulsions is more complex than that of microemulsions, an important advantage of nano-emulsions from a practical viewpoint is that they require lower amounts of surfactants for their formation.

Nano-emulsion droplets are generally stabilized by surfactants. Although it is considered that surfactant molecules are adsorbed at the oil–water interface in the form of monolayers, other surfactant self-organizing structures such as multilayers may play an important role in nano-emulsion stability. In this context, the results of studies of the relation between nano-emulsion formation, stability, and phase behavior are very illustrative [14–16,47].

In this chapter, different methods for nano-emulsion formation, with special emphasis on low-energy emulsification methods, are discussed in Section II. This is followed by a description of nano-emulsion stability (Section III). Finally, the most relevant applications of nano-emulsions are reviewed in Section IV.

II. NANO-EMULSION FORMATION

A. High-Energy Emulsification Methods

Nano-emulsion formation using energy input is generally achieved by applying mechanical shear such as that produced by high-shear stirring, high-pressure homogenizers, and ultrasound generators.

Various processes take place during emulsification [45]: breakup of droplets, adsorption of surfactant molecules, and droplet collisions (which may lead to coalescence and larger droplets). These processes may occur simultaneously during emulsification, as the time scale for each step is very small (microseconds). Breaking of drops is feasible if the deforming force exceeds the Laplace pressure, p_L (the difference between the pressure inside and outside the droplet), which is the interfacial force that acts against droplet deformation:

$$p_L = \gamma \left(\frac{1}{R_1} + \frac{1}{R_2} \right) \tag{1}$$

where R_1 and R_2 are the smaller and the larger radii of curvature of a deformed emulsion drop and γ is the interfacial tension.

From Eq. (1) it can be readily inferred that the smaller the droplet size for a given system, the more energy input and/or surfactant is required. Consequently, nano-emulsion production would cost more than that for conventional emulsions (macroemulsions). The effect of energy input on droplet size of nano-emulsions of the system water/$C_{18}E_{30}$/liquid paraffin, prepared with a high-pressure homogenizer, is illustrated in Fig. 1 [10]. The droplet size is reduced with decreasing oil/surfactant ratio (increasing surfactant concentration) or increasing pressure of homogenization. At high oil/surfactant

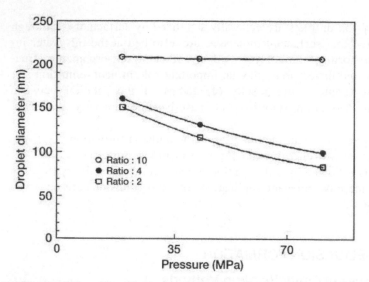

FIG. 1 Effect of applied pressure on the droplet diameter of emulsions having oil/surfactant weight ratios of 2, 4, and 10. (From Ref. 10, p 191.)

ratio, the droplet size is independent of the energy input because the surfactant concentration is insufficient to stabilize smaller droplets [10].

It has been shown [45,48] that the apparatus supplying the available energy in the shortest time and having the most homogeneous flow produces the smallest sizes. High-pressure homogenizers meet these requirements. Because of this, they are the most widely used emulsifying machines to prepare nano-emulsions. Although ultrasonic emulsification is also very efficient in reducing droplet size, as shown in Fig. 2, it is appropriate only for small batches [48].

Considering only mechanical energy aspects, nano-emulsion formation should be considerably costly. However, it is well known that by taking advantage of the physicochemical properties of the system, dispersions can be produced almost "spontaneously" [3,6,14]. This is the case with the so-called low-energy emulsification methods that are described next. In practice, the two types of methods are often combined.

B. Low-Energy Emulsification Methods

These methods make use of the phase transitions that take place during the emulsification process. The so-called phase inversion temperature (PIT) method is widely used in industry [49,50]. This method, introduced by Shi-

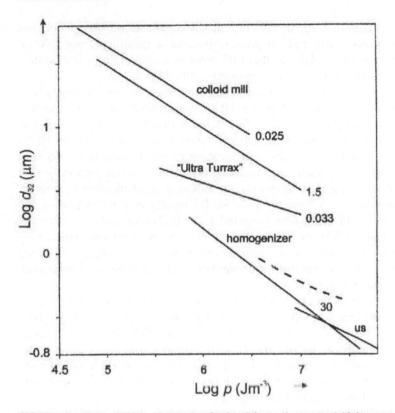

FIG. 2 Average droplet diameters obtained in various emulsifying machines as a function of the energy consumption p; us means ultrasonic generator. The numbers near the curves denote the viscosity ratio λ. The results with the homogenizer are for $\phi = 0.04$ (solid line) and $\phi = 0.3$ (dashed line). (From Ref. 48, by permission of the Royal Society of Chemistry.)

noda and Saito [51], is based on the changes in solubility of polyoxyethylene-type nonionic surfactants with temperature.

These types of surfactants become lipophilic with increasing temperature because of dehydration of the polyoxyethylene chains. At low temperature, the surfactant monolayer has a large positive spontaneous curvature forming oil-swollen micellar solution phases (or O/W microemulsions), which may coexist with an excess oil phase. At high temperatures, the spontaneous curvature becomes negative and water-swollen reverse micelles (or W/O microemulsions) coexist with excess water phase. At intermediate temperatures, the hydrophile–lipophile balance (HLB) temperature, the spontaneous curvature becomes close to zero and a bicontinuous, D phase, microemulsion containing comparable amounts of water and oil phases coexists with both excess water and oil phases. Because a transition from O/W to W/O emulsions takes place at this intermediate temperature, it is also designated as the phase inversion temperature, PIT [51–53].

The PIT emulsification method takes advantage of the extremely low interfacial tensions achieved at the HLB temperature [54,55] to promote emulsification (droplet breakup is facilitated with a low energy input). The interfacial tensions between the different phases are of the order of 10^{-2}–10^{-5} mN m^{-1}, and as a result emulsification is greatly facilitated and very small droplets can be formed [56]. However, coalescence is extremely fast. Consequently, at the HLB temperature, although emulsification is favored, the emulsions are very unstable [57,58]. By rapidly cooling or heating (by about 25–30°C) the emulsions prepared at the HLB temperature, kinetically stable emulsions (O/W or W/O, respectively) can be produced with a very small droplet size and narrow size distribution. If the cooling or heating process is not fast, coalescence predominates and polydisperse coarse emulsions are formed [59,60].

The factors that affect the HLB temperature have been extensively studied and are at present well known [53,61–63]. With decreasing alkyl chain length of the surfactant, increasing ethylene oxide (EO) units, or increasing alkyl chain length of the oil, the HLB temperature increases. Electrolytes with a salting-out effect (NaCl, Na_2SO_4, etc.) decrease the HLB temperature. This allows preparation of a wide variety of emulsions with different components and additives [11–13,49,50].

Other low-energy emulsification methods take advantage of the phase transitions that take place on changing the composition during emulsification at constant temperature [6,14–17,47]. As an example, a recent study of the relation between nano-emulsion formation, phase behavior, and stability [15,16] is described here.

In this study, the system water/Brij 30/decane was chosen as a model system (Brij 30 is an industrial grade ethoxylated lauryl alcohol with an average number of ethylene oxide units of 4). The surfactant concentration was kept constant (5.0 wt%) and the oil weight fraction, $R = O/(O + W)$, varied between 0.2 and 0.8. Emulsification was performed at 25°C by three low-energy methods: (A) stepwise addition of oil to a water–surfactant mixture, (B) stepwise addition of water to a solution of the surfactant in oil, and (C) mixing all the components in the final composition and pre-equilibrating the samples prior to emulsification. A schematic representation of the experimental paths followed in methods A and B is shown in Fig. 3. The results showed [15,16] that nano-emulsions were formed only at low R values when water was added to mixtures of surfactant and oil (emulsification method B). The droplet size of the nano-emulsions obtained was of the order of 50

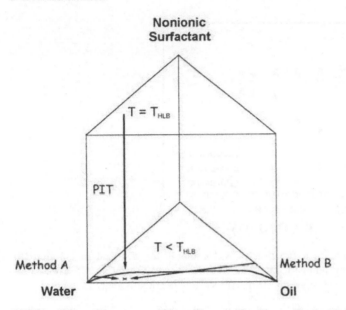

FIG. 3 Schematic representation of emulsification methods: A, B (at constant temperature) and PIT (phase inversion temperature). (From Ref. 16 by permission of Langmuir, Copyright 2001, American Chemical Society.)

nm (Fig. 4). In contrast, emulsification methods A and C lead only to coarse emulsions.

The phase diagram of the system at 25°C (Fig. 5) showed that nano-emulsions were obtained in compositions falling in the (W_m + L_α + O) region. These compositions, at equilibrium, consist of three phases: O/W microemulsion, lamellar liquid crystal, and oil. Their HLB temperature is close to 25°C and their equilibrium interfacial tensions reach very low values, of the order of 10^{-3} mN m^{-1} [15,16]. However, the equilibrium properties cannot explain nano-emulsion formation. Low interfacial tensions are probably necessary but not sufficient to form nano-emulsions. The key factor is the kinetics of the emulsification process. The change in the natural curvature of the surfactant during the emulsification process may play a major role in achieving emulsions with a small droplet size. In the emulsions obtained by method A, initially a dispersion of liquid crystals in water (vesicles) is formed. On adding decane to the system, O/W emulsions are obtained. In emulsification method B, the natural curvature of the surfactant during the emulsification process changes from negative (W/O) to positive (O/W): there is a transition from an isotropic oil-continuous phase (W/O microemulsion) through a multiphase region including lamellar liquid crystal

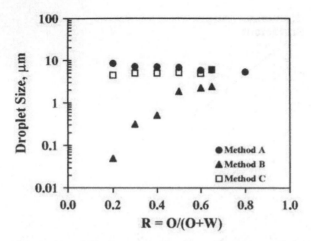

FIG. 4 Droplet size as a function of R for emulsions obtained in a water/Brij 30/decane system by emulsification methods A, B, and C. (From Ref. 16 by permission of Langmuir, Copyright 2001, American Chemical Society.)

(L_α) and a shear birefringent isotropic phase (D') before the O/W emulsion is formed.

III. NANO-EMULSION STABILITY

The main mechanisms of instability that are involved in leading to complete phase separation of emulsions are creaming [64], flocculation [65,66], coalescence [67], and Ostwald ripening [68,69]. However, nano-emulsions do not cream (or sediment) because the Brownian motion is larger than the small creaming rate induced by gravity. Practically, the creaming of droplets smaller than 1 μm is stopped by their faster diffusion rate.

With respect to flocculation of nano-emulsion droplets, it is not clear whether such droplets can adhere and form a thin flat film, as do large drops. On the one hand, because of their small size, the curvature is very high and the Laplace pressure opposes deformation. On the other hand, thermal agitation of small droplets (Brownian motion) can increase collisions and enhance deformation [70]. Anyway, flocculation is achieved spontaneously if the profile of the interaction energy as a function of the separation distance has a minimum deep enough to overcome the thermal energy of the droplets.

Two main interaction potentials are considered in systems stabilized by nonionic surfactants. The emulsion droplets are attracted by van der Waals interaction, which can be counteracted by an energy barrier because of steric repulsion. These potentials are represented schematically in Fig. 6.

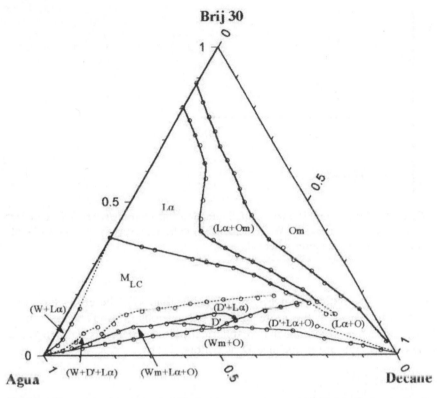

FIG. 5 Phase behavior of water/Brij 30/decane system at 25°C. O_m, isotropic liquid phase; L_α, lamellar liquid crystalline phase; D', shear birefringent liquid phase; W_m, bluish liquid phase (O/W microemulsion); W, aqueous liquid phase; O, oil liquid phase; M_{LC}, multiphase region including lamellar liquid crystal. (From Ref. 16 by permission of Langmuir, Copyright 2001, American Chemical Society.)

The steric repulsion, W_s, has been studied in some detail [71]. The repulsion of emulsion droplets, highly covered with grafted polymer molecules or head groups that attain a brushlike conformation in a good solvent, was described by de Gennes [72,73]. It can be simplified as follows, according to an overlap model;

$$W_s \propto kTe^{-\pi D/L} \tag{2}$$

where k is the Boltzmann constant, D is the separation distance between droplet surfaces, and L is the film thickness of the grafted polymer. Nano-emulsions, with Brownian motion $\approx kT$, stabilized by nonionic surfactants, can remain unflocculated if the minimum in the total interaction energy as

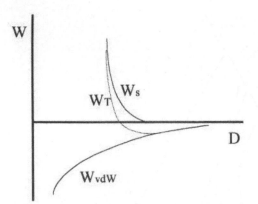

FIG. 6 Schematic representation of interaction potentials between two emulsion droplets stabilized sterically. W_{vdw}, W_s, and W_T indicate, respectively, van der Waals, steric, and total interaction potentials.

presented schematically in Fig. 6 is smaller than kT. Therefore, the larger the film thickness L, the more stable the emulsions. Moreover, the van der Waals attractive potential of two spherical particles depends on their radius R and their separation distance D as follows [71]:

$$W_{vdw} \approx \frac{-AR}{12D} \tag{3}$$

where A is the Hamaker constant. Therefore, the smaller the radius, the smaller the van der Waals potential. Emulsions with droplet size small enough and with surfactant film thick enough can be stable against flocculation because the minimum in the total interaction potential is overcome by the Brownian motion. In this sense, nano-emulsions may behave differently than conventional large drop emulsions (also called macroemulsions).

The stability of emulsions containing nonionic surfactants is minimum at the HLB temperature where the interfacial tension reaches a minimum. The coalescence is enhanced at low interfacial tensions because deformation of the droplets can occur more easily. Thermal fluctuations on the surfactant monolayers may increase, producing a hole in the thin film that separates the drops. This hole may heal and the droplets will not coalesce, or it may propagate in the film, producing its final rupture, as described by the Kabalnov–Wennerström theory [67,74]. A linear dependence in the Arrhenius plot [logarithm of the macroemulsion lifetime, $\ln(\tau_{1/2})$ as a function of the inverse of temperature] is predicted. The activation energy of the film rupture can be calculated from the slope of such a plot [67,74].

However, some nano-emulsions can be rather stable against coalescence [75,76]. One mechanism could be stabilization by a thick multilamellar surfactant film adsorbed on the interface [14,77]. The phase separation of nano-emulsions can result in three-phase systems containing liquid crystals [14–16]. These liquid crystalline phases could form multilayer film structures if enough surfactant were available.

It has been reported that nano-emulsions can behave as hard spheres [78]. Oil-in-water nano-emulsions in the ternary system composed of water, hexadecane, and heptaethylene glycol dodecyl ether possess a hard sphere interaction potential between droplets, as deduced from the variation of the scattered light intensity with varying dispersed phase volume fraction. Very small droplets would not deform enough to form a thin flat film between flocculated droplets, which could lead to coalescence by thermal fluctuations. In the same work, the thickness of bound water was also estimated. The Kabalnov–Wennerström model assumes that flocculation of deformable drops must be present, as a transient stage, before coalescence occurs [67,74]. Therefore, according to this theory, hard undeformable droplets would be more stable than the deformable ones. A dense adsorbed surfactant monolayer may prevent droplet deformation in nano-emulsions and thinning of the liquid film between the droplets and finally may avoid disruption of the film, thereby preventing coalescence.

Therefore, the only process that may produce coarsening of nano-emulsions is Ostwald ripening. It is described by the LSW theory, formulated by Lifshitz and Slezov [68] and independently by Wagner [69]. Several authors have indicated that this theory can be applied to macroemulsions with reasonable accuracy [79,80]. It has also been reported that the presence of microemulsion droplets in the continuous phase accelerates the Ostwald ripening rate by increasing the diffusion coefficient [80,81]. However, this effect is relatively small because microemulsion droplets have much smaller diffusion coefficients than molecules.

The LSW theory assumes that the droplets are separated by distances much larger than their diameters, the transport of the dispersed component is due to molecular diffusion, and the concentration of the dissolved species is constant except when adjacent to the droplet boundaries. These assumptions may not be completely valid for nano-emulsions because the strong Brownian motion may induce convective diffusion accelerating the diffusion rate, which would be slower if it were due only to molecular diffusion. However, it has been shown that convective contributions do not change the fundamental nature of Ostwald ripening processes [82].

The Ostwald ripening rate, ω, as described by the LSW theory, is expressed as follows:

$$\omega = \frac{dr_c^3}{dt} = \frac{8c(\infty)\gamma V_m D}{9RT} \tag{4}$$

where r_c is the critical radius of droplets that are neither growing nor decreasing in size, $c(\infty)$ is the bulk phase solubility, γ is the interfacial tension, V_m is the solute molar volume, D is the diffusion coefficient, R is the gas constant, and T is the temperature.

This equation shows that r^3 varies linearly with time. Therefore plotting r^3 versus time makes it possible to determine Ostwald ripening rates [83,84]. An example is shown in Fig. 7. The linear relationship indicates that the emulsion instability is due to Ostwald ripening.

Several reports show that Ostwald ripening can play the main role in the instability of nano-emulsions and that the LSW theory can also be applied to such systems [75,76], despite the fact that nano-emulsion droplets are not fixed in space and that convective contributions can be very important in the total diffusion coefficient. Two different Ostwald ripening regimes have been detected in nano-emulsions [76]. The ripening rate increased after an induction period, and it was dependent on the volume fraction of the nano-emulsion droplets. Such behavior was not observed in macroemulsions formed in the same system. The ripening rates were slower for nano-emulsions than for macroemulsions. It has been suggested that the Ostwald ripening rate is strongly affected by the initial state of the emulsion, for example, the polydispersity and the interaction between droplets [76]. The nano-emulsions would have slower rates because of narrower polydispersity. An example of the influence of the initial size distribution on Ostwald ripening rates is shown in Fig. 8. It compares the stability of emulsions obtained

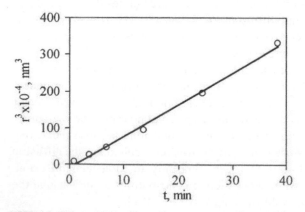

FIG. 7 Cube of droplet radius as a function of time for the system $H_2O/C_{12}E_4/$ $C_{12}E_6/$decane (3 wt% $C_{12}E_4$, 2 wt% $C_{12}E_6$, $H_2O/$decane = 80:20, $T = 25°C$).

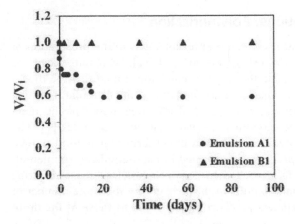

FIG. 8 Stability of emulsions obtained by two low-energy emulsification methods (methods A and B, described in Section II). The composition of both emulsions is the same (water/decane = 80:20, 5 wt% Brij 30). V_i and V_f are initial and final emulsion volumes, respectively. (From Ref. 16 by permission of Langmuir, Copyright 2001, American Chemical Society.)

by two low-energy emulsification methods (methods A and B, described in Section II). The composition of both emulsions is the same (water/decane = 80:20, 5 wt% Brij 30). The ratio V_f/V_i is represented as a function of time (V_f and V_i are the final and the initial emulsion volumes, respectively). The emulsion obtained by emulsification method A showed phase separation in less than 1 h. In contrast, the nano-emulsion obtained by method B (with the same composition) was kinetically stable and did not show phase separation within the measuring time (1 year). The difference in emulsion stability could be explained because the emulsions obtained by method B have lower polydispersity than those obtained by method A [15,16].

IV. APPLICATIONS

Nano-emulsions have found increasing use in many different applications. The advantages of nano-emulsions over conventional emulsions (or macro-emulsions) are a consequence of their characteristic properties, namely small droplet size, high kinetic stability, and optical transparency. In addition, nano-emulsions offer the possibility of using microemulsion-like dispersions without the need for high surfactant concentrations. In the following, the most relevant applications of nano-emulsions in the chemical, pharmaceutical, and cosmetic fields are summarized.

A. Chemical Applications: Polymerization

One of the earliest applications of nano-emulsions was in the preparation of polymer latexes [1,5,9,36,40–43]. Ugelstad et al. [1], who introduced the term miniemulsion to designate this type of emulsion, found that the mechanism involved in miniemulsion polymerization was quite different from that of macroemulsion polymerization (Fig. 9). They suggested that the main locus of nucleation was the monomer droplets instead of micelles [1]. The so-called miniemulsion polymerization is a broad term that is used to designate all polymerization processes performed in nano-emulsion (miniemulsion) media. However, it is also used in a more restrictive sense referring to the polymerization of nano-emulsion droplets giving the same number of polymer particles with particle size distributions equal to those of the droplets [9a].

Several advantages of miniemulsion polymerization over conventional emulsion polymerization have been reported [85]. It is considered to be a process more insensitive to variations in the composition or to the presence of impurities. The wide variations in the conversion rate and particle size obtained in a continuous macroemulsion polymerization process are highly reduced when performing continuous miniemulsion polymerization [85]. It

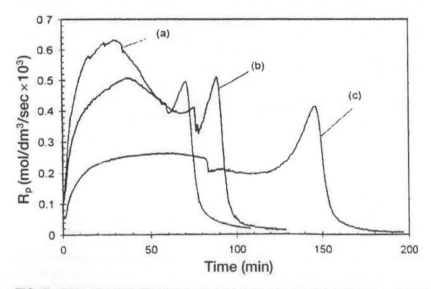

FIG. 9 Rate of polymerization versus time for (a) conventional styrene emulsion, 10 mM SLS; (b) homogenized styrene emulsion, 10 mM SLS; and (c) styrene miniemulsion, 10 mM SLS/30 mM CA. [KPS] = 1.33 mM, T_r = 70°C. (From Ref. 9b, with permission from Elsevier Science.)

also allows better control of the process as depicted in Fig. 10. Moreover, the latexes are more stable under shear and allow higher solid contents. Miniemulsion polymerization allows the encapsulation of many kinds of substrates such as drugs, perfumes, and inorganic pigments in a polymeric matrix [86–88]. Grafted polymers have been developed in these media, producing materials with more uniform composition. In so-called hybrid miniemulsion polymerization, acrylic monomers have been grafted with different kinds of resins: polyester, alkyd, or urethane type. The materials obtained exhibit better properties than those prepared by emulsion polymerization [37,89–92].

Many kinds of polymerizable miniemulsion recipes have been described [9,85]. In the majority of the described systems, the emulsification is achieved by high-energy methods. The emulsifier in the earlier formulations [1] consisted of ionic surfactant/fatty alcohol (cosurfactant) mixtures. It was thought that the stabilizing mechanism was due to the presence of a protective interfacial complex. Later, it was shown that the replacement of the fatty alcohol by a highly hydrophobic compound (e.g., hexadecane) decreased more effectively the Ostwald ripening without the existence of any interfacial complex [9]. Different types of molecules such as reactive co-

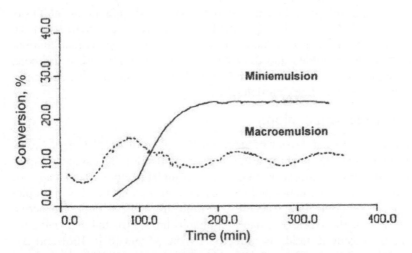

FIG. 10 Continuous macroemulsion and miniemulsion polymerization of methyl methacrylate. Continuous (stirred tank) macro- and miniemulsion polymerization of methyl methacrylate at 40°C in a surfactant (SLS) concentration of 0.67 wt% (based on monomer). Cosurfactant (miniemulsion only): 2 wt% (based on monomer). Initiator: potassium persulfate; 0.01 M. Total solids: 31 wt%. (From Ref. 85, with permission from Elsevier Science.)

monomers, e.g., alkylmethacrylates [93]; block and diblock copolymers [94,95]; initiators, e.g., persulfates [96]; chain transfer agents, e.g., mercaptans [97]; or dyes, e.g., Blue 70 [98] have been used to reduce the monomer diffusion through the continuous phase (Ostwald ripening).

The surfactants used in the preparation of miniemulsions can be very varied. Formulations with anionic [9], single cationic and gemini [99], and nonionic and polymeric surfactants [100] have been reported. The use of one or other surfactant type depends on the final use of the latex.

The most studied polymerization reactions were free radical polyadditions. Therefore, the vast majority of the monomers used in miniemulsion polymerization are of vinylic type. However, polyaddition of expoxides to various diamines, diols or bisphenols [38] and an anionic polymerization of phenyl glycidyl ether [101] in nano-emulsion media have now been reported.

The initiator is another important component in a polymerization system. In the early stages, the initiation was started by applying thermal energy to a free radical generator of the persulfate type. Oil-soluble initiators such as 2,2'-azobis(2-methylbutyronitrile) (AMBN) have also been used and have made it possible to explain thoroughly the kinetics of the process [102]. Nano-emulsion polymerization in the presence of stable radicals, so-called "living radical polymerization," has been reported to reduce the polydispersity of the final latex [103,104].

Other compounds such as chain transfer agents, retardants, or inhibitors can also be included in order to control the molecular weight (MW) of the final latex [105]. In a seeded miniemulsion polymerization, a small amount of latex particles is added and the system is then polymerized. It has been shown that this process enhances the control of the number, size, and polydispersity of the final latex particles [9].

B. Cosmetic Applications

The transparent visual aspect of nano-emulsions with droplet sizes below 200 nm makes them especially attractive for application in cosmetics. Apart from the appearance, similar to that of microemulsions, other advantages of nano-emulsions for cosmetic applications are their kinetic stability, a droplet size that can be controlled, and the possibility to achieve improved active delivery. For all these reasons, nano-emulsions have attracted increasing interest in the cosmetic field, as reflected by the papers [8,10,106] and numerous patents [19–21,23,24,27,107–111] that have appeared in the last few years. Oil-in-water nano-emulsions with a droplet size lower than 100 nm have been described in patents as hair- and skin-care [19,21,23,24,107–109], makeup [110], and sunscreen [20,111] formulations.

Cosmetic emulsions are generally formulated with water concentrations higher than 70 wt%. The active ingredients are dissolved in the aqueous and/or the oil phase of the nano-emulsion. The aqueous phase usually contains components such as glycerol, urea, amino acids, α-hydroxy acids, and water-soluble vitamins. The oil phase, apart from its functional role in preventing water loss from the skin, serves as the carrier for perfumes, oil-soluble vitamins, etc. [112]. A wide range of surfactants is used to stabilize nano-emulsions. These include nonionic surfactants (alkyl polyoxyethylene ethers, POE-POP-POE block copolymers, alkyl sugar derivatives, silicone derivatives, etc.), as well as ionics surfactants (alkylsulfates, alkylsulfonates, phospholipids, lipoamino acid derivatives, etc.).

Condensation (low-energy) and dispersion (high-energy) methods have been used to prepare nano-emulsions for cosmetic applications. The former are mainly based on Shinoda's PIT method [51–53], and the latter generally make use of high-pressure homogenization. Nano-emulsion formation by cooling a single-phase W/O microemulsion to a temperature lower than the HLB temperature, without shaking, is well established [10,39]. It was reported that the droplet sizes of nano-emulsions of systems with ethoxylated nonionic surfactants and oils such as squalane, paraffin oil, hexadecane, pentadecane, tetradecane, and dodecane were 10–100 nm [113,114]. Figure 11 shows the results corresponding to systems with the surfactant octa-ethyleneglycol n-hexadecyl ether ($C_{16}E_8$). Nano-emulsions with liquid paraffin oil and squalane remained unchanged for more than 1 year at 25°C. However, this method is limited to nonionic surfactant systems that can form

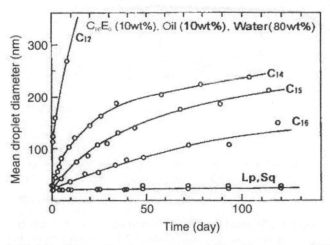

FIG. 11 Effect of carbon number of the oil on the stability of alkane ultrafine emulsions at 25°C. L_p, liquid parafin; S_q, squalane. (From Ref. 10.)

single-phase microemulsions at temperatures above the storage or use temperature.

Examples of different emulsification routes based on the PIT method and the corresponding droplet sizes obtained are shown in Fig. 12. In this example [7,115], the emulsification of a polar oil, cetyl isononanoate, was performed with a surfactant mixture consisting of a long-chain ethoxylated alcohol ($C_{16/18}E_2$) and glyceryl monostearate (GMS) nonionic surfactants. It was shown that in order to obtain nanometer-size droplets in the system studied, either a liquid crystalline phase or a bicontinuous microemulsion should be formed during emulsification [7].

Concerning the use of dispersion (or high-energy) methods for nano-emulsion formation, it should be noted that nano-emulsions with sizes below 50 nm could not be prepared. The addition of water-soluble solvents, such as glycerol or butanediol, etc. proved to be effective for the preparation of fine emulsions [2,39,116]. In this context, a method consisting of the homogenization of a coarse emulsion having large quantities of water-soluble solvents (WSSs) in the aqueous phase has been developed [39].

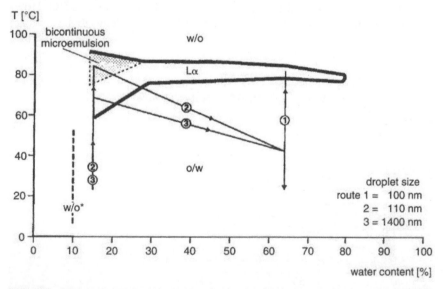

FIG. 12 Water-phase map for O/W emulsions in the system water/$C_{16/18}E_{12}$: glyceryl monostearate (GMS)/cetyl isononanoate. $C_{16/18}E_{12}$, GMS = 2:1; oil/mixed emulsifier = 4.5:1; W/O*, unstable emulsion; numbers refer to the emulsification routes. (From Refs. 7 and 115 with permission from Elsevier Science.)

C. Pharmaceutical Applications

The use of traditional disperse systems, e.g., macroemulsions, in the pharmaceutical industry has been limited due to manufacturing complexity and stability problems [117]. The characteristic properties of nano-emulsions (kinetic stability, small and controlled droplet size, etc.) make them interesting systems for pharmaceutical applications. Indeed, nano-emulsions are used as drug delivery systems for administration through various systemic routes. There are numerous publications on nano-emulsions as drug delivery systems for parenteral [17,18,28,29,118–124], oral [25,125–129], and topical administration, which includes the administration of formulations to the external surfaces of the body skin [32,130,131] and to the body cavities nasal [30,132] as well as ocular administration [31,133–136]. Moreover, many patents concerning pharmaceutical applications of nano-emulsions have been registered [17,18,25,137–145]. An application of nano-emulsions in this field has been in the development of vaccines [33,146–147].

Parenteral (or injectable) administration can be performed intravenously, intramuscularly, or subcutaneously. This administration route is employed for a variety of purposes, namely nutrition (e.g., administration of fats, carbohydrates, vitamins), controlled drug release, and targeting of drugs to specific sites in the body [148,149]. There are strict requirements for emulsions for their use in parenteral administration. Emulsions intended for this use are required to be sterile, isotonic, nonpyrogenic, nontoxic, biodegradable, and stable (physically and chemically) with a droplet size lower than 1 μm [149,150]. In practice, commercial formulations usually have a mean droplet size of about 200–500 nm, with 90 wt% or more particles below 1 μm. Droplet sizes larger than 5 μm could give rise to blockages in the fine capillaries (embolism) [151]. Therefore, using nano-emulsions with a narrow size distribution for parenteral administration is advantageous.

The effect of oil and emulsifiers on the emulsion droplet size and stability of parenteral emulsions has been widely studied. Different oils and their mixtures have been studied in order to obtain emulsions with good long-term stability and small droplet size using a phospholipid surfactant. It has been reported [118] that 20 wt% of a mixture of castor oil with either soybean oil or middle-chain triglycerides (MCTs) forms a very stable nano-emulsion, with a droplet size of about 130–140 nm. At a higher oil concentration (30 wt%), emulsions with a mixture of castor oil and MCT with a weight ratio of 1:1 also have a very small droplet size, depending on the homogenization pressure, as shown in Fig. 13. As for the emulsifier, ultrafine lipid emulsions for intravenous administration with soybean oil and lecithin have been described [17,18]. It has also been reported that a mixture of a nonionic surfactant and phospholipid (Tween 80® and egg phosphatidylcholine, 0.3:0.4) leads to stable emulsions with a small droplet size [152].

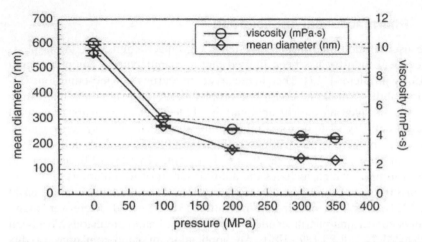

FIG. 13 Effect of the homogenization pressure on mean particle size and viscosity of 30% emulsions containing castor oil and middle chain triglycerides (MCTs), 1:1. (From Ref. 118 with permission from Elsevier Science.)

Self-emulsifying systems suitable for parenteral drug delivery have also been described. One of them consists of a mixture of lecithin and Span 20® as primary and secondary emulsifiers, respectively, which are mixed with the oil phase containing soybean oil [119]. The addition of glycerol allowed spontaneous emulsification at a concentration of 30 wt%. The main emulsion droplet size was 400 nm. A parenteral self-emulsifying drug delivery system containing 0.5 wt% lidocaine as a model drug showed similar spontaneous emulsification with particle size of 390 nm.

The drugs incorporated in nano-emulsions for parenteral administration are numerous. One representative example is paclitaxel (Taxol®; Bristol-Myers), a promising antineoplasic agent, poorly soluble in water and orally inactive, that requires intravenous administration. It has been shown that nano-emulsion droplets coated with a hydrophilic polymer (polyethylene glycol–modified phosphatidylethanolamine) have a prolonged circulation lifetime and accumulate in tumors (Fig. 14), resulting in an enhancement of the antitumor activity [28,120]. Other examples of drugs incorporated in nano-emulsions for parenteral administration include antimalaria drugs such as mefloquine and halofantrine [29,121], the anxiolytic drug diazepam [122], a free radical scavenger (tirilazad) [123], and the antifungal agent amphotericin B [124].

Among the various systemic drug delivery routes, oral administration is considered to be the most popular. Oil-in-water emulsions are already considered very interesting formulations for oral drug administration of poorly

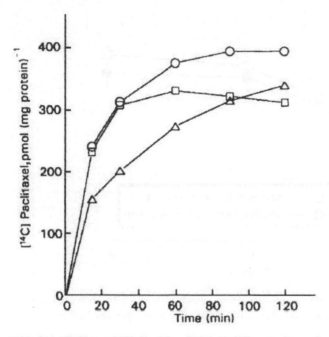

FIG. 14 Uptake of [^{14}C]paclitaxel into T-47D cells (breast cancer cells) from lipid emulsions (○), liposomes (△) and Diluent 12 (□). Time course curves obtained with a drug concentration of 10 μM. (From Ref. 28 with permission from the Royal Pharmaceutical Society of Great Britain.)

water-soluble drugs in terms of bioavailability [153–156] because of en-hancement of intestinal absorption [157–160] and therefore enhanced activ-ity. All the formulations that increase drug solubility and decrease enzymatic attack in intestinal washings are appropriate for oral administration. The absorption of the emulsion in the gastrointestinal tract after oral administra-tion is correlated to the droplet size of the formulation. The smaller droplet size of the emulsion causes greater absorption [161]. Thus, nano-emulsions are good candidates for oral administration.

The hormone calcitonin has been formulated in a nano-emulsion contain-ing Carbopol® 940 (BFGoodrich). This adhesive polymer, located on the droplet surface, is thought to increase the time of emulsion adhesion to the intestinal mucosa and, consequently, the absorption of the drug [125]. The cationic polysaccharide chitosan has also been used in this context because of its mucoadhesivity. It has been reported that nano-emulsions with chitosan release drugs for a prolonged period of time (Fig. 15) [126].

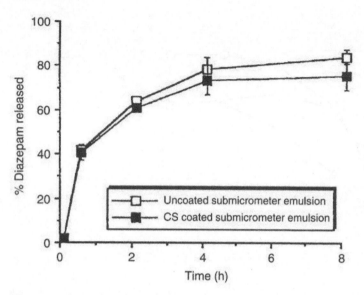

FIG. 15 Diazepam release profiles from uncoated and chitosan-coated submicrometer emulsion. (From Ref. 126 with permission from Springer Verlag.)

Self-emulsifying systems for oral administration have also been developed. Cyclosporin (an immunosuppressing agent) has been formulated in a solution of a polar lipid self-emulsifying drug delivery system filled in soft gelatin capsules [25]. A self-emulsifying system containing indomethacin has been shown to increase the bioavailability of the drug significantly [127]. It is thought that these mixtures of surfactant and oil form a fine emulsion with gentle agitation when exposed to aqueous media, and the gastrointestinal motility can provide the agitating effect necessary for emulsification.

Other examples of drugs incorporated in nano-emulsions for oral administration are the cephalosporin cefpodoxime proxetil [128] and the hormone desmopressin acetate [129].

The oral route has certain limitations for some drugs, such as drug degradation in the gastrointestinal tract, gastrointestinal tissue irritation, and/or gut wall and first-pass metabolism. In this context, the nasal route has received a great deal of attention because of the many advantages of nasal delivery over oral and parenteral administration.

A submicrometer emulsion with testosterone for nasal administration has been formulated and tested in rabbits [132]. It is noteworthy that testosterone, the male sex hormone, is ineffective when administered orally. Other examples are human immunodeficiency virus (HIV) vaccines that have been formulated as nano-emulsions for nasal administration [30]. A study of the response of the immunological system after the nasal administration of a proteosome-rgp 160 vaccine to which a bioadhesive nano-emulsion has been added revealed an increase of specific immunoglobulin G (IgG) and IgA in serum and secretions. These results are very promising for mucosal vaccine development to help control the spread of HIV transmission and the acquired immunodeficiency syndrome (AIDS).

Another application of nano-emulsions in the pharmaceutical field is in ocular administration (a topical administration). Nano-emulsions are used as ocular delivery systems to sustain the pharmacological effect of drugs in comparison with their respective aqueous solutions [133].

Pilocarpine, a drug used to produce miosis, has been widely studied in order to increase its low bioavailability when applied topically, which is due to its low lipophilicity and the rapid loss of the drug from the precorneal area through drainage and conjunctival absorption. This drug is generally used as an aqueous solution that leads to a formulation that must be administered three or four times per day. The administration of a submicrometer emulsion containing a prodrug of pilocarpine that is enzymatically converted to the active parent drug within the cornea, in a dose equivalent to 0.5 wt% pilocarpine base, produces a prolonged miotic effect compared with the pilocarpine-containing aqueous solution (Fig. 16) [133]. However, the bioavailability was not improved. The administration of pilocarpine as an ion pair with monododecylphosphoric also did not increase the bioavailability of the drug [134].

A better result was reported with indomethacin, an anti-inflammatory agent used to reduce postoperative inflammation after cataract surgery. The incorporation of this drug of low water solubility into a submicrometer emulsion stabilized by a combination of phospholipids and small amounts of amphoteric surfactant resulted in higher bioavailability and higher corneal permeability of the drug [31,135,136].

Nano-emulsions are also interesting candidates for the delivery of drugs through the skin (topical administration). Positively and negatively charged submicrometer emulsions containing antifungal drugs (econazole nitrate and miconazole nitrate) have been described [32]. The positively charged submicrometer emulsions were more effective in terms of skin penetration of econazole or miconazole nitrate than negatively charged emulsions. Other nano-emulsions described for topical administration contain diazepam [130] as well as steroidal and nonsteroidal anti-inflamatory drugs [131].

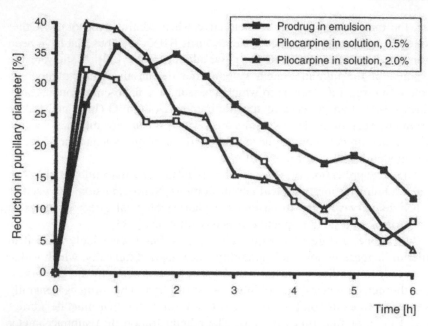

FIG. 16 Comparison of the miotic effect after ocular administration of a submicrometer emulsion containing 1.2% (w/v) pilocarpine prodrug or aqueous solutions containing 0.5% (w/v) or 2.0% (w/v) pilocarpine HCl. (From Ref. 133 with permission from Elsevier Science.)

V. CONCLUSIONS

In this chapter, the characteristic properties of nano-emulsions and relevant applications have been described. A great deal of research effort in recent years has been focused toward the conditions required for nano-emulsion formation. Low interfacial tension values and the presence of lamellar liquid crystalline phases are among the factors that have been shown to be important for their formation. However, it has also been shown that the kinetics of the emulsification process plays a key role. Comprehensive knowledge of the fundamental aspects related to nano-emulsion formation and stability will allow improvement of established applications, such as those described in this chapter, and development of new ones.

ACKNOWLEDGMENTS

The financial support by CICYT (Grant QUI99-0997-CO2-01) and "Comissionat per a Universitats i Recerca, Generalitat de Catalunya" (grant 1999SGR-00193) is gratefully acknowledged.

REFERENCES

1. J Ugelstad, MS El-Aasser, JW Vanderhoff. J Polym Sci Polym Lett 11:503, 1973.
2. H Sagitani. J Am Oil Chem Soc 58:738, 1981.
3. GWJ Lee, ThF Tadros. Colloids Surf 5:105–115, 1982.
4. H Nakajima, S Tomomasa, M Kochi. J Soc Cosmet Chem Jpn 34:335, 1983.
5. MS El-Aasser, CD Lack, YT Choi, TI Min, JW Vanderhoff. Colloids Surf 12: 79–97, 1984.
6. H Sagitani. In: SE Friberg, B Lindman, eds. Organized Solutions. New York: Marcel Dekker, 1992, pp 259–271.
7. A Wadle, T Föester, W Von Rybinski. Colloids Surf A 76:51–57, 1993.
8. Z Zeevi, S Klang, V Alard, F Brossard, S Benita. Int J Pharm 108(1):57–58, 1994.
9. (a) ED Sudol, MS El-Aasser. In: PA Lovell, MS El-Aasser, eds. Emulsion Polymerization and Emulsion Polymers. New York: Wiley, 1997, pp 699–722. (b) PI Blyte, A Kleim, JA Phillips, ED Sudol, MS El-Aasser. J Polym Sci A Polym Chem 37:4449–4457, 1999.
10. H Nakajima. In: C Solans, H Kunieda, eds. Industrial Applications of Micro-emulsions. New York: Marcel Dekker, 1997, pp 175–197.
11. M Miñana-Pérez, C Gutron, C Zundel, JM Andérez, JL Salager. J Dispersion Sci Technol 20:893–905, 1999.
12. E Sing, A Graciaa, J Lachaise, P Brochette, J Salager. Colloids Surf A 152: 31–39, 1999.
13. Y Katsumoto, H Ushik, B Mendiboure, A Graciaa, J Lachaise. Colloid Polym Sci 278:905–909, 2000.
14. MJ Rang, CA Miller. J Colloid Interface Sci 209:179–192, 1999.
15. A Forgiarini, J Esquena, C González, C Solans. Prog Colloid Polym Sci 115: 36–40, 2000.
16. A Forgiarini, J Esquena, C González, C Solans. Langmuir 17:2076–2083, 2001.
17. US patent 5,098,606, 1992.
18. European patent 0363928 B1, 1994.
19. M Petrescu, M Lupulet, GV Pintilie, S Paraschiv. Romanian patent RO 108842 B1 30, 1994.
20. T Föester, S Heinen, B Heide. Ger Offen DE 19532543 A1, 1997.
21. JT Simonnet. Eur patent appl EP 780114 A1, 1997.
22. K Fontenot, FJ Schork, JL Reimers. US patent 5686518 A, 1997.
23. S Restle, D Cauwet-Martin. Eur patent appl EP 842652 A1, 1998.
24. H Lorenz, HR Wagner, A Kawamata. Ger Offen DE 19735851 A1, 1999.
25. VT Bhalani, SP Satishchandra. US patent 5858401 A, 1999.
26. M Antonietti, K Landfester. Ger Offen DE 19852784 A1, 2000.
27. JT Simonnet, O Sonneville, S Legret. Eur patent appl EP 1018363 A1, 2000.
28. BB Lundberg. J Pharm Pharmacol 49:16–21, 1997.
29. TKM Mbela, E Deharo, A Haemers, A Ludwig. J Pharm Pharmacol 50:1221–1225, 1998.

30. GH Lowell, RW Kaminski, TC VanCott, B Silkie, K Kersey, E Zawoznik, L Loomis-Price, G Smith, DL Birx. J Infect Dis 175:292–301, 1997.
31. P Calvo, MJ Alonso, J Vila-Jato. J Pharm Sci 85:530–536, 1996.
32. MP Youenang Piemi, D Korner, S Benita, JP Marty. J Controlled Release 58(2):177–187, 1999.
33. D O'Hagan, G Van Nest, GS Ott, M Singh. PCT Int patent appl WO 9930737 A1, 1999.
34. DI Jon, DI Prettypaul, MJ Benning, KS Narayanan, RM Ianniello. In: JD Nalewaja, GR Gross, RS Tann, eds. Pesticide Formulations and Application Systems. ASTM STP 1347. West Conshohocken, PA: ASTM, 1998, pp 228–241.
35. DI Jon, DI Prettypaul, MJ Benning, KS Narayanan, RM Ianniello. Int patent appl WO 9919256 A2, 1999.
36. MS El-Aasser, CM Miller. In: JM Asua, ed. Polymeric Dispersions: Principles and Applications. NATO ASI Ser. Dordrecht: Kluwer Academic, 1997, pp 109–126.
37. ST Wang, FJ Schork, GW Poehlein, JW Gooch. J Appl Polym Sci 60:2069–2076, 1996.
38. K Landfester, F Tiarks, HP Hentze, M Antonietti. Macromol Chem Phys 201(1):1–5, 2000.
39. H Nakajima, S Tomomasa, M Okabe. Proceedings of First World Emulsion Conference, EDS, Paris, 1993, vol 1, pp 1–11.
40. MS El-Aasser, C Lack, Y Choi, T Min, J Vanderhoff, FM Fowkes. Colloids Surf 12:79–97, 1984.
41. CD Lack, MS El-Aasser, JW Vanderhoff, FM Fowkes. In: DO Shah, ed. Macro and Microemulsions: Theory and Applications. ACS Symp Ser No 272. Washington, DC: American Chemical Society, 1985, pp 345–355.
42. W Brouwer, MS El-Aasser, JW Vanderhoff. Colloids Surf 21:69–86, 1986.
43. MS El-Aasser, CD Lack, JW Vanderhoff. Colloids Surf 29:103–118, 1988.
44. US patent 4177177, 1979.
45. P Walstra. In: P Becher, ed. Encyclopedia of Emulsion Technology. Vol 1. New York: Marcel Dekker, 1983, pp 57–127.
46. ESR Gopal. In: P Sherman, ed. Emulsion Science. New York: Academic Press, 1968, pp 1–75.
47. A Forgiarini, J Esquena, C González, C Solans. Prog Colloid Polym Sci 118:184–189, 2001.
48. P Walstra, PEA Smulders. In: BP Binks, ed. Modern Aspects of Emulsion Science. Cambridge: The Royal Society of Chemistry, 1998, pp 56–59.
49. T Förster. In: M Rieger, LD Rhein, eds. Surfactants in Cosmetics. New York: Marcel Dekker, 1997, pp 105–125.
50. T Förster, WV Rybinski. In: BP Binks, ed. Modern Aspects of Emulsion Science. Cambridge: The Royal Society of Chemistry, 1998, pp 395–426.
51. K Shinoda, H Saito. J Colloid Interface Sci 26:70, 1968.
52. K Shinoda, H Kunieda. J Colloid Interface Sci 42:381, 1973.
53. K Shinoda, H Kunieda. In: P Becher, ed. Encyclopedia of Emulsion Technology. Vol. 1. New York: Marcel Dekker, 1983, pp 337–367.

54. H Kunieda, SE Friberg. Bull Chem Soc Jpn 54:1010, 1981.
55. H Kunieda, K Shinoda. Bull Chem Soc Jpn 55:1777, 1982.
56. S Friberg, C Solans. J Colloid Interface Sci 66:367, 1978.
57. JL Salager, I Loaiza-Maldonado, M Miñana-Pérez, F Silva. J Dispersion Sci Technol 3:279, 1982.
58. BP Binks, W-G Cho, PDI Fletcher, DN Petsev. Langmuir 16:1025–1034, 2000.
59. H Kunieda, H Fukui, H Uchiyama, C Solans. Langmuir 12:2136, 1996.
60. K Ozawa, C Solans, H Kunieda. J Colloid Interface Sci 188:275, 1997.
61. K Shinoda, S Friberg. Emulsions and Solubilization. New York: Wiley, 1986.
62. H Kunieda, M Yamagata. Langmuir 9:3345–3351, 1993.
63. M Kahlweit. J Phys Chem 99:1281–1284, 1995.
64. GG Stokes. Philos Mag 1:337, 1851.
65. EJW Verwey, JThG Overbeek. Theory of the Stability of Lyophobic Colloids. Amsterdam: Elsevier, 1948.
66. DN Petsev, ND Denkov, P Kralchevsky. J Colloid Interface Sci 176:201–213, 1995.
67. A Kabalnov, H Wennerström. Langmuir 12:276–292, 1996.
68. IM Lifshitz, VV Slezov. J Phys Chem Solids 19:35, 1961.
69. C Wagner. Z. Elektrochem 65:581, 1961.
70. GK Batchelor. J Fluid Mech 74:1, 1976.
71. JN Israelachvili. Intermolecular and Surface Forces. San Diego: Academic Press, 1991.
72. PG de Gennes. C R Acad Sci (Paris) 300:839–843, 1985.
73. PG de Gennes. Adv Colloid Interface Sci 27:189–209, 1987.
74. AS Kabalnov. In: BP Binks, ed. Modern Aspects of Emulsion Science. Cambridge: The Royal Society of Chemistry, 1998, pp 205–257.
75. P Taylor, RH Ottewill. Colloids Surf A 88:303–316, 1994.
76. Y Katsumoto, H Ushiki, B Mendiboure, A Graciaa, J Lachaise. J Phys Condens Matter 12:3569–3583, 2000.
77. T Suzuki, H Takei, S Yamazaki. J Colloid Interface Sci 129:491–500, 1989.
78. Y Katsumoto, H Ushiki, B Mendiboure, A Graciaa, J Lachaise. J Phys Condens Matter 12:249–264, 2000.
79. EP Ludwig, J Schmelzer, J Bartels. J Mater Sci 29:4852, 1994.
80. P Taylor. Colloids Surf A 99:175–185, 1995.
81. AS Kabalnov. Langmuir 10:680–684, 1994.
82. K Parbhakar, L Lewandowski, LH Dao. J Colloid Interface Sci 174:142, 1995.
83. AS Kabalnov. J Colloid Interface Sci 118:590–597, 1987.
84. Y De Smet, J Malfait, L Deriemaeker, C De Vos, R Finsy. Bull Soc Chim Belg 105:789–792, 1996.
85. (a) FJ Schork, GW Poehlein, S Wang, J Reimers, J Rodrigues, C Samer. Colloids Surf A 153:39–45, 1999. (b) DT Barnette, FJ Schork. Chem Eng Prog 83:25, 1987.
86. E Bedri, ED Sudol, VL Dimonie, MS El-Aasser. Polym Mater Sci Eng 80: 583–584, 1999.

87. E Bedri, ED Sudol, VL Dimonie, MS El-Aasser. Macromol Symp 155:181–198, 2000.
88. T Clemens, A Boehm, S Kielhorn-Bayer. Polym Prep (Am Chem Soc Div Polym Chem) 41(1):24–25, 2000.
89. JG Tsavalas, JW Gooch, FJ Schork. J Appl Polym Sci 75:916–927, 2000.
90. JW Gooch, H Dong, FJ Schork. J Appl Polym Sci 76:105–114, 2000.
91. XQ Xu, FJ Schork. J Polym Sci A Polym Chem 37:4159–4168, 1999.
92. JW Gooch, ST Wang, FJ Schork, GW Poehlein. Proceedings of the 24th Waterborne, High Solids, Powder Coatings Symposium, 1997, pp 336–377.
93. CS Chern, YC Liou. Macromol Chem Phys 199:2051–2061, 1998.
94. J Reimers, FJ Schork. J Appl Polym Sci 59:1833–1841, 1996.
95. I Aizpurua, JI Amalvy, MJ Barandiaran. Colloids Surf A 166:59–66, 2000.
96. JA Alduncin, J Forcada, JM Asua. Macromolecules 27:2256–2261, 1994.
97. D Mouran, J Reimers, FJ Schork. J Polym Sci A Polym Chem 34:1073–1981, 1996.
98. CS Chern, YC Liou. Polymer 40:3763–3772, 1999.
99. K Landfester, N Bechthold, F Tiarks, M Antonietti. Macromolecules 32:2679–2683, 1999.
100. K Landfester, M Willert, M Antonietti. Macromolecules 33:2370–2376, 2000.
101. C Maitre, F Ganachaud, P Hemery. Macromolecules 33:7730, 2000.
102. PJ Blythe, A Klein, JA Phillips, ED Sudol, MS El-Aasser. J Polym Sci A Polym Chem 37:4449–4457, 1999.
103. T Prodpran, VL Dimonie, ED Sudol, MS El-Aasser. Macromol Symp 155:1–14, 2000.
104. A Butte, G Storti, M Morbidelli. Macromolecules 33:3485–3487, 2000.
105. D Kukulj, TP David, RG Gilbert. Macromolecules 30:7661, 1997.
106. G Marti-Mestres, F Nielloud, R Fortune, C Fernandez, H Maillols. Drug Dev Ind Pharm 26:349–356, 2000.
107. T Förster, D Prinz, M Hollenbrock, B Mueller. PCT int patent appl WO 9952496 A1, 1999.
108. M Beckman, B Eckhardt, B Kaiser, S Goering. Ger Offen DE 196118809 C1, 1997.
109. T Förster, M Claas, B Banowski, B Heide, A Wadle. PCT int patent appl WO 9959537 A1, 1999.
110. JT Simonnet, O Sonneville, S Legret. Eur patent appl EP 1010413 A1, 2000.
111. A Stehlin, G Kreyer, H Luther. PCT int patent appl WO 9703642 A1, 1997.
112. MM Breuer. In: P Becher, ed. Encyclopedia of Emulsion Technology. Vol 2. New York: Marcel Dekker, 1985, pp 385–424.
113. S Tomomasa, M Kochi, H Nakajima. Yukagaku 37:1012, 1988.
114. H Nakajima, S Tomomasa, M Okabe. J Soc Cosmet Chem Jpn 23:288, 1990.
115. T Föester, W Von Rybinski, A Wadle. Adv Colloid Interface Sci 58:119–149, 1995.
116. T Suzuki. Japan patent 57-29213, 1982.
117. N Weiner. In: HA Lieberman, MM Rieger, GS Banker, eds. Pharmaceutical Dosage Forms: Disperse Systems. Vol 1. New York: Marcel Dekker, 1996, pp 1–15.

118. M Jumaa, BW Müller. Int J Pharm 163:81–89, 1998.
119. G Krishna, BB Sheth. J Pharm Sci Technol 53:167–176, 1999.
120. BB Lundberg, BC Mortimer, TG Redgrave. Int J Pharm 134:119–127, 1996.
121. TKM Mbela, A Ludwig. STP Pharma Sci 5(3):225–231, 1995.
122. MY Levy, S Benita. J Parenteral Sci Technol 45(2):101–117, 1991.
123. Y Wang, GM Mesfin, CA Rodríguez, JG Slatter, MR Schuette, AL Cory, MJ Higgins. Pharm Res 16:930–938, 1999.
124. ES Tabosa do Egito, H Fessi, M Appel, F Puisieux, J Bolard, JP Devissaguet. STP Pharma Sci 4(2):155–162, 1994.
125. M Baluom, DI Friedman, A Rubinstein. Int J Pharm 154:235–243, 1997.
126. P Calvo, C Remuñán-López, JL Vila-Jato, MJ Alonso. Colloid Polym Sci 275:46–53, 1997.
127. JY Kim, YS Ku. Int J Pharm 194:81–89, 2000.
128. S Crauste-Manciet, D Brossard, MO Decroix, R Farinotti, JC Chaumeil. Int J Pharm 165:197–106, 1998.
129. E Ilan, S Amselem, M Weisspapir, J Schwarz, A Yogev, E Zawoznik, D Friedman. Pharm Res 13:1083–1087, 1996.
130. JS Schwartz, MR Weisspapir, DI Friedman. Pharm Res 12:687–692, 1995.
131. DI Friedman, JS Schwarz, MR Weisspapir. J Pharm Sci 84:324–329, 1995.
132. KT Ko, TE Needham, II Zia. J Macroencapsulation 15:197–205, 1998.
133. M Sznitowska, K Zurowska-Pryczkowska, S Janicki, T Järvinen. Int J Pharm 184:115–120, 1999.
134. M Sznitowska, K Zurowska-Pryczkowska, E Dabrowska, S Janicki. Int J Pharm 202:161–164, 2000.
135. P Calvo, MJ Alonso, J Vila-Jato, JR Robinson. J Pharm Pharmacol 48:1147–1152, 1996.
136. S Muchtar, M Abdulrazik, J Frucht-Pery, S Benita. J Controlled Release 44:55–64, 1997.
137. R Palacios Peláez, C Ruiz-Bravo López. Spanish patent 2118033 A1, 1998.
138. R Palacios Peláez, C Ruiz-Bravo López. PCT int patent appl WO 9916424 A1, 1999.
139. A Toledo. Eur patent appl EP 770387 A1, 1997.
140. PE Heide. Ger Offen DE 19810655 A1, 1999.
141. A Burek-Kozlowska, R Bolli, R Moudry, HG Weder. Eur patent appl EP 865792 A1, 1998.
142. Ger Offen DE 19825856 A1, 1999.
143. FJ Galan Valdivia, A Coll Dachs, N Carreras Perdiguer. Eur patent appl EP 6696452 A1, 1996.
144. H Aviv, D Friedman, A Bar-Ilan, M Vered. Int. patent appl WO 9405298 A1, 1994.
145. Pharmos Corporation USA. Israeli patent 104328 A1, 1997.
146. JE McVormak, J Douglas, G Van Nest. Int patent appl WO 9902123 A2, 1999.
147. GH Dowell, S Amselm, D Friedman, H Aviv. Int patent appl WO 9511700 A1, 1995.

148. MR Gasco. In: C Solans, H Kunieda, eds. Industrial Applications of Micro-emulsions. New York, Marcel Dekker, 1997, pp 97–122.
149. SS Davis, J Hadgraft, KJ Palin. In: P Becher, ed. Encyclopedia of Emulsion Technology. Vol 2. New York: Marcel Dekker, 1985, pp 159–238.
150. S Benita, MY Levy. J Pharm Sci 82:1069–1079, 1973.
151. AJ Wretlind. Acta Chir Scand (Suppl) 325:31–42, 1964.
152. P Kan, ZB Chen, RY Kung, CHJ Lee, IM Chu. Colloids Surf B 15:117–125, 1999.
153. JG Wagner, ES Gerard, DG Kaiser. Clin Pharmacol Ther 7:610–619, 1966.
154. PJ Carrigan, TR Bates. J Pharm Sci 62:1476–1479, 1973.
155. TR Bates, PJ Carrigan. J Pharm Sci 64:1475–1481, 1975.
156. TR Bates, JA Sesqueira. J Pharm Sci 64:793–797, 1975.
157. KJ Palin, AJ Phillips, A Ning. Int J Pharm 33:99–104, 1986.
158. J Drewe, R Meier, J Vonderscher, D Kiss, U Polanski, T Kissel, K Gyr. J Clin Pharmacol 34:60–64, 1992.
159. PP Constantinides, JP Scalart, C Lancaster, J Marcello, G Marks, H Ellens, PL Smith. Pharm Res 11:1385–1390, 1994.
160. PP Constantinides. Pharm Res 12:1561–1572, 1995.
161. H Toguchi, Y Pgawa, T Shimamoto. Chem Pharm Bull 38:2797–2800, 1990.

26

Surface Modifications of Liposomes for Recognition and Response to Environmental Stimuli

JONG-DUK KIM, SOO KYOUNG BAE, JIN-CHUL KIM, and EUN-OK LEE KAIST, Daejeon, Korea

ABSTRACT

Surface modifications of bilayers and preparations of liposomes are introduced for applications to stimuli-sensitive delivery systems. Modified or intermediate liposomes can be obtained by direct mixing of lipids and receptor-modified lipids, by direct reaction of intermediate-modified liposomes with receptor or ligands, or by insertion of modified receptors into the liposomal bilayer. Proteinaceous receptors can be modified with alkyl chains or lipids positioned along the hydrophobic part of bilayers. The enhanced release of poly(NIPAM)-coated liposomes is attributed to the collapse of hydrogel on bilayers, destroying the order of lipids in the membrane. Target-sensitive immunoliposomes are designed to destabilize upon binding to the target cell and to release their contents at the cell surface. The improved efficacy of liposome-associated adjuvants has been observed at hepatitis B surface antigen (HbsAg) incorporated in negatively charged liposomes. The synthetic peptides of epitopes in HbsAg have been used in the development of the hepatitis B virus vaccine by incorporating poorly immunogenic peptides in lipid A.

I. INTRODUCTION

Many therapeutically active agents are limited in their clinical use because of the obscure delivery processes and methods for the specific sites of action, but there are many off-patent safe drugs whose usefulness and strength would be extended by a new type of formulation. Further, the efforts to develop delivery systems would be much less than those involved in developing a new drug [1]. Therefore, there have been continuous attempts to develop new delivery methods and to design special carriers that would uniquely guide drugs specific to target cells and tissues as well.

Liposomes consist of concentric bilayers of fatty acids, predominantly phospholipids, in the range of 50 nm to several micrometers in diameter. The properties and preparation methods as well as their applications have been widely reported [2–4], including cancer chemotherapy [5,6], antibiotic and antifungal agents [7,8], gene transfer [9,10], immunological adjuvants [11,12], and angiomarkers and diagnostic agents [13,14]. However, regardless of their compositions, sizes, and charges, liposomes are quantitatively captured by cells of the reticuloendothelial system (RES) within the first hour after their intravenous administration.

The specific applications require different bulk and surface properties of liposomes. Among others, for example, the unique properties of stealth, targeted, and cationic liposomes could be achieved by surface modification. Figure 1 illustrates examples of liposomal surface modifications for such specific applications. The goals of these surface modifications are (1) to increase liposome longevity and stability in the circulation, (2) to change liposome biodistribution, (3) to achieve targeting effect, and (4) to impart to liposomes some "unusual" properties such as pH or thermal sensitivity.

We have investigated the surface modification of liposomes for functionally mediated delivery systems including the modification of carbohydrates, proteins, and polymers. The fluidity and membrane state of liposomes will be discussed in terms of promoting their recognition and response to environmental stimuli.

II. MEMBRANE STATES AND INTERACTIONS WITH CELLS

The size and surface properties of liposomes vary with types of lipids, their compositions, their modification, and methods of preparation. For example, multilamellar vesicles (MLVs) several hundred nanometers in size can be produced by a reverse phase evaporation and extrusion, but smaller unilamellar vesicles (SUVs), whose size is less than 100 nm, can be produced by a sonication process [15]. Further, the membrane state of a bilayer is of primary interest not only for surface treatment but also for recognition of a cell surface and delivery of active ingredients. We will briefly review the microfluidity of bilayers and the interaction of liposomes with a cell surface.

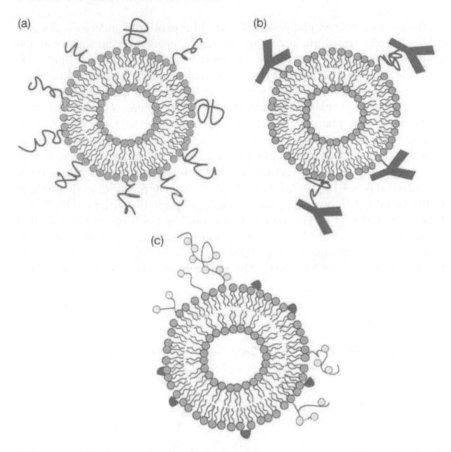

FIG. 1 Surface-modified liposomes with polymer, protein, and ionic ligands for specific applications to (a) stealth liposome, (b) targeted liposome, and (c) cationic liposome.

A. Phase Transition and Fluidity

The functions of liposomes, such as interaction, incorporation, recognition, and stabilization, are attributed to the microfluidity of a membrane and its transitional state [2]. Lipids dispersed in water can form a variety of structures, for example, the liposome-type structure at low lipid/water ratios. As temperature increases, the lipid phase shifts from a crystalline to a condensed gel-like state and then to a fluidic, expanded state, and such a transition state at the corresponding temperature can be determined by various methods [2,16].

Bilayer membranes composed of pure phospholipids undergo a discrete order–disorder transition involving primarily an increase in the rotational freedom of fatty acid side chains and an increase in the area per lipid molecule in a bilayer. In differential scanning calorimetry (DSC) measurements [16], most lipids show two peaks, which indicates the transition of the membrane state. One is related to the melting of head groups, T_{m1}, and the other to the melting of chains, T_{m2}. The former is broader and shows a smaller enthalpy change than the latter. The chain melting usually occurs 5–10°C higher than that of the head groups melting. Lipid molecules have higher polarization in a gel-like structure than in a sol-like structure. Figure 2 shows the polarization of lipid bilayers with respect to temperature, as determined by the polarized fluorescence method using diphenylhexatriene (DPH) as a fluorescent probe [17,18]. The membranes of egg phosphatidylcholine (PC) are apparently in a sol-like state in the range of temperature, whereas those of distearoylphosphatidylcholine (DSPC; T_{m1} = 51.5°C, T_{m2} = 54.9°C), are in a crystalline or gel-like state. The polarities of dipalmitoylphosphatidyl-

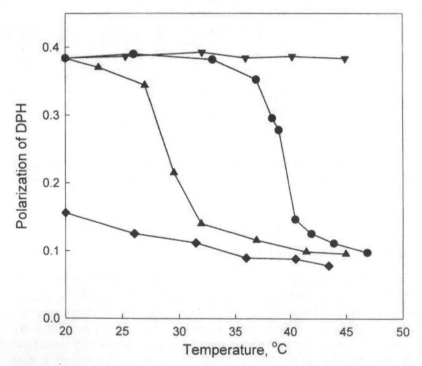

FIG. 2 Polarization of DPH embedded in egg PC (♦), DPPC/DMPC (5:5, wt/wt) (▲), DPPC (●), and DSPC (▼) liposomal membranes with temperature.

choline (DPPC; $T_{m1} = 35.3°C$, $T_{m2} = 41.4°C$) and 50:50 mixtures of DPPC and dimyristoylphosphatidylcholine (DMPC; $T_{m1} = 14.2°C$, $T_{m2} = 23.9°C$) show the transition from a gel-like state to a sol-like state in the range of temperature, and the polarities of the mixtures are reduced by the effect of fluidic DMPC.

Cholesterol in membrane bilayers has an important modulatory effect on the bilayer phase of phospholipids [2,15]. The sterol interacts strongly with phospholipids and keeps them in an "intermediate fluid" condition. Thus, above its transition temperature, the presence of cholesterol tends to increase the packing and rigidity of bilayers [19], and below its transition temperature, it expands and fluidizes the bilayers [20].

The fluidity and packing of a bilayer greatly affect the bilayer–ligand interaction as shown in Fig. 3 [18]. The coupling efficiency of liposomes with peptides (dotted line) increases as the fluidity of membrane (solid line) increases. From monolayer studies [18,21,22], it is known that proteins and

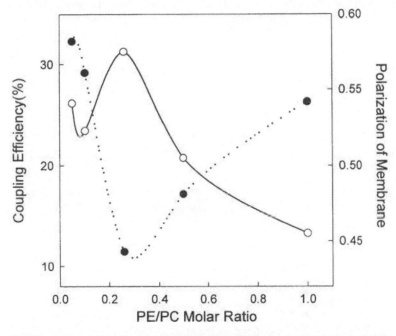

FIG. 3 Polarization of mixed liposomal membrane as a function of PE/PC molar ratio. Liposomes were composed of 2:1 phospholipid and cholesterol. Liposomes were prepared by extrusion through a polycarbonate filter. Lipid concentration was 110 μM and lipid/DPH molar ratio was 50:1. The dotted line represents the polarization of the membrane and the solid line is the coupling efficiency.

other molecules penetrate most readily when the film pressure is low. The fluid-to-solid phase transition increases the packing of lipid molecules, thus tending to prevent ligands from penetrating the lipid film.

B.   Interaction Between Liposomes and Target Cells

Figure 4 shows that the interactions of liposomes with cells fall into four categories [3,15,23]: (1) exchange of lipids or proteins with cell membranes, (2) stable adsorption or binding of liposomes to the cell surface, (3) internalization such as by endocytosis or phagocytosis, or (4) fusion of bound liposome bilayers with the cell membrane.

A number of lipid transfer proteins similar to liposomes have been detected in plasma, and lipid exchange can also occur in the absence of enzymatic activity and in liposome fusion with cells [3,15,24]. Lipid transfer occurs by two separate processes associated with transfer proteins: either by direct contact of solubilized molecules in the aqueous phase or upon liposome collisions with cells. During lipid exchange there is practically no mixing of liposomes and cell contents.

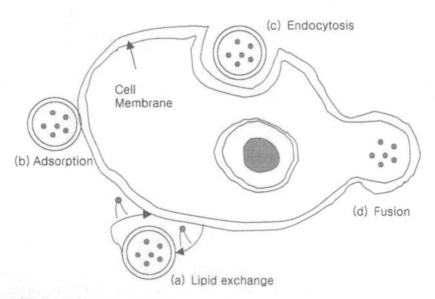

FIG. 4 Interactions of liposomes with the cell membrane. (a) Exchange of lipids or proteins with cell membranes. (b) Stable adsorption or binding of liposomes to a cell surface. (c) Internalization by endocytosis or phagocytosis. (d) Fusion of bound liposome bilayers with cell membrane.

Endocytosis is the most common mechanism for delivery of liposome contents into cells, but only a few types of cells, derived from bone marrow, can effectively phagocytose, especially with large liposomes [3,15,25]. Liposomes can be absorbed on a cell surface, engulfed into phagosomes, or transported to lysosomes. After the lipids are digested, the encapsulated molecules are released into the surrounding. If the molecules are not affected by the pH or by an intercalation and enzymatic activity in lysosomes, the molecules can be delivered into the cytoplasm.

The fusion of liposomes with cells is envisioned to deliver their contents directly to the cytoplasm [15,26]. However, whereas the fusion is an essential cellular process in endocytosis, it appears that the liposome fusion with the cells occurs very rarely and is enhanced by reconstitution of viral surface proteins. Therefore, it is apparent that this process is largely controlled by membrane protein of a cell or virus. This can be done not by a simple fusion of bilayers with cells but by incorporating fusogenic proteins or, in vitro, addition of fusogens.

It is observed that both endocytosis and adsorption are less affected in a membrane state but that fusion with cells is significantly affected in rigid cells. Therefore, both the membrane state and the surface interaction with a cell play a key role in engineering the liposomal transfer.

III. SURFACE MODIFICATION

A. Modification Methods for Liposomes

The most evident approaches to modify the surface properties of liposomes are (1) to vary liposome compositions (resulting in a variation of liposome charges and phase states) and (2) to attach some nonphospholipid compounds to the liposome surface. Various modifiers have been suggested for controlling the distribution and in vivo properties of liposomes [27–30]. The most important and well-studied modifiers are as follows:

1. Antibodies and their fragments
2. Proteins
3. Mono-, oligo-, and polysaccharides
4. Chelating compounds (such as EDTA or DTPA)
5. Soluble synthetic polymers

Liposome modification with antibodies or specific ligands leads to a drastic change in distribution, which is the result of specific recognition between the liposome-immobilized substances and the appropriate target within the body. In addition, it is known that the permeability of the liposomal membrane changes (reflecting intramembrane phase separation, variations of membrane components, lateral diffusion, and some other phenomena) when a liposome interacts with polyelectrolytes [31,32].

Figure 5 shows three methods for the formation and surface modification of liposomes.

1. *Direct mixing of lipids and receptor-modified lipids.* Receptor- or ligand-modified lipids [33,34] are mixed with normal phospholipids in a small portion (Fig. 5a). Because receptors are usually proteins that bind sugars and proteins exposed on a cell surface, receptors can be covalently attached to phospholipids by a chemical reaction. A succinyl group or glutaraldehyde [35,36] may be used as a cross-linking agent for the covalent binding of a protein receptor, and some representative ligands attached are oligo- and polysaccharides, gangliosides, immunoglobulins, viral epitopes, and so on.

2. *Direct reaction of intermediate-modified liposomes with receptors or ligands.* Lipids can also be modified with a variety of intermediates [15,37], which bind to carbohydrates and proteins. After liposomes are

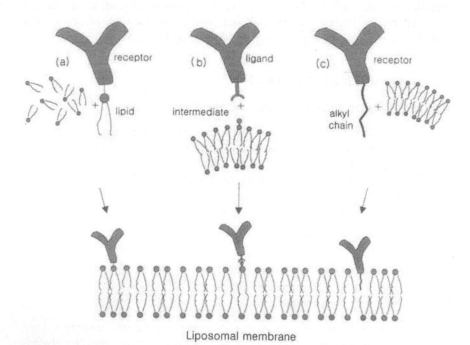

Liposomal membrane

FIG. 5 Methods for surface modifications of liposomes. (a) Direct mixing of lipids and receptor-modified lipids. (b) Direct reaction of intermediate-modified liposomes and ligands. (c) Insertion of modified receptors into the liposomal bilayer.

formed with a mixture of intermediate-modified lipids and normal lipids, ligands or receptors are mixed with liposomes and attached to the intermediates covalently (Fig. 5b).

3. *Insertion of modified receptors into liposomal bilayers.* Bilayers of liposomes consist of phospholipid assemblies that hold individual lipid molecules by weak van der Waals forces. Ligands or receptors with surface-active groups can be inserted into liposomal bilayers (Fig. 5c). Proteinaceous receptors are usually nonamphiphilic, and hence they can be modified with alkyl chains or lipids that can be positioned along the hydrophobic part of bilayers. Then the relatively hydrophilic receptor part is exposed to the liposomal surface and can interact with ligands of cell surfaces [37,38].

Described in the following are reported examples of compounds that have been used in the surface modification of liposomes.

B. Compounds Used for Surface Modification of Liposomes

1. Carbohydrates

A liposome surface modified with carbohydrates attached either to proteins or to small hydrophobic anchors can be used to recognize lectins and lectinlike receptors on mammalian cell membranes. The recognition of carbohydrate by lectin is highly dependent on the exposure of carbohydrate to the aqueous region [39,40]. When sialoglycoproteins of mammalian erythrocytes were incorporated into SUVs, the carbohydrate portion of glycoprotein was exposed on the external surface of vesicles [39]. However, appropriate hapten sugars on liposome surfaces may inhibit the binding of lectin to liposomes or the attachment of liposomes to erythrocytes.

SUVs containing two different mannosyl-pyranoside derivatives can be reversibly aggregated in the presence of concanavalin A [41]. The maximal concentration of these glycolipids in a bilayer is about 14 mol%, and the size, permeability to sucrose, and fluidity of a bilayer are not affected by their insertion, but its analogue with a longer spacer arm is more sensitive to concanavalin A–mediated agglutination. It is suggested that steric constraints will be of major importance for recognition by lectinlike proteins. Such protein-induced aggregation and fusion appear and undergo a maximum at the gel phase transition temperature of lipids.

2. Polymers

Vesicles can interact with a variety of polymers, such as natural polysaccharides [42], poly(amino acid)s, or hydrophilic synthetic polymers. The interaction of a lipid membrane with polymers depends on membrane constituents; e.g., the insertion of cholesterol significantly increases the interaction with dextran.

The interactions of poly(amino acid)s with liposomes have long been used as models for lipid–protein interactions. Basic polypeptides [e.g., poly(L-lysine), poly(L-ornithine)] form complexes with negatively charged liposomes such as phosphatidylserine [43] and undergo a conformational change from a random to an α-helical configuration [44]. Copolymers of lysine and phenylalanine show a behavior similar to that of pure polylysine in the presence of phosphatidylserine SUV but remain in the "random coil" configuration and alter the distribution of liposomes in vivo.

While attempting to prepare biologically stable liposomes, an important breakthrough was achieved by constructing long-circulating liposomes [45,46] coated with poly(ethylene glycol), PEG [47–49]. The possible mechanisms of the PEG protective effect on liposomes involve the participation of PEG in the repulsive interactions between PEG-grafted membranes and other blood moieties [50], the role of surface charge and hydrophilicity of PEG-coated liposomes [51], and the decreased rate of plasma protein (opsonin) adsorption on the hydrophilic surface of PEGylated liposomes [52]. The flexibility of polymer molecules in solution causes a dense polymeric "cloud" over a liposome surface even at relatively low polymer concentrations [53,54].

To reduce the liposome affinity for the reticuloendothelial system (RES), ganglioside G_{M1}, hydrogenated phosphatidylinositol (PI), or poly(ethylene glycol) phosphatidylethanolamine (PEG-PE) was added to standard egg PC: cholesterol liposomes. Such liposomes are not taken up so readily by macrophases of RES and hence stay in circulation streams longer. It may also depend on the size of G_{M1}-containing liposomes (diameter >300 nm) [55]. PEG-PE has a similar effect [56,57] because PEG-PE increases the hydrophilicity of a liposome surface. These PEGylated liposomes modified with antibodies are efficient in both long circulation and targeting and hence are called third-generation liposomes.

In addition, amphiphilic poly(acrylamide) (PAA) and poly(vinyl pyrrolidone) (PVP) are considered candidates among others [58]. Their protective activities are much lower than those of longer acyl anchors. A long-chain anchor binds firmly to liposomes and thus sterically stabilizes the liposomes.

3. Proteins

The attachment of proteins, particularly antibodies, to a liposome surface has been an impetus for the development of target delivery. The earliest attempts to insert antibodies into liposomes were based on the simple expedient of rehydrating dried lipid films in the presence of antibody [59].

Such noncovalently associated antibodies have not been successful in achieving a measure of antibody targeting [60]. Anionic phospholipids such as phosphatidylglycerol (PG) and phosphatidylserine (PS), but not cholesterol, enhance binding by about 50% over that obtained with a neutral PC [57].

Covalent coupling methods have been attempted to bind proteins to functional groups on a liposome surface [56,62–70] or to attach a hydrophobic residue covalently to proteins and allow it to intercalate noncovalently into a bilayer during or after the liposome formation [71,72]. The earliest methods used various bifunctional cross-linking reagents, such as dimethyl suberimidate, glutaraldehyde, and carbodiimide or periodate to oxidize carbohydrates to aldehyde. Protein conjugation can be achieved by the different processes of carboxyl groups: (1) with amino groups to produce amide bonds [35], (2) with a pyridyl-dithio derivative of phosphatidylethanolamine (PE) to produce disulfide bonds [73], and (3) with maleimide derivatives to produce thioether bonds [74]. The thiol-reactive phospholipids are synthesized using N-succinimidyl pyridyl dithio propionate (SPDP) and N-succinimidyl(4-[p-maleimidophenyl]) butyrate (SMPB) as shown in Fig. 6. The former approach results in reversible coupling of protein via a disulfide bond; the latter produces an irreversible thioether linkage.

IV. APPLICATIONS OF STIMULI OR TARGET SENSITIVITY

The surface modification of liposomes is a useful way to impart functionality, especially target sensitivity, to liposomes. In physical targeting, some characteristic of the environment is used either to direct the liposomes to a particular anatomical location or to cause a selective release of its contents as shown in Table 1, but we limit the discussion here to temperature- and pH-sensitive liposomes and immunoadjuvants.

A. Temperature-Sensitive Liposomes

A temperature-sensitive liposome can be produced in two ways: direct transition of lipid bilayers or incorporation of temperature-sensitive triggers. A liposome applied to tumors [75,76] can be made to release its contents rapidly and almost completely at the phase transition temperature, T_m. Temperature-sensitive liposomes have achieved a selectivity greater than 10-fold between heated and nonheated tumors in the delivery of methotrexate to tumors implanted in mice [77–79] or with cisplatin [80]. Hyperthermal targeting [81,82] was used in combination with radiation or chemotherapy with masked and temperature-sensitive liposomes [76,83–87].

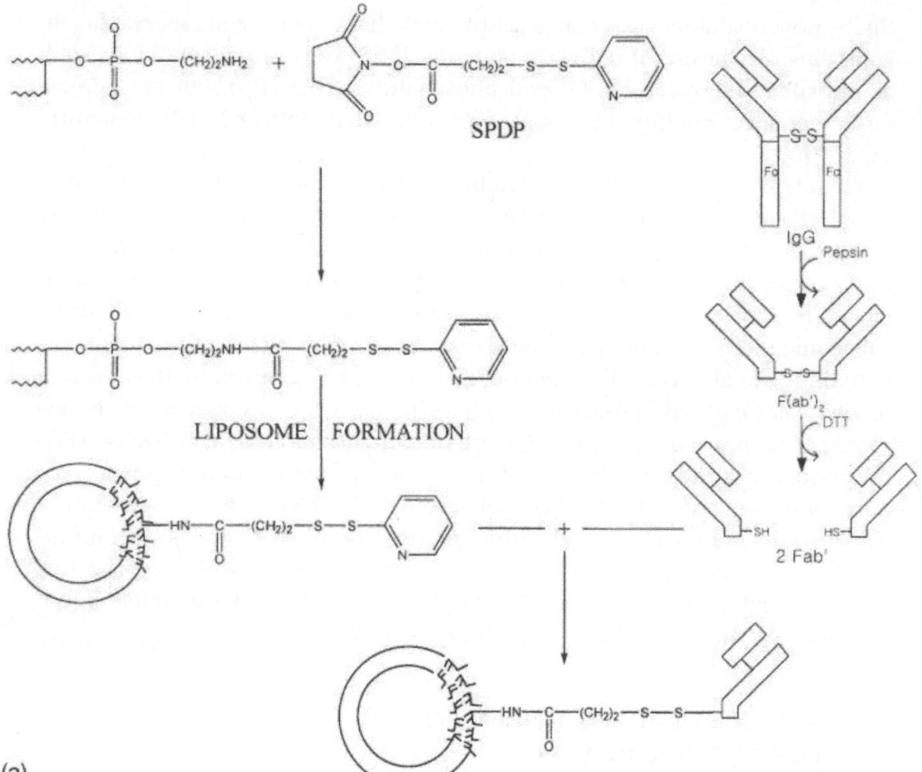

(a)

FIG. 6 Covalent coupling of Fab' fragments to vesicles. (a) Synthesis route of N-[3-(2-pyridyldithio)propionyl]phosphatidylethanolamine (PDP-PE) vesicle using N-succinimidyl 3-(2-pyridyldithio)propionate (SPDP) and coupling of Fab'. The Fab' monomers are generated from F(ab')₂ dimers by reduction with dithiothreitol at low pH and coupled by a disulfide exchange reaction between the thiol group on each Fab' fragment and the pyridyldithio moiety of PDP-PE molecules present in vesicle membranes. (b) Synthesis route of N-[4-(p-maleimidophenyl)butyryl] phosphatidylethanolamine (MPB-PE) vesicle using N-succinimidyl 4-(p-maleimidophenyl) butyrate (SMPB) and coupling of Fab'. The Fab' monomers are generated from F(ab')₂ dimers by reduction with dithiothreitol at low pH. Addition of the Fab'-SH to the double bond of the maleimide moiety of MPB-PE molecules present in vesicle membranes results in a stable thioether cross-linkage.

(b)

TABLE 1 Environmentally Sensitive Ligands or Hydrogels

Stimulus	Ligand or hydrogel
pH	Acidic or basic hydrogel
Ionic strength	Ionic hydrogel
Chemical species	Electron-accepting groups
Enzyme/substrate	Immobilized enzymes
Magnetic	Particles in alginate
Thermal	Thermoresponsive hydrogel
Electrical	Polyelectrolyte hydrogel
Ultrasonic	Poly(vinyl alcohol)

The surfaces of liposomes have been coated [17] with thermosensitive polymers such as poly(N-isopropylacrylamide) [poly(NIPAM)] by taking advantage of the phase transition of polymers. The molecular structure of a hydrophobically modified poly(NIPAM), which has been studied in the preparation of temperature-sensitive liposomes [17], is depicted in Fig. 7. Poly(NIPAM) exhibits a low critical solution temperature (LCST) around 32°C, and the LCST can be altered toward the body temperature by copolymerization [56]. The polymer is in an expanded form at low temperature, but above the critical temperature it is in a contracted form. The interactions of SUVs and hydrophobically modified poly(NIPAM) were studied by fluorescence spectroscopy [88]. More recently, sonicated DPPC and egg PC liposomes coated with a copolymer of NIPAM and octadecylacrylate in a molar ratio of 100:1 were prepared [17,89]. It was shown that above the LCST of the copolymer, the release of calcein and carboxyfluorescein from

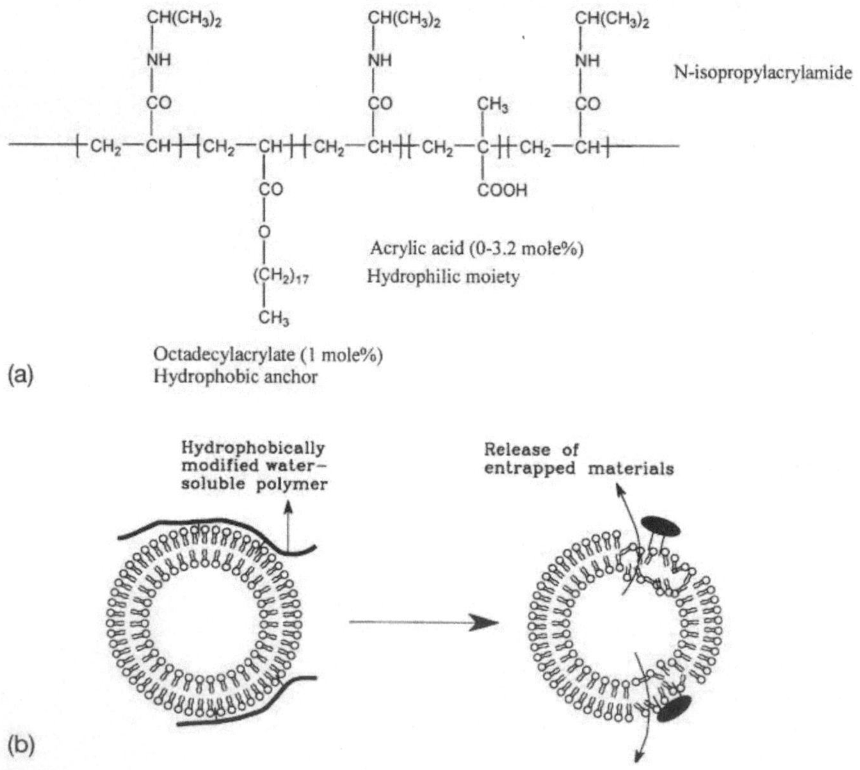

(a)

(b)

FIG. 7 (a) Structure of hydrophobically modified poly(NIPAM) and (b) mechanism of enhanced release of poly(NIPAM)-coated liposomes.

coated liposomes was significantly enhanced as temperature increased [88,90] as illustrated in Fig. 8. The enhanced release is attributed to the collapse of hydrogel on bilayers, resulting in destruction of the order of lipids in the membrane.

As shown in Fig. 2, the polarization of lipid bilayers provides information on membrane states, i.e., gel-like or fluidlike. Therefore, if we match the transition temperatures of both membrane and polymer, the release of a fluorescence dye, in fact of a delivered drug, can be maximized in the transition temperature [90].

B. pH-Sensitive Liposomes

Phosphatidylethanolamine bilayers with acidic head groups were utilized in the pH-sensitive liposomes, which contained negatively charged head

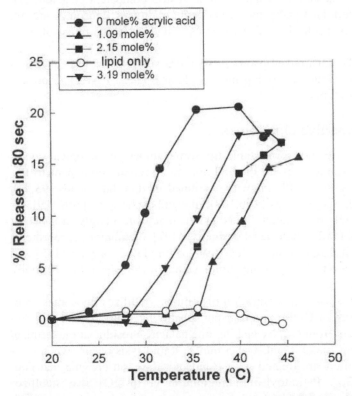

FIG. 8 Release of calcein from liposomes coated with hydrophobically modified poly(NIPAM). Alteration of the LCST of poly(NIPAM) by copolymerizing with acrylic acid can be seen by the significant increase of release efficiency.

groups. Palmitoylhomocysteine (PHC), which possesses a titrable carboxyl group, was combined with dioleoyl-PE (DOPE) to generate pH-sensitive liposomes [91], and rapid fusion between these liposomes occurred when the medium pH was lowered from 7 to 5. Intermixing of bilayer lipids indicated the fusion activity of pH <7, with the maximum fusion occurring at pH 4.4 or below [92]. Palmitic acid (PA) [93] and oleic acid (OA) [93–96], each possessing a carboxylic head group, can also stabilize PE bilayers under physiological conditions.

Another well-characterized pH-sensitive liposome includes cholesterol hemisuccinate (CHEMS) as a stabilizer [97–100], and a carboxylated derivative of PE, N-succinyldioleoyl-PE, has been combined with DOPE to generate pH-sensitive liposomes [101]. pH-sensitive liposomes have been constructed using titrable double-chain glycerol-based amphiphiles as a PE vesicle stabilizer [102]. The pH-sensitive liposomes are stable at neutral or basic pH but are destabilized and become fusion competent at acidic pH. Following the cellular uptake by an endocytic pathway, liposomes are exposed to a mildly acidic pH of endosomes, which is in the range of 5.0–6.5 [103,104].

Therefore, pH-sensitive liposomes are more suitable for the functional delivery of many drugs, including nucleoside analogues, DNA, and protein delivery to the cytoplasm.

C. Target-Sensitive Liposomes

Unsaturated PE does not form stable liposomes under physiological conditions but forms stable ones on the addition of a second component to PE. Several "stabilizers" for PE bilayers examined include fatty acids [93,94], detergents [105], gangliosides [106,107], diacylphospholipids [108,109], lysolipids [105,109], haptenated lipids [110], diacylsuccinylglycerols [102, 111], cholesterol [112], cholesterol derivatives [97], membrane glycoproteins [113,114], palmitylated amino acids [89], palmitylated enzyme [115] and palmitylated antibody [116]. Many lipids stabilize PE bilayers at 20 mol% or higher [108].

Ganglioside G_{M1} at a concentration of 5 mol% stabilizes liposomes composed of a mixture of DOPE and transphosphatidylated-PE (TPE) [101], but G_{M2}, which differs from G_{M1} by lacking one galactose residue at the terminal position, does not make it stable at 5 mol%. Rapid lysis occurs when G_{M1}-stabilized liposomes are treated with β-galactosidase, an enzyme that converts G_{M1} to G_{M2}. Palmitoyl-immunoglubulin G (p-IgG) also stabilizes DOPE bilayers, where the hydrophilic Fab portion of a derivative antibody increases the interfacial hydration and, further, prevents interbilayer contact [115,117,118]. These PE immunoliposomes could be lysed by treatment with papain, which digested the hydrophilic Fab protein domains at a liposome surface [119,120].

The stabilization of a PE bilayer phase also depends upon the hydrophobic portion of amphiphiles. The introduction of a double bond into a carbon chain of diacylglycerol further decreased the transition temperature of saturated lipid. These results can be explained by the relationship v/al, where v is the volume of the hydrophobic portion of the molecule, l is the effective length of the hydrophobic portion, and a is the hydrophilic surface area [121]. Molecules with a large v/al value have a high tendency to aggregate into a hexagonal phase. Lysolecithin possesses a large head group (a) and a small hydrophobic volume (v) and therefore stabilizes the bilayer phase of PE [105,122].

It is thus apparent that the most efficient stabilizers are molecules that have bulky, polar, and/or highly charged head groups. Cholesterol, which merely possesses a 3-β-OH as a polar head group, begins to stabilize the PE lamella only when included at higher than 30 mol% [112]. The dynamic shape concept may be applicable in predicting the molecule's ability to promote PE bilayer formation. Amphiphilic molecules that have large head groups in comparison with the hydrophobic moiety, having the shape of an inverted cone, would probably pack into a bilayer with a proper curvature complementing the cone-shaped PE.

Target-sensitive immunoliposomes are designed to be unstable upon binding to the target cell and to release their contents in high concentrations at the cell surface. Delivery by target-sensitive immunoliposomes may be applicable to a wide variety of target cells, which express sufficient and specific antigen from which antibodies can be generated. However, drug delivery is limited to small molecules, which are rapidly transported into the cells or macromolecules, and bind to cell surface receptors.

D. Liposomes as Immunological Adjuvants

The immunological carrier of protein antigens produces humoral or cellular immunity in animals [123,124], for example, viral antigens, malarial antigens, and bacterial toxins [64,125,126]. In a formulation, the most popular example of the conjugated phospholipid antigens is in series having dinitrophenyl (DNP) as a hapten, particularly DNP-aminocaproyl-phosphatidylethanolamine (DNP-Cap-PE) [64,127,128]. The importance of spacer groups between liposome-associated hapten and phospholipid head groups of liposomes was recognized in the degree of the immune response to hapten.

The effect of bilayer fluidity on the immunogenicity of membrane-soluble antigens showed that liposomes made of phospholipids with a T_c higher than ambient temperature (37°C) (solid liposomes) [129,130] provoked strong antibody responses to antigens. Although strong responses were seen with fluid liposomes (i.e., made of phospholipids with a low T_c), responses were negligible when phospholipids were substituted [125].

Lipopolysaccharide (LPS) and lipid A or muramyl dipeptide (MDP) strongly simulate immune responses [70,131]. Figure 9 shows the improved efficacy of liposome-associated adjuvants [18]. Hepatitis B surface antigen (HbsAg) incorporated in negatively charged liposomes was greater than those obtained with similar amounts of the free antigen [18,56,64–67]. The synthetic peptides of T- and B-cell epitopes in HbsAg have been used in the development of the hepatitis B virus (HBV) vaccine [55,132,133]. Incorporation of poorly immunogenic peptides in lipid A–containing liposomes was a successful adjuvant strategy in humans for including high levels of specific antibody production [67]. The optimal conditions for component proportions and spatial arrangement within the liposome structure are largely

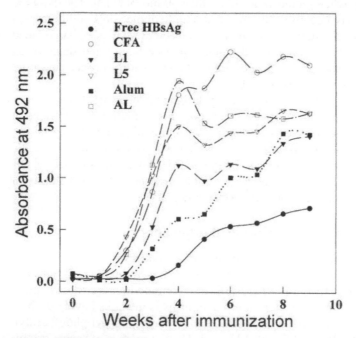

FIG. 9 Effects of different adjuvants on immune response to hepatitis B surface antigen (HbsAg). Antibody production was observed with maximum absorbance at 492 nm. Antigens injected were as follows: (●) 5 μg of free HbsAg; (○) 5 μg of HbsAg in complete Freund adjuvant (CFA); (▼) 1 μg of liposomal HbsAg (L1); (▽) 5 μg of liposomal HbsAg (L5); (■) 5 μg of alum adsorbing HbsAg (Alum); (□) 5 μg of alum adsorbing liposomal HbsAg (AL).

unknown and present a great challenge to those involved in this area of vaccinology.

V. CONCLUSIONS

The surface modifications of bilayers were investigated for stimuli-sensitive applications of liposomes. The preparations of temperature-, pH-, target-, and immunosensitive liposomes and their results were discussed.

The membrane phase of the gel-like or fluidic phase has been recognized as important for liposome delivery and surface modification. Both endocytosis and adsorption are less affected by the membrane state, but the fusion with the cells is significantly influenced.

Modified or intermediate liposomes can be obtained by direct mixing of lipids and receptor-modified lipids, by direct reaction of intermediate-modified liposomes with receptor or ligands, or by insertion of modified receptors into the liposomal bilayer. Proteinaceous receptors may be modified with alkyl chains or lipids positioned along the hydrophobic part of the bilayers. Covalent coupling methods have been attempted to bind proteins to functional groups on a liposome surface or to attach a hydrophobic residue covalently to the proteins and allow it to intercalate noncovalently into the bilayer during or after the formation of the liposome.

The enhanced release of poly(NIPAM)-coated liposomes is attributed to the collapse of the hydrogel on the bilayers, resulting in destruction of the order of lipids in the membrane. The polarization of lipid bilayers provides information on the membrane states, i.e., whether gel-like or fluidlike. By matching the transition temperatures of membranes and polymers, the maximum release of a fluorescent dye was achieved.

Target-sensitive immunoliposomes are designed to destabilize upon binding to the target cell and to release their contents in high concentrations at the cell surface. Target-sensitive delivery could be applicable to a wide variety of target cells but is limited to small molecules, which are rapidly transported into the cells, and to macromolecules, which can bind to cell surface receptors.

The improved efficacy of liposome-associated adjuvants has been observed with hepatitis B surface antigen (HbsAg) incorporated in negatively charged liposomes. The synthetic peptides of T- and B-cell epitopes in HbsAg have been used in the development of the HBV vaccine by incorporating poorly immunogenic peptides in lipid A.

In summary, both the fluidity and packing state of lipid bilayers play significant roles in increasing the coupling efficiency at the surface of and the transport through a membrane.

ACKNOWLEDGMENT

This work was supported by the Korean Science and Engineering Foundation, Dongkook Pharmaceutical Co., and BK project. We also thank Dr. G.Y. Kim, KRIBB and Dr. W.G. Ji, Chungnam University for valuable discussions and help.

REFERENCES

1. The Boston Consulting Group, http://www.bcg.com.
2. RRC New, ed. Liposomes: A Practical Approach. New York: Oxford University Press, 1990.
3. MJ Ostro, ed. Liposomes. New York: Marcel Dekker, 1983.
4. G Gregoriadis. N Engl J Med 295:704–765, 1976.
5. IJ Fidler, S Sone, WE Fogler, ZL Barnes. Proc Natl Acad Sci USA 78:1680, 1981.
6. GL Scherphof, T Daemen, HJH Spanjer, FH Roerdink. Lipids 22:891, 1987.
7. RT Mehta. Adv Drug Delivery Rev 3:283, 1989.
8. G Lopez-Berenstein, V Rainstein, R Hopter, K Mehta, MP Sullivan, M Keating, MG Rosenblum, R Mehta, M Luna, EM Hersch. J Infect Dis 151:704, 1985.
9. X Gao, L Huang. Biophys Biochem Res Commun 179:280, 1991.
10. D Litzinger, L Huang. Biochim Biophys Acta 1113:201, 1992.
11. G Gregoriadis. Immunol Today 11:89, 1990.
12. WE Vannier, SL Snyder. Immunol Lett 19:59, 1988.
13. VJ Caride. Biochem Soc Trans 12:346, 1984.
14. SE Seltzer. Radiology 171:19, 1989.
15. DD Lasic. Liposomes: From Physics to Applications. Amsterdam: Elsevier, 1993.
16. CG Knight. Liposomes: From Physial Structure to Therapeutic Applications. Oxford: Elsevier, 1981.
17. JC Kim, SK Bae, JD Kim. J Biochem 121:15–19, 1997.
18. EO Lee. PhD dissertation, KAIST, Taejon, Korea, 1994.
19. SH Kim. MS thesis, KAIST, Taejon, Korea, 2001.
20. D Papahadjopoulos, M Moscarello, EM Eyler, T Isaac. Biochim Biophys Acta 40:317, 1975.
21. T Araki, S Oinuma, K Iriyama. Langmuir 7:738, 1991.
22. T Araki, Y Sugawara, N Hirao. Chem Lett 2:329, 1989.
23. D Lasic. Am Sci 80(1):20–31, 1992.
24. H Ruterjans, T Maurer, C Lucke. J Biol Chem 370:629, 1988.
25. JN Verma, NM Wasset, RA Wirtz. Biochim Biophys Acta 1066:229, 1991.
26. G Knoll, KNJ Burger, R Bron. J Cell Biol 107:2511, 1988.
27. A Suloria, BK Bachhawat, SK Podder. Nature 257:802, 1975.
28. RL Juliano, D Stamp. Nature 261:335, 1976.
29. TD Heath, RT Fraley. Science 210:539, 1980.

30. RJY Ho, BT Rouse, L Huang. Biochemistry 25:5500, 1986.
31. K Seki, DA Tirrell. Macromolecules 17:1692, 1984.
32. OO Petrukhina, NN Ivanov, MM Feldstein, AE Vasil'ev, NA Plate, VP Torchilin. J Controlled Release 3:137, 1986.
33. G Gregoriadis, D Neerunjun, R Hunt. Life Sci 21:357, 1977.
34. B Wolff, G Gregoriadis. Biochim Biophys Acta 802:259, 1984.
35. A Huang, L Huang, SJ Kennel. J Biol Chem 255:8015, 1980.
36. A Huang, YS Tsao, SJ Kennel. Biochim Biophys Acta 716:140, 1982.
37. R Helmut, S Bernhard, V Joachim. Angew Chem 27:113, 1988.
38. H Masayuki, M Masao, L Sandayo. Supramol Sci 5:777, 1998.
39. RL Juliano, D Stamp. Nature 261:235–237, 1976.
40. WR Redwood, TG Polefka. Biochim Biophys Acta 455:631–643, 1976.
41. CD Muller, F Schuber. Biochim Biophys Acta 986:97–105, 1989.
42. D Schachter. Biochem Biophys Res Commun 84:840–844, 1978.
43. HK Kimelberg, D Papahadjopoulos. J Biol Chem 246:1142–1148, 1971.
44. GG Hammes, SE Schullery. Biochemistry 9:2555–2563, 1970.
45. G Blume, G Cevc. Biochim Biophys Acta 1029:91, 1990.
46. D Papahadjopoulos, TM Allen, A Gabizon, E Mayhew, SK Huang, KD Lee, MC Woodle, DD Lasic, C Redemann, FJ Martin. Proc Natl Acad Sci USA 88:11460, 1991.
47. AL Klibanov, K Maruyama, VP Torchilin, L Huang. FEBS Lett 277:263, 1990.
48. A Mori, AL Klibanov, VP Torchilin, L. Huang. FEBS Lett 284:263, 1991.
49. MC Woodle, DD Lasic. Biochim Biophys Acta 1113:171, 1992.
50. D Needham, TJ McIntosh, DD Lasic. Biochim Biophys Acta 1108:40, 1992.
51. A Gabizon, D Papahadjopoulos. Biochim Biophys Acta 1103:94, 1992.
52. DD Lasic, FG Martin, A Gabizon, SK Huang, D Papahadjopoulos. Biochim Biophys Acta 1070:187, 1991.
53. VP Torchilin, VS Trubetskoy, MI Papisov, AA Bogdanov, VG Omelyanenko, J Narula, BA Khaw. Proceedings of the 20th International Symposium on Controlled Release of Bioactive Materials, The Controlled Release Society, Washington, DC, 1993, p 194.
54. VP Torchilin, MI Papisov. J Liposome Res 4:725, 1994.
55. A Mori, AL Klibanov, VP Torchilin, L Huang. FEBS Lett 284:263–266, 1991.
56. AR Neurath, SBH Kent, N Strick, P Taylor, CE Stevens. Nature 315:154–156, 1985.
57. DC Litzinger, L Huang. Biochim Biophys Acta 1127:249–254, 1992.
58. VP Torchilin, MI Shtilman, VS Trubetskoy, K Whiteman, AM Milstein. Biochim Biophys Acta 1195:181, 1994.
59. WE Magee, OV Miller. Nature 235:339–341, 1972.
60. G. Gregoriadis, ED Neerunjun. Eur J Biochem 47:179–185, 1974.
61. L Huang, SJ Kennel. Biochemistry 18:1702–1707, 1979.
62. TD Heath, D Robertson, MSC Birberk, AJS Davies. Biochim Biophys Acta 599:42–62, 1980.
63. TD Heath, RT Fraley, D Papahadjopoulos. Science 210:539–541, 1980.
64. EK Manesis, CH Cameron, G Gregoriadis. FEBS Lett 102:107–111, 1979.

65. Y Sanchez, I Ionescu-Matiu, GR Dreesman, W Kramp, HR Six, FB Hollinger, JL Melnick. Infect Immun December:728–733, 1980.
66. GR Dreesman, Y Sanchez, I Ionescu-Matiu, JT Sparrow, HR Six, DL Peterson, FB Hollinger, JL Melnick. Nature 295:158–160, 1982.
67. AR Neurath, SB Kent, N Strick. Science 224:392–395, 1984.
68. DR Milich. Immunol Rev 99:71–103, 1987.
69. DR Milich, JE Jones, A McLachlan, G Bitter, A Moriarty, JL Hughes. J Immunol 144:3544–3551, 1990.
70. LF Fries, DM Gordon, RL Richards, JE Egan, MR Hollingdale, M Gross, C Silverman, CR Alving. Proc Natl Acad Sci USA 89:358–362, 1992.
71. A Huang, L. Huang, SJ Kennel. J Biol Chem 255:8015–8018, 1980.
72. D Sinha, F Karush. Biochem Biophys Res Commun 90:554–560, 1979.
73. FJ Martin, WL Hubbell, D Papahadjopoulos. Biochemistry 20:4229–4238, 1981.
74. FJ Martin, D Papahajopoulos. J Biol Chem 257:286–288, 1982.
75. MB Yatvin, JN Weinstein, WH Dennis, R Blumenthal. Science 202:1290–1293, 1978.
76. JN Weinstein, RL Magin, MB Yatvin, DS Zaharko. Science 204:188–191, 1979.
77. JN Weinstein, RL Magin, RL Cysyl, DS Zaharko. Cancer Res 40:1388–1392, 1980.
78. JB Basset, RU Anderson, JR Tacker. J Urol 135:612–615, 1986.
79. JR Tacker, RU Anderson. J Urol 127:1211–1214, 1982.
80. MB Yatvin, H Muhlensiepen, W Porschen, JN Weinstein, LE Feinendegen. Cancer Res 41:1602, 1981.
81. JH Kim, EW Hahn, FJ Benjamin. Clin Bull 9:13–16, 1979.
82. GM Hahn, J Braun, I Har-Kedar. Proc Natl Acad Sci USA 72:937–940, 1975.
83. RL Magin, MR Neisman. Cancer Drug Delivery 1:109, 1984.
84. RP Liburdy, RL Magin. Radiat Res 103:266–275, 1985.
85. JN Weinstein. In Rational Basis for Chemotherapy. New York: McGraw-Hill, 1983, pp 441–473.
86. SM Sullivan, L Huang. Biochim Biophys Acta 812:116–126, 1985.
87. JR Bertino, CD Kowal, ME Klein, J Dombrowski, E Mini. Front Radiat Ther Oncol 18:162–170, 1984.
88. H Ringsdorf, E Sackmann, J Simon, FM Winnik. Biochim Biophys Acta 1153:335–344, 1993.
89. K Kono, H Hayashi, TJ Takagishi. J Controlled Release 30:69–75, 1994.
90. JC Kim, MS Kim, JD Kim. Korean J Chem Eng 16(4):28–33, 1999.
91. J Connor, MB Yatvin, L Huang. Proc Natl Acad Sci USA 81:1715–1718, 1984.
92. DK Struck, D Hoekstra, RE Pagano. Biochemistry 20:4093–4099, 1981.
93. L Huang, SS Lui. Biophys J 45:72a, 1984.
94. J Duzgunes, RM Straubinger, PA Baldwin, DS Friend, D Papahadjopoulos. Biochemistry 24:3091–3098, 1985.
95. RM Straubinger, N Duzgunes, D Papahadjopoulos. FEBS Lett 179:148–154, 1985.

96. J Connor, L Huang. Cancer Res 46:3431–3435, 1986.
97. H Ellens, J Bentz, FC Szoka. Biochemistry 23:1532–1538, 1984.
98. H Ellens, J Bentz, FC Szoka. Biochemistry 24:3099–3106, 1985.
99. J Bentz, H Ellens, MZ Lai, FC Szoka. Proc Natl Acad Sci USA 82:5742–5745, 1985.
100. MZ Lai, N Duzgunes, FC Szoka. Biochemistry 24:1654–1661, 1985.
101. R Nayar, AJ Schroit. Biochemistry 24:5967–5971, 1985.
102. R Leventis, T Diacovo, JR Silvius. Biochemistry 26:3267–3276, 1987.
103. S Ohkuma, B Poole. Proc Natl Acad Sci USA 80:3334–3338, 1978.
104. B Tycko, FR Maxfield. Cell 28:643–651, 1982.
105. TD Madden, PR Cullis. Biochim Biophys Acta 684:149–153, 1982.
106. YS Tsao, L Huang. Biochemistry 24:1092–1098, 1985.
107. P Pinnaduwage, L Huang. Biochim Biophys Acta 939:375–382, 1988.
108. MJ Hope, DC Walker, PR Cullis. Biochem Biophys Res Commun 110:15–22, 1983.
109. A Tari, L Huang. Biochemistry 28:7708–7712, 1989.
110. RJY Ho, L Huang. J Immunol 134:4035–4040, 1985.
111. D Collins, DC Litzinger, L Huang. Biochim Biophys Acta 1025:234–242, 1990.
112. RM Epand, R Bottega. Biochemistry 26:1820–1825, 1987.
113. TF Taraschi, TM van der Steen, B de Kruijff, C Tellier, AJ Verkleij. Biochemistry 21:5756–5764, 1982.
114. P Pinnaduwage, L Huang. Biochemistry 31:2850–2855, 1992.
115. EO Lee, JD Kim. J Biochem 117:54–58, 1995.
116. RJY Ho, BT Rouse, L Huang. Biochemistry 25:5500–5506, 1986.
117. EO Lee, JG Kim, JD Kim. J Korean Pharm Sci 20(3):135–144, 1990.
118. CS Kim, EO Lee, JD Kim. J Korean Pharm Sci 21(3):161–170, 1991.
119. JH Park, EO Lee, JD Kim. J Korean Pharm Sci 22(2):115–125, 1992.
120. EO Lee, JG Kim, JD Kim. J Biochem 112:671–676, 1992.
121. JN Israelachvili, S Marcelja, RG Horn. Q Rev Biophys 13:121–200, 1980.
122. RM Epand. Biochemistry 24:7092–7095, 1985.
123. CR Alving. Pharmacol Ther 22:407, 1983.
124. N van Rooijen, R van Nieuwmengen. In G Gregoriadis, J Senior, A Trouet, eds. Targeting of Drugs. New York: Plenum, 1982, p 301.
125. D Davis, G Gregoriadis. Immunology 61:229–234, 1987.
126. D Davis, G Gregoriadis. Immunology 68:277–282, 1989.
127. GF Dancey, T Yasuda, SC Kinsky. J Immunol 119:1868–1873, 1977.
128. GF Dancey, T Yasuda, SC Kinsky. J Immunol 120:1109–1113, 1978.
129. SC Kinsky, T Yasuda, T Tadakuma. Immunol Today 3:308–310, 1982.
130. Q Bakouch, F David, D Gerlier. Eur J Immunol 17:1839–1842, 1987.
131. M Friede, S Muller, JP Briand, MHV van Regenmortel, F Scheber. Mol Immunol 30:539–547, 1993.
132. DR Milich. Immunol Today 9:380–386, 1988.
133. JP Tam, YA Lu. Proc Natl Acad Sci USA 86:9084–9088, 1989.

27

Specific Partition of Surface-Modified Liposomes in Aqueous PEO/Polysaccharide Two-Phase Systems

EUI-CHUL KANG and KAZUNARI AKIYOSHI Kyoto University, Kyoto, Japan

JUNZO SUNAMOTO Niihama National College of Technology, Ehime, Japan

ABSTRACT

First, the partition of hydrophobized polysaccharide (HP)–coated liposomes was investigated in aqueous two-phase systems such as poly(ethylene oxide) (PEO) (top phase)/pullulan (bottom phase) and poly(ethylene oxide) (top)/dextran (bottom phase). HPs such as cholesterol-bearing pullulan, dextran, and mannan nicely coated the liposomal surface. When conventional uncoated liposomes were added to the aqueous two-phase system, they mostly located at the interface between the two polymer phases. The HP-coated liposomes, on the other hand, were significantly partitioned into the bottom polysaccharide phase, depending on the structure of the HP on the liposomal surface.

Second, the partition of ganglioside (GM_3, GD_{1a}, GD_{1b}, or GT_{1b})-reconstituted liposomes was investigated using the same system as that used for the case of the HP-coated liposomes. The partition of the ganglioside-reconstituted liposomes was strongly affected by the buffer composition employed. In 10 mM sodium phosphate containing 150 mM sodium chloride, for instance, the conventional liposome without ganglioside mostly stayed at the interface between the two phases, whereas the ganglioside-reconstituted and negatively charged liposomes were significantly partitioned into the bottom polysaccharide phase. The extent of the partition increased with increasing the ganglioside density on the liposomal surface. This partition to the polysaccharide-rich phase also depended on the chemical structure of the ganglioside on the liposomal surface. However, these ganglioside-reconstituted liposomes hardly partitioned into the top PEO-rich phase and were even insensitive to the buffer composition of the system.

I. INTRODUCTION

Aqueous two-phase systems have been widely used for separation and purification of proteins, enzymes, and cells [1,2]. The partition of conventional liposomes in the aqueous two-phase system of poly(ethylene oxide) (PEO)/ dextran has been investigated as a model of cell separation [3–6]. The partition is affected by the size [3,4] and lipid composition of the liposome employed [5,6]. The surface property of the liposome was, of course, predominant in affecting the partition [7,8].

First, to understand carbohydrate–carbohydrate interaction in water, partition of hydrophobized polysaccharide (HP)–coated liposomes in an aqueous two-phase system [PEO (top phase)/pullulan (bottom phase) or PEO (top)/dextran (bottom)] was investigated. HPs such as cholesterol-bearing pullulan (CHP), dextran (CHD), and mannan (CHM) coat the liposomal surface nicely. When the liposomes were added into the aqueous two-phase system, the HP-coated liposomes were significantly partitioned to the bottom polysaccharide-rich phase. However, conventional liposomes without any HP coating mostly stayed at the interface between the two polymer phases. This specific partition depended on the chemical structure of the HP on the liposomal surface. The affinity between HP on the liposomal surface and the phase-forming polysaccharide in the bulk bottom phase controls the partition efficiency. Second, the partition of ganglioside (GM$_3$, GD$_{1b}$, GD$_{1a}$, or GT$_{1b}$)-reconstituted liposomes was also investigated using the same aqueous two-phase system. The partition of the ganglioside-reconstituted liposomes was strongly affected by the composition of the buffer employed. With decreasing concentration of sodium phosphate, the partition of the negatively charged liposomes largely increased in the bottom dextran-rich phase. This partition to the dextran-rich phase also depended on the chemical structure of ganglioside on the liposomal surface. The partition of the liposome to the top PEO-rich phase was negligibly small.

II. PARTITION OF HP-COATED LIPOSOMES IN AQUEOUS TWO-PHASE SYSTEMS

A. Materials and Methods

The dl-α-Dipalmitoyl phosphatidylcholine (DPPC) (purity, 99%; Sigma, St. Louis, MO), cholesterol (Nacalai Tesque, Kyoto, Japan), pullulan-50 [mo-

lecular weight (MW) 50,000; Hayashibara Biochemical Laboratory, Oka-
yama, Japan], dextran-40 (MW 40,000; Tokyo Kasei Kogyo Co. Ltd., Tokyo,
Japan), mannan (MW 85,000; Nacalai Tesque, Kyoto, Japan), PEO-20 (MW
20,000; Nacalai Tesque, Inc., Kyoto, Japan), and 1,6-diphenyl-1,3,5-hexa-
triene (DPH) (Wako Pure Chemicals Ltd., Osaka, Japan) were used without
further purification. For CHP-108-1.3, 1.3 cholesteryl groups per 100 glucose units are substituted on the parent pullulan (MW 108,000); for CHP-
55-1.7 and CHP-55-2.5, 1.7 or 2.5 cholesteryl groups per 100 glucose units
are substituted on the parent pullulan (MW 55,000); for CHD-70-1.7, 1.7
cholesteryl groups per 100 glucose units are substituted on the parent dextran
(MW 70,000); for CHM-85-2.3, 2.3 cholesteryl groups per 100 mannose
units are substituted on the parent mannan (MW 85,000); for $C_{16}P$-55-2.4,
2.4 hexadecyl groups per 100 glucose units are substituted on the parent
pullulan (MW 55,000); for $2C_{12}P$-55-2.3, 2.3 α,α'-dodecyl diglyceryl diether
groups per 100 glucose units are substituted on the parent pullulan (MW
55,000); for FITC-0.55-pullulan-50, 0.55 fluorescein isothiocyanate (FITC)
groups per 100 glucose units are substituted on the parent pullulan (MW
50,000); and for FITC-0.49-CHP-50-1.6, 0.49 FITC groups per 100 glucose
units are substituted on the parent pullulan (MW 50,000). All HPs were
synthesized by a method previously reported [9]. The chemical structures of
HPs used in this work are given in Fig. 1.

A lipophilic fluorescent (DPH) was employed as the lipid-soluble probe
for the investigation of liposome partitioning in aqueous two-phase systems
[10–12]. The DPH-loaded liposomes and the HP-coated liposomes were
prepared according to methods described previously [13,14]. The final concentration of the liposomal suspension was adjusted to 1.0×10^{-3} M as the
total liposomal lipid concentration.

Based on the phase diagram, 6.0% (w/w) PEO-20/8.0% (w/w) pullulan-
50 and 6.0% (w/w) PEO-20/8.0% (w/w) dextran-40 systems were selected
for these partition experiments. An aqueous polymer stock solution (20.0
wt%) was prepared by dissolving 2.0 g of the polymer in 8.0 g of 20 mM
aqueous Tris-HCl buffer containing 200 mM NaCl (pH 7.0). An HP-coated
liposome solution (0.05 g) was added to a two-phase system of 0.30 g of
20.0% (w/w) PEO-20 and 0.4 g of 20.0% (w/w) pullulan-50 or 20.0% (w/
w) dextran-40. Tris-HCl buffer (0.25 g) was added to give a total sample
weight of 1.00 g. The resulting mixture was centrifuged at $2000 \times g$ (Capsule HF-120, Tomy Seiko Co.) for 10 min. After reaching equilibrium, a
0.2-mL sample solution was pipetted out from both the PEO-rich top phase
and the polysaccharide-rich bottom phase, and each solution was diluted
with 0.8 mL of the same buffer. To determine partitioning of the DPH-loaded
liposome, the fluorescence intensity at 430 nm (excited at 360 nm) was

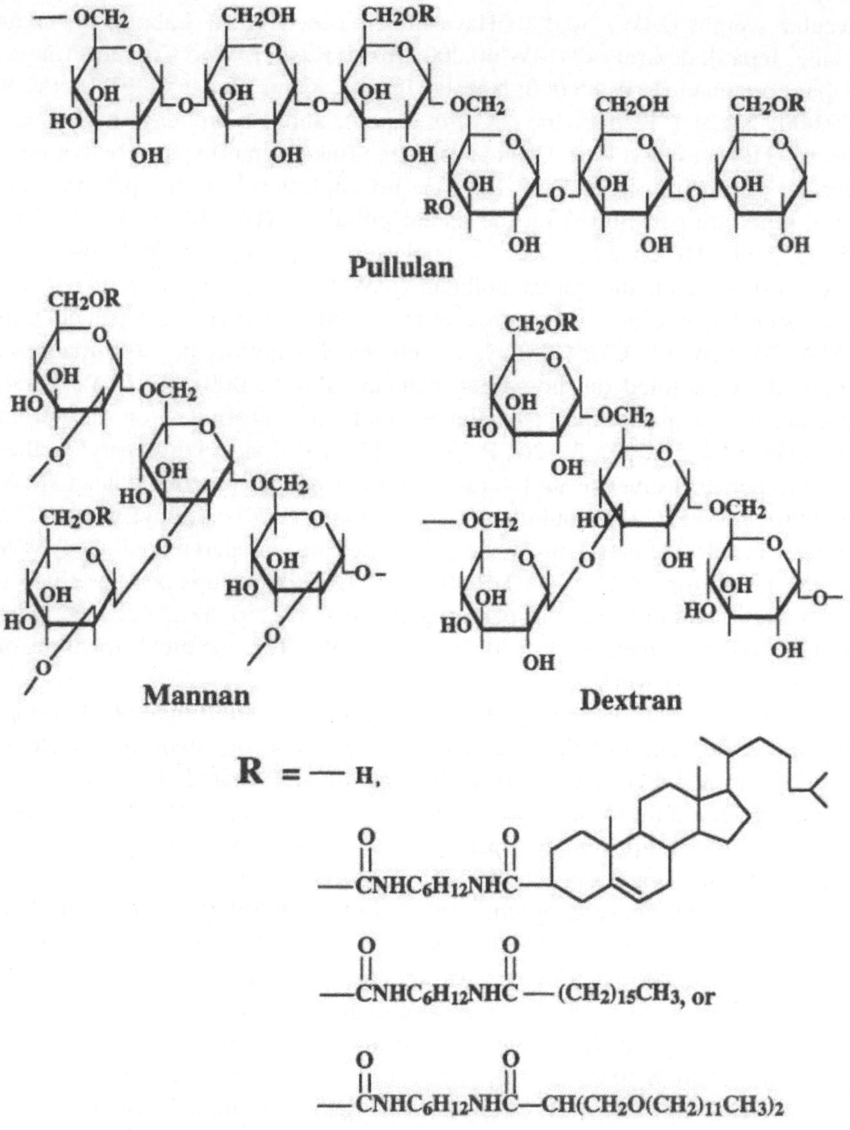

FIG. 1 Chemical structure of polysaccharide derivatives used in this work.

measured for both the top and bottom phases on a fluoroescence spectro-photometer (F-3010, Hitachi, Tokyo, Japan).

B. Partition of HP-Coated Liposomes

The effects of coating the liposomes with HPs on the partition in two different aqueous two-phase systems are summarized in Tables 1 and 2 and Fig. 2. The conventional liposomes without any HP coat locate mostly at the interface between the two polymer phases as reported previously [5–7]. On the other hand, the HP-coated liposomes were significantly partitioned to the bottom polysaccharide-rich phase. The extent of the partition depended on the chemical structure and density of HP on the liposomal surface.

The interfacial adsorption of relatively large particles such as cells or liposomes is a general phenomenon in aqueous two-phase systems of polymers. The higher the interfacial tension becomes, the less the particles present at the interface are partitioned to the two polymer phases [1,2]. The phase diagram reflects the relationship between interfacial tension and the tie-line length (TLL) in a two-phase system [14,15]. In higher TLL systems, the interfacial adsorption of the particles increases. The interfacial tension in the 6.0% (w/w) PEO-20/8.0% (w/w) pullulan-50 system (TLL = 21.9) is higher than that in the 6.0% (w/w) PEO-20/8.0% (w/w) dextran-40 system (TLL = 18.4). Therefore, the slight difference in adsorption of conventional liposomes between these two different polysaccharide systems must be due to the difference in interfacial tension.

The partition of the liposomes drastically changed on coating the surface with HPs (Tables 1 and 2). Partition to the bottom phase increased with an increase in the density of the HP on the liposomal surface; nevertheless, partition to the top PEO-rich phase did not change at all. The partitioning efficiency leveled off above the point where the weight ratio of polysaccharide to lipid was approximately 0.5 (Fig. 2).

Sharpe and Warren [7] have investigated the partition of glycolipid-containing liposomes in the PEO/dextran system. Surface modification of the liposome by carbohydrates markedly increased partition of the liposome to the top PEO-rich phase. Considering the higher hydrophobicity of the top PEO-rich phase compared with the bottom polysaccharide-rich phase, they ascribed the larger partition of the glycolipid-containing liposomes into the top phase to the hydrophobicity of the carbohydrates on the liposomal surface [7].

In contrast to the previous report, in this study partition of the HP-coated liposomes to the bottom polysaccharide-rich phase occurred significantly. This suggests that a specific interaction takes place between the liposomal polysaccharide and the phase-forming polysaccharide in the bulk bottom

TABLE 1 Partition Efficiency (%) of Hydrophobized Polysaccharide–Coated Liposomes into an Aqueous 6.0% (w/w) PEO-20/8.0% (w/w) Pullulan-50 System

Liposomal polysaccharide	R	Partition (%) Top phase	Interface	Bottom phase
None		0.5 (0.1)	97.7 (1.1)	1.8 (0.9)
CHP-55-1.7	0.05	0.4 (0.1)	85.7 (4.8)	13.9 (4.7)
	0.1	0.4 (0.1)	80.6 (2.8)	19.0 (2.8)
	0.2	0.4 (0.1)	69.9 (1.0)	29.7 (1.1)
	0.3	0.4 (0.1)	55.8 (5.1)	43.8 (5.1)
	0.5	0.4 (0.1)	43.8 (4.0)	55.8 (4.0)
	1.0	0.4 (0.1)	49.4 (6.4)	50.2 (6.4)
CHP-55-2.5	0.05	0.4 (0.1)	86.7 (3.4)	12.9 (3.4)
	0.1	0.5 (0.1)	81.3 (2.1)	18.2 (2.0)
	0.2	0.4 (0.1)	70.1 (3.8)	29.5 (3.8)
	0.3	0.4 (0.1)	56.6 (6.3)	40.0 (6.3)
	0.5	0.4 (0.1)	53.2 (4.6)	46.4 (4.7)
	1.0	0.3 (0.1)	48.3 (4.6)	51.4 (4.4)
CHP-108-1.3	0.05	0.4 (0.1)	87.5 (2.2)	12.1 (2.1)
	0.1	0.4 (0.1)	84.0 (1.7)	15.6 (1.7)
	0.2	0.3 (0.1)	70.9 (3.8)	28.8 (3.8)
	0.3	0.4 (0.1)	52.2 (8.8)	47.4 (8.8)
	0.5	0.4 (0.1)	44.6 (7.3)	55.0 (7.3)
	1.0	0.4 (0.1)	43.1 (6.6)	56.5 (6.6)
CHD-70-1.7	0.05	0.3 (0.1)	88.0 (2.0)	11.7 (2.0)
	0.1	0.3 (0.1)	83.6 (1.6)	16.1 (1.4)
	0.2	0.3 (0.1)	76.2 (0.7)	23.5 (0.7)
	0.3	0.3 (0.1)	72.4 (7.3)	27.3 (7.3)
	0.5	0.3 (0.1)	69.9 (2.8)	29.8 (2.8)
	1.0	0.3 (0.1)	72.4 (3.4)	27.3 (3.4)
CHM-85-2.3	0.05	0.7 (0.1)	90.3 (1.1)	9.0 (1.1)
	0.1	0.8 (0.1)	86.5 (2.8)	12.7 (2.8)
	0.2	0.7 (0.1)	82.1 (1.7)	17.2 (1.7)
	0.3	0.7 (0.1)	72.8 (1.4)	26.5 (1.4)
	0.5	0.8 (0.1)	72.0 (3.3)	27.2 (3.4)
	1.0	0.7 (0.1)	72.2 (1.9)	27.1 (1.9)
$2C_{12}P$-55-2.3	0.05	0.4 (0.1)	88.9 (0.7)	10.7 (0.7)
	0.1	0.4 (0.1)	83.5 (3.2)	16.1 (3.2)
	0.2	0.4 (0.1)	73.6 (3.5)	26.0 (3.5)
	0.3	0.5 (0.1)	54.9 (5.3)	44.7 (5.4)
	0.5	0.4 (0.1)	42.1 (1.1)	57.5 (1.2)
	1.0	0.4 (0.1)	43.1 (4.7)	56.5 (4.7)
$C_{16}P$-55-2.4	0.05	0.6 (0.3)	88.2 (3.1)	11.2 (3.0)
	0.1	0.8 (0.1)	84.1 (4.8)	15.1 (4.8)
	0.2	0.6 (0.2)	66.8 (1.9)	32.6 (1.8)
	0.3	0.6 (0.1)	62.0 (1.7)	37.4 (1.7)
	0.5	0.4 (0.1)	53.0 (2.5)	46.6 (2.5)
	1.0	0.4 (0.1)	54.9 (2.0)	44.7 (2.0)

All experiments were performed in triplicate. The values in parentheses are standard deviations. R indicates the weight ratio of hydrophobized polysaccharides to lipids (mg/mg).

TABLE 2 Partition Efficiency (%) of Hydrophobized Polysaccharide–Coated Liposomes into an Aqueous 6.0% (w/w) PEO-20/8.0% (w/w) Dextran-40 System

Liposomal polysaccharide	R	Partition (%)		
		Top phase	Interface	Bottom phase
None		0.6 (0.1)	95.2 (0.2)	4.3 (0.1)
CHP-55-1.7	0.05	0.6 (0.1)	80.2 (4.3)	19.2 (4.3)
	0.1	0.6 (0.1)	66.5 (3.1)	32.9 (3.1)
	0.2	0.6 (0.1)	49.3 (1.0)	50.1 (1.0)
	0.3	0.6 (0.1)	39.6 (6.3)	59.8 (6.3)
	0.5	0.6 (0.1)	28.6 (6.2)	70.8 (6.2)
	1.0	0.4 (0.1)	34.2 (6.2)	65.2 (5.7)
CHP-55-2.5	0.05	0.6 (0.1)	80.8 (1.0)	18.6 (1.0)
	0.1	0.6 (0.1)	60.1 (3.4)	39.3 (3.3)
	0.2	0.6 (0.1)	47.8 (3.8)	51.6 (3.8)
	0.3	0.6 (0.1)	36.0 (4.8)	63.4 (4.8)
	0.5	0.6 (0.1)	30.3 (2.8)	69.1 (2.8)
	1.0	0.5 (0.1)	29.4 (9.8)	70.1 (9.6)
CHP-108-1.3	0.05	0.5 (0.1)	81.7 (1.3)	17.8 (1.3)
	0.1	0.4 (0.1)	62.8 (2.2)	36.8 (2.2)
	0.2	0.4 (0.1)	36.9 (7.0)	62.7 (7.2)
	0.3	0.5 (0.1)	32.1 (5.9)	67.4 (5.9)
	0.5	0.6 (0.1)	27.6 (9.4)	71.8 (9.4)
	1.0	0.6 (0.1)	31.5 (5.7)	67.9 (5.7)
CHD-70-1.7	0.05	0.6 (0.1)	69.7 (1.3)	29.7 (1.1)
	0.1	0.5 (0.1)	51.1 (6.3)	48.4 (6.3)
	0.2	0.5 (0.1)	8.9 (5.5)	90.6 (5.5)
	0.3	0.5 (0.1)	4.8 (4.7)	94.7 (5.2)
	0.5	0.5 (0.1)	4.9 (4.9)	94.6 (5.3)
	1.0	0.5 (0.1)	9.1 (5.8)	90.4 (5.7)
CHM-85-2.3	0.05	0.7 (0.1)	93.4 (0.6)	5.9 (0.6)
	0.1	0.7 (0.1)	82.3 (1.9)	17.0 (1.9)
	0.2	0.7 (0.1)	62.5 (1.2)	36.8 (1.2)
	0.3	0.7 (0.1)	53.1 (6.5)	46.2 (6.5)
	0.5	0.7 (0.1)	39.6 (4.0)	59.7 (4.0)
	1.0	0.7 (0.1)	18.5 (3.1)	80.8 (3.1)
$2C_{12}P$-55-2.3	0.05	0.4 (0.1)	78.1 (3.7)	21.5 (3.7)
	0.1	0.4 (0.1)	68.0 (2.4)	31.6 (2.4)
	0.2	0.4 (0.1)	58.4 (2.3)	41.2 (2.3)
	0.3	0.5 (0.1)	35.8 (7.0)	63.8 (7.0)
	0.5	0.4 (0.1)	33.9 (7.9)	65.7 (7.9)
	1.0	0.4 (0.1)	31.8 (6.4)	67.8 (6.4)
$C_{16}P$-55-2.4	0.05	0.7 (0.3)	88.9 (3.0)	10.4 (2.7)
	0.1	0.8 (0.2)	84.0 (2.3)	15.2 (2.2)
	0.2	0.5 (0.1)	51.9 (6.7)	47.6 (6.7)
	0.3	0.5 (0.1)	44.7 (4.0)	54.8 (4.0)
	0.5	0.5 (0.1)	41.4 (1.7)	58.1 (1.7)
	1.0	0.5 (0.1)	41.1 (1.6)	58.4 (1.6)

All experiments were performed in triplicate. The values in parentheses are standard deviations. R indicates the weight ratio of hydrophobized polysaccharides to lipids (mg/mg).

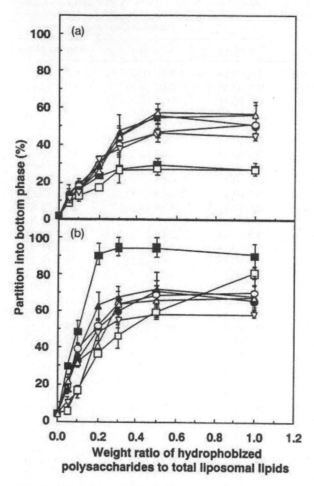

FIG. 2 Partition of the polysaccharide-coated liposomes in (a) the 6.0% (w/w) PEO-20/8.0% (w/w) pullulan-50 system and (b) the 6.0% (w/w) PEO-20/8.0% (w/w) dextran-40 system. The liposomes were coated with (●) CHP-55-1.7, (○) CHP-55-2.5, (▲) CHP-108-1.3, (■) CHD-70-1.7, (□) CHM-85-2.3, (▽) $C_{16}P$-55-2.4, and (△) $2C_{12}P$-55-2.3.

phase. This is a more important factor than the hydrophobicity of the polysaccharide. Even when the liposomes were coated with pullulans of different molecular weights and substitution degrees of the cholesterol moiety, no significant differences were observed in the partition behavior. This was also observed in the different two-phase systems using pullulan-50 or dextran-40. Only the density of the polysaccharide on the liposomal surface seemed to be important. However, a more significant difference was observed when the liposomes were coated with structurally different polysaccharides.

C. Quantitative Analysis of Partition of Liposomes in Aqueous Two-Phase Systems

The partition of the HP-coated liposomes from the interface into the bottom polysaccharide-rich phase can be described by specific binding between the two carbohydrates [Eq. (1)].

$$PS_{1(liposome)} + PS_{2(bulk)} \overset{K}{\rightleftharpoons} PS_{1(liposome)}PS_{2(bulk)} \tag{1}$$

where PS_1 is the HP on the liposomal surface and PS_2 is the polysaccharide in the bottom phase of the two polymer phases. The binding constant (K), or the partition coefficient, is calculated based on a Langmuir-type adsorption equation:

$$K = \frac{[PS_{1(liposome)}PS_{2(bulk)}]_b}{[PS_{1(liposome)}]_f [PS_{2(bulk)}]_f} \tag{2}$$

where $[PS_{1(liposome)}PS_{2(bulk)}]_b$ is the amount of complex produced between the liposomal polysaccharide and the polysaccharide in the bulk bottom phase, $[PS_{1(liposome)}]_f$ is the amount of liposomal polysaccharide remaining at the interface, and $[PS_{2(bulk)}]_f$ is the amount of free polysaccharide in the bulk bottom phase. The following assumptions are made:

$$[PS_{1(liposome)}]_f = [PS_{1(liposome)}]_t - [PS_{1(liposome)}]_b$$
$$[PS_{1(liposome)} \cdot P_{S2(bulk)}]_b = [PS_{1(liposome)}]_b$$
$$[PS_{2(bulk)}]_f = [PS_{2(bulk)}]_t - [PS_{2(bulk)}]_b$$

where subscripts b, f, and t represent bound, free, and total, respectively. However, because $[PS_{2(bulk)}]_t$ is much larger than $[PS_{2(bulk)}]_b$, $[PS_{2(bulk)}]_f$ is almost equal to $[P_{S2(bulk)}]_t$. Therefore, Eq. (2) can be converted to Eq. (3):

$$[PS_{1(liposome)}]_b = \frac{[PS_{1(liposome)}]_t K [PS_{2(bulk)}]_t}{(1 + K[PS_{2(blk)}]_t)} \tag{3}$$

where $[PS_{1(liposome)}]_b$ is the amount of liposomal polysaccharide partitioned to the bottom phase, $[PS_{1(liposome)}]_t$ is the initial amount of liposomal polysaccharide added, $[PS_{2(bulk)}]_t$ is the total amount of polysaccharide in the bottom phase, and $[PS_{2(bulk)}]_b$ is the amount of polysaccharide interacting with the liposomal polysaccharide in the bottom phase. A plot of $[PS_{1(liposome)}]_b$ against $[PS_{1(liposome)}]_t$ yields a straight line (Fig. 3). The binding constant, K, is obtained from the slope of this straight line (Tables 3 and 4).

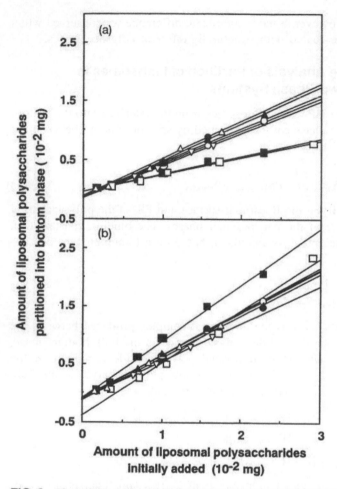

FIG. 3 Plot of the amount of liposomal polysaccharide partitioned into the bottom phase versus the total amount of liposomal polysaccharide initially added in (a) the 6.0% (w/w) PEO-20/8.0% (w/w) pullulan-50 system and (b) the 6.0% (w/w) PEO-20/8.0% (w/w) dextran-40 system. The liposomes were coated with (●) CHP-55-1.7, (○) CHP-55-2.5, (▲) CHP-108-1.3, (■) CHD-70-1.7, (□) CHM-85-2.3, (▽) $C_{16}P$-55-2.4, and (△) $2C_{12}P$-55-2.3.

The affinity between the liposomal polysaccharide (PS_1) and the polysaccharide in the bulk bottom phase (PS_2) affects the partition efficiency. The sequence of the strength of interaction between the two carbohydrates was as follows: for the PEO/dextran two-phase system, $dextran_{(liposome)}$-$dextran_{bulk}$ > $mannan_{(liposome)}$-$dextran_{bulk}$ > $pullulan_{(liposome)}$-$dextran_{(bulk)}$. For the

TABLE 3 Binding Constant, K, for the Interaction Between the Liposomal Polysaccharide and the Bulk Polysaccharide (Pullulan-50) in the Aqueous 6.0% (w/w) PEO-20/8.0% (w/w) Pullulan-50 System

Liposomal polysaccharide	K (mg^{-1})
CHP-55-1.7	$1.7\ (\pm 0.3) \times 10^{-2}$
CHP-55-2.5	$1.8\ (\pm 0.2) \times 10^{-2}$
CHP-108-1.3	$2.2\ (\pm 0.3) \times 10^{-2}$
CHD-70-1.7	$5.5\ (\pm 0.3) \times 10^{-3}$
CHM-85-2.3	$5.3\ (\pm 0.3) \times 10^{-3}$
$2C_{12}P$-55-2.3	$2.3\ (\pm 0.4) \times 10^{-2}$
$C_{16}P$-55-2.4	$1.7\ (\pm 0.5) \times 10^{-2}$

TABLE 4 Binding Constant, K, for the Interaction Between the Liposomal Polysaccharide and the Bulk Polysaccharide (Dextran-40) in the Aqueous 6.0% (w/w) PEO-20/8.0% (w/w) Dextran-40 System

Liposomal polysaccharide	K (mg^{-1})
CHP-55-1.7	$3.8\ (\pm 0.6) \times 10^{-2}$
CHP-55-2.5	$4.5\ (+0.4) \times 10^{-2}$
CHP-108-1.3	$4.1\ (\pm 0.5) \times 10^{-2}$
CHD-70-1.7	$5.6\ (\pm 2.0) \times 10^{-1}$
CHM-85-2.3	$1.3\ (\pm 0.6) \times 10^{-1}$
$2C_{12}P$-55-2.3	$4.0\ (\pm 0.5) \times 10^{-2}$
$C_{16}P$-55-2.4	$2.7\ (\pm 0.4) \times 10^{-2}$

PEO/pullulan system, the sequence of the strength of interaction was pullulan$_{(\text{liposome})}$-pullulan$_{(\text{bulk})}$ > dextran$_{(\text{liposome})}$-pullulan$_{(\text{bulk})}$ $\approx$ mannan$_{(\text{liposome})}$-pullulan$_{(\text{bulk})}$. The interaction between branched polysaccharides such as dextran or mannan seems to be stronger than that between more linear polysaccharides such as pullulan. At present, it is impossible to compare the two polysaccharide systems directly. For this purpose, the two systems must be normalized using a parameter such as the TLL or the interfacial tension.

III. PARTITION OF GANGLIOSIDE-RECONSTITUTED LIPOSOMES IN AQUEOUS TWO-PHASE SYSTEMS

A. Materials and Methods

Gangliosides (GM$_3$, GD$_{1b}$, GD$_{1a}$, and GT$_{1b}$) (purity, 95%; Sigma, St. Louis, MO) were used without further purification. The chemical structures of gangliosides used in this work are given in Fig. 4.

Gangliosides were reconstituted in the liposomal membrane according to a method previously established [16,17]. Both the diameter and the size distribution of the liposomes were determined by the dynamic light scatter-

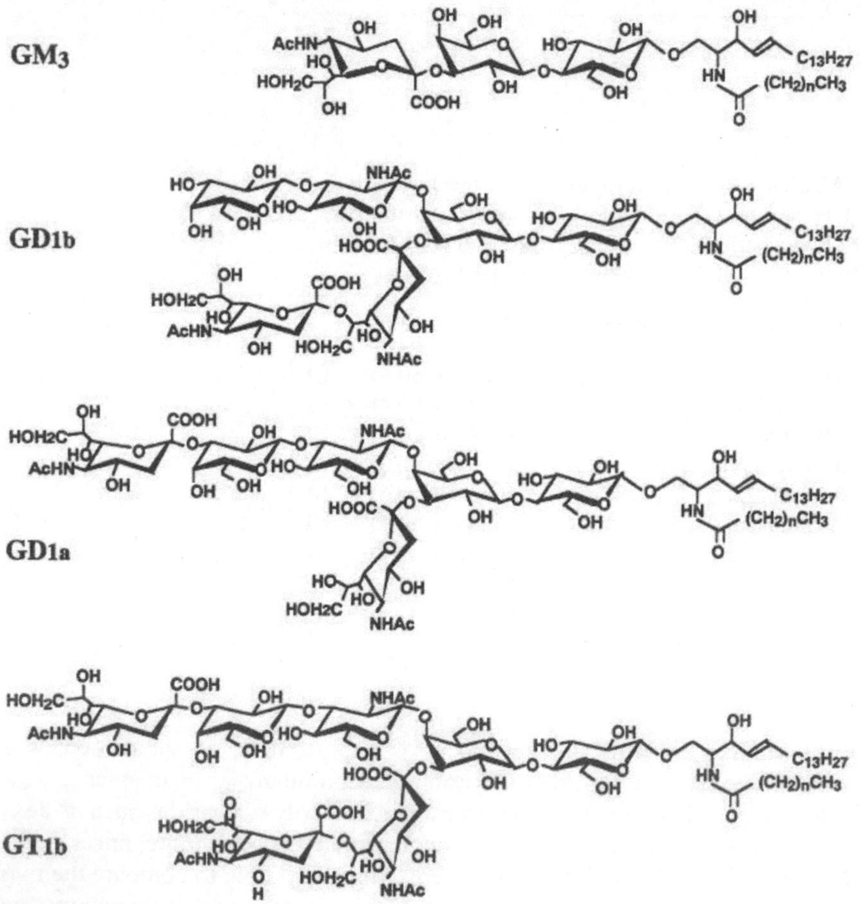

FIG. 4 Chemical structures of gangliosides.

ing (DLS) method on a DLS-700 (Photal Otsuka Electronics, Hirakata, Japan) [17]. The mean diameter of ganglioside-reconstituted liposomes so obtained was approximately 125 ± 5 nm, and the size distribution was rather monodisperse. The concentration of the liposomal phospholipid was determined using a Phospholipid Test Kit (Wako Pure Chemicals Ltd.). The final concentration of the liposomal lipids was adjusted to 1.0×10^{-3} M.

In this system, the actual amount of ganglioside on the outermost surface of the liposome is a very important factor. Therefore, the surface density of the ganglioside on the liposomal surface was determined precisely in advance.

For this purpose, the liposomal lipid concentration was adjusted to 3.0×10^{-4} M, and the liposomal suspension was filtered through a Millipore filter (pore size, 0.45 μm) prior to the DLS measurements to remove any dust.

The system of 4.0% (w/w) PEO-20 and 8.0% (w/w) dextran-40 was prepared in 110 mM sodium phosphate (pH 7.0), in 60 mM sodium phosphate (pH 7.0), in 20 mM sodium phosphate (pH 7.0), and in 10 mM sodium phosphate containing 150 mM sodium chloride (pH 7.0). To a mixture of 0.2 g of 20.0% (w/w) PEO-20 and 0.4 g of 20.0% (w/w) dextran-40 was added a ganglioside-reconstituted liposome suspension (50.0 mg) in a vial, and then a given buffer solution (0.35 g) was added to give a total 1.0-g sample. The sample was mixed well by inversion of the vial 30 times and then centrifuged for 10 min at $2000 \times g$ (Capsule HF-120, Tomy Seiko Co.) at room temperature. After reaching equilibrium, a 0.2-mL sample solution was carefully taken out using a long needle-syringe from both the top PEO-rich and the bottom dextran-rich phases and diluted with 0.8 mL of the same buffer. The fluorescence intensity was measured at 430 nm (excited at 360 nm) for both the top and bottom phases on a fluorescence spectrophotometer (F-3000, Hitachi, Tokyo, Japan).

Prior to the partition studies, the hydrodynamic diameter of various ganglioside-reconstituted liposomes was determined by DLS because the size of liposomes affects the partition in the aqueous two-phase system. In general, the larger particles gather more at the interface than the smaller ones [3]. However, the diameter of the liposomes did not change even when the ganglioside/lipid ratio was changed. The difference in the diameter of the ganglioside-reconstituted liposomes was hardly discernible.

B. Effects of Electrolytes and Ganglioside Density on Liposomal Surface

The partition of liposomes was investigated using different compositions of gangliosides in the system of 4.0% (w/w) PEO-20/8.0% (w/w) dextran-40 as a function of sodium phosphate concentration. The PEO/dextran two-phase system containing sodium phosphate has a so-called positive potential; that is, the top PEO phase is more positively charged than the bottom poly-saccharide phase [2,3,5,18–21]. Figure 5 shows the partition of ganglio-side-reconstituted liposomes in the aqueous 4.0% (w/w) PEO-20/80% (w/w) dextran-40 system (pH 7.0).

With the conventional liposomes without ganglioside, approximately 80% of the liposomes located at the interface between the two phases. This was irrespective of the sodium phosphate concentration. A liposome of neutral surface potential generally locates at the interface of the two phases [3,5,22]. The partition of ganglioside-reconstituted liposomes, which are negatively charged, was drastically changed by changing the sodium phosphate con-centration of the system. In the system of 110 mM sodium phosphate, the partition of GM_3-, GD_{1a}-, and GT_{1b}-reconstituted liposomes to the dextran-rich bottom phase decreased while the interfacial adsorption increased. This increase was related to an increase in the surface density of ganglioside on the liposome. However, their partition to the PEO-rich top phase was not affected much by the ganglioside density. For the system containing 20 mM sodium phosphate, the partition of ganglioside-reconstituted liposomes to the dextran-rich bottom phase significantly increased with an increase in the surface density of ganglioside on the liposomes. On increasing the gangli-oside density, their interfacial adsorption decreased considerably. However, the partition to the PEO-rich top phase changed slightly (Fig. 5, column I). In 60 mM sodium phosphate (Fig. 5, column II), the partition of the lipo-somes showed behavior in between the two cases of 20 mM and 110 mM sodium phosphate.

Interestingly, the partition of the conventional liposomes without any gan-glioside was not much affected by the sodium phosphate concentration. However, the partition of the ganglioside-reconstituted liposomes was largely affected by the buffer concentration. In addition, the more negatively charged liposomes, the GT_{1b}-reconstituted liposomes, were partitioned more to the bottom dextran-rich phase. Another interesting finding is that this effect of the buffer concentration was not observed at all in the partition to the top PEO-rich phase. At the low buffer concentration, the ganglioside bearing a large number of anionic moieties partitioned more to the bottom phase, not to the top PEO-rich phase. When the sodium phosphate concen-tration was increased to 110 mM, even ganglioside-reconstituted and nega-tively charged liposomes mostly located at the interface (Fig. 6).

The system containing 150 mM sodium chloride, which is almost com-parable to the physiological condition, does not lead to an electrostatic po-tential difference [2,3,5,18–21]. As previously reported [19–21], an increase in sodium chloride concentration up to 150 mM in the PEO/dextran system with 10 mM sodium phosphate decreased the electrostatic potential differ-

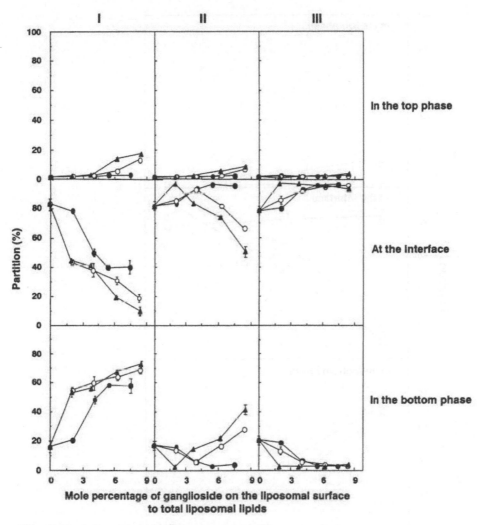

FIG. 5 Partition of ganglioside-reconstituted liposomes in aqueous 4.0% (w/w) PEO-20/8.0% (w/w) dextran-40 system at pH 7.0: I, in 20 mM sodium phosphate; II, in 60 mM sodium phosphate; and III, in 110 mM sodium phosphate. The liposomes were reconstituted with GM_3 (●), GD_{1a} (○), and GT_{1b} (▲).

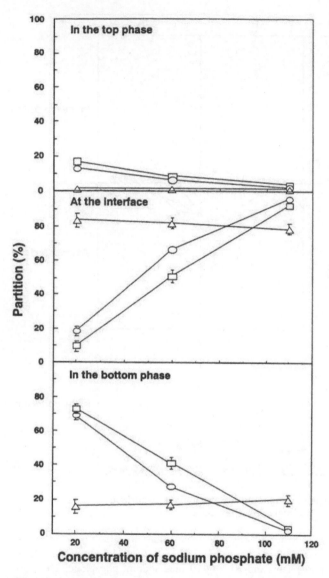

FIG. 6 Partition of ganglioside-reconstituted liposomes as a function of the sodium phosphate concentration in the top phase, at the interface, and in the bottom phase: (○) GD_{1a}-reconstituted liposome, (□) GT_{1b}-reconstituted liposome, and (△) conventional liposome. The amount of ganglioside initially added to the total lipid was 20 mol%.

ence virtually to zero. Figure 7 shows the effect of 150 mM sodium chloride in 10 mM sodium phosphate on the partition of GT_{1b}-reconstituted liposomes. Figure 8 shows the effect of the ganglioside density and the differences in the structures of gangliosides on the partition of the liposomes. The presence of sodium chloride significantly affected the partition of the liposomes. Especially at the higher density of ganglioside, more liposomes were partitioned into the bottom dextran-rich phase. In addition, the extent of the partition was strongly related to the structure and the conformation of the ganglioside (Fig. 8). Another interesting finding is that the partition of these liposomes to the top PEO-rich phase was almost negligible and was not affected at all by the density and the structure of the ganglioside. The conventional liposome locates mostly at the interface between the two phases. This is certainly consistent with previous findings [3,5,18].

C. Partition of Liposomes to Top PEO Phase

An unequal distribution of the cationic and anionic species of the added salt between the top and bottom phases causes a difference in the interfacial potential between the two polymer phases. This largely affects the partition behavior of charged substances present in the system. Tilcock et al. [3] reported that the negatively charged liposomes were partitioned more to the PEO-rich top phase when 110 mM sodium phosphate was present. Zaslavsky et al. [20] studied the electrostatic potential difference between the two polymer phases of an aqueous PEO/dextran system when sodium phosphate was employed and found that the PEO-rich top phase was more positively charged than the dextran-rich bottom phase. An increase in the concentration of sodium phosphate reduces the electrostatic potential difference between the two phases. Ballard et al. [21] also reported that the potential difference became optimal in the PEO/dextran system containing 22 mM sodium phosphate. The addition of more electrolytes presumably provides more mobile ions, which may diminish this potential difference. Figure 6 shows the partition of ganglioside-reconstituted liposomes as a function of the sodium phosphate concentration in the PEO/dextran system. Increase in the sodium phosphate concentration slightly decreased the partition of the liposomes to the top PEO-rich phase.

D. Partition of Liposomes to Bottom Polysaccharide Phase

Contrary to what would be expected, the partition of ganglioside-reconstituted liposomes to the dextran-rich bottom phase largely increased in spite of an increase in the electrostatic potential difference. Bamberger et al. [22] investigated the partition of sodium phosphate in an aqueous PEO/dextran

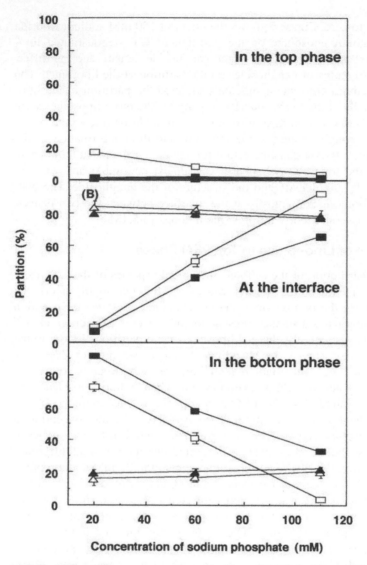

FIG. 7 Effect of the concentration of sodium chloride in the partition of liposomes in an aqueous 4.0% (w/w) PEO-20/8.0% (w/w) dextran-40 system at pH 7.0. The liposomal surface was reconstituted with (■, □) or without (▲, △) GT_{1b}. Open symbols indicate the partition in the sodium phosphate without sodium chloride; closed symbols indicate those with 150 mM sodium chloride.

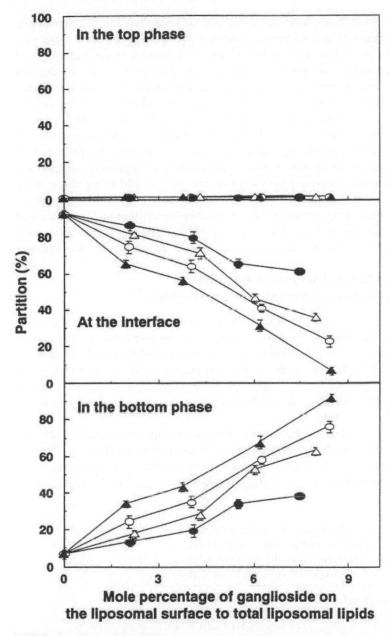

FIG. 8 Partition of ganglioside-reconstituted liposomes in an aqueous 4.0% (w/w) PEO-20/8.0% (w/w) dextran-40 system with 10 mM sodium phosphate containing 150 mM sodium chloride (pH 7.0). The liposomes were reconstituted with GM_3 (●), GD_{1b} (△), GD_{1a} (○), or GT_{1b} (▲).

two-phase system and found that more anionic phosphate ion was partitioned to the dextran-rich bottom phase. Johansson [23] also reported a large partition of the phosphate ions to the dextran-rich bottom phase and proposed specific binding of the phosphate ions to the hydroxyl groups of dextran through hydrogen bonding. Watanabe et al. [24] studied the binding of saccharide molecules at the surface of organized phosphate-containing amphiphiles. They confirmed that the hydroxy group of the saccharide bonded to the phosphate group of the lipid at the air–water interface via specific hydrogen bonding. Considering these previous findings, the phosphate ion of the buffer specifically binds to both saccharides of the liposomal surface and the bottom phase–forming polysaccharides. This would bring about more partition of the phosphate anion to the bottom phase. In addition, this causes a weaker interaction between the ganglioside-reconstituted liposomes and the phase-forming polysaccharide of the bottom phase. The phosphate ion would interfere with the saccharide–saccharide interaction in this system.

E. Quantitative Analysis of Partition of Ganglioside-Reconstituted Liposomes

The partition of ganglioside-reconstituted liposomes was also quantitatively analyzed by the binding isotherm according to the method described earlier. In this system PS_1 is ganglioside on the liposomal surface and PS_2 is polysaccharide in the bottom phase of the two polymer phases [refer to Eqs. (1) and (2)]. The plot of $[PS_{1(liposome)}]_b$ against $[PS_{1(liposome)}]_t$ yields a straight line, and the binding constant, K, is obtained from the slope of this straight line [refer to Eq. (3)] (Table 5).

The affinity between the ganglioside on the liposomal surface and the dextran in the bulk bottom phase controls the partition efficiency. The sequence of the strength of the interaction between the two carbohydrates was the following: $GT_{1b(liposome)} > GD_{1a(liposome)} > GD_{1b(liposome)} > GM_{3(liposome)}$. Both

TABLE 5 Binding Constant, K, for the Interaction Between the Liposomal Ganglioside and the Bulk Polysaccharide (Dextran-40) in the Aqueous 4.0% (w/w) PEO-20/8.0% (w/w) Dextran-40 System at 24°C

Liposomal ganglioside	K (mg^{-1})
GM_3	8.1 (± 1.6) $\times 10^{-3}$
GD_{1b}	1.4 (± 0.3) $\times 10^{-2}$
GD_{1a}	2.3 (± 0.4) $\times 10^{-2}$
GT_{1b}	4.9 (± 1.3) $\times 10^{-2}$

GD_{1a} and GD_{1b} are digangliosides with structures in which only the positions of sialic acid moieties are different. The chemical structure of ganglioside on the liposomal surface seems to be dominant in determining the partition of the liposomes. The chemical structure of the saccharide moiety of ganglioside should be important in the specific carbohydrate–carbohydrate interaction in water (Fig. 4).

Another interesting finding in this work is the more significant partition of GT_{1b}- and GD_{1a}-reconstituted liposomes to the bottom polysaccharide-rich phase compared with other liposomes. From the conformations of the glycoparts of gangliosides on the liposomal surface, it is clear that the glycoparts of GT_{1b} and GD_{1a} are more extended to the bulk aqueous phase (Fig. 9). This may also affect the efficiency of the specific interaction with the phase-forming polysaccharides of the bottom phase.

Gangliosides are prominent cell surface constituents of tumors, such as melanoma. The monosialogangliosides (GM_2 and GM_3) and the disialogangliosides (GD_2 and GD_3) are of particular interest because of their potential as targets for passive immunization with monoclonal antibodies and for ac-

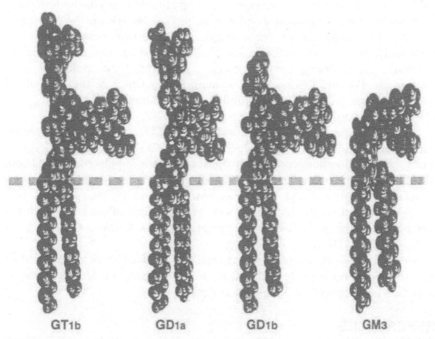

FIG. 9 Schematic drawings of conformations of the glycoparts of gangliosides. The three-dimensional structure was drawn computationally.

tive immunization related to cancer vaccines [25,26]. Sunamoto and Shiku [27,28] found that the growth of B16 melanoma in vivo was 100% suppressed when C57BL/6 mice were immunized by a GT- or GQ-containing egg-phosphatidylcholine liposome. However, GM- or GD-containing liposomes showed no significant immunogenicities. The results obtained in this work are not inconsistent with these previous findings in in vitro studies [27–29].

IV. SUMMARY

In this work, we revealed the existence of the specific carbohydrate–carbohydrate interaction in water using an aqueous two-phase system consisting of poly(ethylene oxide) and polysaccharide.

First, the partition of hydrophobized polysaccharide (HP)–coated liposomes into the poly(ethylene oxide) (PEO)/polysaccharide system was studied. The polysaccharides employed were pullulan, dextran, and mannan, and the HPs used to coat the liposomes were cholesteryl pullulan, cholesteryl dextran, and cholesteryl mannan. Conventional liposomes without any HP coating mostly located at the interface between the two polymer phases, but the HP-coated liposomes were significantly partitioned into the bottom polysaccharide phase depending on the structure of the HP on the liposomal surface. The sequence of the strength of interaction between the two carbohydrates was as follows: for the PEO/dextran two-phase system, $\text{dextran}_{(liposome)}$-$\text{dextran}_{(bulk)}$ > $\text{mannan}_{(liposome)}$-$\text{dextran}_{(bulk)}$ > $\text{pullulan}_{(liposome)}$-$\text{dextran}_{(bulk)}$. For the PEO/pullulan system, the sequence of the strength of interaction was $\text{pullulan}_{(liposome)}$-$\text{pullulan}_{(bulk)}$ > $\text{dextran}_{(liposome)}$-$\text{pullulan}_{(bulk)}$ $\approx$ $\text{mannan}_{(liposome)}$-$\text{pullulan}_{(bulk)}$.

Second, the partition of ganglioside (GM_3, GD_{1a} GD_{1b}, or GT_{1b})-reconstituted liposomes was investigated using the PEO/dextran two-phase system. The ganglioside-reconstituted liposomes were largely partitioned into the dextran-rich bottom phase. The specific carbohydrate–carbohydrate interaction was also found in this system even though the partition was strongly affected by the buffer and salt. The sequence of the strength of the interaction between the two carbohydrates was as follows: $GT_{1b(liposome)}$ > $GD_{1a(liposome)}$ > $GD_{1b(liposome)}$ > $GM_{3(liposome)}$.

So even the weak interactions between polysaccharides in water could be quantitatively and directly detected by using the partition of HP-coated liposomes in the aqueous two-phase system.

REFERENCES

1. PÅ Albertsson. Partition of Cell Particles and Macromolecules. New York: Wiley, 1986.

2. H Walter, DE Brooks, D Fisher. Partitioning in Aqueous Two-Phase Systems. Theory, Methods, Uses and Applications to Biotechnology. New York: Academic Press, 1985.
3. C Tilcock, P Cullis, T Dempsey, BN Youens, D Fisher. Biochim Biophys Acta 979:208–214, 1989.
4. C Tilcock, P Cullis, T Dempsey, D Fisher. In: D Fisher, IA Sutherland, eds. Applications in Cell Biology and Biotechnology. New York: Plenum, 1989, pp 179–189.
5. E Eriksson, PÅ Albertsson. Biochim Biophys Acta 507:425–432, 1978.
6. C Tilcock, R Chin, J Veiro, P Cullis, D Fisher. Biochim Biophys Acta 986: 167–171, 1990.
7. PT Sharpe, GS Warren. Biochim Biophys Acta 772:176–181, 1984.
8. J Senior, C Delgado, D Fisher, C Tilcock, G Gregoriadis. Biochim Biophys Acta 1062:77–82, 1991.
9. K Akiyoshi, S Deguchi, N Moriguchi, S Yamaguchi, J Sunamoto. Macromolecules 26:3026–3068, 1993.
10. LA Chen, RE Dale, S Roth, L Brand. J Biol Chem 252:2163–2169, 1977.
11. S Kawato, K Kinosita, A Ikegami. Biochemistry 17:5026–5031, 1978.
12. JR Lakowicz, FG Prendergast, D Hogen. Biochemistry 18:508–519, 1979.
13. M Takada, T Yuzuriha, K Katayama, K Iwamoto, J Sunamoto. Biochim Biophys Acta 802:237–244, 1984.
14. J Sunamoto, T Sato, M Hirota, K. Fukushima, K Hiratani, K Hara. Biochim Biophys Acta 898:323–330, 1987.
15. J Ryden, PÅ Albertsson. J Colloid Interface Sci 37:219–222, 1971.
16. E Kato, K Akiyoshi, T Furuno, M Nakanishi, A Kikuchi, K Kataoka, J Sunamoto. Biochem Biophys Res Commun 203:1750–1755, 1994.
17. EC Kang, K Akiyoshi, J Sunamoto. J Bioactive Compatible Polym 12:14–26, 1997.
18. PT Sharpe, GS Warren. Biochim Biophys Acta 772:176–182, 1984.
19. R Reitherman, SD Flanagan, SH Barondes. Biochim Biophys Acta 297:193–202, 1973.
20. BY Zaslavsky, LM Miheeva, NM Mestechkina, SV Rogozhin. J Chromatogr 253:149–158, 1982.
21. CM Ballard, JP Dickinson, JJ Smith. Biochim Biophys Acta 582:89–101, 1979.
22. S Bamberger, GVF Seaman, KA Sharp, DE Brooks. J Colloid Interface Sci 99:187–193, 1984.
23. G Johansson. Biochim Biophys Acta 221:387–390, 1970.
24. E Watanabe, N Kimizuka, T Kunitake. Polym Prepr Jpn 45:2480–2481, 1996.
25. T Tai, JC Paulson, LD Cahan, RF Irie. Proc Natl Acad Sci USA 80:5392–5396, 1983.
26. PO Livingston, EJ Natoli, MG Calves, E Stockert, HF Oettgen, LJ Old. Proc Natl Acad Sci USA 84:2911–2915, 1987.
27. J Sunamoto, H Shiku. Proceedings of the 3rd Japanese–French Biomedical Technologies Symposium held in Himeji, Japan, 1989, pp 82–85.
28. J Sunamoto, H Shiku. Ann NY Acad Sci 613:116–127, 1990.
29. E Kato, A Taguchi, S Sakashita, K Akiyoshi, J Sunamoto. Proc Jpn Acad 76: 63–67, 2000.

28

Novel Cationic Transfection Lipids for Use in Liposomal Gene Delivery

RAJKUMAR BANERJEE,* PRASANTA KUMAR DAS,† and GOLLAPUDI VENKATA SRILAKSHMI   Indian Institute of Chemical Technology, Hyderabad, India

NALAM MADHUSUDHANA RAO   Centre for Cellular and Molecular Biology, Hyderabad, India

ARABINDA CHAUDHURI   Indian Institute of Chemical Technology, Hyderabad, India

ABSTRACT

A novel series of nontoxic and non–glycerol-based simple monocationic transfection lipids containing one or two hydroxyethyl groups directly linked to the positively charged nitrogen atom were synthesized. The in vitro transfection efficiencies of these new liposomal gene delivery reagents were better than that of lipofectamine, a transfection agent widely used in cationic lipid-mediated gene transfer. The most efficient transfection formulation was observed to be a 1:1:0.3 mole ratio of DHDEAB (N,N-di-n-hexadecyl-N,N-dihydroxyethylammonium bromide)/cholesterol/HDEAB (N-n-hexadecyl-N, N-dihydroxyethylammonium bromide) using a DHDEAB-to-DNA charge ratio (+/−) of 0.3:1. Observation of good transfection at charge ratios lower than one suggests that the amphiphile/DNA complex may have a net negative charge. Our results reemphasize the important point that in cationic lipid–mediated gene delivery, the overall charge of the lipid–DNA complex

Current affiliation:

*University of Pittsburgh, Pittsburgh, Pennsylvania, U.S.A.

†Massachusetts Institute of Technology, Cambridge, Massachusetts, U.S.A.

Reprinted, in part, with permission from J Med Chem 42:4292–4299, 1999. Copyright 1999 American Chemical Society.

need not always be positive. In addition, our transfection results imply that favorable hydrogen-bonding interactions between the lipid head groups and the cell surface of biological membranes may have some role in improving the transfection efficiency in cationic lipid–mediated gene delivery.

I. INTRODUCTION

In gene therapy, patients carrying identified defective genes are supplemented with copies of the corresponding normal genes [1]. Many gene delivery reagents (also known as transfection vectors) including retrovirus [2], adenovirus [3], positively charged polymers and peptides [4–6], and cationic amphiphilic compounds [7,8] are currently being used as carriers of genes in combating hereditary diseases by gene therapy. Reproducibility, low cellular and immunological toxicities, and the ease of preparation and administration associated with cationic transfection lipids are increasingly making them the transfection vector of choice in gene therapy.

Since the first report [7] on cationic liposome–mediated gene delivery by Felgner et al. in 1987, an upsurge of global interest has been witnessed in synthesizing efficient cationic transfection lipids [9–28]. Many of the reported liposomal transfection vectors, e.g., DOTMA [7], DMDHP [17], DMRIE [24], and DOTAP [28], have a common element in their molecular structures, namely the presence of a glycerol backbone. Interestingly, among the glycerol-based cationic transfection lipids, the polar head group domains of the most efficient lipids, such as DMRIE [24] and DMDHP [17], contain one or two hydroxyethyl groups directly linked to the positively charged nitrogen atoms. The development of efficient non–glycerol-based liposomal transfection lipids has been reported, e.g., DC-Chol synthesized by Gao and Huang [27] and the long chain alkyl acyl carnitine esters designed by Szoka and colleagues [12]. These non–glycerol-based liposomal gene delivery reagents have no hydroxyethyl groups present in their polar head group regions. Except for the patent report by Nantz et al. [20] on the development of 1,4-diaminobutane–based dicationic transfection lipids, a detailed investigation of the transfection efficiencies of non–glycerol-based monocationic liposomal transfection vectors containing hydroxyethyl groups directly attached to the positively charged nitrogen atoms has not been reported. Toward this end, we have developed [29] a highly efficient novel series of non–glycerol-based and nontoxic simple monocationic transfection lipids containing a hydroxyethyl group(s) directly attached to the positively charged quaternized nitrogen atom (lipids 1–5, Chart 1). The present chapter reviews the design, synthesis, and transfection biology of these novel transfection lipids.

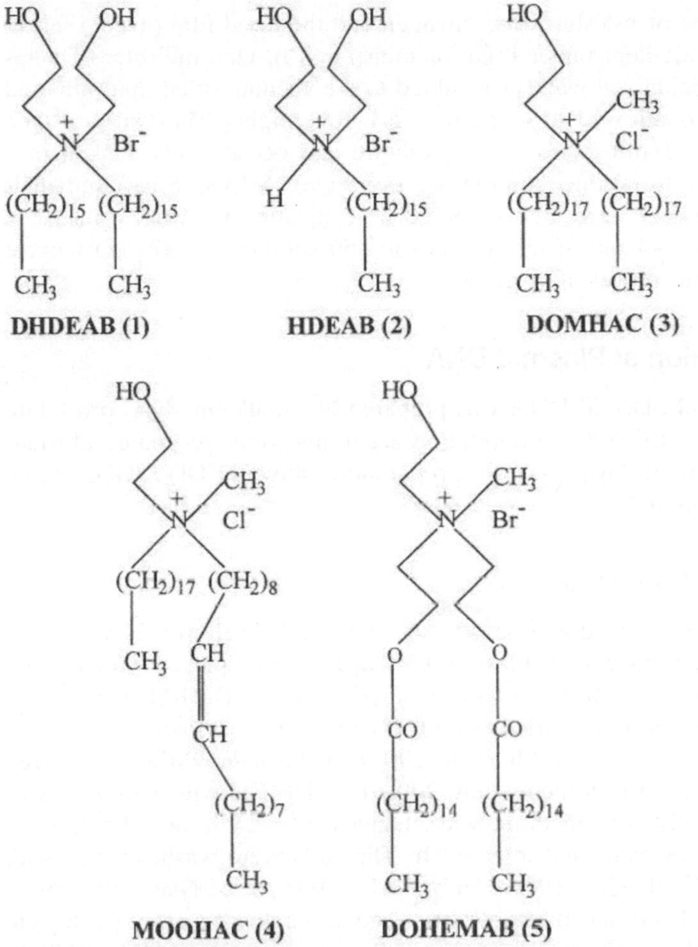

CHART 1 Structures of new cationic lipids 1–5. (Reprinted, in part, with permission from J Med Chem 42:4292–4299, 1999. Copyright 1999 American Chemical Society.)

II. METHODS

The details of the synthetic procedures for all the novel transfection lipids shown in Chart 1 have already been described [29].

A. Liposome Preparation

Mixtures of cationic amphiphiles and cholesterol in the appropriate ratio were dissolved in chloroform in a glass vial. The chloroform was removed with a thin flow of moisture-free nitrogen and the dried film of lipid left in the vial was then kept under high vacuum for 8 h. One milliliter of autoclaved sterile deionized water was added to the vacuum-dried lipid film and the mixture was allowed to swell for 15 h (overnight). The vial was then vortexed for 2–3 min at room temperature and occasionally shaken in a 45°C water bath to produce multilamellar vesicles (MLVs). Small unilamellar vesicles (SUVs) were then prepared by sonicating the MLVs placed in an ice bath for 3–4 min using a Branson 450 sonifier at 100% duty cycle and 25 W output power until clarity.

B. Preparation of Plasmid DNA

The pRSV-β-gal plasmid DNA was prepared by an alkaline lysis procedure and purified by PEG-8000 precipitation according to the procedure of Maniatis and coworkers [30]. Plasmid preparations showing OD_{260}/OD_{280} more than 1.8 were used.

C. Transfection Assay

COS-1 cells were seeded at a density of 50,000 cells/well in a 24-well plate 18 h before the transfection. Plasmid (0.3 μg) was complexed with varying amounts of lipid (0.1–0.9 nmoles) in 25 μL of plain DMEM medium for 30 min. The charge ratio varied from 0.1:1 to 9:1 (+/−) over this range of the lipid. The complex was diluted to 200 μL with plain DMEM and added to the wells. After 3 h of incubation, 200 μL of DMEM with 10% FCS was added to the cells. The medium was changed after 24 h, and the reporter gene activity was estimated after 48 h. The cells were washed twice with phosphate-buffered saline (PBS) and lysed in 100 μL of lysis buffer (0.25 M Tris-HCl, pH 8.0 and 0.5% NP40). Care was taken to ensure complete lysis. The β-galactosidase activity per well was estimated by adding 50 μL of 2× substrate solution (1.33 mg/mL of ONPG, 0.2 M sodium phosphate, pH 7.15 and 2 mM magnesium chloride) to 50 μL of lysate in a 96-well plate. Absorption at 405 nm was converted to β-galactosidase units by using the calibration curve obtained each day using pure commercial β-galactosidase enzyme. The values of β-galactosidase units in replicate plates assayed on the same day varied by less than 20%. The transfection efficiency values shown in Figs. 1–3 are the average values from two replicate transfection plates assayed on the same day. Each transfection experiment was repeated three times on three different days and the day-to-day variation in average transfection efficiency values for identically treated two-replicate transfection plates was approximately twofold and was dependent on the cell density and conditions of the cells.

D. Toxicity Assay

Cytotoxicity of amphiphiles was assessed using a 3-(4,5-dimethyl*t*hiazol-2-yl)-2,5-diphenyl*t*etrazolium bromide (MTT) reduction assay as described earlier [31]. The assay was performed in 96-well plates by maintaining the ratio of the number of cells to the amount of cationic amphiphile constant in cytotoxicity and transfection experiments. The MTT was added 3 h after adding the cationic amphiphile to the cells. The results were expressed as percent viability = $[OD_{540}$(treated cells) $-$ background]$/[OD_{540}$(untreated cells) $-$ background] $\times$ 100.

III. RESULTS AND DISCUSSION

A. Chemistry

The key structural elements common to all the transfection lipids 1–5 (Chart 1) described in the present investigation include (1) the presence of a hydrophobic group either directly linked to the positively charged nitrogen atom or linked to the positively charged nitrogen via an ester group, (2) the presence of at least one hydroxyethyl group directly linked to the positively charged nitrogen atom, and (3) absence of glycerol backbone in the molecular architecture of the monocationic amphiphiles. As delineated in Schemes 1–3, the chemistries involved in preparing these new lipids are straightforward. Scheme 1 outlines the one-step synthetic procedure for preparing DHDEAB and HDEAB. Diethanolamine was initially refluxed with *n*-hexadecyl bromide in the presence of potassium carbonate in methanol. The resulting intermediate tertiary amine (*N-n*-hexadecyldiethanolamine, not isolated) was then refluxed in a mixed solvent containing 80:15:5 (v/v) aceto-

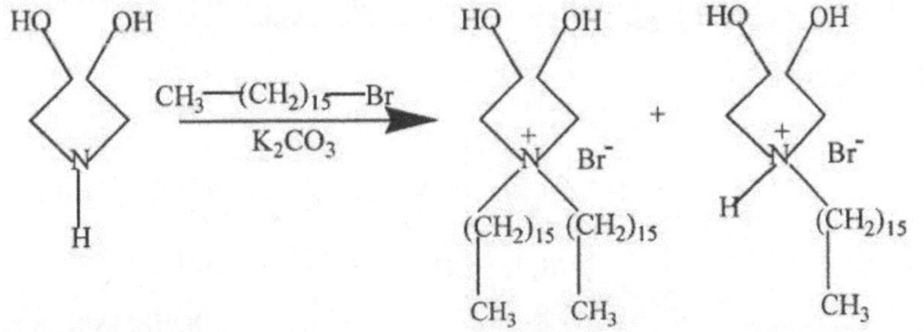

SCHEME 1 Synthesis of DHDEAB and HDEAB. (Reprinted, in part, with permission from J Med Chem 42:4292–4299, 1999. Copyright 1999 American Chemical Society.)

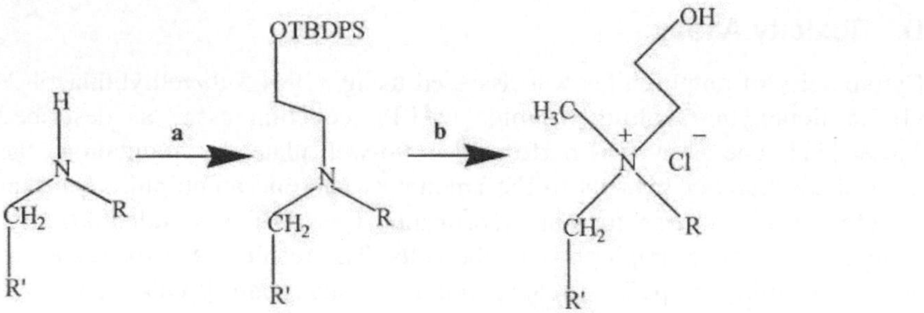

DOMHAC, $R = R'CH_2- = n-C_{18}H_{37}$
MOOHAC $R = n-C_{18}H_{37}$; $R'CH_2- = n-C_{18}H_{35}$

Reagents: (a) $Br-(CH_2)_2-OTBDPS$ (1equiv)/K_2CO_3 (1.1 equiv)/ethyl acetate

(b) TBAF (2.5 equiv)/THF, MeI (huge excess)/$CHCl_3$/MeOH,
Amberlyst A-26 chloride ion-exchange resin.

SCHEME 2   Synthesis of MOOHAC and DOMHAC. (Reprinted, in part, with permission from J Med Chem 42:4292–4299, 1999. Copyright 1999 American Chemical Society.)

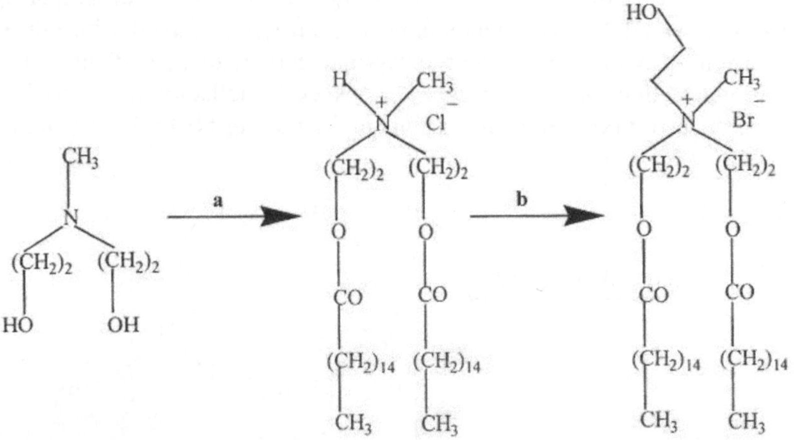

DOHEMAB,

Reagents: (a) n-hexadecanoyl chloride (2.2 equiv)/DMF

(b) 1.0 M aq NaOH/DCM (biphasic system); 2-bromoethanol (1.5 equiv)/85°C/4h

SCHEME 3   Synthesis of DOHEMAB. (Reprinted, in part, with permission from J Med Chem 42:4292–4299, 1999. Copyright 1999 American Chemical Society.)

nitrile, ethyl acetate, and methanol. Finally, column chromatographic purification of the product mixture afforded pure DHDEAB and HDEAB.

The steps used in synthesizing MOOHAC and DOMHAC (Scheme 2) include (1) coupling the appropriate aliphatic saturated or unsaturated aldehyde with the appropriate long-chain aliphatic amine followed by reduction of the resulting imine to obtain the corresponding secondary amine, (2) conversion of the secondary amine obtained in step (1) to N-hydroxyethyl-N,-N-dialkyl amine (tertiary amine) by reacting with the hydroxyl-protected 2-bromoethanol followed by removal of the hydroxyl protecting group, and (3) quaternizing the tertiary amine obtained in step (2) with excess methyl iodide followed by chloride ion-exchange chromatography on the resulting intermediate quaternary amphiphilic iodide. Synthesis of DOHEMAB (Scheme 3) essentially consists of (1) reacting n-hexadecanoyl chloride with N-methyldiethanolamine to obtain the hydrochloride salt of the di-O-acylated intermediate, (2) neutralizing the hydrochloride salt obtained in step 1 with alkali, and (3) quaternizing the resulting tertiary amine obtained in step (2) with 2-bromoethanol [32].

B. Transfection Biology

The transfection efficiencies of the cationic amphiphiles 1–5 (Chart 1) were tested in COS-1 cells using pCH 110 plasmid carrying a β-galactosidase reporter gene under the control of an RSV promoter. Initially, we tested the transfection efficiencies of all the novel transfection lipids using the widely used auxiliary lipid DOPE. All the amphiphiles with DOPE showed very poor transfection. Interestingly, the amphiphiles 1 and 3–5 (Chart 1) showed remarkable transfection efficiencies with varying amounts of cholesterol as helper lipid (Fig. 1). Amphiphile 2 did not show any transfection even with cholesterol as the helper lipid at any ratio, probably because of a single acyl chain, which might interfere with the proper formation of a bilayer. Amphiphiles 4 and 5 showed the highest transfection efficiency in the presence of 60 mol% cholesterol (with respect to the cationic lipid), whereas amphiphile 3 showed the highest efficiency in the presence of 40 mol% of cholesterol. The transfection efficiencies of amphiphiles 3–5 having a single hydroxyethyl group in the head group regions were poorer than that of amphiphile 1. Amphiphile 1 with two hydroxyethyl functionalities directly linked to the positively charged nitrogen atom in combination with an equimolar amount of cholesterol was clearly the most efficient transfection lipid among all the lipids tested (Fig. 1). Among amphiphiles 3–5 with single hydroxyethyl groups directly linked to the positively charged nitrogen atom, amphiphile 3 was the most efficient one (Fig. 1). The transfection efficiency of the most efficient amphiphile DHDEAB was observed to be two to three

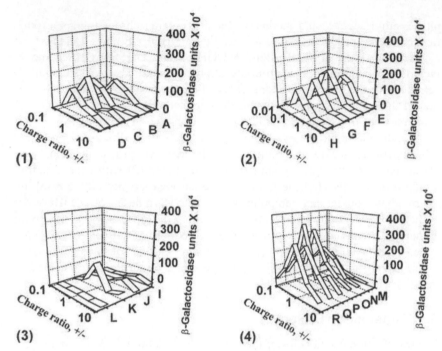

FIG. 1 Transfection efficiencies of (1) DOMHAC, (2) MOOHAC, (3) DOHEMAB, and (4) DHDEAB on COS-1 cells. The transfection efficiencies of the four lipids were tested by varying both the charge ratio (x-axis) and cholesterol (z-axis). The following mole ratios of cholesterol to cationic lipids were used in the z-axis: 0.2:1 (A, E, I, M); 0.4:1 (B, F, J, N); 0.6:1 (C, G, K, O); 1:1 (D, H, L, P); 1.2:1 (Q); 1.5:1 (R). In each well of a 24-well plate, a fixed amount of plasmid DNA (0.3 μg) was used to complex with 0.01 to 9 nmoles of cationic lipid to vary the charge ratio (+/−) from 0.01 to 9. (Reprinted, in part, with permission from J Med Chem 42: 4292–4299, 1999. Copyright 1999 American Chemical Society.)

times more in COS-1 cells than that of Lipofectamine, one of the most widely used commercially available transfection lipids (Fig. 2).

An interesting observation with these non–glycerol-based hydroxyethyl head group amphiphiles was that the optimal transfection efficiencies were in most cases observed with formulations containing lipid-to-DNA charge ratios (+/−) less than one (Fig. 1). Amphiphiles **1** and **3** were most efficient at lipid-to-DNA charge ratios of 0.3:1 and 0.1 to 1 respectively (Fig. 1). Formulation with lipid-to-DNA charge ratios less than one for optimal transfection have previously been reported for cationic lipids with hydroxyethyl groups directly attached to positively charged nitrogen atoms [20,24]. How-

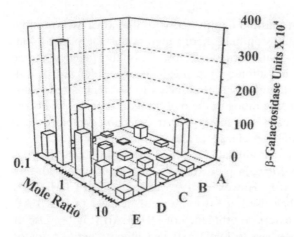

FIG. 2 Transfection efficiencies of DHDEAB and DDAB with cholesterol and DOPE as helper lipids on COS-1 cells. The charge ratios and amount of DNA used were as in Fig. 1. Lipofectamine (A) was used for comparison. DHDEAB (C, E) and DDAB (B, D) were used in combination with DOPE (B, C) and cholesterol (D, E) at a mole ratio of 1:1. The *x*-axis is given as a mole ratio ([cationic amphiphile]/ [DNA]) instead of charge ratio to compare the Lipofectamine with cationic amphiphiles on the same scale. The charge ratio and the mole ratio are the same for our cationic amphiphiles (because they carry one charge per molecule), whereas Lipofectamine carries five positive charges on one molecule. (Reprinted, in part, with permission from J Med Chem 42:4292–4299, 1999. Copyright 1999 American Chemical Society.)

ever, in cationic lipid–mediated gene delivery, it is generally believed that the overall positive charge of the cationic lipid–DNA complex plays a key role in their interaction with the negatively charged biological membranes. Thus, the remarkable efficiencies of the presently described transfection complexes prepared using lipid-to-DNA charge ratios significantly less than one (Figs. 1 and 2) convincingly indicate that the overall charge of the lipid–DNA complex for efficient gene delivery need not be positive. Such improved transfection efficiencies with lower lipid-to-DNA ratios have also been previously observed for the 1,4-diaminobutane–based dicationic amphiphile, *N,N,N',N'*-tetramethyl-*N,N'*-bis(hydroxyethyl)-2,3-di(oleoyloxy)-1,4-butanediammonium iodide [20]. The positive charge may be important for condensation of DNA and/or it may reduce the electrostatic repulsion between the negatively charged biological cell surface and the polyanionic naked DNA, thereby improving the uptake efficiency of the cationic lipid–DNA complex by the cells. Because the plasmid DNA has to interact with a variety of environments and membranes before it is expressed in the nucleus, higher expression of plasmid complexed with the novel hydroxyethyl group–containing amphiphiles outlined here suggests that the plasmid longevity is enhanced on its route to the nucleus. Thus, the function of the positive charge on lipids in cationic lipid–mediated gene delivery is still open to investigation. An important point deserves to be emphasized at this stage of discussion. The charge ratios in the present work refer to the charge ratios of the lipid to DNA used in preparing the transfection complexes, and these may or may not be the net charge of the resulting complexes.

Amphiphils **2** with only one aliphatic chain in the hydrophobic tail did not show any transfection either in pure form or in combination with any helper lipids (data not shown). However, at 0.3 charge ratio of DHDEAB to DNA, amphiphile **2** when used at 30 mol% with respect to DHDEAB modestly enhanced the transfection efficiency of DHDEAB (Fig. 3). Use of a higher mol% of **2** with respect to DHDEAB and higher charge ratios of DHDEAB to DNA (i.e., >0.3) did not improve the transfection further (Fig. 3). Given that the single-chain micelle-forming surfactants are known to destroy the bilayer structures of liposomes, the observed modest increase in the transfection efficiency of DHDEAB in the presence of 30 mol% of HDEAB is intriguing. Clearly, detailed investigations using a host of known

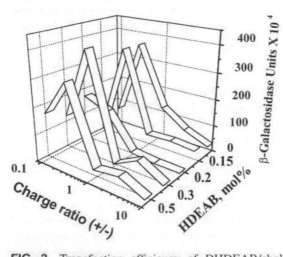

FIG. 3 Transfection efficiency of DHDEAB/cholesterol (1:1 mole ratio) with HDEAB (2) on COS-1 cells. HDEAB, which carries a single hydrophobic chain, was added at a mole ratio of zero (A); 0.15 (B); 0.20 (C); 0.30 (D); and 0.5 (E) with respect to DHDEAB. The charge ratio on the x-axis is based on the charge of DHDEAB only. (Reprinted, in part, with permission from J Med Chem 42:4292–4299, 1999. Copyright 1999 American Chemical Society.)

transfection lipids and varying the mole percent of HDEAB need to be carried out to ensure possible future use of HDEAB as a new helper lipid.

The cytotoxicities (Fig. 4) of the amphiphiles as 1:1 amphiphile–cholesterol preparations were tested in COS-1 cells by using reduction of MTT as described previously [31]. The cytotoxicity assays were performed under conditions identical to those in transfection experiments. In most cases, the cell viabilities were more than 80% up to 9 nmoles of lipid, which is the highest concentration of the lipid used in transfections. Low cytotoxicities of the amphiphiles and good transfection efficiencies indicate that the formulations may be used in a variety of cell lines.

Toward understanding any key role played by the hydroxyethyl head groups of the present transfection lipids, we compared the transfection efficiency of the most efficient amphiphile DHDEAB with that of DDAB having two methyl groups directly linked to nitrogen instead of two hydroxyethyl groups. Transfection results shown in Fig. 2 demonstrate that in the presence of an equimolar amount of cholesterol as the auxiliary lipid, DDAB was two to three times less efficient than DHDEAB. Interestingly, both DHDEAB and DDAB were also observed to show their optimal transfection efficiencies at the lipid-to-DNA charge ratio of 0.3:1 (Fig. 2). Such modestly improved transfection efficiency of DHDEAB compared with DDAB (Fig. 2) implies that the presence of hydroxyethyl functionalities in

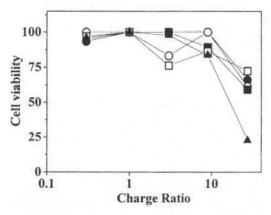

FIG. 4 Cell viabilities of cationic amphiphiles on COS-1 cells. The amphiphiles DOMHAC (filled circle), MOOHAC (open circles), DOHEMAB (filled square), and DHDEAB (open square) were tested in combination with cholesterol at a 1:1 mole ratio. The DHDEAB/cholesterol/HDEAB (1:1:0.5) mole ratio was also tested (filled triangle). (Reprinted, in part, with permission from J Med Chem 42:4292–4299, 1999. Copyright 1999 American Chemical Society.)

the head group regions of the monocationic non–glycerol-based amphiphiles may have some role in enhancing the transfection efficiencies. Similar observations were made earlier with glycerol-based amphiphiles such as DMRIE, DORIE, DORI, and DMDHP, with one or two hydroxyethyl groups linked to the nitrogen atom [17,24], and in the case of 1,4-diaminobutane–based dicationic transfection lipid [20]. Perhaps hydrogen-bonding interactions with the cell surface of biological membranes play some role in transfection lipids containing hydroxyethyl groups in their head group structures.

Previous reports have also indicated that the enhanced transfection efficiencies of glycerol-based cationic lipids containing polar hydroxyethyl head groups might originate from improved interactions of such functionalized cationic lipids or lipid–DNA complexes with cellular membranes via hydrogen bonding [17,24]. However, given the modestly (two- to threefold) enhanced transfection efficiency of DHDEAB compared with DDAB with no hydroxyalkyl functionalities in the head group regions (Fig. 2), the role of hydrogen-bonding interactions between the lipid head groups and the cell surface of biological membranes is not likely to be a key issue in cationic lipid–mediated gene delivery. In sharp contrast to most other reported transfection results, transfection capabilities of both DHDEAB and DDAB were observed to be virtually lost when used in combination with DOPE as the helper lipid (Fig. 2). It is worth mentioning here that the commercially available DDAB-containing transfection reagent LipofectACE (manufactured by Life Technologies Inc. USA) also contains DOPE as the auxiliary lipid. However, to our knowledge, except for the present comparative study (Fig. 2), the relative transfection efficiencies of DDAB–cholesterol and DDAB–DOPE combinations have not been reported so far.

IV. CONCLUSIONS

We have synthesized a novel series of nontoxic and non–glycerol-based simple monocationic liposomal transfection lipids containing one or two hydroxyethyl groups directly linked to the positively charged nitrogen atom. The in vitro transfection efficiency of DHDEAB, the most efficient transfection lipid described herein, is better than that of Lipofectamine, one of the most widely used transfection vectors in cationic lipid–mediated gene transfer. Unlike the findings in most of the reported liposomal transfection studies, cholesterol instead of DOPE needs to be used as the helper lipid with the presently described amphiphiles. Interestingly, the most efficient formulations contained cationic lipid-to-DNA charge ratio of 0.3:1. Thus, our results reemphasize the important point that in cationic lipid–mediated gene delivery, the overall charge of the lipid–DNA complex need not always be positive. In addition, our transfection results imply that favorable hydrogen-bonding interactions between the lipid head groups and the cell surface of biological membranes may have some role in improving the transfection efficiency in cationic lipid–mediated gene delivery. However, the transfection results delineated in the present investigation also indicate that such hydrogen-bonding interactions are not likely to be a key controlling parameter in liposomal transfection.

V. FUTURE RESEARCH

The promising in vitro transfection efficiencies of the presently described novel series of non–glycerol-based simple monocationic transfection lipids containing one of two hydroxyethyl head groups justify the immediate launching of systematic structure–activity investigations using a wide array of structural analogues of these lead cationic lipids. Investigations using novel cationic transfection lipids with multiple hydroxyl functionalities in the head group regions will yield new insights on how important the number of head group hydroxyl functionalities is in cationic liposome–mediated gene delivery. The in vivo transfection efficiencies and in vivo cytotoxicities of these new series of cationic lipids need to be evaluated in the near future for their eventual use in nonviral gene therapy. Investigations toward these ends are in progress in our laboratories.

ABBREVIATIONS

DOTMA, 1,2-dioleyl-3-N,N,N-trimethylaminopropane chloride; DC-Chol, 3-β-[N-(N',N'-dimethyl-ethane)carbamoyl]cholesterol; DMDHP, ($\pm$)-N,N-[bis(2-hydroxyethyl)]-N-[2,3-bis(tetradecanoyloxy)propyl]ammonium chloride; DMRIE, 1,2-dimyristyloxypropyl-3-dimethyl-hydroxyethyl ammonium bromide; DOTAP, 1,2-dioleoyloxy-3-(trimethylamino)propane; DHDEAB, N,N-di-n-hexadecyl-N,N-dihydroxyethylammonium bromide; HDEAB, N-n-hexadecyl-N,N-dihydroxyethylammonium bromide; MOOHAC, N-methyl-N-n-octadecyl-N-oleyl-N-hydroxyethylammonium chloride; DOMHAC, N-methyl-N,N-di-n-octadecyl-N-hydroxyethylammonium chloride; DOHEMAB, N,N-di[O-hexadecanoyl]hydroxyethyl-N-hydroxyethyl-N-methylammonium bromide; DOSPA, 2,3-dioleoyloxy-N-[2-(sperminecarboxamido)ethyl]-N,N-dimethyl-1-propanaminium trifluoroacetate; DDAB, dioctadecyldimethylammonium bromide; DOPE, 1,2-dioleoyl-propyl-3-phosphatidylethanolamine.

ACKNOWLEDGMENTS

Financial support from the Department of Biotechnology, Government of India, New Delhi (to A.C.) for this work is gratefully acknowledged. Financial supports in the form of doctoral research fellowships from the Council of Scientific and Industrial Research (CSIR), Government of India, New Delhi (to R.K.B. and G.V.S.) and from the University Grant Commission (UGC), Government of India, New Delhi (to P.K.D.) are gratefully acknowledged.

REFERENCES

1. AD Miller. Nature 357:455–460, 1992.
2. C Dunbar, D Kohn, S Karlsson, N Barton, R Brady, M Cottler-Fox, O Crooks, R Emmons, J Esplin, S Leitman, C Lenarsky, J Nolta, R Parkman, M Pensiero, R Schifmann, P Tolstoshev, K Weinberg. Hum Gene Ther 7:231–253, 1996.
3. JF Engelhardt, Y Yang, LD Stratford-Perricaudet, ED Allen, K Kozarsky, M Perricaudet, JR Yankaskas, JM Wilson. Nat Genet 4:27–34, 1993.
4. PL Felgner. Adv Drug Delivery Rev 5:163–187, 1990.
5. JP Behr. Bioconj Chem 5:382–389, 1994.
6. TB Wyman, F Nicol, O Zelphati, PV Scaria, C Plank, FC Szoka Jr. Biochemistry 36:3008–3017, 1997.
7. PL Felgner, TR Gadek, M Holm, R Roman, W Chan, M Wenz, JP Northorp, GM Ringold, M Danielsen. Proc Natl Acad Sci USA 84:7413–7417, 1987.
8. N Zhu, D Liggitt, Y Liu, R Debs. Science 261:209–211, 1993.
9. FH Camerol, MJ Moghaddam, VJ Bender, RG Whittaker, M Mott, TJ Lockett. Biochim Biophys Acta 1417:37–50, 1999.
10. V Floch, GL Bolc'h, C Gable-Guillaume, NL Bris, J-J Yaouanc, HD Abbayes, C Fe'rec, J-C Cle'ment. Eur J Med Chem 33:923–934, 1998.
11. AD Miller. Angew Chem Int Engl Ed 37:1768–1785, 1998.
12. J Wang, X Guo, Y Xu, L Barron, FC Szoka Jr. J Med Chem 41:2207–2215, 1998.
13. F Tan, JA Hughes. Biochem Biophys Res Commun 242:141–145, 1998.
14. T Hara, Y Tan, L Huang. Proc Natl Acad Sci USA 94:14547–14552, 1997.
15. NS Templeton, DD Lasic, PM Frederik, HH Strey, DD Roberts, GN Pavlakis. Nat Biotechnol 15:647–652, 1997.
16. Y Liu, LC Mounkes, HD Liggitt, CS Brown, I Solodin, TD Heath, RJ Debs. Nat Biotechnol 15:167–173, 1997.
17. MJ Bennett, AM Aberle, RP Balasubramaniam, JG Malone, RW Malone, MH Nantz. J Med Chem 269:4069–4078, 1997.
18. HEJ Hofland, L Shephard, SM Sullivan. Proc Natl Acad Sci USA 93:7305–7309, 1996.
19. CJ Wheeler, PL Felgner, YJ Tasi, J Marshall, L Sukhu, G Doh, J Hartikka, J Nietupski, M Manthorpe, M Nichols, M Piewe, X Liang, J Norman, A Smith, SH Cheng. Proc Natl Acad Sci USA 93:11454–11459, 1996.
20. MH Nantz, MJ Bennet, RW Malone. US patent 5,527,928, 1996.
21. NJ Caplen, EW Alton, PG Middleton, JR Dorin, BJ Stevenson, X Gao, SR Durham, PK Jeffery, ME Hodson, C Coutelle, L Huang, DJ Porteous, R Williamson, DM Geddes. Nat Med 1:39–46, 1995.

22. I Solodin, C Brown, M Bruno, C Chow, E-H Jang, R Debs, T Health. Biochemistry 34:13537–13544, 1995.
23. Y Liu, D Liggitt, G Tu, W Zhong, K Gaensler, R Debs. J Biol Chem 270: 24864–24870, 1995.
24. JH Felgner, R Kumar, CN Sridhar, CJ Wheeler, Y-J Tsai, R Border, P Ramsey, M Martin, PL Felgner. J Biol Chem 269:2550–2561, 1994.
25. EW Alton, PG Middleton, NJ Caplen, SN Smith, DM Steel, FM Munkonge, PK Jeffery, BJ Stevenson, G McLachlan, JR Dorin, DJ Porteous. Nat Genet 5: 135–142, 1993.
26. MJ Stewart, GE Plautz, L Del Buono, ZY Yang, L Xu, X Gao, L Huang, EG Nabel, GJ Nabel. Hum Gene Ther 3:267–275, 1992.
27. X Gao, L Huang. Biochem Biophys Res Commun 179:280–285, 1991.
28. R Leventis, JR Silvius. Biochim Biophys Acta 1023:124–132, 1990.
29. R Banerjee, PK Das, GV Srilakshmi, A Chaudhuri, NM Rao. J Med Chem 42: 4292–4299, 1999. U.S. Patent 6,346,516 B1 (2002); U.S. Patent 6,333,433 B1 (2001).
30. J Sambrook, EF Fritsch, T Maniatis. Molecular Cloning: A Laboratory Manual, Vol 1, 2nd ed. New York: Cold Spring Harbor Laboratory Press, 1989, pp 1.40–1.41.
31. MB Hansen, SE Neilsen, K Berg. J Immunol Methods 119:203–210, 1989.
32. P Tundo, DJ Kippenberger, PL Klahn, NE Prieto, T-C Jao, JH Fendler. J Am Chem Soc 104:456–461, 1982.

29

Combinatorial Surface Chemistry: A Novel Concept for Langmuir and Langmuir–Blodgett Films Research

QUN HUO North Dakota State University, Fargo, North Dakota, U.S.A.

ROGER M. LEBLANC University of Miami, Coral Gables, Florida, U.S.A.

ABSTRACT

In the recent past, combinatorial chemistry has revolutionized medicinal chemistry and this approach has emerged as a powerful technique to discover novel materials. For the first time, we have attempted to combine Langmuir monolayer and combinatorial chemistry techniques to create proteinlike supramolecular structures. We synthesized a peptide lipid library and three sublibraries and studied their monolayer properties at the air–water interface. It was found that the peptide lipid libraries readily formed stable monolayers at the air–water interface and exhibited different binding activities toward carbohydrate molecules from the aqueous subphase. Our study suggests that combinatorial surface chemistry is a possible novel technique in the design and creation of artificial proteins.

I. INTRODUCTION

A. Molecular Recognition in Langmuir Monolayer

Since the first systematic study of monolayers of amphiphilic molecules at the air–water interface published by Langmuir in 1917 [1], Langmuir monolayers have served mainly as model systems to mimic biological membranes. With the development of nanotechnology in the last two decades, the Langmuir monolayer technique has become an efficient tool to make nanoscale materials, especially as thin films for chemical and biosensor development [2–5].

In a traditional concept of a Langmuir monolayer, amphiphilic molecules are spread at the air–water interface. After the evaporation of organic solvent, the amphiphilic molecules stay at the interface. When compressed at this interface, the amphiphilic molecules start to reorient themselves and eventually form a compact monolayer with the hydrophilic moieties embedded in the water phase and the hydrophobic tails extruded into the air phase [6]. During this orientation and organization process, the amphiphilic molecules have the freedom to move around at the interface. This freedom has provided a unique opportunity for supramolecular chemists.

Supramolecular chemistry is a branch of chemical research aimed at developing molecular and supramolecular systems by using noncovalent bonding [7,8]. Generally, small molecular species are designed with binding elements carefully positioned in the appropriate parts of the molecules. When mixed in solution, complementary molecular species are expected to bind together to form larger supramolecular complexes. However, this remains a significant challenge to supramolecular and synthetic chemists. Even when the binding mechanism of a protein toward its ligand is well known, it is still a tremendous challenge for a synthetic chemist to synthesize a molecule with all the binding units appropriately incorporated into the right parts of the molecule. It is a very common situation that after a careful design and synthesis, it turns out that the artificial receptors bind with the ligand with a completely different mode than expected. This results from the inadequate capability of chemists to control the complicated noncovalent binding.

In the late 1980s and early 1990s, some pioneer work from Kunitake et al. opened a new door in the creation of supramolecular species using the Langmuir monolayer technique [9–11]. A multiple molecular recognition system was designed based on the complementary binding of diaminotriazine (T), guanidinium (G), and orotate (O) moieties to the barbituric acid, phosphate, and adenine functional groups, respectively, from guest molecules such as adenosine monophosphate (AMP), adenosine diphosphate (ADP), flavin mononucleotide (FMN), and flavin adenine dinucleotide (FAD) (Fig. 1) [12–15]. The greatest significance of this artificial molecular recognition system can be found in the fact that with the right combination of lipid molecules, these different lipids formed appropriate multiple binding sites for different guest molecules. For example, diaminotriazine lipid T and guanidinium lipid G were combined to form a multiple binding site for FMN by using the complementary hydrogen bonding and electrostatic interactions of triazine with barbituric acid and guanidine with phosphate groups of FMN. Furthermore, if an orotate lipid O is added, the combination of these three lipids can form a binding site for the FAD molecule, which is composed of one isoalloxazine unit, two phosphate units, and one adenine unit.

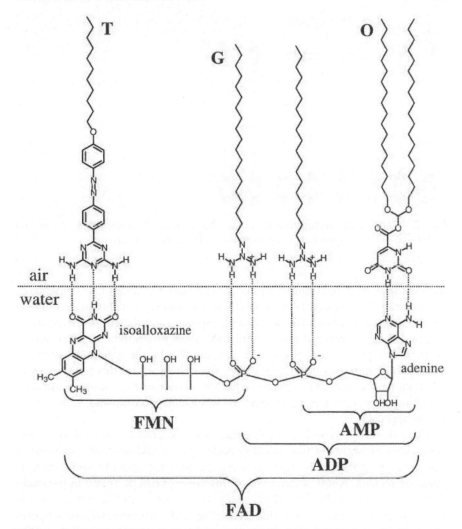

FIG. 1 The combination of small lipid molecules in the Langmuir monolayers to form multiple binding sites for different guest molecules from the aqueous subphase.

The aforementioned work provides an invaluable clue to the supramolecular chemist. Here, instead of incorporating all the necessary binding units into one molecule, different binding units can exist in different molecules and later assemble together to form the desired binding sites under an external force, as illustrated in a cartoon picture (Fig. 2). Compared with synthesizing one molecule with all the necessary binding units incorporated,

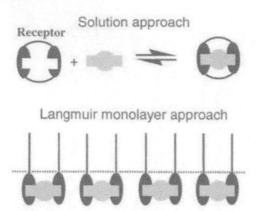

FIG. 2 Cartoon illustrating the difference between the Langmuir monolayer and traditional solution approaches for supramolecular chemistry study.

the synthesis of a few simple lipids is much more efficient and convenient, considering the popularity of combinatorial synthesis techniques. Furthermore, the compression process of the Langmuir monolayers functions as an external force to drive the two lipid molecules to combine to form united artificial receptors. As a result, the van der Waals interaction between the hydrophobic moieties helps to stabilize the system. From this point of view, the Langmuir monolayer approach for the design and creation of artificial molecular receptors has an incomparable advantage compared with the solution approach.

Kunitake et al. [16–19] have further demonstrated the feasibility of this novel idea by showing that small peptide lipids could also be combined in Langmuir monolayers to form binding pockets for specific small peptide guest molecules. Another example of this concept can be found in the study of hydrogen bond direct self-assembly at the air–water interface. A few research groups have shown that melamine and barbiturates form a hydrogen-bonding network at the air–water interface, as illustrated in Fig. 3 [20–26]. This specific molecular recognition can be envisaged as the combination of two molecules using their binding units to complex with one guest molecule. All these previous studies point to one fact that as a classical surface chemistry technique, the Langmuir monolayer approach could be a very useful assembly tool in supramolecular chemistry research.

B. Combinatorial Library Techniques

Since the early 1990s, the combinatorial library technique has revolutionized medicinal chemistry and materials science [27–31]. In traditional synthetic

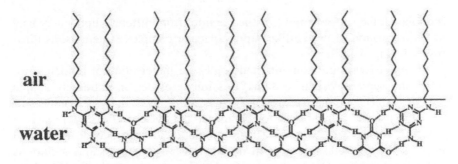

FIG. 3 Hydrogen-bonding network formed at the air–water interface between complementary barbiturate and triaminotriazine lipid.

chemistry, one target molecule is synthesized each time followed by its activity testing. This is a very lengthy procedure, and the cost of discovering one drug molecule for a pharmaceutical company can be as high as millions of dollars. In contrast to this classical approach, the combinatorial technique allows the synthesis of large amounts of diverse molecules such as hundreds, thousands, of even millions of compounds in a row. This technique largely reduces the cost of drug discovery and has been welcomed by the pharmaceutical industry, research institutes, and university laboratories. Following the tremendous success in the medicinal area, the combinatorial library technique has attracted increased interest from materials chemists [32–34]. Novel materials are continuously being discovered through combinatorial library synthesis.

C. The Concept of Combinatorial Surface Chemistry

However, the basic essence of combinatorial chemistry has not yet been completely employed. Nature is the best combinatorial chemist by showing how the four deoxyribonucleotides and 20 amino acids make a whole biological world through the "combination" of these small molecular species. Indeed, one may think a long polypeptide chain is a "combination" of different amino acids connected through amide bonds. The long polypeptide chains then fold into the unique three-dimensional structures of proteins through the noncovalent bonds between the amino acid residues. As a result, the "active site" of a protein can be envisaged merely as a combination of amino acids in the three-dimensional space. This fundamental feature of proteins has provided a very interesting clue to creating protein mimics. If a peptide lipid library sample with hundreds or even thousands of different peptides is spread at the air–water interface and compressed into a Langmuir monolayer, the self-assembly of the peptides from different lipids may lead to the formation of proteinlike supramolecular complex structures, as illustrated in Fig. 4 [35].

Compared with previous methodologies for the creation of artificial proteins and supramolecular systems, this idea provides an extremely simple and convenient alternative approach. It is well known that the synthesis of proteins itself is a huge task for chemists, and even after a protein is successfully synthesized according to the amino acid sequence of the natural protein, the synthetic protein will not always maintain the same activity as the natural protein. This is because the folding process of the synthetic long peptides may lead to a three-dimensional structure completely different from that of the natural protein. From this combinatorial surface chemistry approach, the proteinlike structure is created through the combination of different small peptides into a three-dimensional structure. We think that peptide lipids can be three to five amino acids long and these small peptides can be easily synthesized by solid-phase combinatorial peptide synthesis methods [27]. If this novel approach is proved to work, it could become an extremely efficient approach to making artificial proteins. The artificial proteins generated on the surface of the thin films can be readily adapted for

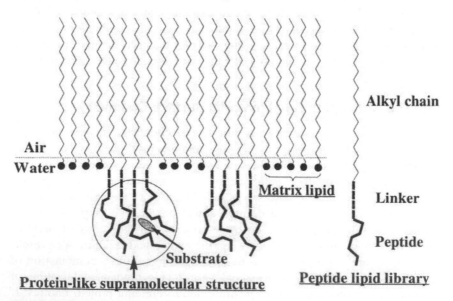

FIG. 4 Illustration of the self-assembly of peptide lipid library components at the air–water interface to form proteinlike supramolecular structures.

biomimetic sensor development or as a novel coating technology for use in many areas of chemistry and medical science, as discussed at the end of this chapter. In the following we will present our results on using this method to make Langmuir monolayers with binding activity for carbohydrate maltose [35].

It is known that the binding site of maltose binding protein (MBP) is exceptionally rich in polar and aromatic amino acid residues [36,37]. The polar charged side chains are involved in the hydrogen bonding with the maltose hydroxyl group, and the stacking of the aromatic residues provides a majority of the van der Waals contacts with maltose. On the basis of this information, we designed and synthesized a peptide lipid library and three sublibraries by including five amino acids, Gly, Glu, Ser, His, and Tyr, as building blocks (Fig. 5). Because these amino acid residues are present in different positions in the peptide lipid library components, we used the library and sublibraries to examine whether the spatial combination of these amino acid residues in the lipid library monolayers could lead to the formation of specific binding sites for maltose, similar to the binding site of MBP.

II. METHODS

A. The Synthesis of Peptide Lipid Libraries

FMOC-protected amino acids and Wang resins were obtained from Advanced ChemTech (Louisville, KY). Other reagents, solvents, and stearic acid for the synthesis of peptide lipids were purchased from Aldrich Chemical Co. (Milwaukee, WI). The peptide lipid library and sublibraries were constructed by using the splitting library synthesis technique. Solid-phase 9-fluorenylmethoxycarbonyl (FMOC) chemistry with the diisopropylcarbodiimide and 1-hydroxybenzotriazole in situ activation method was used for amino acid coupling. The loading of the first glycine to the resin followed the literature procedure [38]. The coupling and deprotection cycle followed the reported procedure [39–41]. FMOC-protected amino acids in N,N-dimethylformamide (DMF) with a concentration of 0.3 M were added to the glycine-loaded resin in a fivefold molar excess relative to the amino groups on the resins. After the coupling of the last amino acid residues, the resins were incubated overnight in a dichloromethane solution of succinamidyl ester of stearic acid at a concentration of 0.1 M. After washing, the peptide lipids were cleaved from the resin by incubating the resin in CF_3COOH and H_2O (95:5, v/v) for 2 h. After filtration, the filtrate was concentrated en vacuo into an oily residue. The product was precipitated out from cold deionized water, washed with cold deionized water five to eight times, and centrifuged. After lyophilization, the product was used for surface chemistry study without further purification.

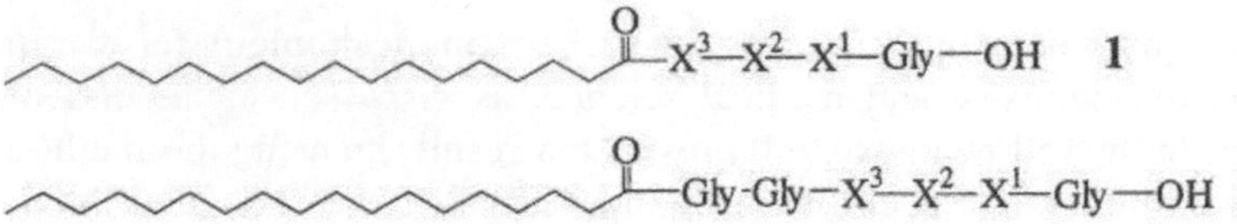

Library	X^1, X^2, X^3	Components
LIB	Gly, Glu, Ser, His, Tyr	250
SUB1	Gly, Ser, Tyr	54
SUB2	Gly, Glu, Tyr	54
SUB3	Glu, Ser, His	54

FIG. 5 The structures and building blocks for the synthesis of the peptide lipid library and sublibraries.

B. Experimental Conditions for Surface Chemistry Studies

High-performance liquid chromatography (HPLC)-grade chloroform and methanol were obtained from Fisher Scientific Co. Peptide lipid library samples were dissolved in a mixed solvent of chloroform, methanol and CF_3COOH (5:1:0.01, v/v/v) to a concentration of 1.0 mM. The injected volume was 40 μL for all samples. After spreading the sample, the solvent was allowed to evaporate for 15 min. The water used for the monolayer study was purified by a Modulab 2020 water purification system (Continental Water Systems Corp., San Antonio, TX). The water had a resistance of 18 MΩ·cm and a surface tension of 72.6 mN/m at 20°C. The D-maltose, D-glucose, and sucrose used for subphase preparation were purchased from Aldrich Chemical Co. and were dissolved in deionized water to a concentration of 10 mM. All these subphases had a pH of 5.8. The compression rate was set at 4 $\mathring{A}^2$ molecule^{-1} min^{-1} for the surface pressure–area isotherm measurements.

All the experiments were conducted in a Class 1000 cleanroom where the temperature (20 $\pm$ 1°C) and the humidity (50 $\pm$ 1%) were controlled. The Langmuir trough used for the surface pressure measurements was a KSV minitrough, model 2000. The trough dimensions were 7.5 cm $\times$ 30.

cm. The surface pressure was measured by the Wilhelmy method, and the sensitivity of the Wilhelmy plate was ±0.01 mN m^{-1}. All the isotherm measurements were repeated three times and the isotherms presented are the average of three measurements. The difference between the average isotherm and any of the three individual isotherms is within ±1 Å^2 molecule^{-1}. Ultraviolet–visible (UV-Vis) spectra of the monolayers at the air–water interface were measured using a modified Hewlett Packard 8452A diode array spectrophotometer with a resolution of ±1 nm through the quartz window in the center of the KSV minitrough.

III. RESULTS AND DISCUSSION

The surface chemistry of the lipid libraries was studied using surface pressure–area isotherm measurements and spectroscopic techniques. In contrast to traditional Langmuir monolayer studies, the present study used a lipid library instead of one or a few lipids at the air–water interface. Surface pressure–area isotherm measurements show that as a whole, these library and sublibrary samples formed monolayers at the air–water interface (Fig. 6). A question was raised during the isotherm measurements of these lipid

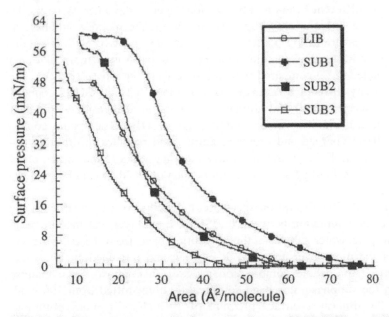

FIG. 6 Surface pressure–area isotherms for the peptide lipid library LIB and sublibrary SUB1, SUB2, and SUB3 monolayers on pure water subphase (pH 5.8, 20°C).

library and sublibrary samples. Because the library components are mixed together through the mix-splitting synthesis, library components are inseparable and are not strictly purified. The exact molecular weight of a library sample cannot be calculated. Therefore, when preparing the spreading solution, all the library and sublibrary samples were given a pseudomolecular weight of 600 g mol^{-1} (a molecule with a C_{18} alkyl hydrocarbon chain and a peptide chain with three to five amino acid residues has an average molecular weight of 600 g mol^{-1}). The molecular areas in the isotherms may not reflect the real average molecular areas of the library components. Therefore, it is not appropriate to compare the molecular areas of monolayers from one library to another library. However, this will not affect the molecular recognition and binding studies of the peptide lipid library, as discussed later, because only the molecular area changes upon binding of the substrates are important.

Using surface pressure–area isotherm measurements, the binding activity of lipid library monolayers toward maltose was then tested. It was found that on a 10 mM D-maltose subphase, the molecular areas of the library and sublibrary monolayers were all expanded (Fig. 7). The molecular area expansion indicates the binding of maltose to library monolayers. However, the molecular area expansions caused by the presence of maltose in the subphase are different from one library monolayer to the other. Whereas the LIB monolayer exhibits only slight expansion on the maltose subphase at the surface pressure lifting point ($\sim$4 Å^2 molecule^{-1}), SUB2 with Gly, Glu, and Tyr as building blocks exhibits the largest expansion of molecular area ($\sim$12 Å^2 molecule^{-1}) compared with the pure water subphase (Fig. 8). The binding activity difference is attributed to the fact that only SUB2 contains both polar charged (Glu) and aromatic (Tyr) amino acid residues, the necessary structural elements in the binding site of MBP. The spatial combination of these charged and aromatic amino acid residues in different positions of peptide lipids at the air–water interface leads to the formation of more appropriate binding sites for maltose than other library and sublibrary monolayers.

From the UV-Vis absorption spectroscopic study, we can clearly see the intermolecular interaction between the SUB2 monolayer and maltose (Fig. 9). On the pure water subphase, the absorption band from the aromatic Tyr residue of the monolayer appeared at 296 nm and remained at this wavelength during the whole compression process. In contrast, when maltose existed in the subphase, this band significantly blue shifted from 296 gradually to 280 nm (at a surface pressure of 30 mN m^{-1}) upon continuous compression, similar to the absorption spectral change observed from the maltose–MBP complex [36,37].

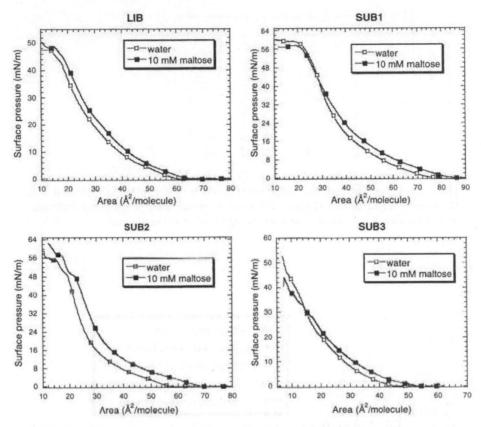

FIG. 7 Surface pressure–area isotherms for the peptide lipid library LIB and sub-library SUB1, SUB2, and SUB3 monolayers on maltose (10 mM) subphase compared with pure water subphase (pH 5.8, 20°C).

Despite only a small library with only three variable amino acid building blocks, the monolayer made by the SUB2 library already exhibits certain specificity as an artificial receptor. Surface pressure–area isotherm measurements show that the presence of glucose or sucrose in the subphase caused almost invisible expansion of the SUB2 monolayer (Fig. 10), indicating less efficient binding between glucose or sucrose and the monolayer. Furthermore, the UV-Vis absorption spectra of the SUB2 monolayer taken from the 10 mM sucrose subphase show that the absorption band arising from Tyr residues appeared at 278 nm on this subphase and remained unchanged at this wavelength during the whole compression period.

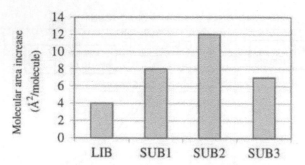

FIG. 8 Increases of the molecular area (in Å²/molecule) of the library and sublibrary monolayers on maltose subphase (10 mM in water, pH 5.8) compared with the pure water subphase. The area increases were calculated based on the molecular area increases at the surface pressure lifting point of the isotherms.

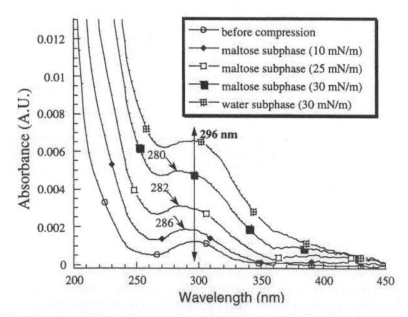

FIG. 9 UV-Vis absorption spectra of SUB2 monolayer on 10 mM maltose subphase compared with the pure water subphase. Spectra of the monolayer on the pure water subphase at a surface pressure lower than 30 mN/m are not shown. The maximum absorption of the monolayer on pure water subphase remains at 296 nm from the beginning of the compression until the collapse.

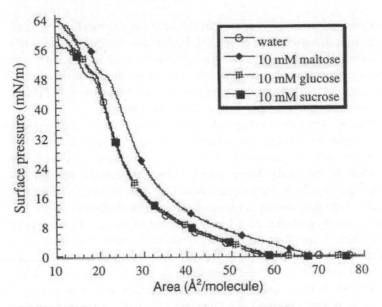

FIG. 10 Surface pressure–area isotherms for SUB2 monolayer on water, 10 mM maltose, glucose, and sucrose subphases (pH 5.8).

IV. CONCLUSION

From the experimental studies presented here, we conclude that it is possible to introduce the combinatorial technique in the Langmuir monolayer research. The application of lipid libraries instead of the traditional one or a few lipids for monolayer formation provides a unique approach to generating artificial proteins or other molecular receptors. The supramolecular species with proteinlike structures located on the surface of the monolayer can be readily used for biomimetic sensor development after the deposition of the film on a transducer such as an optic fiber. This combinatorial surface chemistry research may become a very important research area in Langmuir and Langmuir–Blodgett film studies.

V. FUTURE RESEARCH

This research is currently in a very early stage. Much deeper and further investigation is needed. Above all, the feasibility of this novel combinatorial surface chemistry technique needs further experimental evidence. The study of Langmuir monolayers made from lipid library samples is unprecedented in itself and, therefore, requires considerable work in this respect. For example, one technical problem involved in this study is the handling of samples for the reason discussed in Section III. Furthermore, although the solid-phase combinatorial synthesis has shown significant progress to ensure that each synthetic step is as clean as possible, i.e., to produce a product with the highest yield and to avoid as much as possible any by-product, the purity of library samples is still a problem that cannot be ignored.

Second, because the proposed novel approach relies on assembling different peptide lipids to organize into desired proteinlike structures, the control and characterization of the assembly process and the aggregate structures are critical steps of the study. More detailed characterizations through microscopic and spectroscopic techniques should be able to bring further important insights into this unique molecular assembly technique.

It should be mentioned that as a long-term goal of this study, we intend to use these library thin films for research related to cells. Fields and co-workers [42–46] have demonstrated that peptide amphiphiles can be very attractive model systems to mimic biointerfaces and to study the cell adhesion and growth at these interfaces. The cell–surface interactions play a crucial role in tissue engineering [47]. Our interest is in screening the peptide lipid library thin films for their biocompatibility and bioactivity toward stem cells. Immobilized functional thin films, which can promote cell growth and organization into desired tissue structures, are expected to be discovered through this approach. This biomimetic coating technology can also be used for the study of cell receptors leading to disease diagnosis and drug discoveries.

Furthermore, this technique can lead to the development of a very efficient system to model natural proteins. It is known that purification and characterization of membrane proteins are normally very difficult because of the conformational change of these proteins in the purification process [48]. The artificial proteins generated in this approach can be used to study the ligand binding and signaling of bioactive molecules. One direction we have started to pursue is to mimic the binding site of metalloproteins [49]. Metal–protein complexation plays a crucial role in the function and activity of proteins and enzymes. Model systems that can mimic the structure and function of metalloproteins are of primary interest to the bioinorganic chemist. A critical factor in determining the binding of metals within a specific structure of a protein is the appropriate positioning of amino acid residues in the three-dimensional space to form a specific pocket that the metal cations can enter and form an energy-minimized complex. We have attempted to incorporate amino acids such as histidine, which is a well-known ligand for transition metals such as copper and zinc, into different peptide lipids. Our study has shown that when assembled at the air–water interface, these different peptide lipids can form structures exhibiting binding activity toward copper cation with a binding constant comparable to that of natural proteins.

REFERENCES

1. I Langmuir. J Am Chem Soc 39:1848, 1917.
2. G Monkman. Sensor Rev 20:127, 2000.
3. T Dubrovsky, S Vakula, C Nicolini. Sensors Actuators B 22:69, 1994.
4. DH Charych, JO Nagy, W Spevak, MD Bednarski. Science 261:585, 1993.
5. J Anzai, T Osa. Selective Electrode Rev 12:3, 1990.
6. GL Gaines Jr. Insoluble Monolayers at Liquid–Gas Interface. New York: Interscience, 1966, p 73.
7. DN Reinhoudt, JF Stoddart, R Ungaro. Chem Eur J 4:1349, 1998.
8. J-M Lehn. Supramolecular Chemistry. New York: VCH, 1995.
9. K Ariga, T Kunitake. Acc Chem Res 31:371, 1998.
10. T Kunitake. Pure Appl Chem 69:1999, 1997.
11. K Kurihara. Colloids Surf A 123–124:425, 1997.
12. DY Sasaki, K Kurihara, T Kunitake. J Am Chem Soc 114:10994, 1992.
13. K Taguchi, K Ariga, T Kunitake. Chem Lett 701, 1995.
14. Y Oishi, T Kato, M Kuramori, K Suehiro, K Ariga, A Kamino, H Koyano, T Kunitake. Chem Lett 857, 1996.
15. Y Oishi, Y Torii, T Kato, M Kuramori, K Suehiro, K Ariga, K Taguchi, A Kamino, H Koyano, T Kunitake. Langmuir 13:519, 1997.
16. X Cha, K Ariga, M Onda, T Kunitake. J Am Chem Soc 117:11833, 1995.
17. X Cha, K Ariga, T Kunitake. J Am Chem Soc 118:9545, 1996.
18. X Cha, K Ariga, T Kunitake. Bull Chem Soc Jpn 69:163, 1996.
19. X Cha, K Ariga, T Kunitake. Chem Lett 73, 1996.
20. TM Bohanon, S Denzinger, R Fink, W Paulus, H Ringsdorf. Angew Chem Int Ed 32:1033, 1993.
21. M Weck, R Fink, H Ringsdorf. Langmuir 13:3515, 1997.
22. H Koyano, P Bissel, K Yoshihara, K Ariga, T Kunitake. Chem Eur J 3:1077, 1997.
23. Q Huo, KC Russell, RM Leblanc. Langmuir 14:2174, 1998.
24. Q Huo, L Dziri, B Desbat, KC Russell, RM Leblanc. J Phys Chem B 103:2929, 1999.
25. Q Huo, KC Russell, RM Leblanc. Langmuir 15:3972, 1999.
26. Q Huo, R Stoyan, T Hasegawa, J Nishijo, J Umemura, G Puccetti, KC Russell, RM Leblanc. J Am Chem Soc 122:7890, 2000.
27. AW Czatnik, SH DeWitt, eds. A Practical Guide to Combinatorial Chemistry. Washington, DC: American Chemical Society, 1997.
28. G Jung, ed. Combinatorial Peptide and Nonpeptide Library. New York: VCH, 1996.
29. IM Chaiken, KD Janda, eds. Molecular Diversity and Combinatorial Chemistry. Washington, DC: American Chemical Society, 1996.
30. S Borman. Chem Eng News 78:53–65, May 15, 2000.

31.   N Terrett. Drug Discovery Today 3:46, 1998.
32.   HE Tuinstra, CH Cummins. Adv Mater 12:1819, 2000.
33.   IE Maxwell. Nature 394:325, 1998.
34.   RF Service. Science 280:1670, 1998.
35.   Q Huo, GD Sui, P Kele, RM Leblanc. Angew Chem Int Ed 39:1854, 2000.
36.   JA Hall, K Gehring, H Nikaido. J Biol Chem 272:17605, 1997.
37.   JC Spurlino, GY Lu, FA Quiocho. J Biol Chem 266:5202, 1991.
38.   P Sieber. Tetrahedron Lett 28:6147, 1987.
39.   RN Zuckermann, JM Kerr, MA Siani, SC Banville, DV Santi. Proc Natl Acad Sci USA 89:4505, 1992.
40.   GB Fields, RL Noble. Int J Pept Protein Res 35:161, 1990.
41.   GB Fields. Methods in Enzymology. Vol 289: Solid Phase Peptide Synthesis. New York: Academic Press, 1997.
42.   GHR Rao, GC Fields, JG White, GB Fields. J Biol Chem 269:13899, 1994.
43.   AJ Miles, APN Skubitz, LT Furcht, GB Fields. J Biol Chem 269:30939, 1994.
44.   B Grab, AJ Miles, LT Furcht, GB Fields. J Biol Chem 271:12234, 1996.
45.   H Nagase, GB Fields. Biopolymers 40:399, 1996.
46.   C Li, JB McCarthy, LT Furcht, GB Fields. Biochemistry 36:15404, 1997.
47.   P Bongrand, PM Claesson, ASG Curtis. Studying Cell Adhesion. New York: Springer-Verlag, 1994.
48.   RK Scope. Protein Purification: Principles and Practice. New York: Springer-Verlag, 1994.
49.   Q Huo, G Sui, Y Zheng, P Kele, T Hasegawa, J Nishijo, J Umemura, RM Leblanc. Chem Eur J 7:4796, 2001.

30

Oscillating Structural Forces Reflecting the Organization of Bulk Solutions and Surface Complexes

PER M. CLAESSON   Royal Institute of Technology and Institute for Surface Chemistry, Stockholm, Sweden

VANCE BERGERON   Ecole Normale Superieure, Paris, France

ABSTRACT

This contribution focuses on structural forces in micellar solutions, in polyelectrolyte solutions, and between adsorbed layers consisting of polyelectrolyte–surfactant complexes. The force measurements have been carried out with different surface force techniques. We can distinguish between two types of structural forces. The first type is due to changes in the organization in the bulk liquid separating the two interacting interfaces occurring as the separating liquid film is thinning. The second one is due to changes in adsorbed layer structure occurring as a result of decreasing the film thickness. The amplitude of the latter type of structural force is significantly larger than that of the former. For both types of structural forces, the periodicity obtained from the force curve shows good agreement with correlation distances observed using scattering techniques.

I. INTRODUCTION

A range of surface force methods have been used during the last 25 years for accurate measurements of classical DLVO forces (electrostatic double-layer and van der Waals forces), polymer-induced forces (steric, bridging, depletion) under a range of solvency conditions, as well as short-range hydration/protrusion forces and long-range attractive forces between nonpolar surfaces in polar solvents. In particular, the interferometric surface force apparatus has been successfully used to probe the liquid structure in the gap between two molecularly smooth surfaces. It is found that the arrangement of the solvent molecules in the gap between the surfaces changes as the surface separation is reduced. Hence, the liquid density in the gap varies and this results in a decaying oscillatory force profile [1] that is detected up to about 10 molecular diameters away from the surface in the case of rigid and spherical solvent molecules. The range of the structural force is considerably less in liquids composed of more flexible molecules [2,3].

In the same manner, the structures of liquid crystalline phases in the gap between two surfaces can be probed by studying surface forces as first demonstrated by Horn et al. [4]. Since then, the relation between the structural forces in concentrated lyotropic liquid crystalline systems trapped between two solid supports has been determined and the perturbing effect of the surface has been clearly demonstrated [5]. The surface may induce a surfactant phase at the solid–liquid interface that is different than that found in the bulk. Related phenomena, induced by the preferential interaction between the surface and one of the components in the environment, are capillary condensation [6], capillary evaporation [7] (an important mechanism behind some of the reports concerned with long-range "hydrophobic interactions" [8]), and surface-induced phase separation in polymer mixtures [9].

In this chapter we focus on structural forces observed in aqueous systems containing surfactants, or polyelectrolytes, or mixtures of surfactants and polyelectrolytes. We will argue that two fundamentally different situations should be distinguished. In some cases the oscillating forces reflect the organization in the bulk solution, whereas in others they reflect the internal structure of the adsorbed layer. In both cases, neutron scattering experiments provide information that facilitates the correct interpretation of the structural forces.

II. SURFACE FORCE METHODS

In this chapter we review some data on the interactions between two solid–liquid or two air–liquid interfaces obtained with a range of surface force techniques. It is beyond the purpose of this chapter to describe the merits and drawbacks of the various methods and the interested reader is referred to the original articles describing the surface force apparatus (SFA) [10], the atomic force microscope (AFM) colloidal probe [11], the thin film balance (TFB) [12] and total internal reflection microscopy (TIRM) [13] as well as a more recent review [14]. It is, however, important to be aware that the different techniques use different interaction geometries, and the results can be compared only by using the Derjaguin approximation [15,16]:

$$G_f(D) = \frac{F_c(D)}{2\pi R} = \frac{F_{sf}(D)}{2\pi R} = \frac{F_{ss}(D)}{\pi R}$$

where the subscripts c, f, and s stand for crossed cylinder, flat surface, and sphere, respectively, and D is the separation, F the force, R the mean radius of the interacting surfaces, and G the free energy of interaction per unit area. Note that in TFB measurements one measures the pressure between flat interfaces, which according to the Derjaguin approximation is proportional to the gradient of the force determined with the SFA or AFM. The Derjaguin approximation is valid provided the range of the force is much smaller than the radius of the surfaces and provided no surface deformation occurs.

III. RESULTS AND DISCUSSION

A. Micellar Solutions

Although observations of mesoscopic layering in thin films date back to the turn of the 20th century, it was not until 1992 that force measurements quantified the oscillatory interactions these layers produce when confined between surfaces [12,17]. These first results were obtained from surfactant solutions well above the critical micelle concentration (cmc) and were reported independently for thin-liquid foam films [12] and between mica surfaces [17]. The former study used anionic sodium dodecyl sulfate (SDS), and the latter investigated cationic cetyltrimethylammonium bromide (C_{16}TAB) solutions. In both cases oscillatory force curves were obtained (Fig. 1), which displayed oscillation periods, Δh, equal to the effective diameter of the surfactant micelle:

$$D_{eff} = d_{mic} + 2\lambda_d$$

where D_{eff} is the effective diameter, d_{mic} the molecular diameter of a micelle, and λ_d the solution Debye length. Figure 2 provides the measured oscillation period, Δh, as a function of SDS micelle concentration, C_{mic}, from which the following relation can be deduced:

$$\Delta h \sim C_{mic}^{-1/3}$$

The exponential dependence of $-1/3$ is consistent with geometric scaling arguments for close-packed spheres of diameter D_{eff} [18,19]. Thus, models used to describe these forces are based on the successive removal of spherical micelles as they are progressively confined between approaching interfaces [17,20,21]. Later, the TIRM technique was utilized to study similar oscillating force profiles at lower surfactant concentrations but, of course, still above the cmc [22].

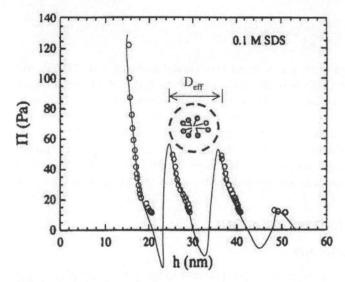

FIG. 1 Oscillatory disjoining pressure isotherm for a 0.1 M solution of sodium dodecyl sulfate.

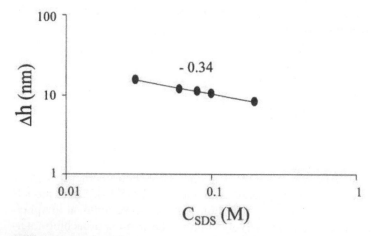

FIG. 2 Oscillation period as a function of SDS concentration. The concentrations are well above the cmc and the total SDS concentration is approximately equal to the concentration of SDS in micelles.

B. Polyelectrolytes

Another important year for the advancement of oscillatory structural forces at mesoscopic length was 1996. Three independent studies emerged that year which reported the observation of oscillatory force interactions involving polyelectrolytes. Two of these studies dealt with polymer–surfactant mixtures [23,24], and one used only polyelectrolyte solutions [25]. Also noteworthy is that one study dealt with flexible foam films [23] and the other two measured forces between rigid solid surfaces [24,25]. Two different phenomena were revealed: oscillatory force interactions originating from bulk solution organization and from surface-specific complexation. What follows is an overview of these different force–structure relationships.

1. Bulk Solution Effects

Oscillatory force measurements involving polyelectrolytes were reported for the first time in foam films made from surfactant solutions containing low levels of polyelectrolyte and surfactant. At nearly the same time, a thorough study by Milling [25] independently showed that oscillatory forces were also present between repulsive silica surfaces in the presence of fully charged polystyrenesulfonate (PSS) solutions with no added surfactant. By systematically investigating various polymer concentrations, C_p, Milling was able to establish that the period of the force oscillations, Δh, followed the same scaling law as the correlation length, ξ, of the polymer solution, namely $\Delta h \sim C_p^{-0.5}$. Moreover, it was shown that the oscillatory forces were highly dependent on the ionic strength. Milling's findings suggested that the oscillatory forces were closely related to the macromolecular structuring of the polyelectrolyte in the bulk. Subsequently, studies following up on the initial foam film observations showed that the same basic features as observed by Milling occurred in foam films, suggesting that although surfactant was present in the foam film system, the phenomena had the same origin [26,27].

One complication that arises with thin-liquid foam film studies is the need to have surface-active components present in order to stabilize the films. Without adequate film stability, measurement of the interactions between the two air–water interfaces cannot be accomplished. These surface-active species provide film stability via surface elasticity and repulsive force interactions between the interfaces (i.e., DLVO-type interactions). In addition, surfactants may interact with polymers added to the system, which can mediate and change the polymer configuration, surface adsorption, and thin-film interactions. Therefore, to determine the role of a polyelectrolyte one must understand independently the various interfacial and polymer–surfactant interactions. Theodoly and colleagues [18,19] have accomplished this through a judicious choice of combined polymer–surfactant mixtures. Two systems for which the effects of the polymer–surfactant association and polymer adsorption are negligible were chosen, thus allowing them to isolate the bulk polyelectrolyte behavior from the possible effects of polyelectrolyte adsorption and complexation with surfactant. The more complex situation involving such interactions is addressed in Section III.B.2.

The two model systems studied by Theodoly and colleagues [18,19] included one nonionic surfactant–polyelectrolyte mixture and one anionic surfactant–polyelectrolyte mixture:

Nonionic surfactant: hexaethylene glycol monododecyl ether
                                  ($C_{12}E_6$)
Anionic surfactant: sodium dodecyl sulfonate ($C_{12}SO_3Na^+$)
Anionic polyelectrolyte: poly (2-acrylamido-2-propane sulfonate)
                                  (PAMPS)

The surfactant in both cases adsorbs at the air–water interface and provides the interfacial properties required for film stability, while the non–surface-active polymer remains dissolved in the solution. Thus, in thin-liquid foam films, under nonionic surfactant conditions the polyelectrolyte solution is confined between neutral film walls, whereas confinement between repulsive charged walls is achieved using an anionic surfactant having the same charge as the polyelectrolyte. In both cases, an oscillatory force interaction superimposed on the native thin-film interactions seen when polymer is not present is observed. An example of each system is shown in Fig. 3. The force oscillations show the same periodic behavior originally found by Milling, $\Delta h \sim C_p^{-0.5}$, regardless of the condition at the surface. We also observe that the magnitude of the forces is comparable and rather weak when both nonionic and anionic surfactants stabilize the film (Fig. 3). The observed decrease in the force magnitude with increased polymer concentration is consistent with the electrostatic nature of the interactions involved and related to the corresponding decrease in Debye length with increased polyelectrolyte concentration. Similarly, the addition of salt can diminish the interaction to such an extent that no force oscillations are observed above 0.1 M NaCl.

In addition to thin-film measurements, solution properties of these mixed polymer/surfactant systems were investigated. For the PAMPS/$C_{12}E_6$ and PAMPS/$C_{12}SO_3^-Na^+$ solutions small-angle X-ray scattering (SAXS) measurements provided a clear correspondence between the polymer bulk correlation length and polymer concentration, $\xi \sim C_p^{-0.5}$, in agreement with standard polyelectrolyte studies. Previous observations of the similar dependence of ξ and Δh on polymer concentration led to the speculation that Δh is related to the structure of the polymer network [25–27]. Most noteworthy in the work of Theodoly and colleagues is that direct independent measure-

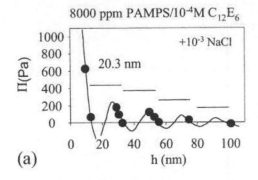

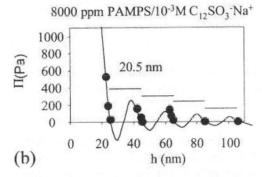

FIG. 3 Oscillatory disjoining pressure isotherms for (a) PAMPS/$C_{12}E_6$ and (b) PAMPS/$C_{12}SO_3^-Na^+$ solutions.

ments of ξ and Δh show for the first time that indeed $\xi = \Delta h$ for the non-associative polymer–surfactant systems tested; see Fig. 4.

As noted previously, spherical structures in the bulk such as micelles also result in oscillatory force behavior in thin films with a periodicity that can be traced to a characteristic distance in the bulk solution films leading to $\xi = \Delta h \sim C_{mic}^{-1/3}$ [18]. By comparison, a simple geometric scaling argument for close-packed cylindrical objects provides a scaling dependence of $C^{-0.5}$, as observed in the polyelectrolyte solutions in the semidilute concentration regime. Thus, these polyelectrolytes appear to be behaving as cylindrical rods, which implies that the persistence length of the polymer is larger than the distance between the chains. Hence, a close analogy between charged micellar and polyelectrolyte systems regarding their correspondence with bulk correlation lengths and induced force oscillations exists. The major difference between these two systems arises from the spherical or cylindrical symmetry of the structures involved. Common to both systems is a bulk correlation length that upon a break in symmetry via the presence of an

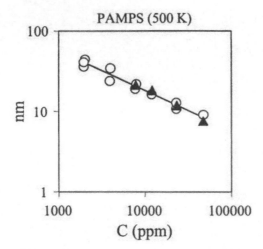

FIG. 4 Comparison between force oscillation period (○) and bulk correlation length (▲) of the PAMPS solution as a function of polymer concentration.

interface, produces density oscillations that subsequently generate oscillatory force interactions in thin films. This type of phenomenon is very similar to the structural forces seen at the molecular level [1]. For nonassociating polyelectrolytes in the semidilute regime, this particular type of oscillatory force behavior can be described in a broad way by these arguments and analogies; however, the detailed nature of the forces remains elusive until a more comprehensive understanding of polyelectrolytes in general is developed.

2. Surface Complexes of Polyelectrolytes and Surfactants

Polyelectrolytes and oppositely charged surfactants associate in bulk aqueous solutions. The association process leads to formation of complexes that can be used for, e.g., controlling rheology, creating gels, or solubilizing hydrophobic molecules. The association process is conveniently characterized by measuring the binding isotherm of the surfactant to the polyelectrolyte; see, e.g., Ref. 28. For charged systems, the literature results convincingly demonstrate that the amount of bound surfactant is low until a critical bulk surfactant concentration has been reached, the critical association concentration (cac), after which the amount of surfactant in the complex increases rapidly. This behavior can be understood by considering the two most important driving forces for the association process: electrostatic and hydrophobic. For example, in the case of oppositely charged polyelectrolyte–surfactant mixtures, the electrostatic force between the surfactant head group and the charged polyelectrolyte segment is attractive, whereas the electrostatic interactions between two head groups and between two segments are repulsive. The attractive hydrophobic interaction, on the other hand, is always present between the nonpolar tails of the surfactant and can also be present between the polyelectrolyte and the surfactant provided the polyelectrolyte contains some hydrophobic regions. Together, these interactions can lead to cooperative association that is sensitive to surfactant concentration.

From the preceding discussion it can be easily understood why the cac is lower than the cmc of the surfactant and increases with increasing salt concentration. On the contrary, the cmc of ionic surfactants decreases with increasing salt concentration. Thus, the difference between the cac and the cmc decreases as more salt is added [29]. The cac in bulk solution also decreases with increasing surfactant chain length and increasing polyelectrolyte charge density. Likewise, the cooperativity of the association process increases with surfactant chain length and with the charge density and flexibility of the polyelectrolyte. These effects can be understood by realizing that any property of the polyelectrolyte that facilitates surfactant tail interactions between two bound surfactants will promote hydrophobic interactions and is, therefore, favorable

The cac concept is less useful when considering the association between hydrophobically modified (HM) polyelectrolytes and surfactants. The reason is that the HM polyelectrolytes themselves associate and form hydrophobic microdomains into which added surfactants can be incorporated [30], a process akin to mixed micelle formation [31]. Hence, even at low concentrations the surfactants added to the polymer solution can be incorporated in the hydrophobic microdomains present. A consequence of this is that polymer–surfactant complexation occurs over a broader range of surfactant concentration and it is difficult to determine a critical concentration.

A question that naturally arises is how the presence of a surface influences the association process. This has been discussed in a review [32]. One can distinguish two broad cases. First, the complexes can be formed in bulk solution and then be adsorbed onto surfaces. In this case it is of interest to learn about the relation between the size of the aggregates present in solution and the thickness of the adsorbed layer, as well as whether the chemical composition of the aggregates changes upon adsorption. Only a few studies devoted to these problems have been reported [33–35]. In a second approach one may precoat the surfaces with a polyelectrolyte layer and then investigate how the addition of surfactants influences, e.g., the adsorbed amount of polyelectrolyte and the layer structure. It is also of interest to learn how the cac is affected by the presence of the surface. These questions have been addressed in quite a few studies [24,36,37]. It has been shown that when a cationic polyelectrolyte is preadsorbed onto a negatively charged surface and later an anionic surfactant is added, the cac at the surface is higher than that in the bulk, particularly for highly charged polyelectrolytes [38]. We note that as yet there is no systematic study of how the cac at the surface is influenced by the nature of the adsorbing substrate.

Here we focus on one aspect of the association between polyelectrolytes and surfactants at a solid surface, namely the structure of the adsorbed layer and its relation to the internal structure of the aggregates formed in bulk solution. The system we will discuss consists of a cationic polyelectrolyte, PCMA (Fig. 5), having one charge per segment that is preadsorbed onto negatively charged mica surfaces. The effect of addition of an anionic surfactant, SDS, was explored. It should be noted that no polyelectrolyte was present in the solution during these experiments, whereas the surfactant was present in bulk solution and was also incorporated in the adsorbed layer.

The forces measured between mica surfaces precoated with PCMA across dilute SDS solutions [24] are illustrated in Fig. 6. The data were obtained using the SFA. The addition of SDS to a concentration of 0.01 cmc or 0.02 cmc (cmc = 8.3 mM) does not result in any change in the long-range interaction or pull-off force (65 mN/m). The surfaces remain uncharged and a bridging attraction [39] acts from a separation of about 15 nm. Hence, at these low SDS concentrations the incorporation of surfactant into the layer is very limited. However, as the SDS concentration is increased further to 0.1 cmc (8.3×10^{-4} M), a long-range repulsive double-layer force appears. The repulsive force is overcome by an attraction at a separation of 11 nm. This attraction pulls the surfaces inward to a separation of 4 nm. A further increase in the compressive force hardly affects the surface separation, indicating a dense layer structure that contains both the polyelectrolyte and SDS. The pull-off force in this case is 25–30 mN/m, i.e., significantly lower than at the lower SDS concentrations.

A further increase in SDS concentration to 0.2 cmc (1.7×10^{-3} M) results in the appearance of pronounced oscillations in the force curve on both the

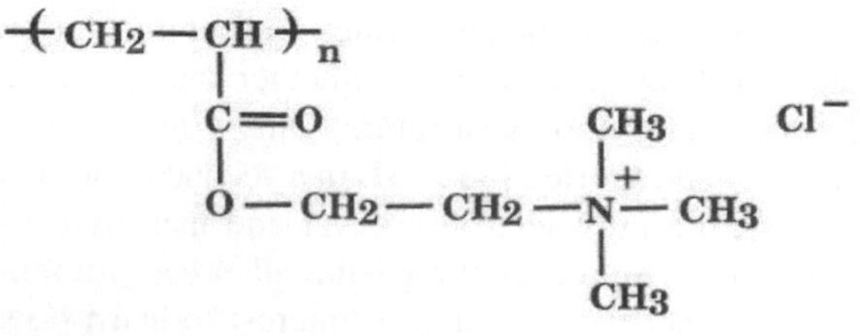

FIG. 5 The monomer structure of PCMA, poly[(3-methacrylamidopropyl)-tri-methylammonium chloride].

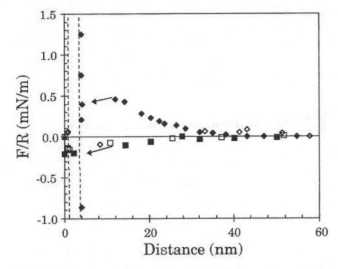

FIG. 6 Force normalized by radius as a function of surface separation between mica surfaces precoated with PCMA. The forces were measured across an aqueous 10^{-4} M KBr solution. The SDS concentration was zero (□), 0.01 cmc (■), 0.02 cmc (◇), and 0.1 cmc (◆). The arrows represent inward jumps and the vertical lines the layer thicknesses.

first and subsequent approaches (Fig. 7). It may be noted that the oscillations are slightly more pronounced when the surfaces have been separated from contact for the first time. The innermost force barrier is located at a separation between 4 and 5 nm, i.e., at the same position as in 0.1 cmc SDS solution. The next force barrier is observed at the distance interval 7–9 nm and the outermost one at a separation of 12–13 nm. The oscillations thus have a periodicity of about 4 nm, and it is observed that both the repulsive and the attractive branches increase in magnitude as the surfaces are moved from an outer to an inner oscillation. A repulsive double-layer force dominates the interaction at separations larger than 13 nm. It should be stressed that these oscillating forces are determined without any polyelectrolyte in the solution and well below the cmc of SDS in bulk solution. Hence, the structural force responsible for the oscillating force profile reflects the structure of the adsorbed layer rather than the organization of the bulk solution. This is different compared with the situation discussed in previous sections.

The forces measured at higher surfactant concentration, up to 2 cmc (1.7 $\times$ 10^{-2} M), also display oscillations with a periodicity of 4 nm (Fig. 8). Clearly, the periodicity of the oscillations remains unchanged when the SDS concentration is increased but the number of oscillations and their magni-

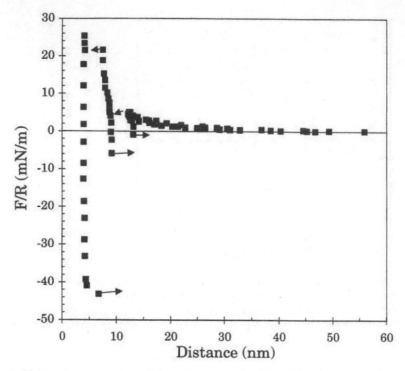

FIG. 7 Force normalized by radius as a function of surface separation between mica surfaces precoated with PCMA. The forces were measured across an aqueous 10^{-4} M KBr solution containing 0.2 cmc SDS. The arrows illustrate inward and outward jumps.

tudes differ. The magnitude of the repulsive force branches increases up to an SDS concentration of 0.5 cmc and decreases again at higher surfactant concentrations. The reduction observed at these high SDS concentrations is most likely due to some desorption of the polyelectrolyte. We also note that the range of the force is somewhat larger at 2 cmc than at lower SDS concentrations, which indicates an increased length of the longest tails.

The data displayed in Figs. 6–8 show that the preadsorbed PCMA layers are strongly swelled by association with SDS when the SDS concentration has reached 0.1 cmc. This indicates that a part of the polyelectrolyte chain is desorbed from the surface. However, most of the polyelectrolytes remain attached to the mica surface for a period of at least several days [35]. When the swelled layers are pushed together, oscillating force curves are observed. The reason is that the internal structure of the adsorbed layer changes in order to minimize the free energy of the system. It is, however, not clear

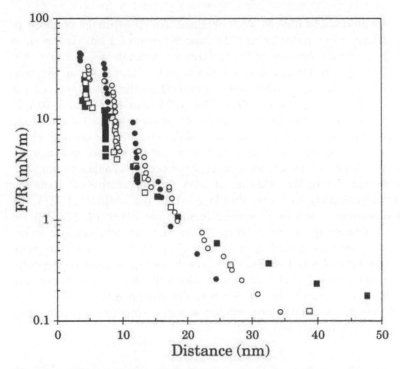

FIG. 8 Force normalized by radius as a function of surface separation between mica surfaces precoated with PCMA. The forces were measured across an aqueous 10^{-4} M KBr solution containing SDS at the following concentrations: (○) 0.2 cmc, (●) 0.5 cmc, (□) 1 cmc, (■) 2 cmc.

how the material in the adsorbed layer is redistributed when going from an outer to an inner oscillation. Is the whole complex deformed laterally along the surface or are surfactants desorbed to the bulk solution? There are indications that the latter process occurs when complexes between a 10% charged polyelectrolyte and SDS on negatively charged surfaces are compressed [40].

The structure of the adsorbed layer can also be visualized by AFM imaging. The images obtained for mica precoated with PCMA before the addition of surfactant are featureless [35], confirming a flat and homogeneous coverage of the surfaces. However, after addition of SDS large features become visible. To obtain reproducible images some care should be taken not to apply too high force. In fact, we noted that reproducible surface features could be obtained only when scanning in the double-layer force mode (i.e., no direct tip–layer contact), whereas as soon as the tip came into contact with the layer irreproducible images were obtained. An AFM image obtained in the double-layer force mode at an SDS concentration of 1 cmc is shown in Fig. 9. The lateral dimensions of the features are typically about 200×200 nm, and the height difference between peaks and valleys is 4 nm. We note that the height differences observed correspond to the periodicity of the structural force observed by the SFA. The AFM scan is, as stated before, carried out away from contact in the double-layer force mode and the image does not provide any information on the total thickness of the coating. However, the force curve measured between the tip and the surface indicates the presence of at least two oscillations, and the periodicity is again 40 nm [35].

We note that before the addition of SDS the polyelectrolyte coats the surface homogeneously in a very thin layer. After the addition of SDS, the material redistributes and large surface features are observed. This can be viewed as a dewetting of the polyelectrolyte from the surface. The incorporation of an anionic surfactant reduces the affinity between the complex and the negatively charged surface. As a result, certain regions of the polyelectrolyte chain desorb and instead associate with the surfactant, forming large complexes as seen in Fig. 9. We note that the polyelectrolyte–surfactant complexes formed are poorly soluble in water, which may explain why complete desorption does not occur.

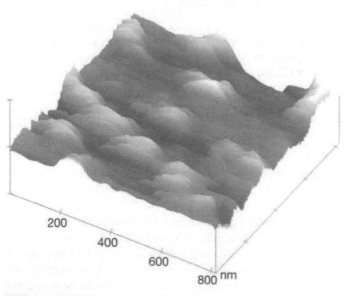

FIG. 9 AFM image of preadsorbed PCMA layers swelled with a 1 cmc SDS solution. The image is taken in liquid using the double-layer repulsion between the tip and the sample. The height scale is 10 nm/div.

Let us now address the question concerning the similarity between the internal structure of the aggregates formed at the solid–liquid interfaces and the internal structure of the polyelectrolyte–surfactant aggregates formed in bulk solution. In order to do so, we return to the complexes formed in bulk solution and apply small-angle neutron scattering (SANS). The details of the study can be found in Ref. 41, and here only some relevant findings are recapitulated. The polyelectrolyte PCMA forms an isotropic and clear solution with water, and the scattering behavior of the samples containing pure polyelectrolyte in D_2O displays a clear peak that is displaced toward higher scattering vector (q) values when the polyelectrolyte concentration is increased. The peak corresponds to a characteristic distance ($\sim 2\pi/q_{max}$) in the polyelectrolyte solution. The peak is rather broad as a result of a comparatively large standard deviation in the distribution of distances between the structural units. Figure 10 shows the characteristic distance as a function of the inverse square root of the polyelectrolyte concentration. The points fall on a straight line as expected for a semidilute polyelectrolyte solution. In fact, it is this structural feature of polyelectrolyte solutions that has been

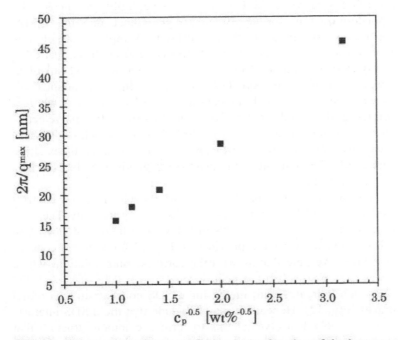

FIG. 10 Characteristic distance ($=2\pi/q_{max}$) as a function of the inverse square root of the PCMA concentration in D_2O.

probed by the AFM [25] and TFB [23,26,27] studies discussed in previous sections. However, this structural feature has no relevance for the oscillating forces observed between preadsorbed PCMA in the presence of SDS.

The SANS data after addition of deuterated SDS (d-SDS) to the PCMA solution demonstrate the appearance of a new structural feature. The polyelectrolyte–surfactant mixture was obtained by adding a small amount of a concentrated surfactant solution to the polyelectrolyte solution under stirring. When a sufficient amount of d-SDS was added to the PCMA solution, it became cloudy and large objects could be seen with the naked eye. Some aggregates remained dispersed in the aqueous phase whereas others precipitated. Away from charge stoichiometry the dispersed aggregates were quite stable (particularly at low polyelectrolyte concentrations), whereas rapid sedimentation occurred when the ratio of SDS to charged segments was close to one. Hence, the scattering data after the addition of sufficient d-SDS were obtained in a two-phase system consisting of aggregates with a high concentration of polyelectrolyte and surfactant dispersed in an aqueous solution containing a low concentration of free surfactant. The amount of precipitate under our measuring conditions was zero or small. Hence, the scattering is due to the dispersed aggregates. The scattering length density for d-SDS relative to the D_2O solvent is negligible compared with the corresponding quantity for the polyelectrolyte. Hence, only the PCMA, and not the d-SDS, contributes to the scattering for samples in which pure D_2O is used as a solvent. Moreover, the sulfate head group has a scattering length density very similar to D_2O [42] so that both head and tail of d-SDS are contrast matched in pure D_2O. When a sufficient amount of d-SDS is added, a new sharp peak appears at $q \approx 0.16$–0.17 Å^{-1}, whereas the peak corresponding to the mesh size in the surfactant-free polyelectrolyte solution disappears. The intensity of the new peak, which is shown in Fig. 11, increases with increasing surfactant concentration, but the position remains unaltered. The position of the peak corresponds to a characteristic distance of 3.7–3.9 nm.

In a mixture of 80% H_2O and 20% D_2O, the polyelectrolyte is contrast matched so that only d-SDS contributes to the scattering intensity. The scattering data at high q values for d-SDS in 0.1 wt% solutions of PCMA in the 80:20 H_2O–D_2O mixtures are provided in Fig. 12 for different concentrations of d-SDS. We note that as with the corresponding solutions where pure D_2O was used as a solvent, a peak located at $q = 0.16$–0.17 Å^{-1}, the intensity of which increases with increasing d-SDS concentration, is found for all samples (Fig. 12). Hence, we may conclude that the d-SDS interacts with parts of the polyelectrolyte chains to form a common structure that contributes similarly to the behavior of the scattering data at high q values for samples in which either the polyelectrolyte or the oppositely charged

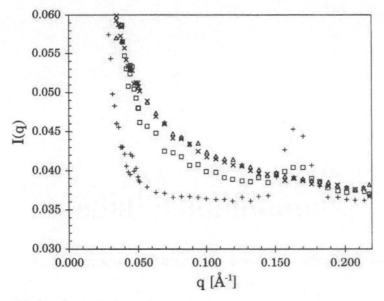

FIG. 11 Scattering intensity at high q values as a function of scattering vector for a 0.1 wt% PCMA solution in D_2O. The d-SDS concentration was 0 ($\triangle$), 0.005 wt% ($r = 0.03$, $\times$), 0.05 wt% ($r = 0.3$, $\square$), and 0.5 wt% ($r = 3$, +), where r is the ratio of d-SDS to charged polyelectrolyte segments.

surfactant is contrast matched. The marked increase in scattering intensity toward the low-q region observed in Fig. 12 for the highest d-SDS concentration is due to formation of free d-SDS micelles.

We note that the characteristic distance describing the PCMA–SDS complexes formed in solution is very similar to the periodicity of the forces measured between mica surfaces precoated with PCMA and swollen by SDS. Hence, it appears that both features are due to the same molecular arrangement. The SANS data are not consistent with a bead-and-necklace structure (i.e., a polyelectrolyte chain decorated with adsorbed micelles). Hence, the interpretation of the force data given in the original article [24] is not correct. Instead, it appears that the structures responsible for the oscillating force curve (Figs. 7 and 8) are similar to the mesomorphous phases characterized by Antonietti and coworkers [43–47] as suggested in the later work by Claesson and colleagues [35,41]. New small-angle X-ray scattering data that are not yet published indicate that the internal arrangement for the PCMA–SDS complex is hexagonal.

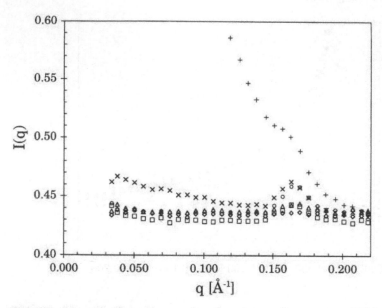

FIG. 12 Scattering intensity as a function of scattering vector at high q values for a 0.1 wt% PCMA solution in an H_2O–D_2O (80:20) mixture. The d-SDS concentration was 0.02 wt% ($r = 0.12$, $\diamond$), 0.05 wt% ($r = 0.3$, $\triangle$), 0.1 wt% ($r = 0.6$, $\square$), 0.2 wt% ($r = 1.2$, $\circ$), 0.5 wt% ($r = 3$, $\times$), and 2 wt% ($r = 12$, $+$), where r is the ratio of d-SDS to charged polyelectrolyte segments.

IV. CONCLUSIONS

A range of surface force techniques has been utilized by several research groups to probe the structures in thin films separating two solid surfaces or two air–water interfaces. Structural forces may arise from changes in packing of the solvent molecules, changes in packing of micelles in the gap between the surfaces, or rearrangement in semidilute polyelectrolyte solutions. Oscillating forces may also appear as a result of disturbance of the internal structure of polyelectrolyte–surfactant complexes attached to the solid surface. In several cases, nice agreement between bulk structures, as probed by SAXS and SANS, and the periodicity of the oscillating forces has been demonstrated. Hence, in some cases the organization in complex solutions is reflected in the force profile between two surfaces. In other cases the internal structure of polymer–surfactant complexes is probed by the force measuring techniques.

ACKNOWLEDGMENT

P. C. acknowledges financial support from the Swedish Natural Science Research Council (NFR) and the SSF program Nanochemistry.

REFERENCES

1. RG Horn, JN Israelachvili. J Chem Phys 75:1400–1411, 1981.
2. HK Christenson, DWR Gruen, RG Horn, JN Israelachvili. J Chem Phys 87:1834–1841, 1987.
3. PM Claesson, A Dedinaite, B Bergenståhl, B Campbell, HK Christenson. Langmuir 13:1682–1688, 1997.
4. RG Horn, JN Israelachvili, E Perez. J Phys 42:39–52, 1981.
5. P Petrov, S Mikalvic, U Olsson, H Wennerström. Langmuir 11:3928–3936, 1995.
6. HK Christenson. Phys Rev Lett 73:1821–1824, 1994.
7. JL Parker, PM Claesson, P Attard. J Phys Chem 98:8468–8480, 1994.
8. HK Christenson, PM Claesson. Adv Colloid Interface Sci 91:391–406, 2001.
9. H Wennerström, K Thuresson, P Linse, E Freyssingeas. Langmuir 14:5664–5666, 1998.
10. JN Israelachvili, GE Adams. J Chem Soc Faraday Trans 1 74:975–1001, 1978.
11. WA Ducker, TJ Senden, RM Pashley. Nature 353:239–241, 1991.
12. V Bergeron, CJ Radke. Langmuir 8:3020–3026, 1992.
13. JY Walz, DC Prieve. Langmuir 8:3073–3082, 1992.
14. PM Claesson, T Ederth, V Bergeron, MW Rutland. Adv Colloid Interface Sci 67:119–183, 1996.
15. B Derjaguin. Kolloid 69:155–164, 1934
16. JN Israelachvili. Intermolecular and Surface Forces. London: Academic Press, 1991.
17. P Kékicheff, P Richetti. Prog Colloid Polym Sci 88:8–17, 1992.
18. O Théodoly. PhD dissertation, College de France, Paris, 1999.
19. O Theodoly, JS Tan, R Ober, CE Williams, V Bergeron. Langmuir 17:4910–4918, 2001.
20. DT Wasan, AD Nikolov, PA Kralchevsky, IB Ivanov. Colloids Surf 67:139–145, 1992.
21. PA Kralchevsky, N Denkov. Chem Phys Lett 240:385–392, 1995.
22. DL Sober, JY Walz. Langmuir 11:2352–2356, 1995.
23. V Bergeron, D Langevin, A Asnacios. Langmuir 12:1550–1556, 1996.
24. PM Claesson, A Dedinaite, E Blomberg, VG Sergeyev. Ber Bunsenges Phys Chem 100:1008–1013, 1996.
25. AJ Milling. J Phys Chem 100:8986–8993, 1996.
26. A Asnacios, A Espert, A Colin, D Langevin. Phys Rev Lett 78:4974–4977, 1997.
27. R v.Klitzing, A Espert, A Asnacios, T Hellweg, A Colin, D Langevin. Colloids Surf A 149:131–140, 1999.

28. K Hayakawa, JCT Kwak. J Phys Chem 86:3866–3870, 1982.
29. B Lindman, K Thalberg. In: ED Goddard, KP Ananthapadmanabhan, eds. Interactions of Surfactants with Polymers and Proteins. Boca Raton, FL: CRC Press, 1993, pp 203–276.
30. O Anthony, R Zana. Langmuir 12:3590–3597, 1996.
31. P Linse, L Piculell, P Hansson. In: JCT Kwak, ed. Models of Polymer–Surfactant Complexation. New York: Marcel Dekker, 1998, pp 193–237.
32. PM Claesson, A Dedinaite, E Poptoshev. In: T Radeva, ed. Physical Chemistry of Polyelectrolytes. Surfactant Science Series Vol 99. New York: Marcel Dekker, 2001, pp 447–507.
33. PM Claesson, M Fielden, A Dedinaite, W Brown, J Fundin. J Phys Chem B 102:1270–1278, 1998.
34. A Dedinaite, PM Claesson. Langmuir 16:1951–1959, 2000.
35. A Dedinaite, PM Claesson, M Bergström. Langmuir 16:5257–5266, 2000.
36. V Shubin, P Petrov, B Lindman. Colloid Polym Sci 272:1590–1601, 1994.
37. O Anthony, CM Marques, P Richetti. Langmuir 14:6086–6095, 1998.
38. PM Claesson, A Dedinaite, M Fielden, URM Kjellin, R Audebert. Prog Colloid Polym Sci 106:24–33, 1997.
39. MAG Dahlgren, Å Waltermo, E Blomberg, PM Claesson, L Sjöström, T Åkesson, B Jönsson. J Phys Chem 97:11769–11775, 1993.
40. URM Kjellin, PM Claesson, R Audebert. J Colloid Interface Sci 190:476–484, 1997.
41. PM Claesson, M Bergström, A Dedinaite, M Kjellin, J Legrand. J Phys Chem B 104:11689–11694, 2000.
42. M Bergström, JS Pederson. Phys Chem Chem Phys 1:4437–4446, 1999.
43. M Antonietti, J Conrad, A Thünemann. Macromolecules 27:6007–6011, 1994.
44. M Antonietti, C Burger, J Effing. Adv Mater 7:751–753, 1995.
45. M Antonietti, A Kaul, A Thünemann. Langmuir 11:2633–2638, 1995.
46. M Antonietti, A Wenzel, A Thünemann. Langmuir 12:2111–2114, 1996.
47. M Antonietti, M Maskos. Macromolecules 29:4199–4205, 1996.

31

Effect of Polymeric Surfactants on the Behavior of Polycrystalline Materials with Special Reference to Ammonium Nitrate

ARUN KUMAR CHATTOPADHYAY   United States Bronze Powders Group of Companies, Haskell, New Jersey, U.S.A.

ABSTRACT

The particles of inorganic polycrystalline materials, for example, the nitrates of potassium, sodium, and ammonium in either their spherical or granular forms, consist of aggregations of irregular forms of crystals, which provide free variable space between the crystals. During the formation and growth of crystals below the crystallization temperature, a progression of crystal growth occurs during both cooling and drying stages. The study discussed here relates to the effect of some sulfonated polymeric surfactants on the changes of crystal growth pattern in ammonium nitrate particles. Atomic force microscopy studies on the particles confirm a unique associated migration of water-bound polymeric additives during crystallization, which limits the growth of the individual crystals to almost unit cell dimensions by controlling the overall crystallization pattern.

I. INTRODUCTION

Inorganic nitrates—sodium nitrate, potassium nitrate, and ammonium nitrate—play major roles in the fertilizer, explosives, and propellant industries. Ammonium nitrate has been the material of prime interest for commercial exploitation because of its favored chemical nature and cost-effectiveness. The commercially available forms of these polycrystalline nitrates are generally granular or spherical (popularly known as prill) consisting of aggregated irregular crystalline forms that provide free variable space between the crystals. Depending upon the compactness of the crystallites or available free variable space between the crystallites, the density of the prill differs. The higher the free space, the lower the prill density. For their applications in fertilizers, the prill density is probably of lesser significance. However, in explosives and propellants the size of the crystallites, prill density, and resistance to thermally induced crystal growth due to interparticle bridging are major considerations for their suitability [1,2].

Despite being most suitable, both chemically and economically, ammonium nitrate poses a major challenge for its wider usage in propellants, and this is related to the material properties of the porous prill as opposed to the problems associated with the compound itself. The material deformation and the density changes under pressure and temperature, which are attributed to the uncontrolled crystal growth and bridging, essentially affect the rate of reaction of ammonium nitrate with the fuel binders, resulting in irregular thrusts, inferior performance, and failure. There have been many attempts in the past to reduce the magnitude of these problems associated with ammonium nitrate by using additives to improve its material property. The large number of additives that have been used to influence the material property of ammonium nitrate can be classified in five distinct categories: (1) crystal habit modifiers, (2) desiccants, (3) solid solutions and double salts, (4) nucleating agents, and (5) anticaking agents. The previous work done on the crystal habit modification aided by various additive molecules indicated a mechanistic relationship between the adsorption of the additive molecules and lattice matching. Such a mechanistic relationship provides a predictive tool for the selection of useful additives for ammonium nitrate. The specific adsorption of additive molecules onto the crystal lattices and its effect on overall crystallization (growth and shape) are the basis for invoking changes in the material property of ammonium nitrate.

In this chapter the introduction of a polymeric anionic surfactant, polystyrene sulfonate, to bring about a marked effect on the crystallinity of ammonium nitrate is discussed and the evidence for microcrystalline forms of ammonium nitrate induced by polystyrene sulfonate is presented. There are many additives documented in the literature that are claimed [1–4], with some evidence, to produce porous ammonium nitrate prill with desirable strength and internal stability. There is, however, very little understanding of these systems, particularly in relation to crystallization. The possible mechanisms of polystyrene sulfonate in relation to its water association as well as drying properties in crystallizing ammonium nitrate are discussed in the following sections.

II. POLYMORPHISM

In a commercial manufacturing process, supersaturated solutions of nitrate salts are showered through a porous plate in a tall tower and the droplets of the supersaturated solutions crystallize during their flight from the top of the tower to the bottom as the droplets cool down below the crystallization temperature. A schematic representation is shown in Fig. 1. Further drying and cooling processes take away the remaining moisture to form dried pseudospherical bodies called granules or prill [3,4].

From the onset of crystallization to drying, ammonium nitrate undergoes various crystal phase changes as shown in Fig. 2. Under ambient storage conditions, the phenomena of interparticle bridging and crystal growth are often related to IV↔III phase transition of ammonium nitrate. As a result, the study of IV↔III transition kinetics has historically received a great deal of attention [5–7] and various additives have been used to prevent interparticle bridging particularly due to such phase transitions. It must be noted in this regard that during particle formation the additives can influence ammonium nitrate right from the initiation of the phase I (cubic) crystalline form of ammonium nitrate. However, no systematic study has ever been carried out to understand the effect of various additives (auxiliary host mol-

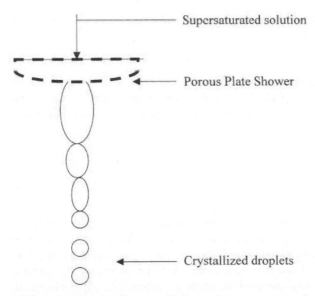

FIG. 1 Schematic diagram of showering supersaturated solutions of inorganic nitrates to produce spherical particles.

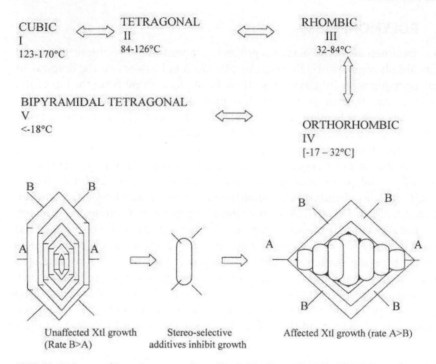

FIG. 2 Polymorphism in ammonium nitrate (top) and effect of additives on crystal habit modifications (bottom).

ecules) on the crystal growth pattern. However, the use of crystal morphology as a determinant of specific interaction between crystal surfaces and auxiliary molecules has received considerable attention in last two decades. Figure 2 demonstrates:

1. A crystal growing in a pure environment with faces B growing faster than faces A
2. Specific adsorption of additive molecules onto B yielding a change in the overall crystal shape

This phenomenon provides a basic understanding of the additive selection principle as well as the basis for rational molecular design of the additive molecules because, as shown here, if the morphology is characterized in terms of the structure of the affected faces, then the nature of the surface binding sites and the key parameters such as geometric, stereochemical, electrostatic, and molecular recognition can be identified [8–14].

Depending upon the nature of additives, the process of crystal habit modification can be initiated at various stages of crystal phase changes, e.g., I (cubic)–II (tetragonal)–III (rhombic)–IV (orthorhombic)–V (bipyramidal tetragonal). Besides modifying the crystal structure by altering the preferences of certain growth planes, the habit modifiers are also known to limit the size of crystals. In search of a suitable habit modifier for ammonium nitrate, a large number of additives have been tried during the last four decades. On looking at the range of compounds studied, it appears that the molecules containing large ionic groups—sulfates, sulfonates, phosphates, phosphonates, etc.—are the most effective for habit modification.

The action of an additive on phase IV of ammonium nitrate is yet another example in this respect (Fig. 3). In the crystal structure of phase IV, the (100), (001), (110), and (011) planes are classified according to the arrangement of NH_4^+ and NO_3^- ions. In the orthorhombic crystal structure of phase IV, the (001) and (110) planes both consist of alternating layers of either NH_4^+ or NO_3^- ions, and the (100) and (011) planes comprise both NH_4^+ and NO_3^- ions. Some dyestuffs (well known as habit modifiers), e.g., Cu-phthalocyanine, bring about habit modifications as a result of their bipolar nature. These molecules adsorb on the (100) and (011) planes via SO_3^- and $—NH_3^+$ groups. Similarly, additive molecules containing anionic groups are found to exert an effect via their adsorption on the (001) and (110) planes containing only NH_4^+.

In order to understand the effect of additives on the material properties of crystals, the examples of crystallization that occur in nature are worth citing. In nature, the structure–activity relationship and recognition factors existing between the molecules, i.e., the crystallizing materials and the host molecules, largely influence crystallization. For example, in animal physiology, crystallization of calcium carbonate occurs in different morphological forms. A great deal of structure variation is observed in the formation of bones, teeth, shells, etc. despite the fact that all of them are principally constituted by the same calcium carbonate. Such variations in morphology are principally guided by the nature of proteins and enzymes involved in the metabolic processes in the growth of desired structural forms. Different proteins influence the crystallization of calcium carbonate to occur in different manners. The nature of the association between an inorganic phase (e.g., calcium carbonate) and an organic film (proteins or enzymes) determines the overall crystal growth pattern, its morphology, and subsequent material properties of the crystallizing materials [8].

III. EXPERIMENTAL

Polystyrene sulfonate (PSS) of molecular weight ~70,000 (Polyscience), polyvinyl sulfonate (PVS) of molecular weight ~25,000 (Air Products), and polyvinyl-co-styrene sulfonate (PVSS) of molecular weight ~50,000 (ob-

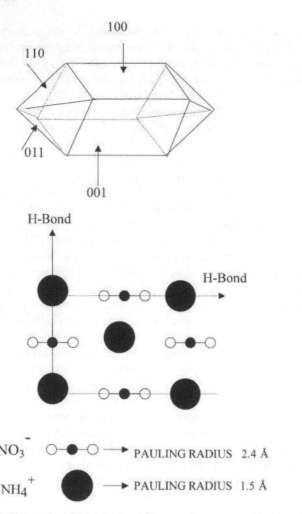

100

110

011

001

H-Bond

H-Bond

NO$_3^-$ ○–●–○ ⟶ PAULING RADIUS 2.4 Å

NH$_4^+$ ● ⟶ PAULING RADIUS 1.5 Å

FIG. 3 Phase IV orthorhombic crystal structure of ammonium nitrate and arrangement of the ionic groups in the lattice.

tained from the Department of Chemistry, McGill University, Montreal, Canada) were used as additives in ammonium nitrate to study their effect on crystallinity. The PVSS was prepared by polymerizing styrene sulfonate and vinyl sulfonate in a 1:1 molar ratio. The Na$^+$, K$^+$, and NH$_4^+$ salts of the polymers were prepared by neutralizing with corresponding alkalis. The concentration of these additives in ammonium nitrate solutions [containing 1.18

moles of ammonium nitrate and 0.28 moles of water (approximately 237.5 molal solution of ammonium nitrate)] was maintained at 500 ppm [3,4].

Crystallization and drying were studied by employing the solutions on a glass hot bed maintained at 70°C fitted with a constant tangential cold air flow unit mounted over the glass bed delivering air at the rate of 0.5 cfm at room temperature, approximately 20°C (Fig. 4). The moisture content of the drying crystal layers on the glass bed was determined by Karl Fischer titration and recorded upon subjecting the ammonium nitrate solution for a certain period of time. These values were compared with those of pure ammonium nitrate solutions of similar concentrations.

Atomic force microscope (AFM) images of the dried crystals of ammonium nitrate (courtesy of the Department of Chemistry, University of Miami, Coral Gables, Florida) were obtained by scanning several samples (at least six samples each) of ammonium nitrate and ammonium nitrate with polymeric additives. These AFM samples were prepared by depositing drops of ammonium nitrate solution on the freshly cleaved mica and drying under the same conditions mentioned before. The films were scanned by the contact mode [15] AFM in a cleanroom of class 1000. The scanning force was set between 5 and 10 nN. The average scanning rate used in this study was 8 Hz.

IV. RESULTS AND DISCUSSION

In all of our experiments the supersaturated solutions of ammonium nitrate, with or without additives, were kept at 145°C prior to placing a constant volume of the solution (0.5 mL per addition) on the glass hotbed. Figures 5 and 6 show the drying profiles of ammonium nitrate solutions. The profiles in Fig. 5 clearly show that despite all being sulfonated polymers, the stereochemical features of the polymers certainly play a very important role in the overall crystallization and drying. Among the three different types of polymers studied, PVS certainly has the least effect on drying compared with the systems without any additive. PSS and PVSS exhibited similar drying trends; however, PSS showed superior results with respect to drying

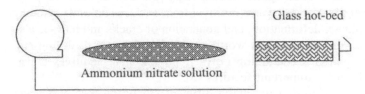

FIG. 4 Schematic diagram of the drying device.

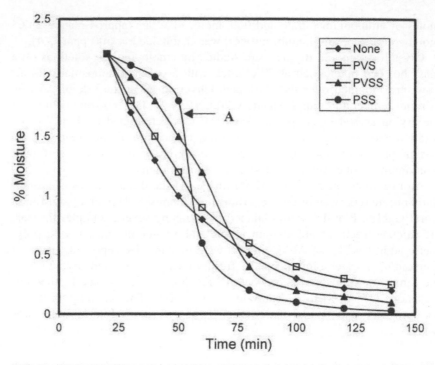

FIG. 5 Drying profiles of ammonium nitrate solutions in the presence of Na salts of the polymeric additives. The critical point, A, during drying of ammonium nitrate in the presence of PSS indicates the rapid change in the rate of moisture loss.

efficiency. In Table 1, the results of the final moisture content of the dried ammonium nitrate in the presence of polymeric additives are given. In all cases the samples were subjected to drying for 3 h. The final moisture contents are indicative of the binding strength of water molecules with the crystal planes and the influence of additives on the water binding energies. The Na salt of PSS (Fig. 6) was found to be the most effective in removing most of the water molecules in the shortest period of time. In a real-life situation, in drying pseudoregular bodies (e.g., granules or prill) of polycrystalline materials, besides water removal from surfaces an understanding of particle shrinkage, deformation, and generation of cracks and flaws is also required [12]. In this respect it is worth noting that the ease of drying ammonium nitrate particles containing polystyrene sulfonate additives is certainly one of the most important features [3].

The mechanism of drying involving different kinetic processes of water removal from a hygroscopic ammonium nitrate body [16,17] is schemati-

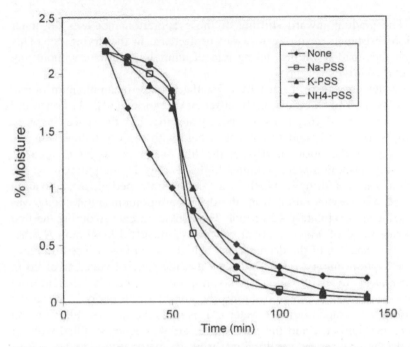

FIG. 6 Drying profiles of ammonium nitrate solutions in the presence of sodium, potassium, and ammonium salts of polystyrene sulfonic acid.

cally represented in Fig. 7. In a regular drying process, when a hygroscopic body with uniformly distributed water is dried under a steady air flow, the water on the outer layer begins to vaporize first and then the internal water moves to the surface. As the drying proceeds, evaporation shifts inward because of the decreasing water transferability. Water vaporized inside the drying body is transferred to the surface and then released to the environ-

TABLE 1 Retained Moisture in Dried Ammonium Nitrate in the Presence of Various Salts of PSS, PVS, and PVSS Polymeric Additives

Additive	None	PSS	PVS	PVSS
% Final moisture	0.2	—	—	—
Polymer counterions	—	—	—	—
Na^+	—	0.02	0.25	0.18
K^+	—	0.05	0.3	0.2
NH_4^+	—	0.04	0.32	0.2

ment. The gradual inward shifting of the evaporation increases the inner diffusion resistance, resulting in a steady decrease in the drying rate. This explains the steadily falling drying rate of ammonium nitrate without any polymeric additives (Fig. 5).

The water present in a partially crystallized body of ammonium nitrate can be classified in three distinctly different categories [14]: (1) film water that surrounds the drying body, (2) pore water that fills the spaces between crystallites, and (3) bound water that is held with the crystallites (Fig. 7). Under a given condition of drying, the film water is easiest to remove, whereas crystal-bound water requires higher energy for its removal. At the very first stage of drying, which is a constant-rate period, the activation energy of water evaporation from the drying ammonium nitrate body was found to be approximately 42 kJ/mol. The activation energy during the first falling-rate stage of water removal was approximately 125 kJ/mol. A schematic representation of the drying kinetics is shown in Fig. 8. The difference in the activation energy values indicates that the state of water involved in the subsequent stages of drying, as shown in Fig. 7, is clearly different from the film water that is released during the constant-rate period. During the first falling-rate stage, the pore water moves to the surface by diffusion and capillary mechanisms, and the pore spaces are subsequently filled with air as the drying progresses, resulting in shrinkage of the drying body.

However, in contrast to the pure ammonium nitrate, ammonium nitrate in the presence of PSS additives offers a completely different feature. The initial drying phase was slower than for the pure ammonium nitrate, whereas the second stage of drying occurred at a remarkably faster rate. The presence of PSS additives probably induces several kinetic processes simultaneously, namely initiation of crystallization, salting out of the polymers from the crystalline phase, solvation of the polymer molecules, and migration of polymer molecules with water as the drying progresses. The PSS additives provide an efficient vehicle to transport water molecules from the core to the surface of a drying ammonium nitrate body. The function of the PSS is shown schematically in Fig. 9. PSS in the form of an Na^+, K^+, or NH_4^+ salt is an anionic polyelectrolyte prepared under conditions that yield a high degree of substitution of sulfonate groups in the polymer molecules (one sulfonate group per styrene moiety). The high degree of substitution imparts properties to the polymer that make it suitable for a variety of applications. The salts of polystyrene sulfonate are highly soluble in water. The solubility of these polymers decreases rapidly with an increase in the concentration of ammonium nitrate. As a result, during crystallization of ammonium nitrate from its aqueous supersaturated solutions, these polymers always remain with the aqueous component of the crystallizing front and migrate as the water leaves the drying body.

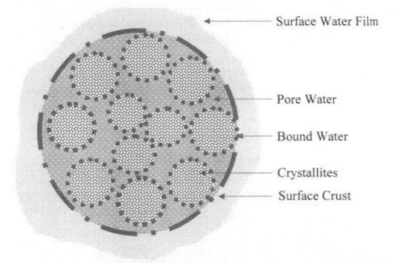

Surface Water Film

Pore Water

Bound Water

Crystallites

Surface Crust

FIG. 7 Mechanism of drying: pictorial description of a drying polycrystalline material featuring three different types of water present in a crystallizing body.

Beyond the critical point, A, as shown in Fig. 5, the loss of moisture takes place with greater rapidity. The initial slow rate is probably due to the capillary saturation of water that occurs near the outer surface of the drying body. Faster crystallization in the presence of PSS keeps the size of the crystallites smaller and the rate of release of the free water molecules is also higher as the crystallization progresses. These water molecules evaporate from the surface by capillary diffusion. The rapid release of water molecules in the presence of PSS affects the steady-state condition of diffusion and evaporation by causing overcrowding of the diffusible free water molecules as opposed to the number of water molecules which can actually diffuse through the capillaries and evaporate. This explains the initial slow rate of moisture loss. As the threshold between the capillary saturation and the internal vapor pressure of water crosses the critical point, A, the liquid initially held up in the capillaries erupts out almost instantaneously (Fig. 10a and b). In this instant, the rate of moisture loss from the drying ammonium nitrate body becomes invariant to the temperature. It appears that the water association to PSS is energetically more favorable than to ammonium nitrate alone. Thus, the water molecules initially bound to the ammonium nitrate crystal lattices are favorably drawn by the PSS molecules, resulting in a unique associated migration of water molecules. The dried forms of these materials were further investigated over a period of 30 days to find out if

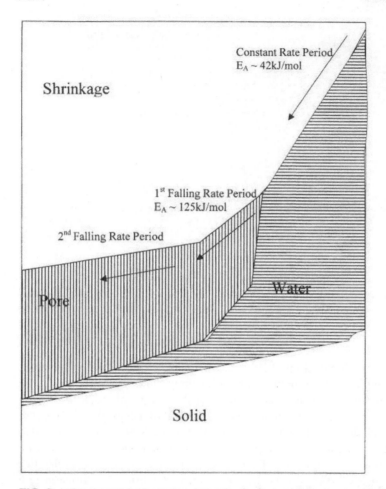

FIG. 8 Kinetics of drying of an irregular body containing water. Deformation and shrinkage occur due to the formation of voids or pores as water leaves the matrix during drying.

there had been any progressive changes in crystal growth. The optical micrographs of dried ammonium nitrate versus ammonium nitrate with PSS additive showed a massive growth of long needlelike crystals in pure ammonium nitrate samples, whereas the sample of ammonium nitrate with PSS additive retained smaller crystals of ammonium nitrate without any noticeable changes in size (Figs. 11 and 12). This motivated us to carry out further studies on these systems by atomic force microscopy (A. K. Chattopadhyay, S. Boussaad, and R. M. Leblanc, unpublished data).

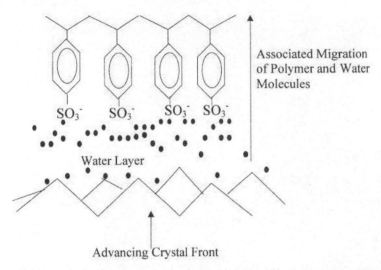

FIG. 9 Schematic description of the action of PSS polymers to enhance the rate of crystallization by withdrawing water molecules from the crystallizing front to themselves.

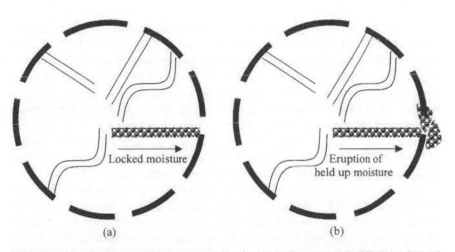

FIG. 10 (a) Capillary saturation (rate of evaporation ≪ rate of diffusion through the capillary) slows down the initial rate of moisture loss in the presence of PSS. (b) Interior liquid initially held up by capillary saturation erupts as soon as the inner vapor pressure rises above the capillary pressure.

FIG. 11 Optical micrograph (500×) showing the large growth of needle-shaped crystals in pure ammonium nitrate.

In Fig. 13, the AFM images, a and b, show the surface structure of dried films of pure ammonium nitrate. In the image of Fig. 13a (15 × 15 μm), one can notice the plane surface formed by the crystals of ammonium nitrate. Considering the image scan size ($Z = 450$ nm) and the dimensions of a unit cell in ammonium nitrate crystals, one can assume that the surface observed by the AFM results from the crystals of ammonium nitrate, indicating a strong intercrystal bridging phenomenon. This has been further clarified in the image of Fig. 13b. This AFM image (6 × 6 μm) represents a magnified view of a region of the image of Fig. 13a. In Fig. 13b there is no clear indication of the presence of any intercrystal distinct boundaries.

However, the study of the films of ammonium nitrate in the presence of the Na salt of PSS showed a direct impact of the additive on the overall crystallinity of ammonium nitrate. In Fig. 13c (15 × 15 μm), the structure of the film clearly shows that the film is composed of clusters of extremely small crystals with distinct boundaries. The large difference in Z range (200 nm) between images 13a and 13c reveals the smoothness of the surface of the films containing the PSS additive. In the higher resolution image in Fig. 13d (4 × 4 μm) the structure of ammonium nitrate crystals in the presence

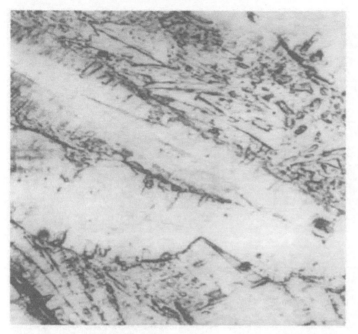

FIG. 12 Optical micrograph (500×) showing clusters of small crystals in ammonium nitrate with PSS additive.

of PSS additive is further revealed. The most important feature of the crystal arrangements is that they are all oriented along the Y axis of the AFM images, substantiating the effect of PSS additive on the associated migration and showing the habits of crystal formation as discussed earlier. The orientation of the crystals of ammonium nitrate in the presence of PSS additive is characteristic of the specific interaction between the polymer molecules and the crystals of ammonium nitrate. It confirms that PSS limits the growth of the rhombic crystals to extremely small sizes. The dark canals that separate the upper and lower clusters of crystals are considered to be void capillaries or pores carrying signatures of the loss of pore water as shown in Fig. 7.

The AFM images presented in Fig. 14 show the crystal shapes and orientation at higher resolutions. In images 14a (2 × 2 μm) and 14b (employing high-pass filters for the AFM images) the crystals appear with a shape similar to the rhombic cells. Also, the AFM images 14c (1 × 1 μm) and 14d (0.5 × 0.5 μm) show the crystals with sharp edge features with their end tips aligned along the Y axis of the AFM images. From the sizes of these crystals determined from the available AFM data, it appears that on

Sz = 15 μm/Zr = 450 nm/Sr = 2 Hz   Sz = 6 μm/Zr = 450 nm/Sr = 2 Hz

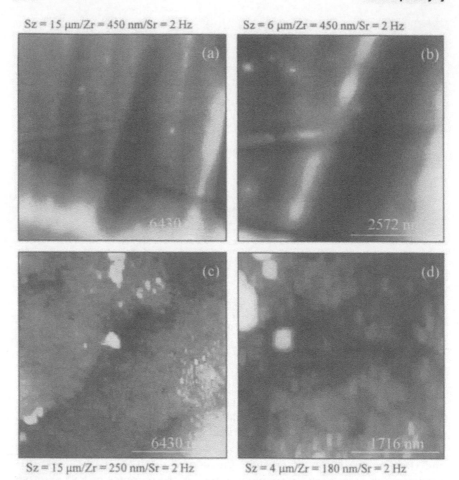

Sz = 15 μm/Zr = 250 nm/Sr = 2 Hz   Sz = 4 μm/Zr = 180 nm/Sr = 2 Hz

FIG. 13   (a) AFM image of pure ammonium nitrate with scanning area Sz = 15 ×
15 μm, image scan size Zr = 450 nm, and scanning rate Sr = 2 Hz. (b) Magnified
view of (a) with scanning area 6 × 6 μm. (c) AFM image of ammonium nitrate
with PSS additive with image scan size Zr = 250 nm. (d) Magnified view of (c)
with Sz = 4 × 4 μm, Zr = 180 nm. (c) and (d) show the directional orientation of
the microcrystals of ammonium nitrate.

average these crystals are 340 nm long, 170 nm wide, and 65 nm thick (Fig.
15), which is surprisingly close to the unit cell dimensions of the rhombic
ammonium nitrate.

Why the anionic PSS performs most effectively compared with PVS and
PVSS can be explained further in terms of their sterochemical configura-

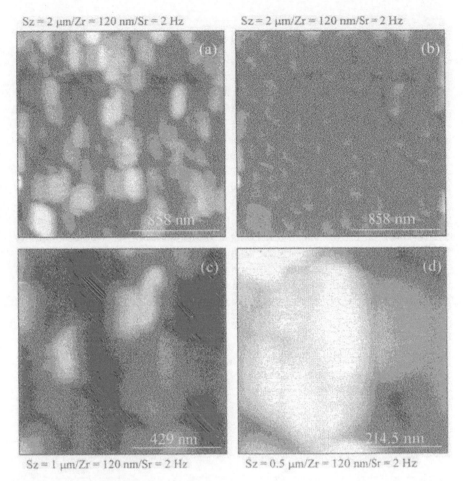

Sz = 2 µm/Zr = 120 nm/Sr = 2 Hz   Sz = 2 µm/Zr = 120 nm/Sr = 2 Hz

(a)   (b)

858 nm   858 nm

(c)   (d)

429 nm   214.5 nm

Sz = 1 µm/Zr = 120 nm/Sr = 2 Hz   Sz = 0.5 µm/Zr = 120 nm/Sr = 2 Hz

FIG. 14 (a) Magnified view of ammonium nitrate crystals in the presence of PSS. (b) Image of (a) employing high-pass filtering. (c) and (d) Further magnified views to obtain the unit crystal dimensions.

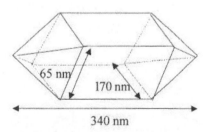

65 nm

170 nm

340 nm

FIG. 15 Average crystal dimensions of ammonium nitrate in the presence of PSS additive as observed by AFM.

tions. In a medium of very high dielectric constant (e.g., water, dielectric constant 78), these polymer molecules remain in random coil forms. However, in a low dielectric situation such as supersaturated ammonium nitrate solution (dielectric constant approximately 16), these polymers take up linear conformations as shown in Fig. 16. Depending upon their sterochemical configurations, the spacing between their sulfonated units varies. In a low dielectric situation the linear conformation of PSS molecules comprises sulfonate groups with a geometric spacing ranging approximately between 6 and 8.5 Å. These values were approximated from the theoretical calculations of minimum energy molecular models [18] (M. A. Whitehead, personal communication, 1995). This indicates that regardless of the crystallizing phases of ammonium nitrate, the PSS, being a multifunctional molecule, is able to interact with various sites on the crystallizing ammonium nitrate planes. For example, in the cubic phase ammonium nitrate has a diagonal spacing of 6.22 Å, whereas in phase III, the a and b axes of the ammonium nitrate unit cell are of 7.14 and 7.7 Å, respectively [5–7,10,19]. As mentioned before,

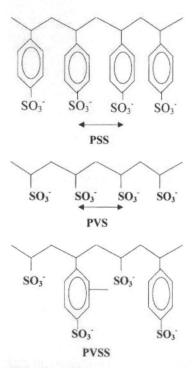

FIG. 16 Differences in stereochemical configurations between PSS, PVS, and PVSS that influence the crystal morphology of ammonium nitrate.

the use of an additive to modify crystal habit is essentially to create surface defects and dislocations in crystals by altering their lattice energies. An additive that lowers the crystal lattice energy (which generally happens with the organic surface-active agents) reduces the bond strength between the crystals and crystal-bound water molecules [19,20]. The changes in the nature of crystal-bound water brought about by the PSS type of additives have a stronger yet definitive influence on the water removal or drying properties compared with any other known conventional additives. Such a property ultimately determines the crystal size, intercrystal spacing, and compactness, which further affect other related material properties such as deformability, fracture, and overall particle strength.

ACKNOWLEDGMENTS

The author thanks his ex-colleagues Dr. Roger Davey, Dr. H. A. Bampfield, and Dr. John Cooper of ICI for their encouragement and fruitful discussions. The author is also grateful to Prof. R. M. Leblanc and Dr. S. Boussaad of the University of Miami for providing the facility to carry out AFM studies.

REFERENCES

1. MA Cook. The Science of Industrial Explosives. Graphic Service and Supply, 1974, pp 1–26.
2. Xuguang Wang. Emulsion Explosives. Beijing: Metallurgical Engineering Press, 1994, chapters 1 and 4.
3. AK Chattopadhyay. US patent 5,597,977.
4. AK Chattopadhyay. European patent 569118.
5. PT Candew, RJ Davey, AJ Ruddick. J Chem Soc Faraday Trans II 43:659, 1984.
6. EJ Griffith. J Chem Eng Data 8(1):22, 1963.
7. JS Ingman, GJ Kearley, SFA Kettle. J Chem Soc Faraday Trans I 78:1817, 1982.
8. AK Chattopadhyay, L Ghaicha, Y Bai, G Munger, RM Leblanc. In: T Provder, MA Winnick, M Urban, eds., Film Formation in Waterborne Coatings. ACS Symp Ser No 648. Washington, DC: American Chemical Society; 1999, chapter 30.
9. Y Bai, G Munger, RM Leblanc, L Ghaicha, AK Chattopadhyay. J Dispersion Sci Technol 17:391, 1996.
10. S Rajam, BR Heywood, JBA Walker, S Mann, RJ Davey, JD Birchall. J Chem Soc Faraday Trans 87:727, 1991.
11. EM Landau, SG Wolf, M Levanon, L Leiserowitz, M Lahav, J Sagiv. J Am Chem Soc 111:1436, 1989.
12. EM Landau, M Levanon, L Leiserowitz, M Lahav, J Sagiv. Nature 318:353, 1985.

13. MW Hosseini, JM Lehn. J Am Chem Soc 109:7047, 1987.
14. PA Bianconi, J Lin, AR Strelecki. Nature 349:6307, 1991.
15. D Sarid, V Elings. J Vac Sci Technol B9:431,1991.
16. S Sarker, AK Chattopadhyay, R Kapadia. Mater Sci Lett 13:983, 1994.
17. S Asami. Drying Technol 11:733, 1993.
18. (a) F Villamagna, MA Whitehead, AK Chattopadhyay. Theochem 343:77, 1995; (b) F Villamagna, MA Whitehead. J Chem Soc Faraday Trans 90:47, 1994.
19. RJ Davey, AJ Ruddick, PD Guy, B Mitchell, SJ Maginn, LA Polwyks. J Phys D Appl Phys 24:176, 1991.
20. T Gacoin, C Train, F Chaput, JP Boilot, P Aubert, M Gandais, Y Wang, A Lecomte. Proc SPIE 1758:565, 1992.

32

Surface Tension Measurements with Top-Loading Balances

BRIAN GRADY, ANDREW R. SLAGLE, LINDA ZHU, EDWARD E. TUCKER, SHERRIL D. CHRISTIAN,[†] **and JOHN F. SCAMEHORN** University of Oklahoma, Norman, Oklahoma, U.S.A.

ABSTRACT

The principles of operation of an inverted force-pull tensiometer are given. The advantages and disadvantages versus other types of tensiometers designed to measure surface tensions are discussed. Some examples of the measurement of surface tensions are presented.

I. INTRODUCTION

Surface tension is perhaps the most important parameter that describes the interfacial properties of a liquid, and its measurement has been the subject of a great many studies. An important subset of these techniques includes methods that rely on measuring the vertical force on a plate, rod, or ring as the surface tension "pulls" the object downward toward the liquid. Currently, the most common method to measure interfacial tension using these pull techniques has the suspended object directly attached by means of a wire to a suspended microbalance so that the force exerted can be measured as the solid is moved in precise increments. Without question this experimental arrangement can give accurate and precise results; however, two substantial problems exist. First, the attachment between the solid object and the balance must not be rigid for accurate measurement of the force. This lack of rigidity has meant that excessive weight cannot be tolerated, and hence very fragile thin rings or plates are typically used. Further, nonzero

---

[†]Deceased.

contact angles will lead to substantial errors in the ring case, and in the Wilhelmy plate case a zero contact angle is preferred to reduce draining times, which can be quite long. Hence, very expensive platinum solids are usually employed because high temperatures can be used to clean the surface thoroughly and ensure a zero contact angle. Second, suspended microbalances are relatively expensive; microbalances must be used because of the small forces involved.

A new principle has been described [1,2], a patent granted [3], and a new product introduced into the market [4] that does not have these two disadvantages. The most significant change has been the use of a top-loading analytical balance, an innovation that has been termed an inverted force-pull tensiometer. An inverted force-pull tensiometer measures the decrease in the weight of a solution as the solid object is pulled out of the solution from above. Because of Newton's third law, the force measured by weighing the solution can, in theory, be no different from that measured by weighing the suspended object.

Determining the surface tension from a pull experiment requires the solution of the Young–Laplace equation:

$$\Delta P = \gamma \left(\frac{1}{R_h} + \frac{1}{R_v} \right) \tag{1}$$

where P is pressure, γ is surface tension, and R_h and R_v are the two radii of curvature for the curved surface that describes the suspended liquid. For a circular rod, the Young–Laplace equation can be rewritten as

$$\frac{\Delta \rho g y}{\gamma} = \frac{y''}{(1 + y'^2)^{3/2}} + \frac{y'}{x(1 + y'^2)^{1/2}} \tag{2}$$

where y is the height of the liquid, $\Delta \rho$ is the density difference between the liquid and gas phases, g is the gravitational constant, and x is the width of the suspended liquid. In order to solve this equation, knowledge of the appropriate contact angles (liquid–solid and liquid–vapor) is required. In theory, any object with arbitrary cross section could be used to measure surface tension; at the very least a solution of the Young–Laplace equation can be numerically generated. However, two geometries have been preferred because the solutions are comparatively straightforward: a circular ring (called a du Noüy ring after the developer of this method) and a very thin plate that is assumed to be infinitely long (called a Wilhelmy plate or slide after the developer of this method). These methods, along with a rod method, are described in this chapter. The latter has not historically been as popular but is well adapted for the inverted force-pull tensiometers described here.

A. du Noüy Method

The du Noüy ring method is unchanged when using a top-loading balance versus the suspended microbalance on most instruments. As with all of the pull methods described in this chapter, to determine surface tension, the general procedure is to move either the solid or liquid up or down to the point of maximum pull, i.e., where the force is a maximum, and record that force. In theory, this movement can be done using steps or continuously as long as the system is at equilibrium when measurements are made. One advantage of using a top-loading balance is that a cylindrical tube rather than a ring can be used; using a tube on instruments with suspended micro-balances would be a problem because of weight and possible torque.

If a ring or tube were infinitely thin, a force balance would lead to the following relationship:

$$W_{balance} = 4\pi R\gamma \tag{3}$$

where $W_{balance}$ is the measured weight, R the radius of the ring, and γ the surface tension. Note that this equation assumes that the balance has been tared with the weight of liquid plus holder and that the weight is defined to be positive. Using a ring or tube of finite thickness introduces substantial deviations from ideality because of the finite difference between inner and outer radii. This nonideality requires a correction in the form $f[R/r, R^3/V]$ that is given in Ref. 5. Here V is the meniscus volume and is given by $W_{balance}/\rho_{liquid}$, and r is half the thickness of the ring. This correction term is valid only when the contact angle between the liquid and solid is zero; if not, then the calculated surface tension will be lower than the actual surface tension.

B. Wilhelmy Plate Method

The Wilhelmy plate method procedurally is identical to the du Noüy method described earlier. Because an infinitely long plate has the same plane of symmetry as the suspended liquid, a correction factor is not necessary. The equation to measure the surface tension is trivial and becomes

$$W_{balance} = tw\gamma \tag{4}$$

where t is the thickness of the plate and w is the width. One advantage of this method is that the contact angle can be any value as long as the meniscus is not broken before the maximum force is reached. Of course, one must be sure that only the bottom is wetted with liquid, or else a correction factor with the appropriate contact angle must be used.

C. Rod-Pull Method

Although using a solid rod has a number of advantages over a ring or plate, a rod has not been used very often in pull methods. The reason for this is twofold. First, the calculations are involved and are most conveniently done on a computer. Second, and more important, measuring the rod weight is difficult because excessive torque can easily ruin a microbalance. However, measuring the weight of liquid means that this concern is eliminated, and a rigid support can be used. By numerical integration of the Young–Laplace equation, it is possible to calculate the mass profile as a function of height. A correction factor is not necessary because the rod has the same infinite rotation axis of symmetry as the suspended liquid profile. Again, the only parameter that must be measured to determine the surface tension is the value of the mass at maximum pull.

Assuming that the vessel is much larger than the radius of the rod, an empirical equation has been determined that relates the radius of the rod [X] to the meniscus constant [k]:

$$
\frac{X}{k} = 2.48573 \left[\frac{X^3}{V}\right]^{1/2} + 0.70985 \left[\frac{X^3}{V}\right] + 4.21654 \left[\frac{X^3}{V}\right]^{3/2}
$$

$$
- 1.94468 \left[\frac{X^3}{V}\right]^2 + 2.30285 \left[\frac{X^3}{V}\right]^3 - 2.77894 \left[\frac{X^3}{V}\right]^4
$$

$$
+ 1.65453 \left[\frac{X^3}{V}\right]^5 - 0.420300 \left[\frac{X^3}{V}\right]^6 + 0.0129372 \left[\frac{X^3}{V}\right]^8
$$

where V has been defined previously and $k = [\gamma/\rho_l g]^{1/2}$, where ρ_l is the density of the liquid phase. As with the Wilhelmy plate, any value of the contact angle can be accommodated with this method. Also as with the Wilhelmy plate, adequate time must be allowed so that the system is at equilibrium during the test, which typically means allowing enough time for the solution to drain from the sides of the rod. The relative magnitude of the error due to incomplete draining is less than for the Wilhelmy plate, however. Both of these constraints mean that some trial and error with the pulling speed or the frequency of movement steps is required.

A further modification of this method is the use of a miniature container; i.e., the radius X is only slightly smaller than the radius of a circular container. Because the object can be held rigid, it is simple to prevent the solid from coming in contact with the sides of the vessel. A theoretical description of this method is given elsewhere along with the relevant equations [6,7]. A numerical integration must be performed to determine the maximum force because the maximum force depends on the contact angle between the liquid and the walls of the container.

II. EXPERIMENTAL

A. Equipment

Schematics of the two types of tensiometers are shown in Fig. 1. A ring, tube, or rod was attached to the metal spindle of a micrometer to conduct the measurement in one type of balance; for the other type, a converter arm was used in conjunction with an electronically controlled stepper motor. A platinum ring furnished by the Fisher Scientific Company was used for the du Noüy ring experiments. The ring had a 9.5366 mm mean radius and a thickness radius of 0.3558 mm. A stainless steel rod of 6.35 mm diameter was used, and two stainless steel tubes with the same value of outside diameter (12.7000 mm) but different wall thicknesses (0.3175 and 0.2591 mm) were used.

B. Materials

All surfactants were obtained from Sigma Chemical Company with purity greater than 99% and were used as received. Sodium bromide (NaBr) was supplied by EM Science with purity greater than 90%. The organic solvents hexadecane (99% purity), ethylene glycol (99.9% purity), and formamide (reagent grade) were supplied by Acros Organics. All the chemicals were used as received. Water was deionized and double distilled. It was used for cleaning and making surfactant solutions.

III. RESULTS AND DISCUSSION

Table 1 lists surface tensions for water measured at 23°C using the inverted rod-pull, inverted tube-pull, and inverted du Noüy ring-pull methods. The reported surface tension results are the average of 7 to 10 measurements. The variance in the measurements was almost certainly due to the ± 1°C temperature fluctuations in the experimental equipment because the magnitude of the variance is comparable to the change in surface tension over this temperature range [8]. Although various surface tension values for water at room temperature are reported in the literature [9,10] the commonly reported value of 72.75 mN/m at 20°C compares favorably with our results [11]. To facilitate this comparison, the measured surface tension values were used to calculate values at 20°C using a temperature correlation [12].

A more quantitative assessment of the accuracy of these measurements can be seen in Fig. 2, which represents measurements for water using the rod. Clearly, numerically integrating the equations gives a result close to the actual measured data, and a slight height correction can be applied (to account for buoyancy effects) that makes the results even closer to the actual

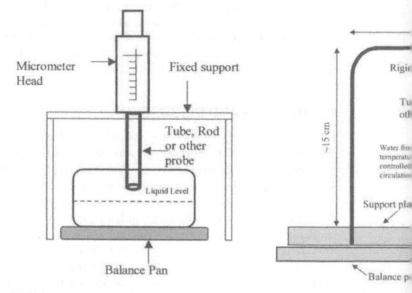

FIG. 1 Schematics of inverted force-pull tensiometers used in our laboratory
the case where a micrometer is directly mounted above a balance; for the ma
was used in conjunction with an electronically controlled stepper motor.

TABLE 1 Surface Tension of Water Measured by Using an Inverted Force-Pull Tensiometer at Room Temperature and Value Corrected to 20°C

Measurement method	Surface tension (γ) at 23°C (mN/m)	Equivalent γ at 20°C (mN/m)	Standard deviation in γ (mN/m)
Inverted rod-pull	71.58	72.15	±0.21
Inverted tube-pull			
Wall thickness: 0.3175 mm	72.33	72.91	±0.18
Wall thickness: 0.3175 mm	72.56	73.07	±0.16
Inverted du Noüy ring-pull	72.13	72.65	±0.10

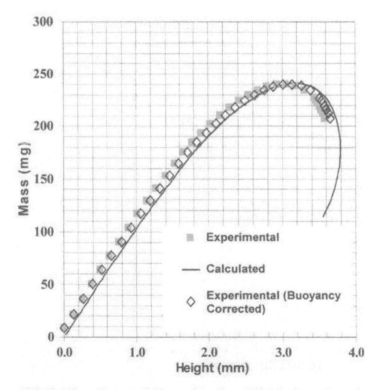

FIG. 2 Force (mass units) as a function of height for rod tensiometer. Corrected measurements represent data adjusted for buoyancy forces.

measured curve. The height corresponding to the maximum pull is clearly smaller in the experimental than in the calculated case, which is presumably due to wetting problems. However, the maximum force agrees almost exactly, which is fortunate because this value is used for surface tension calculations.

One extremely important parameter in the rod-pull measurements is the shape of the end of the rod. Typically, a machinist rounds corners as shown in Fig. 3; i.e., the actual outside diameter is greater than the diameter of the circle at the end. This error in machining can lead to substantial errors in the surface tension, as shown graphically in Fig. 3. Similar concerns apply to the tube as well. In fact, the size of the rod should be checked using water or some other known material before measuring unknown surface tensions.

One important benefit of choosing a container where the radius is only slightly greater than the radius of the rod is that the maximum force measured is much greater. For example, for water and a rod of very similar

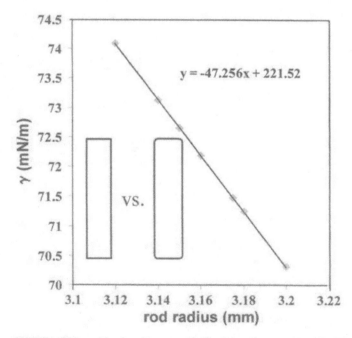

FIG. 3 Effect of rod radius on calculated surface tension. The inset represents common machining problems; corners must be sharp when the rod is manufactured.

diameter (6.23 mm), the force increases by approximately 50% if a container of 10.55 mm is used. This modification also helps to alleviate a disadvantage of the rod-pull method: for a given amount of wetted perimeter, the force measured is roughly half that measured by the du Noüy or Wilhelmy plate method. Because the rod is rigidly held, the liquid can be almost totally encased to reduce evaporation dramatically.

Wetting issues are less important with this device than with the du Noüy ring or Wilhelmy plate, but wetting can still affect results. Table 2 shows the values of surface tension obtained from measurements for selected organic solvents and an aqueous surfactant (cetyl trimethylammonium bromide, CTAB) solution versus literature values at 20°C. Note that the results obtained by using the inverted du Noüy ring-pull method agree better with literature values (average difference of 0.27 mN/m from literature values) than the results obtained by using the inverted tube-pull method (average difference of 1.51 mN/m from literature values). Better precision (±0.03 mN/m) was found for the inverted du Noüy ring-pull method than the inverted tube-pull method (±0.12 mN/m). Lack of cleanliness is almost certainly not the source of the error; the most likely source is the poor wettability of the stainless steel tube by organic solvents, which would lower the value of the experimental maximum-pull force because the contacting area between the liquid and the bottom of the tube would be reduced.

Measuring the surface tension in order to determine the critical micelle concentration (cmc) is also straightforward. Figure 4 shows plots of the surface tension versus surfactant concentration for CTAB and sodium dodecyl sulfate (SDS) solutions. The cmc can be determined as the intersection of the straight lines through the two linear portions of the semilogarithmic plot; the values determined are close to literature values. The familiar dip in surface tension at the cmc for SDS due to the presence of impurities is also clearly apparent. Although the data are not shown, this method can also be used effectively to measure the cmc in the presence of added electrolyte [13].

Similarly to a Wilhelmy plate, the contact angle can easily be measured using a rod. The procedure is very similar except that the rod is below the liquid level and a precise measurement of the relative height is necessary. If the contact angle with a different solid is required, one advantage of this method is that the rod can often be coated with the appropriate material; a new rod is not necessary. The buoyancy effect must be accounted for, and, as with methods for other measurements of contact angle, surface smoothness is critical. Figure 5 shows the determination of the advancing and receding contact angles on stainless steel for water. The region for determination of the surface tension is shown as part of the previous plot.

3LE 2 Surface Tension of Various Substances Measured Using an Inverted Force-Pull Tens

|)stance | Inverted du Noüy ring-pull | | | Inverted tube-p | |
	Surface tension (γ) (mN/m)	Standard deviation (mN/m)	Equivalent γ at 20°C (mN/m)	Surface tension (γ) (mN/m)	Standard deviation in (mN/m)
mamide	57.69	±0.03	57.92	56.9	±0.11
ylene glycol	47.40	±0.03	47.41	46.05	±0.08
‹adecane	27.13	±0.04	27.34	25.54	±0.13
ʻAB] = 0.9 mM	37.84	±0.04	N/A	38.14	±0.17

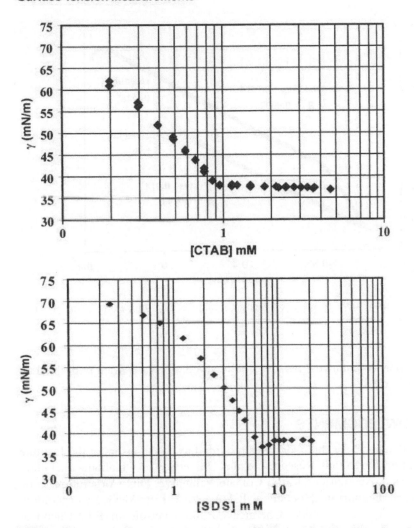

FIG. 4 Representative cmc measurements with inverted rod-pull tensiometer.

IV. CONCLUSIONS

In many situations, "pull" methods plus the use of a top-loading balance simplify the measurement of surface tension. The largest advantages derive from the ability to use a rigid support for the solid, which means that a more mechanically robust object can be used. The measurements are essentially as accurate as those with more complicated and expensive instruments. Further, measuring small volumes of liquids, which is difficult in systems

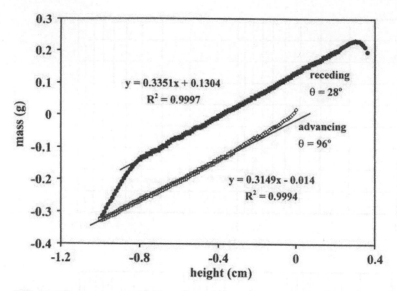

FIG. 5 Measurement of advancing and receding contact angles (q) using inverted rod-pull tensiometer.

with suspended balances, becomes quite straightforward when a top-loading balance is used.

ACKNOWLEDGMENTS

Financial support for this work was provided by the sponsors of the Institute for Applied Surfactant Research at the University of Oklahoma including Akzo Nobel, Abemarle, Clorox, Colgate-Palmolive, Dial Corporation, Dow Chemical, Halliburton, Huntsman, ICI America, Kerr-McGee Corporation, Lubrizol Corporation, Nikko Chemicals, Phillips Petroleum, Pilot Chemical Company, Procter and Gamble, Reckitt & Colman, Schlumberger, Shell Chemical Company, Sun Chemical Corporation, and Unilever. John Scamehorn holds the Asahi Glass Chair in Chemical Engineering at the University of Oklahoma. Much of this chapter was written while BPG was at the Max Planck Institute for Colloid and Interface Science supported by a Humboldt fellowship; this support is gratefully acknowledged.

REFERENCES

1. SD Christian, AR Slagle, EE Tucker, JF Scamehorn. Langmuir 14:3126–3128, 1998.

2. SD Christian, AR Slagle, K. Fujio, EE Tucker, JF Scamehorn. J Colloid Interface Sci 214:224–230, 1999.
3. SD Christian, AR Slagle, EE Tucker, JF Scamehorn, K Fujio. US patent 6,119,511, 2000.
4. For more information see http://www.temco.com/tensiometer.htm.
5. AW Adamson. Physical Chemical of Surfaces. 5th ed. New York: Wiley-Interscience, 1990, pp 23–28.
6. DN Furlong, S Hartland. J Chem Soc Faraday Trans 76:457–466, 1980.
7. DN Furlong, S Hartland. J Chem Soc Faraday Trans 76:467–475, 1980.
8. R Cini, G Loglio, A Ficalbi. J Colloid Interface Sci 41:287–295, 1972.
9. DR Lide, ed. CRC Handbook of Chemistry and Physics. 80th ed. Boca Raton, FL: CRC Press, 1999, p 6-3.
10. AW Adamson. Physical Chemistry of Surfaces. 5th ed. New York: Wiley-Interscience, 1990, p 41.
11. BW Rossiter, RC Baetzold. Physical Methods of Chemistry. 2nd ed. New York: Wiley-Interscience, 1993, p 333.
12. WV Kayser. J Colloid Interface Sci 56:622–629, 1976.
13. L Zhu. MS thesis, University of Oklahoma, 2000.

Index

T - #0195 - 071024 - C0 - 244/170/39 - PB - 9780367395674 - Gloss Lamination